25.00
x

PRIMATE
VISIONS

PRIMATE VISIONS

Gender, Race, and Nature in the World of Modern Science

DONNA HARAWAY

ROUTLEDGE
New York London

FOR E.R.H. AND B.J.M.

First published in 1989 by

Routledge
an imprint of
Routledge, Chapman & Hall, Inc.
29 West 35 Street
New York, NY 10001

Published in Great Britain by

Routledge
11 New Fetter Lane
London EC4P 4EE

© 1989 by Routledge, Chapman & Hall, Inc.

Printed in the United States of America

Library of Congress Cataloging in Publication Data

Haraway, Donna Jeanne.
    Primate visions : gender, race, and nature in the world of modern
  science / Donna J. Haraway.
        p.   cm.
    Bibliography : p.
    Includes index.
    ISBN 0-415-90114-6
      1. Primates.   2. Primates—Research—History.   3. Feminism.
    4. Sociobiology.   I. Title
  QL737.P9H245   1989   89-10079
  599.8′072—dc20   CIP

British Library cataloging in publication data also available

CONTENTS

ACKNOWLEDGMENTS

Primate Visions began to germinate about 1978, and many people and institutions have contributed enormously. Colleagues and students at Johns Hopkins and the Université de Montréal taught me the pleasure of writing and teaching the history of science, particularly William Coleman, Stephen Cross, Robert Kargon, Camille Limoges, and Philip Pauly. I am especially grateful to the academic community in the History of Consciousness Board and in the Women's Studies Program at the University of California at Santa Cruz. Teaching with Jim Clifford on the traffic between "nature" and "culture" has been a special pleasure; and my writing has been stimulated and informed also by the colleagueship of Bettina Aptheker, Norman O. Brown, Teresa de Lauretis, Barbara Epstein, Gary Lease, Helene Moglen, and Hayden White. A warm thanks also to Billie Haris.

The graduate seminars in feminist theory in History of Consciousness have been uniquely valuable. Engaging with students over several years, a faculty member is privileged to read papers and take part in conversations that fundamentally change one's way of seeing. This is a delicate, multidirectional, and collective process. It is impossible adequately to state my debt, and naming a few students leaves unacknowledged many others whose work deeply informs my own. But I must thank especially those with whom I have worked closely who will have finished their doctorates by the time *Primate Visions* is published: Sandra Azeredo, Elizabeth Bird, Lisa Bloom, Catherine Borchert, Mary Crane, Paul Edwards, Elliott Butler-Evans, Ruth Frankenberg, Debbie Gordon, Valerie Hartouni, Caren Kaplan, Katie King, Carole McCann, Lata Mani, Chela Sandoval, Zoe Sofoulis, and Noel Sturgeon. The notes and bibliography only hint at the intellectual generosity of these and other History of Consciousness students.

This book has also drawn significantly from the work of UCSC faculty and students in the Group for the Critical Study of Colonial Discourse and the Feminist Studies Organized Research Activity. Thanks also to Nathalie Magnan and Sarah Williams, who in helping produce a Paper Tiger Television program made me think through my arguments and images.

I have relied extensively on the help of primate scientists, scholars in science studies, photographers, librarians, archivists, and artists. I owe special thanks to those who submitted to interviews, gave me access to unpublished material, supplied photographs, and criticized drafts of chapters. In particular, I am grateful to Jeanne Altmann, Stuart Altmann, Pamela Asquith, Linda Fedigan, Sarah Blaffer Hrdy, Alison Jolly, Lita Osmundsen, Barbara Smuts, Shirley Strum, Sherwood Washburn, and Adrienne Zihlman.

A crucial network of friends and colleagues has read drafts, criticized arguments, offered insights, and generally sustained the intellectual and emotional work of this project. Reading every word critically, Joan Scott has had a major impact. Her friendship and intellectual acumen are a rare gift. I am also fundamentally grateful to Lila Abu-Lughod, Judith Butler, Scott Gilbert, Nancy Hartsock, Dorinne Kondo, Bruno Latour, Rayna Rapp, David Schneider, Dorothy Stein, George Stocking, Marilyn Strathern, and Robert Young.

The heterogeneous, contentious, and wonderful people in science studies who are working in relation to feminist questions have been especially important. *Primate Visions* is particularly in conversation with Ruth Bleier, Anne Fausto-Sterling, Elizabeth Fee, Sandra Harding, Ruth Hubbard, Ludamilla Jordanova, Evelyn Fox Keller, Diana Long, Helen Longino, Hilary Rose, and Sharon Traweek.

Financial help, allowing the time and other resources to write, has been provided by grants from the Academic Senate of the University of California, the Alpha Fund of the Institute for Advanced Study in Princeton, NJ, and the Wenner-Gren Foundation for Anthropological Research. Faced with a monster manuscript, William Germano at Routledge has provided editorial guidance, intellectual stimulus, and a fine sense of humor throughout.

Members of my households in Honolulu, Baltimore, Santa Cruz, and Healdsburg have given intellectual, personal, and political support over two decades. It is they, especially Rusten Hogness and Jaye Miller, who sustain my sense of possibility in the human regions of the primate order.

Sections of *Primate Visions* have been previously or simultaneously published in the following essays. All have been edited for *Primate Visions*, some extensively as noted, and are reprinted with permission. Chapter 3: "Teddy Bear Patriarchy: Taxidermy in the Garden of Eden, New York City, 1908–1936," *Social Text*, no. 11 (winter 1984–1985): 20–64. Chapter 5: Extensively revised from "The High Cost of Information in Post-World War II Evolutionary Biology: Ergonomics, Semiotics, and the Sociobiology of Communications Systems," *Philosophical Forum* XIII, nos. 2–3 (1981–1982): 244–78; and "Signs of Dominance: From a Physiology to a Cybernetics of Primate Society, C.R. Carpenter, 1930–70," *Studies in History of Biology* 6 (1983): 129–219. Chapter 8: "Remodeling the Human Way of Life: Sherwood

Washburn and the New Physical Anthropology, 1950–1980," *History of Anthropology*, ed. George Stocking, 5 (1988): 206–59. Chapter 10: Section on Alison Jolly revised from "Monkeys, Aliens, and Women: Love, Science, and Politics at the Intersection of Feminist Theory and Colonial Discourse," *Women's Studies International Forum*, Ruth Bleier memorial volume, Sue Rosser, ed., forthcoming 1989. Chapter 15: "Sarah Blaffer Hrdy and Sociobiological Feminism: Investment Strategies for the Evolving Portfolio of Primate Females," in Mary Jacobus, Evelyn Fox Keller, and Sally Shuttleworth, eds., *Women, Science, and the Body* (New York: Routledge, forthcoming 1989).

1

INTRODUCTION:
THE PERSISTENCE OF VISION

The names you uncaged primates give things affect your
attitude to them forever after. (Herschberger 1970 [1948])

For thus all things must begin, with an act of love. (Marais
1980)

How are love, power, and science intertwined in the constructions of nature
in the late twentieth century?[1] What may count as nature for late industrial
people? What forms does love of nature take in particular historical contexts? For whom and at what cost? In what specific places, out of which social and
intellectual histories, and with what tools is nature constructed as an object of erotic
and intellectual desire? How do the terrible marks of gender and race enable and
constrain love and knowledge in particular cultural traditions, including the modern
natural sciences? Who may contest for what the body of nature will be? These
questions guide my history of the modern sciences and popular cultures emerging
from accounts of the bodies and lives of monkeys and apes.

The themes of race, sexuality, gender, nation, family, and class have been written
into the body of nature in western life sciences since the eighteenth century. In the
wake of post-World War II decolonization, local and global feminist and anti-racist
movements, nuclear and environmental threats, and broad consciousness of the
fragility of earth's webs of life, nature remains a crucially important and deeply
contested myth and reality. How do material and symbolic threads interweave in
the fabric of late twentieth-century nature for industrial people?

Monkeys and apes have a privileged relation to nature and culture for western
people: simians occupy the border zones between those potent mythic poles. In the
border zones, love and knowledge are richly ambiguous and productive of meanings
in which many people have a stake. The commercial and scientific traffic in monkeys
and apes is a traffic in meanings, as well as in animal lives. The sciences that tie

1

monkeys, apes, and people together in a Primate Order are built through disciplined practices deeply enmeshed in narrative, politics, myth, economics, and technical possibilities. The women and men who have contributed to primate studies have carried with them the marks of their own histories and cultures. These marks are written into the texts of the lives of monkeys and apes, but often in subtle and unexpected ways. People who study other primates are advocates of contending scientific discourses, and they are accountable to many kinds of audiences and patrons. These people have engaged in dynamic, disciplined, and intimate relations of love and knowledge with the animals they were privileged to watch. Both the primatologists and the animals on whose lives they reported command intense popular interest—in natural history museums, television specials, zoos, hunting, photography, science fiction, conservation politics, advertising, cinema, science news, greeting cards, jokes. The animals have been claimed as privileged subjects by disparate life and human sciences—anthropology, medicine, psychiatry, psychobiology, reproductive physiology, linguistics, neural biology, paleontology, and behavioral ecology. Monkeys and apes have modeled a vast array of human problems and hopes. Most of all, in European, American, and Japanese societies, monkeys and apes have been subjected to sustained, culturally specific interrogations of what it means to be "almost human."

Monkeys and apes—and the people who construct scientific and popular knowledge about them—are part of cultures in contention. Never innocent, the visualizing narrative "technology" of this book draws from contemporary theories of cultural production, historical and social studies of science and technology, and feminist and anti-racist movements and theories to craft a view of nature as it is constructed and reconstructed in the bodies and lives of "third world" animals serving as surrogates for "man."

I have tried to fill *Primate Visions* with potent verbal and visual images—the corpse of a gorilla shot in 1921 in the "heart of Africa" and transfixed into a lesson in civic virtue in the American Museum of Natural History in New York City; a little white girl brought into the Belgian Congo in the 1920s to hunt gorilla with a camera, who metamorphosized in the 1970s into a writer of science fiction considered for years as a model of masculine prose; the chimpanzee HAM in his space capsule in the Mercury Project in 1961; HAM's chimp contemporary, David Greybeard, reaching out to Jane Goodall, "alone" in the "wilds of Tanzania" in the year in which 15 African primate-habitat nations achieved national independence; a *Vanity Fair* special on the murdered Dian Fossey in a gorilla graveyard in Rwanda in 1986; the bones of an ancient fossil, reconstructed as the grandmother of humanity, laid out like jewels on red velvet in a paleontologist's laboratory in a pattern to ground, once again, a theory of the origin of "monogamy"; infant monkeys in Harry Harlow's laboratory in the 1960s clinging to cloth and wire "surrogate mothers" at an historical moment when the images of surrogacy began to surface in American reproductive politics; the emotionally wrenching embrace between a young, middle-class, white woman scientist and an adult American Sign Language-speaking chimpanzee on an island in the River Gambia, where white women teach captive apes to "return" to the "wild"; a Hallmark greeting card reversing the images of King Kong with a monstrous blond woman and a cringing silverback gorilla in bed in a drama called "Getting Even"; the anatomical drawings of living and fossil female apes sharing the basic lines of their bodies with a modern human female, in order to teach medical students the functional meaning of human adaptations; ordinary women and men from Africa, the United States, Japan, Europe, India, and elsewhere, with tape recorders and data clipboards transcribing the lives of monkeys and apes into specialized texts that become contested items in political controversies in many cultures.

I am writing about primates because they are popular, important, marvelously varied, and controversial. And all members of the Primate Order—monkeys, apes, and people—are threatened. Late twentieth-century primatology may be seen as part of a complex survival literature in global, nuclear culture. Many people, including myself, have emotional, political, and professional stakes in the production and stabilization of knowledge about the order of primates. This will not be a disinterested, objective study, nor a comprehensive one—partly because such studies are impossible for anyone, partly because I have stakes I want to make visible (and probably others as well). I want this book to be interesting for many audiences, and pleasurable and disturbing for all of us. In particular, I want this book to be responsible to primatologists, to historians of science, to cultural theorists, to the broad left, anti-racist, anti-colonial, and women's movements, to animals, and to lovers of serious stories. It is perhaps not always possible to be accountable to those contending audiences, but they have all made this book possible. They are all inside this text. Primates existing at the boundaries of so many hopes and interests are wonderful subjects with whom to explore the permeability of walls, the reconstitution of boundaries, the distaste for endless socially enforced dualisms.

Fact and Fiction

Both science and popular culture are intricately woven of fact and fiction. It seems natural, even morally obligatory, to oppose fact and fiction; but their similarities run deep in western culture and language. Facts can be imagined as original, irreducible nodes from which a reliable understanding of the world can be constructed. Facts ought to be discovered, not made or constructed. But the etymology of facts refers us to human action, performance, indeed, to human feats (OED). Deeds, as opposed to words, are the parents of facts. That is, human action is at the root of what we can see as a fact, linguistically and historically. A fact is the thing done, a neuter past participle in our Roman parent language. In that original sense, facts are what has actually happened. Such things are known by direct experience, by testimony, and by interrogation—extraordinarily privileged routes to knowledge in North America.

Fiction can be imagined as a derivative, fabricated version of the world and experience, as a kind of perverse double for the facts or as an escape through fantasy into a better world than "that which actually happened." But tones of meaning in fiction make us hear its origin in vision, inspiration, insight, genius. We hear the root of fiction in poetry and we believe, in our Romantic moments, that original natures are revealed in good fiction. That is, fiction can be *true*, known to be true by an appeal to nature. And as nature is prolific, the mother of life in our major myth systems, fiction seems to be an inner truth which gives birth to our actual lives. This, too, is a very privileged route to knowledge in western cultures, including the United States. And finally, the etymology of fiction refers us once again to human action, to the act of fashioning, forming, or inventing, as well as to feigning. Fiction is inescapably implicated in a dialectic of the true (natural) and the counterfeit (artifactual). But in all its meanings, fiction is about human action. So, too, are all the narratives of science—fiction and fact—about human action.

Fiction's kinship to facts is close, but they are not identical twins. Facts are opposed to opinion, to prejudice, but not to fiction. Both fiction and fact are rooted in an epistemology that appeals to experience. However, there is an important difference; the word *fiction* is an active form, referring to a present act of fashioning, while *fact* is a descendant of a past participle, a word form which masks the generative deed or performance. A fact seems done, unchangeable, fit only to be recorded; fiction seems always inventive, open to other possibilities, other fashionings of life. But in this opening lies the threat of merely feigning, of not telling the true form of things.

From some points of view, the natural sciences seem to be crafts for distinguishing between fact and fiction, for substituting the past participle for the invention, and thus preserving true experience from its counterfeit. For example, the history of primatology has been repeatedly told as a progressive clarification of sightings of monkeys, apes, and human beings. First came the original intimations of primate form, suggested in the pre-scientific mists in the inventive stories of hunters, travelers, and natives, beginning perhaps in ancient times, perhaps in the equally mythic Age of Discovery and of the Birth of Modern Science in the sixteenth century. Then gradually came clear-sighted vision, based on anatomical dissection and comparison. The story of correct vision of primate social form has the same plot: progress from misty sight, prone to invention, to sharp-eyed quantitative knowledge rooted in that kind of experience called, in English, experiment. It is a story of progress from immature sciences based on mere description and free qualitative interpretation to mature science based on quantitative methods and falsifiable hypotheses, leading to a synthetic scientific reconstruction of primate reality. But these histories are stories about stories, narratives with a good ending; i.e., the facts put together, reality reconstructed scientifically. These are stories with a particular aesthetic, realism, and a particular politics, commitment to progress.

From only a slightly different perspective, the history of science appears as a narrative about the history of technical and social means to produce the facts. The facts themselves are types of stories, of testimony to experience. But the provocation of experience requires an elaborate technology—including physical tools, an accessible tradition of interpretation, and specific social relations. Not just anything can emerge as a fact; not just anything can be seen or done, and so told. Scientific practice may be considered a kind of story-telling practice—a rule-governed, constrained, historically changing craft of narrating the history of nature. Scientific practice and scientific theories produce and are embedded in particular kinds of stories. Any scientific statement about the world depends intimately upon language, upon metaphor. The metaphors may be mathematical or they may be culinary; in any case, they structure scientific vision. Scientific practice is above all a story-telling practice in the sense of historically specific practices of interpretation and testimony.

Looking at primatology, a branch of the life sciences, as a story-telling craft may be particularly appropriate. First, the discourse of biology, beginning near the first decades of the nineteenth century, has been about organisms, beings with a life history; i.e., a plot with structure and function.[2] Biology is inherently historical, and its form of discourse is inherently narrative. Biology as a way of knowing the world is kin to Romantic literature, with its discourse about organic form and function.

Biology is the fiction appropriate to objects called organisms; biology fashions the facts "discovered" from organic beings. Organisms perform for the biologist, who transforms that performance into a truth attested by disciplined experience; i.e., into a fact, the jointly accomplished deed or feat of the scientist and the organism. Romanticism passes into realism, and realism into naturalism, genius into progress, insight into fact. *Both* the scientist and the organism are actors in a story-telling practice.

Second, monkeys, apes, and human beings emerge in primatology inside elaborate narratives about origins, natures, and possibilities. Primatology is about the life history of a taxonomic order that includes people. Especially western people produce stories about primates while simultaneously telling stories about the relations of nature and culture, animal and human, body and mind, origin and future. Indeed, from the start, in the mid-eighteenth century, the primate order has been built on tales about these dualisms and their scientific resolution.

To treat a science as narrative is not to be dismissive, quite the contrary. But neither is it to be mystified and worshipful in the face of a past participle. I am interested in the narratives of scientific fact—those potent fictions of science—within a complex field indicated by the signifier SF. In the late 1960s science fiction anthologist and critic Judith Merril idiosyncratically began using the signifier SF to designate a complex emerging narrative field in which boundaries between science fiction (conventionally, sf) and fantasy became highly permeable in confusing ways, commercially and linguistically. Her designation, SF, came to be widely adopted as critics, readers, writers, fans, and publishers struggled to comprehend an increasingly heterodox array of writing, reading, and marketing practices indicated by a proliferation of "sf " phrases: speculative fiction, science fiction, science fantasy, speculative futures, speculative fabulation.

SF is a territory of contested cultural reproduction in high-technology worlds. Placing the narratives of scientific fact within the heterogeneous space of SF produces a transformed field. The transformed field sets up resonances among all of its regions and components. No region or component is "reduced" to any other, but reading and writing practices respond to each other across a structured space. Speculative fiction has different tensions when its field also contains the inscription practices that constitute scientific fact. The sciences have complex histories in the constitution of imaginative worlds and of actual bodies in modern and postmodern "first world" cultures. Teresa de Lauretis speculated that the sign work of SF was "potentially creative of new forms of social imagination, creative in the sense of mapping out areas where cultural change *could* take place, of envisioning a different order of relationships between people and between people and things, a different conceptualization of social existence, inclusive of physical and material existence" (1980: 161). This is also one task of the "sign work" of primatology.

So, in part, *Primate Visions* reads the primate text as science fiction, where possible worlds are constantly reinvented in the contest for very real, present worlds. The conclusion perversely reads a sf story about an alien species that intervenes in human reproductive politics as if it were a monograph from the primate field. Beginning with the myths, sciences, and historical social practices that placed apes in Eden and apes in space, at the beginnings and ends of western culture, *Primate Visions* locates aliens in the text as a way to understand love and knowledge among primates on a contemporary fragile earth.

Four Temptations

Analyzing a scientific discourse, primatology, as story telling within several contested narrative fields is a way to enter current debates about the social construction of scientific knowledge without succumbing completely to any of four very tempting positions, which are also major resources for the approaches of this book. I use the image of temptation because I find all four positions persuasive, enabling, and also dangerous, especially if any one position finally silences all the others, creating a false harmony in the primate story.

The first resourceful temptation comes from the most active tendencies in the social studies of science and technology. For example, the French prominent analyst of science, Bruno Latour, radically rejects all forms of epistemological realism and analyzes scientific practice as thoroughly social and constructionist. He rejects the distinction between social and technical and represents scientific practice as the refinement of "inscription devices," i.e., devices for transcribing the immense complexity and chaos of competing interpretations into unambiguous traces, writings, which mark the emergence of a fact, the case about reality. Interested in science as a fresh form of power in the social-material world and scientists as investing "their political ability in the heart of doing science," Latour and his colleague Stephen Woolgar powerfully describe how processes of construction are made to invert and appear in the form of discovery (1979: 213). The accounts of the scientists about their own processes become ethnographic data, subject to cultural analysis.

Fundamentally, from the perspective of *Laboratory Life*, scientific practice is literary practice, writing, based on jockeying for the power to stabilize definitions and standards for claiming something to be the case. To win is to make the cost of destabilizing a given account too high. This approach can explain scientific contests for the power to close off debate, and it can account for both successful and unsuccessful entries in the contest. Scientific practice is negotiation, strategic moves, inscription, translation. A great deal can be said about science as effective belief and the world-changing power to enforce and embody it.[3] What more can one ask of a theory of scientific practice?

The second valuable temptation comes from one branch of the marxist tradition, which argues for the historical superiority of particular structured standpoints for knowing the social world, and possibly the "natural" world as well. Fundamentally, people in this tradition find the social world to be structured by the social relations of the production and reproduction of daily life, such that it is only possible to see these relations clearly from some vantage points. This is not an individual matter, and good will is not at issue. From the standpoint of those social groups in positions of systematic domination and power, the true nature of social life will be opaque; they have too much to lose from clarity.

Thus, the owners of the means of production will see equality in a system of exchange, where the standpoint of the working class will reveal the nature of domination in the system of production based on the wage contract and the exploitation and deformation of human labor. Those whose social definition of identity is rooted in the system of racism will not be able to see that the definition of human has not been neutral, and cannot be until major material-social changes occur on a world scale. Similarly, for those whose possibility of adult status rests on the power to appropriate the "other" in a socio-sexual system of gender, sexism will not look like a fundamental barrier to correct knowledge *in general.* The tradition indebted to marxist epistemology can account for the greater adequacy of some ways of knowing and can show that race, sex, and class fundamentally determine the most intimate details of knowledge and practice, especially where the appearance is of neutrality and universality.[4]

These issues are hardly irrelevant to primatology, a science practiced in the United States nearly exclusively by white people, and until quite recently by white men, and still practiced overwhelmingly by the economically privileged. Much of this book examines the consequences for primatology of the social relations of race, sex, and class in the construction of scientific knowledge. For example, perhaps most primatologists in the field in the first decades after World War II failed to appreciate that the interrelationships of people, land, and animals in Africa and Asia are at least partly due to the positions of the researchers within systems of racism and imperialism. Many sought a "pure" nature, unspoiled by contact with people; and so they sought untouched species, analogous to the "natives" once sought by colonial anthropologists. But for the observer of animals, the indigenous peoples of Africa and Asia were a nuisance, a threat to conservation—indeed, encroaching "aliens"—until decolonization forced white western scientists to restructure their bio-politics of self and other, native and alien. The boundaries among animals and human beings shift in the transition from a colonial to a post- or neo-colonial standpoint. Insisting that there can be less deformed contents and methods in the natural as well as social sciences, the marxist, feminist, and anti-racist accounts reject the relativism of the social studies of science. Explicitly political accounts take sides on what is a more adequate, humanly acceptable knowledge. But these analyses have limits for guiding an exploration of primate studies. Wage labor, sexual and reproductive appropriation, and racial hegemony are structured aspects of the human social world. There is no doubt that they affect knowledge systematically, but it is not clear precisely how they relate to knowledge about the feeding patterns of patas monkeys or about the replication of DNA molecules.

Another aspect of the marxist tradition has made significant progress in answering that kind of question. In the 1970s, people associated with the British *Radical Science Journal* developed the concept of science as a labor process in order to study and change scientific mediations of class domination in the relations of production and reproduction of human life.[5] Like Latour, they leave no holes for a realist or positivist epistemology, the preferred versions of most practicing scientists. Every aspect of scientific practice can be described in terms of the concept of mediation: language, laboratory hierarchies, industrial ties, medical doctrines, basic theoretical preferences, and stories about nature. The concept of labor process seems cannibalistic, making the social relations of other basic processes seem derivative. For example, the complex systems of domination, complicity, resistance, equality, and nurturance in gendered practices of bearing and raising children cannot be accommodated by the concept of labor. But these reproductive practices visibly affect more than a few contents and methods in modern primate studies. But even an extended concept of mediation and systematic social process, one that does not insist on the reduction to labor in a classic marxist sense, leaves out too much.

The third temptation comes from the siren call of the scientists themselves; they keep pointing out that they are, among other things, watching monkeys and apes. In some sense, more or less nuanced, they insist that scientific practice "gets at" the world. They claim that scientific knowledge is not simply about power and control. They claim that their knowledge somehow translates the active voice of their subjects, the objects of knowledge. Without necessarily being compelled by their aesthetic of realism or their theories of representation, I believe them in the strong sense that my imaginative and intellectual life and my professional and political commitments in the world respond to these scientific accounts. Scientists are adept at providing good grounds for belief in their accounts and for action on their basis. Just how science "gets at" the world remains far from resolved. What does seem resolved, however, is that science grows from and enables concrete ways of life, including particular constructions of love, knowledge, and power. That is the core of its instrumentalism and the limit to its universalism.

Evidence is always a question of interpretation; theories are accounts *of* and *for* specific kinds of lives. I am looking for a way of telling a story of the production of a branch of the life sciences, a branch which includes human beings centrally, that listens very carefully to the stories themselves. My story must listen to the practices of interpretation of the primate order in which the primates themselves—monkeys, apes, and people—all have some kind of "authorship." I would suggest that the concept of constrained and contested story-telling allows an appreciation of the social construction of science, while still guiding the hearer to a search for the other animals who are active participants in primatology. I want to find a concept for telling a history of science that does not itself depend on the dualism between active and passive, culture and nature, human and animal, social and natural.

The fourth temptation intersects with each of the other three; this master temptation is to look always through the lenses ground on the stones of the complex histories of gender and race in the constructions of modern sciences around the globe. That means examining cultural productions, including the primate sciences, from the points of view enabled by the politics and theories of feminism and anti-racism. The challenge is to remember the particularity as well as the power of this way of reading and writing. But that is the same challenge that should be built into reading or writing a scientific text. Race and gender are not prior universal social categories—much less natural or biological givens. Race and gender are the world-changing products of specific, but very large and durable, histories. The same thing is true of science. The visual system of this book depends upon a triple filter of race, gender, and science. This is the filter which traps the marked bodies of history for closer examination.[6]

Stories are always a complex production with many tellers and hearers, not all of them visible or audible. Story-telling is a serious concept, but one happily without the power to claim unique or closed readings. Primatology seems to be a science composed of stories, and the purpose of this book is to enter into contestations for their construction. The lens of story-telling defines a thin line between realism and nominalism; but primates seem to be natural scholastics, given to equivocation when pressed. Also, I think there is an aesthetic and an ethic built into thinking of scientific practice as story-telling, an aesthetic and ethic different from capitulation to "progress" and belief in knowledge as passive reflection of "the way things are," and also different from the ironic skepticism and fascination with power so common in the social studies of science. The aesthetic and ethic latent in the examination of story-telling might be pleasure and responsibility in the weaving of tales. Stories are means to ways of living. Stories are technologies for primate embodiment.

Primatology is (Judeo-) Christian Science

Western Jews and Christians or post-Judeo-Christians are not the only practitioners of primate sciences. But this book focuses primarily on the history of studies
of the social behavior of monkeys and apes done in the United States or by Euro-
Americans in the twentieth century. In these stories, there is a constant refrain
drawn from salvation history; primatology is about primal stories, the origin and
nature of "man," and about reformation stories, the reform and reconstruction of
human nature. Implicitly and explicitly, the story of the Garden of Eden emerges
in the sciences of monkeys and apes, along with versions of the origin of society,
marriage, and language.

From the beginning, primatology has had this character in the west. If the eighteenth-century Swedish "father" of modern biological classification, Linnaeus, is
cited at all by twentieth-century scientists, he is noted for placing human beings in
a taxonomic order of nature with other animals, i.e., for taking a large step away
from Christian assumptions. Linnaeus placed "man" in his taxonomic order of
Primates as *Homo sapiens*, in the same genus with *Homo troglodytes*, a dubious and
interesting creature illustrated as a hairy woman in Linnaeus's probable source.
Also in the new primate order in the tenth edition of the *Systema naturae* of 1758
were a genus for monkeys and apes, one for lemurs, and one for bats. But there is
quite another way to see Linnaeus's activity as the "father" of a discourse about
nature. He referred to himself as a second Adam, the "eye" of God, who could give
true representations, true names, thus reforming or restoring a purity of names
lost by the first Adam's sin.[7] Nature was a theatre, a stage for the playing out of
natural and salvation history. The role of the one who renamed the animals was to
ensure a true and faithful order of nature, to purify the eye and the word. The
"balance of nature" was maintained partly by the role of a new "man" who would
see clearly and name accurately, hardly a trivial identity in the face of eighteenth-
century European expansion. Indeed, this is the identity of the modern authorial
subject, for whom inscribing the body of nature gives assurance of his mastery.

Linnaeus's science of natural history was intimately a Christian science. Its first
task, achieved in Linnaeus's and his correspondents' life work, was to announce the
kinship of "man" and beast in the modern order of an expanding Europe. Natural
man was found not only among the "savages," but also among the animals, who
were named primates in consequence, the first Order of nature. Those who could
bestow such names had a powerful modern vocation; they became scientists. Taxonomy had a secular sacred function. The "calling" to practice science has kept this
sacralized character into the late twentieth century, although we will see it at its
strongest in the early part of our century. The stories produced by such practitioners
have a special status in a repressed protestant biblical culture like that of the United
States.

Nature for Linnaeus was not understood "biologically," but "representationally."
In the course of the nineteenth century, biology became a discourse about productive, expanding nature. Biology was constructed as a discourse about nature known
as a system of production and reproduction, understood in terms of the functional
division of labor and the mental, labor, and sexual efficiency of organisms. In this
context, by the twentieth century primates were cast into an *Ecological Theatre and
an Evolutionary Play* (Hutchinson 1965). The drama has been about the origin and

development of many persistent mythic themes: sex, language, authority, society, competition, domination, cooperation, family, state, subsistence, technology, and mobility. There are two major readings of the play adopted in this book: One attends to symbolic meanings, to the primate sciences as a kind of art form making repeated use of the narrative resources of Judeo-Christian myth systems. The second pays particular attention to the ways primate biology is theorized as a material system of production and reproduction, a kind of "materialist" reading. Both interpretations listen for echoes and determinants of race, sex, and class in the stories. The primate body, as part of the body of nature, may be read as a map of power. Biology, and primatology, are inherently political discourses, whose chief objects of knowledge, such as organisms and ecosystems, are icons (condensations) of the whole of the history and politics of the culture that constructed them for contemplation and manipulation. The primate body itself is an intriguing kind of political discourse.

Primatology is Simian Orientalism

The argument of this book is that primatology is about an Order, a taxonomic and *therefore* political order that works by the negotiation of boundaries achieved through ordering differences. These boundaries mark off important social territories, like the norm for a proper family, and are established by social practice, like curriculum development, mental health policy, conservation politics, film making, and book publishing. The two major axes structuring the potent scientific stories of primatology that are elaborated in these practices are defined by the interacting dualisms, *sex/gender* and *nature/culture*. Sex and the west are axiomatic in biology and anthropology. Under the guiding logic of these complex dualisms, western primatology is simian orientalism. [Figure 1.1]

Edward Said (1978) argued that western (European and American) scholars have had a long history of coming to terms with countries, peoples, and cultures in the Near and Far East that is based on the Orient's special place in western history—the scene of origins of language and civilization, of rich markets and colonial possession and penetration, and of imaginative projection. The Orient has been a troubling resource for the production of the Occident, the "East's" other and periphery that became materially its dominant. The West is positioned outside the Orient, and this exteriority is part of the Occident's practice of representation. Said quotes Marx, "They cannot represent themselves; they must be represented" (xiii). These representations are complex mirrors for western selves in specific historical moments. The west has also been positioned mobily; westerners could be *there* with relatively little resistance from the other. The difference has been one of power. The structure has been limiting, of course, but more importantly, it has been *productive*. That productivity occurred within the structured practices and discourses of orientalism; the structures were a condition of having anything to say. There never is any question of having anything truly original to say about origins. Part of the authority of the practices of telling origin stories resides precisely in their intertextual relations.

Without stretching the comparison too far, the signs of orientalist discourse mark primatology. But here, the scene of origins is not the cradle of civilization, but the cradle of culture, of human being distinct from animal existence. If orientalism

Figure 1.1   Tom Palmore, *Reclining Nude*, 1976, acrylic on canvas. The Philadelphia Museum of Art: Purchased with funds given by Marion B. Stroud and the Adele Haas Turner and Beatrice Pastorius Turner Fund. Published with permission.

concerns the western imagination of the origin of the city, primatology displays the western imagination of the origin of sociality itself, especially in the densely meaning-laden icon of "the family." Origins are in principle inaccessible to direct testimony; any voice from the time of origins is structurally the voice of the other who generates the self. That is why both realist and postmodernist aesthetics in primate representations and simulations have been modes of production of complex illusions that function as fruitful generators of scientific facts and theories. "Illusion" is not to be despised when it grounds such powerful truths.

Simian orientalism means that western primatology has been about the construction of the self from the raw material of the other, the appropriation of nature in the production of culture, the ripening of the human from the soil of the animal, the clarity of white from the obscurity of color, the issue of man from the body of woman, the elaboration of gender from the resource of sex, the emergence of mind by the activation of body. To effect these transformative operations, simian "orientalist" discourse must first construct the terms: animal, nature, body, primitive, female. Traditionally associated with lewd meanings, sexual lust, and the unrestrained body, monkeys and apes mirror humans in a complex play of distortions over centuries of western commentary on these troubling doubles. Primatology is western discourse, and it is sexualized discourse. It is about potential and its actualization. Nature/culture and sex/gender are not loosely related pairs of terms; their specific form of relation is hierarchical appropriation, connected as Aristotle taught by the logic of active/passive, form/matter, achieved form/resource, man/animal, final/material cause. Symbolically, nature and culture, as well as sex and gender, mutually (but not equally) construct each other; one pole of a dualism cannot exist without the other.

Said's critique of orientalism should alert us to another important point: neither sex nor nature is the truth underlying gender and culture, any more than the "East" is really the origin and distorting mirror of the "West." Nature and sex are as crafted as their dominant "others." But their functions and powers are different. The task of this book is to participate in showing how the whole dualism is built, what the stakes might be in the architectures, and how the building might be redesigned. It matters to know precisely how sex and nature become natural-technical objects of knowledge, as much as it matters to explain their doubles, gender and culture. It is not the case that no story could be told without these dualisms or that they are part of the structure of the mind or language. For one thing, alternative stories within primatology exist. But these binarisms have been especially *productive* and especially *problematic* for constructions of female and race-marked bodies; it is crucial to see how the binarisms may be deconstructed and maybe redeployed.

It seems nearly impossible for those who produce natural sciences and comment on them for a living really to believe that there is no *given* reality beneath the inscriptions of science, no untouchable sacred center to ground and authorize an innocent and progressive order of knowledge. Maybe in the humanities there is no recourse from representation, mediation, story-telling, and social saturation. But the sciences succeed that other faulty order of knowledge; the proof is in their power to convince and reorder the whole world, not just one local culture. The natural sciences are the "other" to the human sciences, with their tragic orientalisms. But these pleas do not survive scrutiny.

The pleas of natural scientists do not convince because they are set up as the "other." The claims are predictable and seem plausible to those who make them because they are built into the taxonomies of western knowledge and because social and psychological needs are met by the persistent voices of the divided knowledge of natural and human sciences, by this division of labor and authority in the production of discourses. But these observations about predictable claims and social needs do not reduce natural sciences to a cynical "relativism" with no standards beyond arbitrary power. Nor does my argument claim there is no world for which people struggle to give an account, no referent in the system of signs and productions of meanings, no progress in building better accounts within traditions of practice. That would be to reduce a complex field to one pole of precisely the dualisms under analysis, the one designated as ideal to some impossible material, appearance to some forbidden real.

The point of my argument is rather that natural sciences, like human sciences, are inextricably *within* the processes that give them birth. And so, like the human sciences, the natural sciences are culturally and historically specific, modified, involved. They matter to real people. It makes sense to ask what stakes, methods, and kinds of authority are involved in natural scientific accounts, how they differ, for example, from religion or ethnography. It does not make sense to ask for a form of authority that escapes the web of the highly productive cultural fields that make the accounts possible in the first place. The detached eye of objective science is an ideological fiction, and a powerful one. But it is a fiction that hides—and is designed to hide—how the powerful discourses of the natural sciences really work. Again, the limits are *productive*, not reductive and invalidating.

One grating consequence of my argument is that the natural sciences are legitimately subject to criticism on the level of "values," not just "facts." They are subject to cultural and political evaluation "internally," not just "externally." But the evaluation is also implicated, bound, full of interests and stakes, part of the field of practices that make meanings for real people accounting for situated lives, including highly structured things called scientific observations. The evaluations and critiques cannot leap over the crafted standards for producing credible accounts in the natural sciences because neither the critiques nor the objects of their discourse have any place to stand "outside" to legitimate such an arrogant overview. To insist on value and story-ladenness at the heart of the production of scientific knowledge is not equivalent to standing nowhere talking about nothing but one's biases—quite the opposite. Only the pose of disinterested objectivity makes "concrete objectivity" impossible.

Part of the difficulty of approaching the embedded, interested, passionate constructions of science non-reductively derives from an inherited analytical tradition, deeply indebted to Aristotle and to the transformative history of "White Capitalist Patriarchy" (how may we name this scandalous Thing?) that turns everything into a resource for appropriation. As "resource" an object of knowledge is finally only matter for the seminal power, the act, of the knower. Here, the object both guarantees and refreshes the power of the knower, but any status as *agent* in the productions of knowledge must be denied the object. It—the world—must, in short, be objectified as thing, not agent; it must be matter for the self-formation of the only social being in the productions of knowledge, the human knower. Nature is only the raw material of culture, appropriated, preserved, enslaved, exalted, or otherwise made flexible for disposal by culture in the logic of capitalist colonialism. Similarly, sex is only the matter to the act of gender; the productionist logic seems inescapable in traditions of western binarisms. This analytical and historical narrative logic accounts for my nervousness about the sex/gender distinction in the recent history of feminist theory as a way to approach reconstructions of what may count as female and as nature in primatology—and why those reconstructions matter beyond the boundaries of primate studies. It has seemed all but impossible to avoid the trap of an appropriationist logic of domination built into the nature/culture binarism and its generative lineage, including the sex/gender distinction.

Reading in the Borderlands

There are many subjects in the history of biology and anthropology that could sustain the themes discussed in this introduction, so why has this book chosen to explore primate sciences in particular? The principal reason is that monkeys and apes, and human beings as their taxonomic kin, exist on the boundaries of so many struggles to determine what will count as knowledge. Primates are not nicely boxed into a specialized and secured discipline or field. Even in the late twentieth century, many kinds of people can claim to know primates, to the chagrin and dismay of many other contestants for official expertise. The cost of destabilizing knowledge about primates remains within reach not only for practitioners of several fields in the life and human sciences, but for people on the fringes of any science—like science writers, philosophers, historians, and zoo goers. In addition, story telling about animals is such a deeply popular practice that the discourse produced within scientific specialties is appropriated by other people for their own ends. The boundary between technical and popular discourse is very fragile and permeable. Even in the late twentieth century, the language of primatology is accessible in contentious political debate about human nature, history, and futures. This remains true despite a transformation of specialized discourses in primatology into the language of mathematics, systems theories, ergonomic analysis, game theory, life history strategies, and molecular biology.

Some of the interesting border disputes about primates, who and what they are (and who and what they are for), are between psychiatry and zoology, biology and anthropology, genetics and comparative psychology, ecology and medical research, agriculturalists and tourist industries in the "third world," field researchers and laboratory scientists, conservationists and multinational logging companies, poachers and game wardens, scientists and administrators in zoos, feminists and anti-feminists, specialists and lay people, physical anthropologists and ecological-evolutionary biologists, established scientists and new Ph.D.'s, women's studies students and professors in animal behavior courses, linguists and biologists, foundation officials and grant applicants, science writers and researchers, historians of science and real scientists, marxists and liberals, liberals and neo-conservatives. All of these intersections appear in this book.

How might different readers travel with pleasure in the borderlands of *Primate Visions*? This is a large book that may be read from start to finish as a chronological and thematic survey of twentieth-century primatology, with a major boundary at about 1955. But each chapter also stands by itself as an essay in cultural studies. Those most intrigued by popular culture might want to read first "Teddy Bear Patriarchy," focused on museum taxidermy and collecting safaris in colonial Africa, and "Apes in Eden, Apes in Space," examining National Geographic television specials in the context of the space race and decolonization. Primatologists might be most intrigued initially by the account of Robert Yerkes's Yale Laboratories of Primate Biology, C.R. Carpenter's pre-war field work, and the case studies of women working in primatology since the 1970s. Physical anthropologists might want to begin with the debates about fossil hominids and the field studies of monkeys and apes encouraged by Sherwood Washburn from the late 1950s. For questions about the reconstructions of nature in the context of decolonization, the reader might begin with "The Bio-politics of a Multicultural Field." An interest in psychological laboratory modeling of human social problems in the 1960s and 1970s might lead a reader to "Metaphors into Hardware: Harry Harlow and the Technology of Love." People coming to *Primate Visions* from feminist studies might want to begin by reading Part Three, "The Politics of Being Female: Primatology Is a Genre of Feminist Theory."

But each chapter is simultaneously history of science, cultural studies, feminist exploration, and engaged intervention into the constitutions of love and knowledge in the disciplined crafting of the Primate Order. I hope that the readers who begin in the position of one of the intended audiences for this book find themselves invited to become members of all of the audiences. And I hope that readers will not be "audience" in the sense of receivers of a finished story. Conventions within the narrative field of SF seem to require readers radically to rewrite stories in the act of reading them. My placing this account of primatology within SF—the narratives of speculative fiction and scientific fact—is an invitation for the readers of *Primate Visions*—historians, culture critics, feminists, anthropologists, biologists, anti-racists, and nature lovers—to remap the borderlands between nature and culture. I want the readers to find an "elsewhere" from which to envision a different and less hostile order of relationships among people, animals, technologies, and land. Like the actors in the stories that follow, I also want to set new terms for the traffic between what we have come to know historically as nature and culture.

Part One

Monkeys and Monopoly Capitalism: Primatology Before World War II

2
PRIMATE COLONIES AND THE EXTRACTION OF VALUE

Before the Second World War, non-human primates were already the subject of international western interest, with research stations and conservation areas fostered by France, Belgium, Russia, Germany, and the United States.[1] Literally and figuratively, primate studies were a colonial affair, in which knowledge of the living and dead bodies of monkeys and apes was part of the system of unequal exchange of extractive colonialism. Primate bodies grounded the discourses that rested on a flow of value from the lands where monkeys and apes lived to the lands where they were exhibited and textualized. Nonhuman primates were a fundamental part of the apparatus of colonial medicine. Part of the ideological framework justifying this directed flow of knowledge was the great chain of being structuring western imperial imaginations; apes especially were located in a potent place on that chain.

In 1924, the French established "Pastoria," a colonial outpost of the Institut Pasteur at Kindia in French Guinea. They also maintained animals at Tunis, as well as a laboratory colony, or "Singerie," in Paris.[2] The French colonial ideology of the civilizing mission fueled the press releases' lively and racist imagination of what went on behind laboratory walls. The international colonial press heralded "Pastoria" with stories of impending "civilizing" experiments on chimpanzees in an effort to determine the limits of their mental capacity. The Chicago *Tribune Ocean Times* boldly announced, "French to Establish Model Village as Training Grounds for Apes in which Civilizing Experiments Will Be Carried Out. Native Women as Nurses and Guides."[3] [Figure 2.1] Race, sex, and animality were simultaneously constituted

CHICAGO TRIBUNE OCEAN TIMES

FRENCH TO ESTABLISH MODEL VILLAGE, AND TRAINING GROUNDS FOR APES, IN WHICH CIVILISING EXPERIMENTS WILL BE TRIED OUT

Native Women as Nurses and Guides.

It is a good many years since the late Professor Garner attempted unsuccessfully to learn the monkey language in the wilds of Africa. Now the Pasteur Institute, with the support of the French Government, has launched a much more imposing project.

This is nothing less than the establishment of the first ape village, where chimpanzees and gorillas will live and be waited on by native women attendants and studied by scientists.

This unique township is situated in the heart of the jungle at Kindin, French Guinea, and already eighty apes are installed there.

Young apes only are to be captured and brought to the new colony, where serious efforts are to be made to teach them to speak and educate them generally, if possible, to the level of human beings.

In addition they will be inoculated with various serums and efforts will be made to try and discover the precise cause and cure of cancer, tuberculosis, infant paralysis, and other diseases which continue to baffle medical science.

Monkey Blood like Human.

Professor Georges Calmette, head of the Pasteur Institute, amplifying the facts concerning this interesting experiment, says that the blood of the chimpanzee has the same qualities as human blood, and that microbic diseases can therefore be transmitted.

Apart from the medical side, he points out how interesting it will be to watch the intellectual development,

which may result in teaching us that the intelligence of the chimpanzee can be trained to a high point.

Native women will act as nurses to the apes, preside over their games, and assist in their education and provision of their meals. They will take them for walks in the shade of the palm trees, leading them by the hand as if they were little children.

Individual Freedom.

Each ape is to have his own hut built on piles a few feet above the ground. The interiors will be warmed by a hot water system, whilst attached to every little house a fountain of running water provides the inmate with ample opportunities for drinking and bathing.

During the day it is planned to give the apes considerable freedom; but the hut doors will be closed at night.

The project includes covering recreation ground with netting at the top and providing windows at the side, to preserve the apes from the dust of the earth and contagious diseases of the neighbourhood.

The bill of fare covers four meals a day, and great care will be exercised in the choice and preparation of the food provided.

Professor Calmette does not expect any great progress for the first few years; but he hopes that the experiment will result in the gain of considerable knowledge in the prevention and treatment of diseases against which the human race is at present more or less helpless.

Figure 2.1 "French to Establish Model Village," from the Chicago *Tribune Ocean Times*, 1924. Robert M. Yerkes Papers, Manuscripts and Archives, Yale University Library. Published with permission.

in the primate order. In 1924 the International Feature Service, Inc., of Great Britain provided a story about plans for "A Monkey College to Make Chimpanzees Human." [Figure 2.2] The bright chimps and orangs (captured on another continent) would be assigned to the monkey college, while the less-gifted apes and all the monkeys would be assigned to serum experiments in tropical medicine. The college, alleged the international colonial news service, would provide native women as nurses, but European teachers trained in human child psychology. The confusion about whether "native women" would be merely servants to the young animals, or also their guides or teachers, suggests the boundary confusion at the nether regions of the great chain of being. There, the bodies marked by race, sex, and species luxuriated in the tropical colonial endeavors that joined black, female, and animal in fantasy and in enforced social labor.

The inflamed imagination of the press envisioned hundreds of ape students shepherded across the border between human and animal. But Director Calmette's remarks had been far more formal and measured. Photographs of the chimpanzee quarters at Kindia revealed a stony, barred prison, an architecture which was typical

Figure 2.2 "A Monkey College to Make Chimpanzees Human," from International Feature Service, Inc., 1924. Robert M. Yerkes Papers, Manuscripts and Archives, Yale University Library. Published with permission.

for the period of primate culture in the metropolis, as well as the colonies (Yerkes and Yerkes 1929: 596). The animals at Kindia were mostly subjects in the study of tropical medicine, a far more crucial link in the chain of factors sustaining western survival in the tropics than chimpanzee genius.

The Germans had their simian outpost in the Canary Islands, on Tenerife.[4] Belgium established the first colonial national park in the Belgian Congo, to protect the mountain gorilla and to ground international research efforts.[5] Before World War II, the United States had two off-shore sites with wild or free-ranging primate colonies, the Barro Colorado Island in the Panama Canal Zone, later administered by the Smithsonian Institution, and Cayo Santiago, attached to the University of Puerto Rico and the Columbia College of Physicians and Surgeons in New York City. Studies of wild monkeys as vectors for tropical diseases were carried out by Herbert Clark from the Gorgas Memorial Institute for Tropical Medicine in Panama. Those expeditions also included behavioral observations and collecting for anatomical series.

The Russians housed a small domestic colony at the Zoological Laboratory at the Darwinian Museum in Moscow, where Nadia Kohts conducted her elegant observations of chimpanzee mental capacities.[6] The Soviet Union established a permanent primate research station at Sukhumi in 1924. From the 1890s, provision of rhesus monkeys for medical research was made at the St. Petersberg Institute of Experimental Medicine; and in North America, Robert Yerkes established the Yale Laboratories primarily for psychobiological research, with the chimpanzee breeding station located in Florida to be as close to the tropics as possible in the continental United States. George Washington Corner and Carl Hartman established the first U.S. physiological laboratory housing nonhuman primates, in the Embryology Department of the Carnegie Institution in Baltimore, in order to provide animals for their study of reproductive physiology, especially the primate menstrual cycle.[7]

Expressing typical beliefs about the matrix of his studies, Hartman argued that anthropology was the study of the human family tree, phylogeny, while gynecology was the study of human reproduction. From a laboratory which produced basic knowledge of human and animal reproductive physiology in the years of rapid discovery in endocrinology, Hartman (1932) situated his work in evolutionary comparative terms, grounded on the one end by anthropology as evolutionary kinship and on the other end by gynecology as experimental medicine. It is this kind of tying of technical and mythic strands that weaves the scientific objects of knowledge called race and sex.

Despite some efforts at animal husbandry, most of the laboratory colonies were not self-sufficient breeding stations, much less an adequate source of research animals for sale to other laboratories, as demand for the animals expanded. The primate animal trade is a critical and unexamined part of the history of primatology from the start. Perhaps the most successful early breeding colony of macaques used for reproductive physiological research was under the direction and personal care of Gertrude van Wagenen, in the Department of Obstetrics at Yale University Medical School from the 1930s. Van Wagenen was one of the very few women primate researchers before the 1950s.[8]

The importance of monkeys to tropical disease research and to reproductive and nervous physiology accounted for the modest scientific commitments, underwritten by western governments and philanthropic foundations, to understand these animals. Beginning in the late 1800s primates appeared sporadically in medical research on human diseases, e.g., those of Metchnikoff on syphillis and Landsteiner on polio. Maurice Wakeman's famous studies on yellow fever were conducted in the bodies of Nigerian yellow monkeys. For many years before Pastoria was established, Metchnikoff had importuned the French government to fund a primate research resource. Monkeys began to appear in neurophysiological experiments in the second half of the nineteenth century, and small numbers were kept in various institutions for the purpose. But the "standardized" research monkey, a position occupied principally by the rhesus monkey in the latter half of the twentieth century, had not yet put in its historical appearance on the primate stage.

The perceived relevance of monkeys and apes to questions of human evolution was another major basis of primate investigation. Physical anthropologists, who were also often medical men, studied comparative anatomy and phylogeny. Several key English-speaking figures in this aspect of primate studies in the early twentieth century were born in the white settler colonies of the British empire (Solly Zuckerman, Raymond Dart) or were Euro-American men whose writing was fundamental to the constitution of race at the heart of primate physical anthropology (E.A. Hooton, W. Gregory, Carleton Coon, H.F. Osborn, and Ales Hrdlicka). Until the middle of the twentieth century, European production of comparative physical anthropology far exceeded North American.[9]

In the United States before mid-century, primatology was also a psychobiological discipline. The tie to medicine and to social interventions, considered as a social therapeutics, grounded primate studies both technically and ideologically. Close ties linked psychobiology both to neural and reproductive biology and to psychiatric and anthropological theories and practices. For example, G.V. Hamilton, who had studied with Robert Yerkes at Harvard and with the psychiatrist Adolph Meyer at the Phipps Clinic at the Johns Hopkins, had a private collection of primates at his estate at Montecito, California. Both a comparative psychologist and psychopathologist, Hamilton studied the phylogeny of mental disorders, especially those tied to sex and learning, inscribed in gonad and brain, the mirror twins of the biological body. Hamilton was also concerned with the biomedical characterization of homosexuality. His studies on captive and free-ranging primates on his estate were a substantial part of the scientific foundation for the belief that primate females exist in a nearly constant state of sexual "receptivity." That belief fell hard in post-war behavioral and ecological investigations, despite cogent criticism.[10] The belief was from the beginning crucial to the scientific construction of "the family" and its defining function of the cultural regulation of biological resource. Ordered by marriage, the heterosexual pair bond grounded the human nuclear family, and so averted sexual chaos. The phylogenesis of psychopathology of the sexual function was a major concern.[11]

Often emphasizing fauna native to particular national colonial possessions, zoological parks in western cities were foci of primate exhibition. For example, Copenhagen, London, Dusseldorf, Paris, Rotterdam, and Berlin all opened new ape or monkey houses between 1900 and 1930. Less often, the zoological collections were utilized for quasi-systematic behavioral study; and occasional eccentrics made fascinating observations of wild primates, such as those by the Afrikaans-speaking naturalist, Eugene Marais.[12]

Private collections complemented the zoological gardens. Perhaps the most spectacular was Mme. Abreu's in Havana, where the first chimpanzee born in captivity in the western hemisphere appeared in 1915. Yerkes's stunning description of Abreu's estate and practices with her animals is a perfect portrait of the intersecting construction of nonhuman primates as pets, surrogate children, endangered species, research animals, colonial subjects, and wild animals. In all these aspects, Abreu's animals were, in Yerkes's words, "almost human." The more orthodox human children, distressed by the depletion of their patrimony through the upkeep of their illegitimate kin, dispersed the collection as soon as possible after their mother's death. Yerkes got several of the animals for his new research colony.

Intricately connected with Yerkes's scientific networks, Clarence Ray Carpenter was another father of primatology; the field is crowded with origins and with contenders for the paternal function. The spatial distributions (read by sociometry) and behavioral interactions of "undisturbed" animals (read as the semiotics of signal exchanges mediating social bonds) were the text of his practice. Carpenter's filmic techniques were shaped by his belief in unmediated direct vision of the natural animal in its natural setting. Simple and powerful field techniques, such as the census copied from ornithology, allowed him to construct the elements for the production of his principle natural-technical object: the whole animal in the whole social group. Carpenter believed that each primate species had a typical grouping pattern explained by the socionomics of sex, i.e., by the principles of sexual efficiencies. The socionomic sex ratio grounded social cooperation, the balanced resolution of the potentially disruptive forces of sex, dominance, and aggression.

But before World War II, with the crucial exception of the naturalistic field studies of Carpenter and the minor exceptions of Bingham (1932) and Nissen (1931), primatology was overwhelmingly a laboratory and museum-based affair. As subjects of science, living monkeys and apes were in labs and public or private collections, and dead ones were in cabinets and dioramas in universities and museums. Expeditions to the "wild" were made primarily to collect animals for circuses, the pet trade, medical research, zoos, or museums, and only incidentally to record the lives of the animals in their own worlds. In its ethnographic dimensions for animals and scientists, "the field," later to become such a potent scene of primatology, was only dimly discerned in the first half of the twentieth century.

From John Fulton's Physiology Department at the Yale Medical School, Theodore Ruch (1941) compiled a comprehensive bibliography on the primates, beginning with ancient texts and concluding in 1940. The newcomer status of North American contributions in all categories and the international character of primate literature stand out. Ten pages of extremely inhomogeneous entries cover the period before 1800; only the twentieth-century scientific bibliographer could have assembled these texts into a common book. There follow 201 pages for the nineteenth and twentieth centuries, comprising 4090 entries. Anatomy, Physiology, and Pharmacology account for 3299 of them. Pathology was reserved for a planned second volume. Phylogeny accounts for only 92 entries. Experimental Psychobiology, itself medically allied, was covered in 288 entries. Comprising 333 publications, Observational Psychobiology is also a very miscellaneous category from late-twentieth-century points of view. It included hunting and travel narratives, a zoo keeper's observations of captive primates' fear of snakes, accounts of home-reared chimpanzees in suburban New York, an English translation of a German nature lover's account of the marvelous animals in the Roman zoo, expanding the genre of German Italian travel literature, and Carpenter's severely positivist howler monkey monograph (1934).

The defamiliarizing power of this bibliography matches Foucault's use of Borges's " 'certain Chinese encyclopedia' "

> in which it is written that 'animals are divided into: (a) belonging to the Emperor, (b) embalmed, (c) tame, (d) sucking pigs, (e) sirens, (f) fabulous, (g) stray dogs, (h) included in the present classification, (i) frenzied, (j) innumerable, (k) drawn with a very fine camelhair brush, (l) *etcetera*, (m) having just broken the water pitcher, (n) that from a long way off look like flies'. In the wonderment of this taxonomy, the thing we apprehend . . . as the exotic charm of another system of thought, is the limitation of our own, the stark impossibility of thinking *that*. (Foucault 1971; italics in original)

What is it about the animals we call primates that holds together all the authors of Ruch's compendium? What is the implicit question that made the prodigious labor of assembling that list make sense to its sponsor, the Yale Medical Library? Ruch's book introduced the word *primatology* and called those who studied these animals *primatologists*, a term he set off with quotation marks. In the introduction, J.F. Fulton notes the relative paucity of "primatological" literature—a mere 5000 titles compared to the lavish 50,000 catalogued for studies of fishes before 1914. Humans eat fish; we watch primates, and build from them cultural-, racial-, and gender-specific models for spiritual and bodily ills. In 1758, the order Primates was new; at the halfway mark of the twentieth century, the discursive order of "primatology" was christened.

3

Teddy Bear Patriarchy
Taxidermy in the Garden of Eden, New York City, 1908–1936

Nature teaches law and order and respect for property. If these people cannot go to the country, then the Museum must bring nature to the city.[1]

I started my thoughts on the legend of Romulus and Remus who had been suckled by a wolf and founded Rome, but in the jungle I had my little Lord Greystoke suckled by an ape.[2]

Experience

In the heart of New York City stands Central Park—the urban garden designed by Frederick Law Olmsted to heal the overwrought or decadent city dweller with a prophylactic dose of nature. Across from the park the Theodore Roosevelt Memorial presides as the central building of the American Museum of Natural History, a monumental reproduction of the Garden of Eden.[3] In the Garden, Western "man" may begin again the first journey, the first birth from within the sanctuary of nature. Founded just after the Civil War and dedicated to popular education and scientific research, the American Museum of Natural History is the place to undertake this genesis, this regeneration. Passing through the Museum's Roosevelt Memorial atrium into the African Hall, opened in 1936, the ordinary citizen enters a privileged space and time: the Age of Mammals in the heart of Africa, scene origins.[4] A hope is implicit in every architectural detail: in immediate vision of the origin, perhaps the future can be fixed. By saving the beginnings, the end can be achieved and the present can be transcended. African Hall offers a unique communion with nature at its highest and yet most vulnerable moment, the moment of the interface of the Age of Mammals with the Age of Man. This communion is offered through the sense of vision by the craft of taxidermy. Its most ecstatic and skillful moment joins ape and man in visual embrace.

Restoration of the origin, the task of genetic hygiene, is achieved in Carl Akeley's African Hall by an art that began for him in the 1880s with the crude stuffing of

26

P. T. Barnum's elephant, Jumbo, who had been run down by a railroad train, the emblem of the Industrial Revolution. The end of his task came in the 1920s, with his exquisite mounting of the Giant of Karisimbi, the lone silverback male gorilla that dominates the diorama depicting the site of Akeley's own grave in the mountainous rain forest of the Congo, today's Zaire. So it could inhabit Akeley's monument to the purity of nature, this gorilla was killed in 1921, the same year the Museum hosted the Second International Congress of Eugenics. From the dead body of the primate, Akeley crafted something finer than the living organism; he achieved its true end, a new genesis. Decadence—the threat of the city, civilization, machine—was stayed in the politics of eugenics and the art of taxidermy. The Museum fulfilled its scientific purpose of conservation, preservation, and the production of permanence. Life was transfigured in the principal civic arena of western political theory—the natural body of man.[5]

Behind every mounted animal, bronze sculpture, or photograph lies a profusion of objects and social interactions among people and other animals, which can be recomposed to tell a biography embracing major themes for twentieth-century United States. But the recomposition produces a story that is reticent, even mute, about Africa. H. F. Osborn, president of the American Museum from 1908–33, thought Akeley was Africa's biographer. But in a stronger sense, Akeley is America's biographer, at least for part of North America. Akeley thought in African Hall the visitor would experience nature at its moment of highest perfection. He did not dream that he crafted the means to experience a history of race, sex, and class in New York City that reached to Nairobi.

To enter the Theodore Roosevelt Memorial, the visitor must pass by a James Earle Fraser equestrian statue of Teddy majestically mounted as a father and protector between two "primitive" men, an American Indian and an African, both standing, dressed as "savages." The facade of the memorial, funded by the State of New York and awarded to the American Museum of Natural History on the basis of its competitive application in 1923, is classical, with four Ionic columns 54 feet high topped by statues of the great explorers Boone, Audubon, Lewis, and Clark. The coin-like, bas-relief seals of the United States and of the Liberty Bell are stamped on the front panels. Inscribed across the top are the words TRUTH, KNOWLEDGE, VISION and the dedication to Roosevelt as "a great leader of the youth of America, in energy and fortitude in the faith of our fathers, in defense of the rights of the people, in the love and conservation of nature and of the best in life and in man." Youth, paternal solicitude, virile defense of democracy, and intense emotional connection to nature are the unmistakable themes.[6]

The building presents itself in many visible faces. It is at once a Greek temple, a bank, a scientific research institution, a popular museum, a neoclassical theater. One is entering a space that sacralizes democracy, Protestant Christianity, adventure, science, and commerce. Entering this building, one knows that a drama will be enacted inside. Experience in this public monument will be intensely personal; this structure is one of North America's spaces for joining the duality of self and community.

Just inside the portals, the visitor enters the sacred space where transformation of consciousness and moral state will begin.[7] The walls are inscribed with Roosevelt's words under the headings Nature, Youth, Manhood, the State. The seeker begins in Nature: "There are no words that can tell the hidden spirit of the wilderness, that can reveal its mystery. . . . The nation behaves well if it treats its natural resources as assets which it must turn over to the next generation increased and not impaired in value." Nature is mystery and resource, a critical union in the history of civilization. The visitor—necessarily a white boy in moral state, no matter what accidents of biology or social gender and race might have pertained prior to the Museum excursion—progresses through Youth: "I want to see you game boys . . . and gentle and tender. . . . Courage, hard work, self mastery, and intelligent effort are essential to a successful life." Youth mirrors Nature, its pair across the room. The next stage is Manhood: "Only those are fit to live who do not fear to die and none are fit to die who have shrunk from the joy of life and the duty of life." Opposite is its spiritual pair, the State: "Aggressive fighting for the right is the noblest sport the world affords. . . . If I must choose between righteousness and peace, I choose righteousness." The walls of the atrium are full of murals depicting Roosevelt's life, the perfect illustration of his words. His life is inscribed in stone in a peculiarly literal way appropriate to this museum. One sees the man hunting big game in Africa, conducting diplomacy in the Philippines and China, helping boy and girl scouts, receiving academic honors, and presiding over the Panama Canal ("The land divided, the world united").

Finally, in the atrium stand the striking life-size bronze sculptures by Carl Akeley of the Nandi spearmen of East Africa on a lion hunt. These African men and the lion they kill symbolize for Akeley the essence of the hunt, of what would later be named "man the hunter." Discussing the lion spearers, Akeley referred to them as men. In every other circumstance he referred to adult male Africans as boys. Roosevelt, the modern sportsman, and the "primitive" Nandi share in the spiritual truth of manhood. The noble sculptures express Akeley's great love for Roosevelt, his friend and hunting companion in Africa in 1910 for the killing of one of the elephants which Akeley mounted for the Museum. Akeley said he would follow Roosevelt anywhere because of his "sincerity and integrity" (Akeley 1923: 162).

In the Museum shop in the atrium in the 1980s, one may purchase *T.R.: Champion of the Strenuous Life*, a photographic biography of the 26th president. Every aspect of the fulfillment of manhood is depicted, even death is labeled "The Great Adventure." One learns that after defeat in the presidential campaign of 1912, Roosevelt undertook the exploration of the Amazonian tributary, the River of Doubt, under the auspices of the American Museum of Natural History and the Brazilian Government. It was a perfect trip. The explorers nearly died, the river had never before been seen by white men, and the great stream, no longer doubtful, was renamed Rio Roosevelt by the Brazilian State. In the picture biography, which includes a print of the adventurers paddling their primitive dugout canoe (one assumes before starvation and jungle fever attenuated the ardor of the photographer), the former president of a great industrial power explains his return to the wilderness: "I had to go. It was my last chance to be a boy" (Johnson 1958: 138, 126–7).[8]

The joining of life and death in these icons of Roosevelt's journeys and in the architecture of his stony memorial announces the central moral truth of the Museum. This is the effective truth of manhood, the state conferred on the visitor who successfully passes through the trial of the Museum. The body can be transcended. This is the lesson Simone de Beauvoir so painfully remembered in the *Second Sex*; man is the sex which risks life and in so doing, achieves his existence. In the upside down world of Teddy Bear Patriarchy, it is in the craft of killing that life is constructed, not in the accident of personal, material birth. Roosevelt is the perfect *locus genii* for the Museum's task of regeneration of a miscellaneous, incoherent urban public threatened with genetic and social decadence, threatened with the prolific bodies of the new immigrants, threatened with the failure of manhood.[9]

The Akeley African Hall itself is simultaneously a very strange place and an ordinary experience for literally millions of North Americans over more than five decades. The types of display in this hall are spread all over the country, and even the world, partly due to the craftspeople Akeley himself trained. In the 1980s sacrilege is perhaps more evident than liminal experience of nature. What is the experience of New York streetwise kids wired to Walkman radios and passing the Friday afternoon cocktail bar by the lion diorama? These are the kids who came to the Museum to see the high tech Nature-Max films. But soon, for those not physically wired into the communication system of the late twentieth century, another time begins to take form. The African Hall was meant to be a time machine, and it is (Fabian 1983: 144). The individual enters the age of Mammals. But one enters alone, each individual soul, as part of no stable prior community and without confidence in the substance of one's body, in order to be received into a saved community. One begins in the threatening chaos of the industrial city, part of a horde, but here one will come to belong, to find substance. No matter how many people crowd the Great Hall, the experience is of individual communion with nature. The sacrament will be enacted for each worshipper. This nature is not constituted from a probability calculus. This is not a random world, populated by late twentieth-century cyborgs, for whom the threat of decadence is a nostalgic memory of a dim organic past, but the moment of origin where nature and culture, private and public, profane and sacred meet—a moment of incarnation in the encounter of man and animal.

The Hall is darkened, lit only from the display cases which line the sides of the spacious room. In the center of the Hall is a group of elephants so lifelike that a moment's fantasy suffices for awakening a premonition of their movement, perhaps an angry charge at one's personal intrusion. The elephants stand like a high altar in the nave of a great cathedral. That impression is strengthened by one's growing consciousness of the dioramas that line both sides of the main Hall and the spacious gallery above. Lit from within, the dioramas contain detailed and lifelike groups of large African mammals—game for the wealthy New York hunters who financed this experience. Called habitat groups, they are the culmination of the taxidermist's art. Called by Akeley a "peep-hole into the jungle,"[10] each diorama presents itself as a side altar, a stage, an unspoiled garden in nature, a hearth for home and family. As an altar, each diorama tells a part of the story of salvation history; each has its special emblems indicating particular virtues. Above all, inviting the visitor to share in its revelation, each tells the truth. Eachs offers a vision. Each is a window onto knowledge.

A diorama is eminently a story, a part of natural history. The story is told in the pages of nature, read by the naked eye. The animals in the habitat groups are captured in a photographer's and sculptor's vision. They are actors in a morality play on the stage of nature, and the eye is the critical organ. Each diorama contains a small group of animals in the foreground, in the midst of exact reproductions of plants, insects, rocks, soil. Paintings reminiscent of Hollywood movie set art curve in back of the group and up to the ceiling, creating a great panoramic vision of a scene on the African continent. Each painting is minutely appropriate to the particular animals in the foreground. Among the 28 dioramas in the Hall, all the major geographic areas of the African continent and most of the large mammals are represented.

Gradually, the viewer begins to articulate the content of the story. Most groups are made up of only a few animals, usually a large and vigilant male, a female or two, and one baby. Perhaps there are some other animals—a male adolescent maybe, never an aged or deformed beast. The animals in the group form a developmental series, such that the group can represent the essence of the species as a dynamic, living whole. The principles of organicism, that is, of the laws of organic form, rule the composition.[11] There is no need for the multiplication of specimens because the series is a true biography. Each animal is an organism, and the group is an organism. Each organism is a vital moment in the narrative of natural history, condensing the flow of time into the harmony of developmental form. The groups are peaceful, composed, illuminated—in "brightest Africa."[12] Each group forms a community structured by a natural division of function; the whole animal in the whole group is nature's truth. The physiological division of labor that has informed the history of biology is embodied in these habitat groups which tell of communities and families, peacefully and hierarchically ordered. Sexual specialization of function— the organic bodily and social sexual division of labor—is unobtrusively ubiquitous, unquestionable, right. The African buffalo, the white and black rhinos, the lion, the zebra, the mountain nyala, the okapi, all find their place in the differentiated developmental harmony of nature. The racial division of labor, the familial progress from youthful native to adult white man, was announced at the steps leading to the building itself; Akeley's original plan for African Hall included bas-relief sculptures of all the "primitive" tribes of Africa complementing the other stories of natural wild life in the Hall. Organic hierarchies are embodied in every organ in the articulation of natural order in the Museum.[13]

But there is a curious note in the story; it begins to dominate as scene after scene draws the visitor into itself through the eyes of the animals in the tableaux.[14] Each diorama has at least one animal that catches the viewer's gaze and holds it in communion. The animal is vigilant, ready to sound an alarm at the intrusion of man, but ready also to hold forever the gaze of meeting, the moment of truth, the original encounter. The moment seems fragile, the animals about to disappear, the communion about to break; the Hall threatens to dissolve into the chaos of the Age of Man. But it does not. The gaze holds, and the wary animal heals those who will look. There is no impediment to this vision, no mediation. The glass front of the diorama forbids the body's entry, but the gaze invites his visual penetration. The animal is frozen in a moment of supreme life, and man is transfixed. No merely living organism could accomplish this act. The specular commerce between man and animal at the interface of two evolutionary ages is completed. The animals in the dioramas have transcended mortal life, and hold their pose forever, with muscles tensed, noses aquiver, veins in the face and delicate ankles and folds in the supple skin all prominent. No visitor to a merely physical Africa could see these animals. This is a spiritual vision made possible only by their death and literal re-presentation. Only then could the essence of their life be present. Only then could the hygiene of nature cure the sick vision of civilized man. Taxidermy fulfills the fatal desire to represent, to be whole; it is a politics of reproduction.

There is one diorama that stands out from all the others, the gorilla group. It is not simply that this group is one of the four large corner displays. There is something special in the painting with the steaming volcano in the background and Lake Kivu below, in the pose of the enigmatic large silverback rising above the group in a chest-beating gesture of alarm and an unforgettable gaze in spite of the handicap of glass eyes. The painter's art was particularly successful in conveying the sense of limitless vision, of a panorama without end around the focal lush green garden. This is the scene that Akeley longed to return to. It is where he died, feeling he was at home as in no other place on earth. It is where he first killed a gorilla and felt the enchantment of a perfect garden. After his first visit in 1921, he was motivated to convince the Belgian government to make this area the first African national park to ensure a sanctuary for the gorilla. But the viewer does not know these things when he sees the five animals in a naturalistic setting. It is plain that he is looking at a natural family of close human relatives, but that is not the essence of this diorama. The viewer sees that the elephants, the lion, the rhino, and the water hole group—with its peaceful panorama of all the grassland species, including the carnivores, caught in a moment outside the Fall—all these have been a kind of preparation, not so much for the gorilla group, as for the Giant of Karisimbi. This double for man stands in a unique personal individuality, his fixed face molded forever from the death mask cast from his corpse by a taxidermist in the Kivu Mountains. Here is natural man, immediately known. His image may be purchased on a picture postcard at the desk in the Roosevelt atrium. [Figure 3.1]

It would have been inappropriate to meet the gorilla anywhere else but on the mountain. Frankenstein and his monster had Mont Blanc for their encounter; Akeley and the gorilla first saw each other on the lush volcanoes of central Africa. The glance proved deadly for them both, just as the exchange between Victor Frankenstein and his creature froze each of them into a dialectic of immolation. But Frankenstein tasted the bitter failure of his fatherhood in his own and his creature's death; Akeley resurrected his creature and his authorship in both the sanctuary of Parc Albert and the African Hall of the American Museum of Natural History. Mary Shelley's story may be read as a dissection of the deadly logic of birthing in patriarchy at the dawn of the age of biology; her tale is a nightmare about the crushing failure of the project of man. But the taxidermist labored to restore manhood at the interface of the Age of Mammals and the Age of Man. Akeley achieved the fulfillment of a sportsman in Teddy Bear Patriarchy—he died a father to the game, and their sepulcher is named after him, the Akeley African Hall.

The gorilla was the highest quarry of Akeley's life as artist, scientist, and hunter, but why? He said himself (through his ghostwriter, the invisible Dorothy Greene), "To me the gorilla made a much more interesting quarry than lions, elephants, or any other African game, for the gorilla is still comparatively unknown" (Akeley 1923: 190). But so was the colobus monkey or any of a long list of animals. What qualities did it take to make an animal "game"? One answer is similarity to man, the ultimate quarry, a worthy opponent. The ideal quarry is the "other," the natural self. That is one reason Frankenstein needed to hunt down his creature. Hunter, scientist, and artist all sought the gorilla for his revelation about the nature and future of manhood. Akeley compared and contrasted his quest for the gorilla with the French-American Paul du Chaillu's, the first white man to kill a gorilla, in 1855,

Figure 3.1   The Giant of Karisimbi. Negative no. 315077. Published with permission of the Department of Library Services, American Museum of Natural History.

eight years after it was "discovered" to science. Du Chaillu's account of the encounter stands as the classic portrayal of a depraved and vicious beast killed in the heroic, dangerous encounter. Disbelieving du Chaillu, Akeley told his own readers how many times du Chaillu's publishers made him rewrite until the beast was fierce enough. Frankenstein plugged up his ears rather than listen to his awful son claim a gentle and peace loving soul. Akeley was certain he would find a noble and peaceful beast; so he brought his guns, cameras, and white women into the garden to hunt, wondering what distance measured courage in the face of a charging alter-ego.

Like du Chaillu, Akeley came upon a sign of the animal, a footprint, or in Akeley's case a handprint, before meeting face to face. "I'll never forget it. In that mud hole were the marks of four great knuckles where the gorilla had placed his hand on the ground. There is no other track like this in the world—there is no other hand in the world so large. . . . As I looked at that track I lost the faith on which I had brought my party to Africa. Instinctively I took my gun from the gun boy" (Akeley 1923: 203). Later, Akeley told that the handprint, not the face, gave him his greatest thrill. In the hand the trace of kinship writ large and terrible struck the craftsman.

But then, on the first day out from camp in gorilla country, Akeley did meet a gorilla face to face, the creature he had sought for decades, prevented from earlier success by mauling elephants, stingy millionaires, and world war. Within minutes of his first glimpse of the features of the face of an animal he longed more than anything to see, Akeley had killed him, not in the face of a charge, but through a dense forest screen within which the animal hid, rushed, and shook branches. Surely the taxidermist did not want to risk losing his specimen, for perhaps there would be no more. He knew the Prince of Sweden was just then leaving Africa after having shot fourteen of the great apes in the same region. The animals must be wary of new hunters; collecting might be very difficult.

Whatever the rational or fantastic logic that ruled the first shot, precisely placed into the aorta, the task that followed was arduous indeed—skinning the animal and transporting various remains back to camp. The corpse had nearly miraculously lodged itself against the trunk of a tree above a deep chasm. As a result of Herculean labors, which included casting the death mask pictured in *Lions, Gorillas, and their Neighbors* (Akeley and Akeley 1922), Akeley was ready for his next gorilla hunt on the second day after shooting the first ape. The pace he was setting himself was grueling, dangerous for a man ominously weakened by tropical fevers. "But science is a jealous mistress and takes little account of a man's feelings."[15] The second quest resulted in two missed males, a dead female, and her frightened baby speared by the porters and guides. Akeley and his party had killed or attempted to kill every ape they had seen since arriving in the area.

On his third day out, Akeley took his cameras and ordered his guides to lead toward easier country. With a baby, female, and male, he could do a group even if he got no more specimens. Now it was time to hunt with the camera.[16] "Almost before I knew it I was turning the crank of the camera on two gorillas in full view with a beautiful setting behind them. I do not think at the time I appreciated the fact that I was doing a thing that had never been done before" (Akeley 1923: 221). But the photogenic baby and mother and the accompanying small group of other gorillas had become boring after two hundred feet of film, so Akeley provoked an action shot by standing up. That was interesting for a bit. "So finally, feeling that I had about all I could expect from that band, I picked out one that I thought to be an immature male. I shot and killed it and found, much to my regret, that it was a female. As it turned out, however, she was such a splendid large specimen that the feeling of regret was considerably lessened" (Akeley 1923: 222).

Satisfied with the triumphs of his gun and camera, Akeley decided it was time to ask the rest of the party waiting in a camp below to come up to hunt gorillas. He was getting considerably sicker and feared he would not fulfill his promise to his friends to give them gorilla. His whole purpose in taking white women into gorilla country depended on meeting this commitment: "As a naturalist interested in preserving wild life, I was glad to do anything that might make killing animals less attractive."[17] The best thing to reduce the potency of game for heroic hunting is to demonstrate that inexperienced women could safely do the same thing. Science had already penetrated; women could follow.

Two days of hunting resulted in Herbert Bradley's shooting a large silverback, the one Akeley compared to Jack Dempsey and mounted as the lone male of Karisimbi in African Hall. It was now possible to admit another level of feeling: "As he lay at the base of the tree, it took all one's scientific ardour to keep from feeling like a murderer. He was a magnificent creature with the face of an amiable giant who would do no harm except perhaps in self defense or in defense of his family" (Akeley 1923: 230). If he had succeeded in his aborted hunt, Victor Frankenstein could have spoken those lines.

The photograph in the American Museum film archive of Carl Akeley, Herbert Bradley, and Mary Hastings Bradley holding up the gorilla head and corpse to be recorded by the camera is an unforgettable image.[18] The face of the dead giant evokes Bosch's conception of pain, and the lower jaw hangs slack, held up by Akeley's hand. The body looks bloated and utterly heavy. Mary Bradley gazes smilingly at the faces of the male hunters, her own eyes averted from the camera. Akeley and Herbert Bradley look directly at the camera in unshuttered acceptance of their act. Two Africans, a young boy and a young man, perch in a tree above the scene, one looking at the camera, one at the hunting party. The contrast of this scene of death with the diorama framing the giant of Karisimbi mounted in New York is total; the animal came to life again, this time immortal.

There was no more need to kill, so the last capture was with the camera. "The guns were put behind and the camera pushed forward and we had the extreme satisfaction of seeing the band of gorillas disappear over the crest of the opposite ridge none the worse for having met with white men that morning. It was a wonderful finish to a wonderful gorilla hunt" (Akeley 1923: 235). Once domination is complete, conservation is urgent. But perhaps preservation comes too late.

What followed was the return to the United States and active work for an absolute gorilla sanctuary providing facilities for scientific research. Akeley feared the gorilla would be driven to extinction before it was adequately known to science (Akeley 1923: 248). Scientific knowledge canceled death; only death before knowledge was final, an abortive act in the natural history of progress. His health weakened but his spirit at its height, Akeley lived to return to Kivu to prepare paintings and other material for the gorilla group diorama. Between 1921 and 1926, he mounted his precious gorilla specimens, producing that extraordinary silverback whose gaze dominates African Hall. When he did return to Kivu in 1926, he was so exhausted from his exertions to reach his goal that he died on November 17, 1926, almost immediately after he and his party arrived on the slopes of Mt. Mikena, "in the land of his dreams" (M. J. Akeley 1929b: Chpt. XV).

Akeley's was a literal science dedicated to the prevention of decadence, of biological decay. His grave was built in the heart of the rain forest on the volcano, where "all the free wild things of the forest have perpetual sanctuary" (M. J. Akeley 1940: 341). Mary Jobe Akeley directed the digging of an eight-foot vault in lava gravel and rock. The hole was lined with closely set wooden beams. The coffin was crafted on the site out of solid native mahogany and lined with heavy galvanized steel salvaged from the boxes used to pack specimens to protect them from insect and other damage. Then the coffin was upholstered with camp blankets. A slab of cement ten by twelve feet and five inches thick was poured on top of the grave and inscribed with the name and date of death of the father of the game. The cement had been carried on porters' backs all the way from the nearest source in Kibale, Uganda. The men ditched the first load in the face of the difficult trails; they were sent back for a second effort. An eight-foot stockade fence was built around the grave to deter buffalo and elephant from desecrating the site. "Derscheid, Raddatz, Bill and I worked five days and five nights to give him the best home we could build, and he was buried as I think he would have liked with a simple reading service and a prayer" (M. J. Akeley 1929b: 189–90). The grave was inviolate, and reincarnation of the natural self would be immortal in African Hall. In 1979, "grave robbers, Zairoise poachers, violated the site and carried off [Akeley's] skeleton" (Fossey 1983: 3).

Biography

> For this untruthful picture Akeley substitutes a real gorilla." (Osborn, in Akeley 1923: xii)

> Of the two I was the savage and the aggressor. (Akeley 1923: 216)

Akeley sought to craft a true life, a unique life. The life of Africa became his life, his telos. But it is not possible to tell his life from a single point of view. There is a polyphony of stories, and they do not harmonize. Each source for telling the story of Akeley's life speaks in an authoritative mode, but I felt compelled to compare the versions, and then to cast Akeley's story in an ironic mode, the register most avoided by my subject. Akeley wanted to present an immediate vision; I would like to dissect and make visible layer after layer of mediation. I want to show the reader how the experience of the diorama grew from the safari in specific times and places, how the camera and the gun together are the conduits for the spiritual commerce of man and nature, how biography is woven into and from a social and political tissue. I want to show how the stunning animals of Akeley's achieved dream in African Hall are the product of particular technologies, i.e., the techniques of effecting meanings.

Life Stories

In harmony with the available plots in U.S. history, it is necessary that Carl Akeley (1864–1926) was born on a farm in New York of poor, but vigorous, old, (white— the only trait that didn't need to be named), American stock. The time of his birth, near the end of the Civil War, was an end and a beginning for so much in North America, including the history of biology and the structure of wealth and social class. In a boyhood full of hard farm labor, he learned self-reliance and skill with tools and machines. He passed long hours alone watching and hunting the wildlife of New York. By the age of 13, aroused by a borrowed book on the subject, Akeley was committed to the vocation of taxidermy. His vocation's bibliogenesis seems also ordained by the plot. At that age (or age 16 in some versions), he had a business card printed up. No Yankee boy could miss the connection of life's purpose with business, although young Carl scarcely believed he could make his living at such a craft. He took lessons in painting, so that he might provide realistic backgrounds for the birds he ceaselessly mounted. From the beginning Akeley's life had a single focus: the recapturing and representation of the nature he saw. On this point all the versions of Akeley's life concur.

After the crops were in, at the age of 19, Akeley set off from his father's farm "to get a wider field for my efforts" (Akeley 1923: 1). First he tried to get a job with a local painter and interior decorator whose hobby was taxidermy, but this man directed the boy to an institution which changed his life—Ward's Natural Science Establishment in Rochester, where Akeley would spend four years and form a friendship pregnant with consequences for the nascent science of ecology as it came to be practiced in museum exhibition. Ward's provided mounted specimens and natural history collections for practically all the museums in the nation. Several important men in the history of biology and museology in the United States passed through this curious institution, including Akeley's friend, William Morton Wheeler. Wheeler completed his career in entymology at Harvard, a founder of the science of animal ecology (which he called ethology—the science of the character of nature) and a mentor to the great organicists and conservative social philosophers in Harvard's biological and medical establishment (Russett 1966; Evans and Evans 1970; Cross and Albury 1987). Wheeler was then a young Milwaukee naturalist steeped in German "Kultur" who began tutoring the rustic Akeley for entry into Yale's Sheffield Scientific School. However, eleven hours of taxidermy in the day and long hours of study proved too much; so higher education was postponed, later permanently, in order to follow the truer vocation of reading nature's book directly.

Akeley was disappointed at Ward's because business imperatives allowed no room for improvement of taxidermy. He felt animals were "upholstered." Developing his own skill and technique in spite of the lack of encouragement, and the lack of money, he got a chance for public recognition when P. T. Barnum's famous elephant was run down by a locomotive in Canada in 1885. Barnum did not want to forego the fame and profit from continuing to display the giant (who had died trying to save a baby elephant, we are told), so Akeley and a companion were dispatched to Canada from Rochester to save the situation. Six butchers from a nearby town helped with the rapidly rotting carcass. What Akeley learned about very large mammal taxidermy from this experience laid the foundation for his later revolutionary innovations in producing light, strong, life-like pachyderms. The popular press followed the monumental mounting, and the day Jumbo was launched in his own railroad car into his post-mortem career, half the population of Rochester witnessed the resurrection.

In 1885, Wheeler returned to Milwaukee to teach high school and soon took up a curatorship in the Milwaukee Museum of Natural History. Wheeler urged his friend to follow, hoping to continue his tutoring and to secure Akeley commissions for specimens from the museum. Museums did not then generally have their own taxidermy departments, although around 1890 taxidermic technique flowered in Britain and the United States. Akeley opened his business shop on the Wheeler family property, and he and the naturalist spent long hours discussing natural history, finding themselves in agreement about museum display and about the character of nature. The most important credo for them both was the need to develop scientific knowledge of the whole animal in the whole group in nature—i.e., they were committed organicists. Wheeler soon became director of the Milwaukee Museum and gave Akeley significant support. Akeley had conceived the idea for habitat groups and wished to mount a series illustrating the fur-bearing animals of Wisconsin. His completed muskrat group (1889), minus the painted backgrounds, was probably the first mammalian habitat group anywhere.

As a result of a recommendation from Wheeler, in 1894 the British Museum invited Akeley to practice his trade in that world-famous institution. On the way to London, Akeley visited the Field Museum in Chicago, met Daniel Giraud Elliot and accepted his offer of preparing the large collection of specimens the Museum had bought from Ward's. In 1896, Akeley made his first collecting expedition to Africa, to British Somaliland, a trip that opened a new world to him. This was the first of five safaris to Africa, each escalating his sense of the purity of the continent's vanishing wildlife and the conviction that the meaning of his life was its preservation through transforming taxidermy into an art. He was next in Africa for the Field Museum in 1905, with his explorer/adventurer/author wife, Delia, to collect elephants in British East Africa. On this trip Akeley escaped with his life after killing a leopard in hand-to-fang combat.

In Chicago Akeley spent four years largely at his own expense preparing the justly famous Four Seasons deer dioramas. In 1908, at the invitation of the new president, H. F. Osborn, who was anxious to mark his office with the discovery of major new scientific laws and departures in museum exhibition and public education, Akeley moved to New York and the American Museum of Natural History in hope of preparing a major collection of large African mammals. From 1909–11 Carl and Delia collected in British East Africa, a trip marked by a hunt with Theodore Roosevelt and his son Kermit, who were collecting for the Washington National Museum. The safari was brought to a limping conclusion by Carl's being mauled by an elephant, delaying fulfillment of his dream of collecting gorillas. His plan for the African Hall took shape by 1911 and ruled his behavior thereafter. In World War I he was a civilian Assistant Engineer to the Mechanical and Devices Section of the Army. He is said to have refused a commission in order to keep his freedom to speak freely to anyone in the hierarchy.

During the war, his work resulted in several patents in his name. The theme of Akeley the inventor recurs constantly in his life story. Included in his roster of inventions, several of which involved subsequent business development, were a motion picture camera, a cement gun, and new taxidermic processes.

With the close of war, Akeley focused his energy on getting backing for the African Hall. He needed more than a million dollars. Lecture tours, articles, a book, and endless promotion brought him into touch with the major wealthy sportsmen of New York, but sufficient financial commitment eluded him. In 1921, financing half the expense himself, Akeley left for Africa, this time accompanied by a married couple, their 5-year-old daughter, their governess, and Akeley's adult niece whom he had promised to take hunting in Africa. In 1923 in New York, Carl and Delia divorced—an event unrecorded in versions of his life; Delia just disappears from the narratives. In 1924 Akeley married Mary L. Jobe, the explorer/adventurer/

author who accompanied him on his last adventure, the Akeley-Eastman-Pomeroy African Hall Expedition, that collected for ten dioramas of the Great Hall. George Eastman, of Eastman Kodak fortunes, and Daniel Pomeroy, the benefactors, accompanied the taxidermist-hunter to collect specimens. Eastman, then 71 years old, went with his own physician and commanded his own railroad train for part of the excursion.

En route to Africa the Akeleys were received by the conservationist and war hero Belgian king, Albert. He was the son of the infamous Leopold II, whose personal rapacious control of the Congo for profit was wrested away and given to the Belgian government by other European powers in 1908. Leopold II had financed Henry Stanley's explorations of the Congo. Akeley is narrated as a man like the great explorers, Stanley and Livingstone, but also as the man who witnessed, and helped birth, a new "bright" Africa. The "enlightened" Albert, led to his views on national parks by a visit to Yosemite, confirmed plans for the Parc Albert and commissioned the Akeleys to prepare topographical maps and descriptions of the area in cooperation with the Belgian naturalist, Jean Derscheid. There was no room for a great park for the Belgians in Europe, so "naturally" one was established in the Congo. Mandating protection for the Pygmies within park boundaries, the park was to provide sanctuary for "natural primitives," as well as foster scientific study by establishing permanent research facilities. After ten months of collecting, Carl and Mary Jobe set off for the Kivu forest, the heart of remaining unspoiled Africa, where he died and was buried "in ground the hand of man can never alter or profane" (M. J. Akeley 1940: 340).

Taxidermy: From Upholstery to Epiphany

> Transplanted Africa stands before him—a result of Akeley's dream. (Clark 1936: 73)

The vision Carl Akeley had seen was one of jungle peace. His quest to *embody* this vision justified to himself his hunting, turned it into a tool of science and art, the scalpel that revealed the harmony of an organic, articulate world. Let us follow Akeley briefly through his technical contributions to taxidermy in order to grasp more fully the stories he needed to tell about the biography of Africa, the life history of nature.

It is a simple tale: Taxidermy was made into the servant of the "real." Artifactual children, better than life, were birthed from dead matter (Sofoulis 1988). Akeley's vocation, and his achievement, was the production of an organized craft for eliciting unambiguous experience of organic perfection. Literally, Akeley "typified" nature, made nature true to type. Taxidermy was about the single story, about nature's unity, the unblemished type specimen. Taxidermy became the art most suited to the epistemological and aesthetic stance of realism. The power of this stance is in its magical effects: what is so painfully constructed appears effortlessly, spontaneously found, discovered, simply there if one will only look. Realism does not appear to be a point of view, but appears as a "peephole into the jungle" where peace may be witnessed. Epiphany comes as a gift, not as the fruit of merit and toil, soiled by the hand of man. Realistic art at its most deeply magical issues in revelation. This art repays labor with transcendence. Small wonder that artistic realism and biological science were twin brothers in the founding of the civic order of nature at the American Museum of Natural History. It is also natural that taxidermy and biology depend fundamentally upon vision in a hierarchy of the senses; they are tools for the construction, discovery of form.

Akeley's eight years in Milwaukee from 1886 to 1894 were crucial for his working out techniques that served him the rest of his life. The culmination of that period was a head of a male Virginia deer that won first place in the first Sportsman's Show, in New York City in 1895. The judge in that national competition was Theodore Roosevelt, whom Akeley did not meet until they befriended each other on safari in Africa in 1906. The head, entitled "The Challenge," displayed a buck "in the full frenzy of his virility as he gave the defiant roar of the rutting season—the call to fierce combat" (M. J. Akeley 1940: 38). Jungle peace was not a passive affair, nor one unmarked by gender.

The head was done in a period of experimentation leading to the production of the Four Seasons group in Chicago, installed in 1902.[19] In crafting those groups over four years, Akeley worked out his manikin method, clay modeling, plaster casting, vegetation molding techniques, and the organized production system. He hired women and men workers by the hour to turn out the thousands of individual leaves needed to clothe the trees in the scenes. Charles Abel Corwin painted background canvases from studies in the Michigan Iron Mountains where the animals were collected. Akeley patented his vegetation process, but gave rights for its use free of charge to the Field Museum in Chicago. He allowed free, worldwide use of his patented methods of producing light, strong papier-mache manikins from exact clay models and plaster casts. Cooperation in museum development was a fundamental value for Akeley, who did not make much money at his craft and whose inventions were significant for economic survival.

Akeley continued to make improvements in his taxidermic technique throughout his life, and he taught several other key workers, including James Lipsitt Clark, who was the Director of Arts, Preparation, and Installation at the American Museum after Akeley's death when African Hall was actually constructed. While Akeley worked long hours alone, taxidermy as he helped to develop it was not a solitary art. Taxidermy requires a complex system of coordination and division of labor, beginning in the field during the hunting of the animals and culminating in a finished diorama. A minimum list of workers on one of Akeley's projects includes taxidermists, collectors, artists, anatomists, and "accessory men" (M. J. Akeley 1940: 217). Pictures of work in the Museum taxidermy studios show men (males, usually white) tanning hides, working on clay models of sizable mammals (including elephants) or on plaster casts, assembling skeleton and wood frames, consulting scale models of the planned display, doing carpentry, making vegetation, sketching, etc. Clark reports that between 1926 and 1936, when African Hall opened, still unfinished, the staff of the project usually employed about 45 men. Painting the backgrounds was a major artistic specialization, and the artists based their final panoramas on numerous studies done at the site of collection. In the field, the entire operation rested on the organization of the safari, a complex social institution where race, sex, and class came together intensely. Skinning a large animal could employ 50 workers for several hours. Photographs, moving picture records, death masks, extensive anatomical measurements, initial treatment of skins, and sketches occupied the field workers. The production of a modern diorama involved the work of hundreds of people in a social system embracing structures of skill and authority on a worldwide scale.

How can such a system produce a unified biography of nature? How is it possible to refer to Akeley's African Hall when it was constructed after he died? On an ideological level, the answer to these questions connects to the ruling conception of organicism, an organic hierarchy, conceived as nature's principle of organization. Clark stressed the importance of "artistic composition" and described the process as a "recreation" of nature based on the principles of organic form. This process required a base of "personal experience," ideally actual presence in Africa, at the site of the animal's life and death. Technical crafts are always imagined to be subordinated by the ruling artistic idea, itself rooted authoritatively in nature's own life. "Such things must be felt, must be absorbed and assimilated, and then in turn, with understanding and enthusiasm, given out by the creator. . . . Therefore, our groups are very often conceived in the very lair of the animals" (Clark 1936: 71).

The credos of realism and organicism interdigitate; both are systematizations of organization by a hierarchical division of labor, perceived as natural and so productive of unity. Unity must be *authored* in the Judeo-Christian myth system; just as nature has an Author, so does the organism or the realistic diorama. The author must be imagined with the aspects of mind, in relation to the body which executes. Akeley was intent on avoiding lying in his work; his craft was to tell the truth of nature. There was only one way to achieve such truth—the rule of mind rooted in the claim to experience. All the work must be done by men who did their collecting and studies on the spot because "[o]therwise, the exhibit is a lie and it would be nothing short of a crime to place it in one of the leading educational institutions of the country" (Akeley 1923: 265). A single mind infused collective experience: "If an exhibition hall is to approach its ideal, its plan must be that of a master mind, while in actuality it is the product of the correlation of many minds and hands" (Akeley 1923: 261). The "mind" is spermatic.

But above all, this sense of telling a true story rested on the selection of individual animals, the formation of groups of "typical" specimens. What was the meaning of "typical" for Akeley and his contemporaries in the biological departments of the American Museum of Natural History? What are the contents of these stories, and what must one do to see these contents? To respond, we must follow Carl Akeley into the field and watch him select an animal to mount. Akeley's concentration on finding the typical specimen, group, or scene cannot be overemphasized. But how could he know what was typical, or that such a state of being existed? This problem has been fundamental in the history of biology; one effort at solution is embodied in African Hall.

First, the concept includes the notion of perfection. The large bull giraffe in the water hole group in African Hall was the object of a hunt over many days in 1921. Several animals were passed over because they were too small or not colored beautifully enough. Remembering record trophies from earlier hunters undermined satisfaction with a modern, smaller specimen taken from the depleted herds of vanishing African nature. When at last the bull was taken as the result of great skill and daring, the minute details of its preservation and recreation were lovingly described.

Similarly, in 1910–11, the hunt for a large bull elephant provided the central drama of the safari for the entire two years. An animal with asymmetrical tusks was rejected, despite his imposing size. Character, as well as mere physical appearance, was important in judging an animal to be perfect.Cowardice would disqualify the most lovely and properly proportioned beast. Ideally, the killing itself had to be accomplished as a sportsmanlike act. Perfection was heightened if the hunt were a meeting of equals. So there was a hierarchy of game according to species: lions, elephants, and giraffes far outranked wild asses or antelope. The gorilla was the supreme achievement, almost a definition of perfection in the heart of the garden at the moment of origin. Perfection inhered in the animal itself, but the fullest meanings of perfection inhered in the meeting of animal and man, the moment of perfect vision, of rebirth. Taxidermy was the craft of remembering this perfect experience. Realism was a supreme achievement of the artifactual art of memory, a rhetorical achievement crucial to the foundations of Western science (Fabian 1983: 105–41). Memory was an art of reproduction.

There is one other essential quality for the typical animal in its perfect expression: it must be an adult male. Akeley describes hunting many fine females, and he cared for their hides and other details of reconstruction with all his skill. But never was it necessary to take weeks and risk the success of the entire enterprise to find the perfect female. There existed an image of an animal which was somehow *the* gorilla or *the* elephant incarnate. That particular tone of perfection could only be heard in the male mode. It was a compound of physical and spiritual quality judged truthfully by the artist-scientist in the fullness of direct experience. Perfection was marked by exact quantitative measurement, but even more by virile vitality known by the hunter-scientist from visual communion. Perfection was known by natural kinship; type, kind, and kin mutually and seminally defined each other.

Akeley hunted for a series or a group, not just for individuals. How did he know when to stop the hunt? Two groups give his criterion of wholeness, the gorilla group collected in 1921 and the original group of four elephants mounted by Akeley himself after the 1910–11 safari. Akeley once shot a gorilla, believing it to be a female, but found it to be a young male. He was disturbed because he wished to kill as few animals as possible and he believed the natural family of the gorilla did not contain more than one male. When he later saw a group made up of several males and females, he stopped his hunt with relief, confident that he could tell the truth from his existing specimens. Also, the photograph of Akeley's original group of four elephants unmistakably shows a perfect family. Nature's biographical unit, the reproductive group had the moral and epistemological status of truth-tellers.

Akeley wanted to be an artist and a scientist. Giving up his early plan of obtaining a degree from Yale Sheffield Scientific School and then of becoming a professional sculptor, he combined art and science in taxidermy. Since that art required that he also be a sculptor, he told some of his stories in bronzes as well as in dioramas. His criteria were similar; Akeley had many stories to tell, but they all expressed the same fundamental vision of a vanishing, threatened scene. In his determination to sculpt "typical" Nandi lion spearmen, Akeley used as models extensive photographs, drawings, and "selected types of American negroes which he was using to make sure of perfect figures" (Johnson 1936: 47). The variety of nature had a purpose— to lead to discovery of the highest type of each species of wildlife, including human beings outside "civilization."

Besides sculpture and taxidermy, Akeley perfected another narrative tool, photography. All of his story-telling instruments relied primarily on vision, but each caught and held slightly different manifestations of natural history. As a visual art, taxidermy occupied for Akeley a middle ground between sculpture and photography. Both sculpture and photography were subordinate means to accomplishing the final taxidermic scene. But photography also represented the future and sculpture the past. Akeley's practice of photography was suspended between the manual touch of sculpture, which produced knowledge of life in the fraternal discourses of organicist biology and realist art, and the virtual touch of the camera, which has dominated our understanding of nature since World War II. The nineteenth century produced the masterpieces of animal bronzes inhabiting the world's museums. Akeley's early twentieth-century taxidermy, seemingly so solid and material, appears as a brief frozen temporal section in the incarnation of art and science, before the camera technically could pervert his single dream into the polymorphous, absurdly intimate filmic reality we now take for granted. Critics accuse Akeley's taxidermy and the American Museum's expensive policy of building the great display halls in the years before World War II of being armature against the future, of having literally locked in stone one historical moment's way of seeing, while calling this vision the whole (Kennedy 1968: 204). But Akeley was a leader technically and spiritually in the perfection of the camera's eye. Taxidermy was not armed against the filmic future, but froze one frame of a far more intense visual communion to be consummated in virtual images. Akeley helped produce the armature—and armament—that would advance into the future.

Photography: Hunting with the Camera

> Guns have metamorphosed into cameras in this earnest comedy, the ecology safari, because nature has ceased to be what it had always been—what people needed protection from. Now nature—tamed, endangered, mortal—needs to be protected from people. When we are afraid, we shoot. But when we are nostalgic, we take pictures. (Sontag 1977: 15)

Akeley and his peers feared the disappearance of their world, of their social world in the new immigrations after 1890 amd the resulting dissolution of the old imagined hygienic, pre-industrial America. Civilization appeared to be a disease in the form of technological progress and the vast accumulation of wealth in the practice of monopoly capitalism by the very wealthy sportsmen who were trustees of the Museum and the backers of Akeley's African Hall. The leaders of the American Museum were afraid for their health; that is, their manhood was endangered. Theodore Roosevelt knew the prophylaxis for this specific historical malaise: the true man is the true sportsman. Any human being, regardless of race, class, and gender, could spiritually participate in the moral status of healthy manhood in democracy, even if only a few (anglo-saxon, male, heterosexual, Protestant, physically robust, and economically comfortable) could express manhood's highest forms. From about 1890 to the 1930s, the Museum was a vast public education and research program for producing experience potent to induce the fertile state of manhood. The Museum, in turn, was the ideological and material product of the sporting life. As Mary Jobe Akeley realized, "[the true sportsman] loves the game as if he were the father of it" (M. J. Akeley 1929b: 116). Akeley believed that the highest expression of sportsmanship was hunting with the camera: "Moreover, according to any true

conception of sport—the use of skill, daring, and endurance in overcoming difficulties—camera hunting takes twice the man that gun hunting takes" (Akeley 1923:
155). The true father of the game loves nature with the camera; it takes twice the
man, and the children are in his perfect image. The eye is infinitely more potent
than the gun. Both put a woman to shame—reproductively.

At the time of Akeley's first collecting safari in 1896, cameras were a nearly useless
encumbrance, incapable of capturing the goal of the hunt—life. According to
Akeley, the first notable camera hunters in Africa appeared around 1902, beginning
with Edward North Burton. The early books like Burton's were based on still
photographs; moving picture wildlife photography, owing much to Akeley's own
camera, did not achieve anything before the 1920s. On his 1910–11 safari to east
Africa, with the best available equipment, Akeley tried to film the Nandi lion
spearing. His failure due to inadequate cameras, described with great emotional
intensity, led him during the next five years to design the Akeley camera, which was
used extensively by the Army Signal Corps during World War I. Akeley formed
the Akeley Camera Company to develop his invention, which received its civilian
christening by filming Man-o-War win the Kentucky Derby in 1920, and his camera's
innovative telephoto lens caught the Dempsey-Carpentier heavyweight battle. Akeley's first taste of his own camera in the field was in 1921 in the Kivu forest.
Within a few days, Akeley shot his first gorillas with both gun and camera: in these
experiences he saw the culmination of his life. Awarded the John Price Wetherhill
Medal at the Franklin Institute in 1926 for his invention, Akeley succeeded that
year in filming to his satisfaction African lion spearing, on the same safari on which
Rochester's George Eastman, of Eastman-Kodak fortunes, was both co-sponsor and
hunter-collector.[20]

The ambiguity of the gun and camera runs throughout Akeley's work. He is a
transitional figure from the western image of darkest to lightest Africa, from nature
worthy of manly fear to nature in need of motherly nurture. The woman/scientist/
mother of orphaned apes popularized by the National Geographic Society's magazine and films in the 1970s was still half a century away. With Akeley, manhood
tested itself against fear, even as the lust for the image of jungle peace held the
finger on the gun long enough to take the picture and even as the intellectual and
mythic certainty grew that the savage beast in the jungle was human, in particular,
industrial human. The industrialist in the field with Akeley, George Eastman, was
an object lesson in the monopoly capitalist's greater fear of decadence than of death.
The narrative has a septagenarian Eastman getting a close-up photograph at 20
feet of a charging rhino, directing his white hunter when to shoot the gun, while
his personal physician looks on. "With this adventure Mr. Eastman began to enjoy
Africa thoroughly . . ." (M. J. Akeley 1940: 270).

Even at the literal level of physical appearance, "[t]o one familiar with the old
types of camera the Akeley resembled a machine gun quite as much as it resembled
a camera" (Akeley 1923: 166). Akeley said he set out to design a camera "that you
can aim . . . with about the same ease that you can point a pistol" (Akeley 1923:
166). He enjoyed retelling the apocryphal story of seven Germans mistakenly surrendering to one American when they found themselves faced by an Akeley. "The
fundamental difference between the Akeley motion-picture camera and the others
is a panoramic device which enables one to swing it all about, much as one would
swing a swivel gun, following the natural line of vision" (Akeley 1923: 167). Akeley

semi-joked in knowing puns on the penetrating, deadly invasiveness of the camera, naming one of his image machines "The Gorilla." "'The Gorilla' had taken 300 feet of film of the animal that had never heretofore been taken alive in its native wilds by any camera. . . . I was satisfied—more satisfied than a man ever should be—but I revelled in the feeling."[21]

The taxidermist, certain of the essential peacefulness of the gorilla, wondered how close he should let a charging male get before neglecting the camera for the gun. "I hope that I shall have the courage to allow an apparently charging gorilla to come within a reasonable distance before shooting. I hesitate to say just what I consider a reasonable distance at the present moment. I shall feel very gratified if I can get a photograph at twenty feet. I should be proud of my nerve if I were able to show a photograph of him at ten feet, but I do not expect to do this unless I am at the moment a victim of suicidal mania" (Akeley 1923: 197). Akeley wrote these words before he had ever seen a wild gorilla. What was the boundary of courage; how much did nature or man need protecting? What if the gorilla never charged, even when provoked? What if the gorilla were a coward (or a female)? Who, precisely, was threatened in the drama of natural history in the early decades of monopoly capitalism's presence in Africa and America?

Aware of a disturbing potential of the camera, Akeley set himself against faking. He stuffed Barnum's Jumbo, but he wanted no part of the great circus magnate's cultivation of the American popular art form, the hoax (Harris 1973). But hoax luxuriated in early wildlife photography (and anthropological photography). In particular, Akeley saw unscrupulous men manipulate nature to tell the story of a fierce and savage Africa that would sell in the motion picture emporia across America. Taxidermy had always threatened to lapse from art into deception, from life to upholstered death as a poor sportsman's trophy. Photography too was full of philistines who could debase the entire undertaking of nature work, the Museum's term for its educational work in the early 1900s. The Museum was for public entertainment (the point that kept its Presbyterian trustees resisting Sunday opening in the 1880s despite that day's fine potential for educating the new Catholic immigrants, who worked a six-day week); but entertainment only had value if it communicated the truth. Therefore, Akeley encouraged an association between the American Museum and the wildlife photographers, Martin and Osa Johnson, who seemed willing and able to produce popular motion pictures telling the story of jungle peace. Johnson claimed in his 1923 prospectus to the American Museum, "The camera cannot be deceived . . . [therefore, it has] enormous scientific value."[22]

Entertainment was interwoven with science, art, hunting, and education. Barnum's humbug tested the cleverness, the scientific acumen, of the observer in a republic where each citizen could discover the nakedness of the emperor and the sham of his rationality. This democracy of reason was always a bit dangerous. There is a tradition of active participation in the eye of science in America which makes the stories of nature ready to erupt into popular politics. Natural history can be— and has sometimes been—a means for millenial expectation and disorderly action. Akeley himself is an excellent example of a self-made man who made use of the mythic resources of the independent man's honest vision, the appeal to experience the testimony of one's own eyes. He *saw* the Giant of Karisimbi. The camera, an eminently democratic machine, has been crucial to crafting stories in biology. Its control has eluded the professional and the moralist, the official scientist. But in

Martin Johnson, Akeley hoped he had the man who would tame specular entertainment for the social uplift promised by science.

In 1906 Martin Johnson shipped out with Jack London for a two-year south sea voyage. The ship, the *Snark*, was the photographer's *Beagle*. Its name could hardly have been better chosen for the ship carrying the two adventurers whose books and films complemented *Tarzan* for recording the dilemma of manhood in the early twentieth century. Lewis Carroll's *The Hunting of the Snark* parodically anticipates the revelation of men like Johnson, London, and Akeley:

> In one moment I've seen what has hitherto been
>   Enveloped in absolute mystery,
> And without extra charge I will give you at large
>   A Lesson in Natural History. (Carroll 1971: 225)

From 1908–13 Johnson ran five motion picture houses in Kansas. He and Osa traveled in the still mysterious, potent places to film "native life": Melanesia, Polynesia, Malekula, Borneo, Kenya Colony. In 1922 the Johnsons sought Akeley's opinion of their new film, *Trailing African Wild Animals*. Akeley was delighted, and the Museum set up a special corporation to fund the Johnsons on a five-year African film safari. They planned a film on "African Babies." "It will show elephant babies, lion babies, zebra babies, giraffe babies, and black babies . . . showing the play of wild animals and the maternal care that is so strange and interesting a feature of wildlife."[23] African human life had the status of wildlife in the Age of Mammals. That was the logic for "protection"—the ultimate justification for domination. Here was a record of jungle peace.

The Johnsons also planned a big animal feature film. The museum lauded both the commercial and educational values. Osborn enthused, "The double message of such photography is, first, that it brings the aesthetic and ethical influence of nature within the reach of millions of people . . . second, it spreads the idea that our generation has no right to destroy what future generations may enjoy."[24] Johnson was confident that their approach of combining truth and beauty without hoax would ultimately be commercially superior, as well as scientifically accurate. "[T]here is no limit to the money it can make. . . . My past training, my knowledge of showmanship, mixed with the scientific knowledge I have absorbed lately, and the wonderful photographic equipment . . . make me certain that this Big Feature is going to be the biggest money maker ever placed on the market, as there is no doubt it will be the last big Africa Feature made, and it will be so spectacular that there will be no danger of another film of like nature competing with it. For these reasons it will produce an income as long as we live."[25] Africa had always promised gold.

The "naked eye" science advocated by the American Museum perfectly suited the camera, ultimately so superior to the gun for the possession, production, preservation, consumption, surveillance, appreciation, and control of nature. Akeley's aesthetic ideology of realism was part of his effort to bridge the yawning gaps in the endangered self. To make an exact image is to insure against disappearance, to cannibalize life until it is safely and permanently a specular image, a ghost. The image arrested decay. That is why nature photography is so beautiful and so religious—and such a powerful hint of an apocalyptic future. Akeley's aesthetic combined the instrumental and contemplative into a photographic technology providing a transfusion for a steadily depleted sense of reality. The image and the real define each other, as all of reality in late capitalist culture lusts to become an image for its own security. Reality is assured, insured, by the image, and there is no limit to the amount of money that can be made. The camera is superior to the gun for the control of time; and Akeley's dioramas with their photographic vision, sculptor's touch, and taxidermic solidity were about the end of time (Sontag 1977).

Telling Stories

The synthetic story told so far has had three major and many minor sources. Telling a life synthetically masks the tones emerging from inharmonious versions. The single biography, the achieved unity of African Hall, can be unraveled to tie its threads into an imagined heteroglossic narrative of nature yet to be written. A polyphonic natural history waits for its sustaining social history. To probe more deeply into the tissue of meanings and mediations making the specific structure of experience possible for the viewer of the dioramas of African Hall, I would like to tease apart the sources for a major event in Akeley's life, an elephant mauling in British East Africa in 1910. This event leavens my story of the structure and function of biography in the construction of a twentieth-century primate order, with its multiform hierarchies of race, sex, species, and class. Whose stories appear and disappear in the web of social practices that constitute Teddy Bear Patriarchy? Questions about authorized writing enforced by publishing practices and about labor that never issues in acknowledged authorship (never becomes father of the game) make up my story.[26]

Authors and Versions

> She didn't write it.
> She wrote it but she shouln't have.
> She wrote it, but look what she wrote about. (Russ 1983: 76)

In Brightest Africa appears to be written by Carl Akeley. But we learn from Mary Jobe Akeley (1940: 222), a prolific author, that the taxidermist "hated to wield a pen." She elaborates that Doubleday and Page (the men, not the company), were enthralled by Carl's stories told in their homes at dinner and so "determined to extract a book from him." So one evening after dinner Arthur W. Page "stationed a stenographer behind a screen, and without Carl's knowledge, she recorded every-thing he said while the guests lingered before the fire." Editing of this material is credited to Doubleday and Page, and the author is named as Carl. The stenographer is an unnamed hand. Her notes gave rise to articles in a journal called *World's Work*, but the publishers wanted a book. Then Akeley read a newspaper account of his Kivu journey that he liked; it had been written by Dorothy S. Greene while she worked for the director of the American Museum. Akeley hired her as his secretary, to record his stories while he talked with explorers and scientists or lectured to raise funds for African Hall. "She unobtrusively jotted down material which could be used in a book" (M.J. Akeley 1940: 223). Who wrote *In Brightest Africa*? To insist on that question troubles official versions of the relation of mind and body in western authorship.

The physical appearance of the books is itself an eloquent story. The stamp of approval from men like H. F. Osborn in the dignified prefaces, the presence of handsome photographs, a publishing house that catered to wealthy hunters: all compose the authority of the books. The frontispieces are like Orthodox icons; the entire story can be read from them. In *Lions, Gorillas and their Neighbors*, published for young people, the frontispiece shows an elderly Carl Akeley in his studio gazing intently into the eyes of the plaster death mask of the first gorilla he ever saw. Maturity in the encounter with nature is announced. *The Wilderness Lives Again*, the biography that resurrected Carl through his wife's vicarious authorship, displays in the front a young Carl, arm and hand bandaged heavily, standing outside a tent beside a dead leopard suspended by her hind legs. The caption reads: "Carl Akeley, when still in his twenties, choked this wounded infuriated leopard to death with his naked hands as it attacked him with intent to kill."

Carl Akeley's story of his encounter with the elephant that mauled him is in a chapter titled "Elephant Friends and Foes." Moral lessons pervade the chapter, prominently those of human ignorance of the great animals—partly because hunters are only after ivory and trophies, so that their knowledge is only of tracking and killing, not of the animals' lives—and of Akeley's difference because of his special closeness to nature embodied in the magnificent elephants. Akeley witnessed two elephants help a wounded comrade escape from the scene of slaughter, inspiring one of the taxidermist's bronzes. But, the reader also sees Akeley making a table to seat eight people out of elephant ears from a specimen which nearly killed him and Delia, despite each of them shooting into his head about 13 times. In this chapter, the taxidermist is hunting as an equal with his wife. He does not hide stories which might seem a bit seedy or full of personal bravado; yet his "natural nobility" pervaded all these anecdotes, particularly for an audience of potential donors to African Hall, who might find themselves shooting big game in Africa.

His near fatal encounter with an elephant occurred when Akeley had gone off without Delia to get photographs, taking "four days' rations, gun boys, porters, camera men, and so forth —about fifteen men in all" (Akeley 1923: 45). He was tracking an elephant whose trail was very fresh, when he suddenly became aware that the animal was bearing down on him directly:

> I have no knowledge of how the warning came. . . . I only know that as I picked up my gun and wheeled about I tried to shove the safety catch forward. It refused to budge. . . . My next mental record is of a tusk right at my chest. I grabbed it with my left hand, the other one with my right hand, and swinging in between them went to the ground on my back. This swinging in between the tusks was purely automatic. It was the result of many a time on the trails imagining myself caught by an elephant's rush and planning what to do, and a very profitable planning too: for I am convinced that if a man imagines such a crisis and plans what he would do, he will, when the occasion occurs, automatically do what he planned. . . . He drove his tusks into the ground on either side of me. (Akeley 1923: 48–49)

Akeley tells that he lay unconscious and untouched for hours because his men felt he was dead, and they came from groups which refused to touch a dead man. When he came to, he shouted and got attention. He relates that word had been sent to Mrs. Akeley at base camp, who valiantly mounted a rescue party in the middle of the night against the wishes of her guides (because of the dangers of night travel through the bush), whom she pursued into their huts to force their cooperation. Sending word to the nearest government post to dispatch a doctor, she arrived at the scene of the injury by dawn. Akeley attributed his recovery to Delia's fast action, but more to the subsequent speedy arrival of a neophyte Scottish doctor, who sped through the jungle to help the injured man partly out of his ignorance of the foolishness of hurrying to help anyone mauled by an elephant—such men simply didn't survive to pay for one's haste. The more seasoned chief medical officer arrived considerably later.

The remainder of the chapter recounts Akeley's chat with other old hands in Africa about their experiences surviving elephant attacks. Like his thoughts as he swung between the giant tusks, the tone is reasoned, scientific, focused on the behavior and character of those interesting aspects of elephant behavior. The ubiquitous moral concludes the chapter:

> But although the elephant is a terrible fighter in his own defense when attacked by man, that is not his chief characteristic. The things that stick in my mind are his sagacity, his versatility, and a certain comradeship which I have never noticed to the same degree in other animals. . . . I like to think back to the day I saw the group of baby elephants playing with a great ball of baked dirt. . . . They have no enemy but man and are at peace amongst themselves. It is my friend the elephant that I hope to perpetuate in the central group in Roosevelt African Hall. . . . In this, which we hope will be an everlasting monument to the Africa that was, the Africa that is fast disappearing, I hope to place the elephant on a pedestal in the centre of the hall—the rightful place for the first among them. (Akeley 1923: 54–5)

Akeley sees himself as an advocate for "nature" in which "man" is the enemy, the intruder, the dealer of death. His own exploits in the hunt stand in ironic juxtaposition only if the reader evades their true meaning—the tales of a pure man whose danger in pursuit of a noble cause brings him into communion with nature through the beasts he kills. This nature is a worthy brother of man, a worthy foil for his manhood. Akeley's elephant is profoundly male, singular, and representative of the possibility of nobility. The mauling was an exciting tale, with parts for many actors, including Delia, but the brush with death and the details of rescue are told with the cool humor of a man ready for his end dealt by such a noble friend and brother, his best enemy, the object of his scientific curiosity. The putative behavior of the "boys" underlines the confrontation between white manhood and the noble beast. "I never got much information out of the boys as to what did happen, for they were not proud of their part in the adventure. . . . It is reasonable to assume that they had scattered through [the area which the elephant thoroughly trampled] like a covey of quail . . ." (1923: 49). Casual and institutional racism heightens the life story of the single adult man. The action in Akeley's stories focuses on the center of the stage, on the meeting of the singular man and animal. The entourage is inaudible, invisible, except for comic relief and anecdotes about native life. In Akeley's rendering, empowered by class and race, white woman stands without much comment in a similar moral position as white man—a hunter, an adult.

Mary Jobe Akeley published her biography of her husband, *The Wilderness Lives Again*, in 1940, four years after the Akeley African Hall opened to the public. Her purpose was to promote conservation and fulfill her life's purpose—accomplishing her husband's life work. She presents herself as the inspired scribe for her husband's story. Through her vicarious authorship and through African Hall and the Parc Albert, not only the wilderness, but Akeley himself, whose meaning was the wilderness, lives again.

Mary Jobe had not always lived for a husband.[27] An explorer since 1913, she had completed ten expeditions to explore and map British Columbian wilderness; and the Canadian government named a peak Mt. Jobe. She recounts the scene at Carl's death when she accepted his commission for her, that she would live thereafter to fulfill his work. The entire book is suffused with her joy in this task. Her self-construction as the other is breathtaking in its ecstasy. The story of the elephant mauling undergoes interesting emendations to facilitate her accomplishment. One must read this book with attention because Carl's words from his field diaries and publications are quoted at great length with no typographical differentiation from the rest of the text. At no point does the wife give a source for the husband's words; they may be from conversation, lectures, anywhere. It does not matter, because the two are one flesh. The stories of Carl and Mary Jobe blend imperceptibly—until the reader starts comparing other versions of the "same" incidents, even the ones written apparently in the direct words of the true, if absent, author-husband.

The key emendation is an absence; the entire biography of Carl Akeley by Mary Jobe Akeley does not mention the name or presence of Delia. Her role in the rescue is taken by the Kikuyu man Wimbia Gikungu, called "Bill," Akeley's gun bearer and companion on several safaris. Bill roused the recalcitrant guides and notified the government post, thus bringing on the Scotsman posthaste (M. J. Akeley, 1940: Chpt. IX). The long quotation from Carl in which the whole story is told simply lacks mention of his previous wife.

Mary Jobe tells a sequel to the mauling not in Akeley's published stories, and apparently taken from his field diaries or lectures. Because it is not uncommon for a man to lose his nerve after an elephant mauling and decline to hunt elephants again, it was necessary for Akeley to face elephants as soon as possible. Again, the first thing to notice is an absence; there is no question that such courage should be regained. But the explicit story does not ennoble Akeley. He tracked an elephant before he was really healthy, needing his "boys" to carry a chair on the trail for him to sit on as he tired; he wounded the elephant with unsportsmanlike hasty shots; and it was not found before dying. If Akeley's nobility is saved in this story, it is by his humility: "The whole thing had been stupid and unsportsmanlike" (M. J. Akeley 1940: 126).

Mary Jobe Akeley pictures herself as Carl's companion and soul mate, but not really as his co-adventurer and buddy hunter—with one exception. Mary Jobe fired two shots in Africa, and killed a magnificent male lion: "An hour later we came upon a fine old lion, a splendid beast, Carl said, and good enough for me to shoot. And so I shot. . . . Carl considered it a valuable specimen; but I was chiefly concerned that I fulfilled Carl's expectations and had killed the lion cleanly and without assistance" (M. J. Akerley 1940: 303). Mary Jobe's authority as a biographer does not depend on her being a hunter, but her status was enhanced by this most desirable transforming experience.

Delia Akeley pictures herself as a joyous and unrepentant hunter; but, by the publication of *Jungle Portraits* in 1930, her husband has some warts. Delia does not bear the authorial moral status of the artist-scientist, Carl Akeley, or his socially sure second wife. Delia's tales clarify the kind of biography that was to be suppressed in African Hall. In Delia's story of the rescue, "Bill" also appears, and he behaves well. But her own heroism in confronting the superstitions of the "boys" and in saving her endangered husband is the central tale: "Examining and cleansing Mr. Akeley's wounds were my first consideration. . . . The fact that his wounds were cared for so promptly prevented infection, and without doubt saved his life" (D. Akeley 1930: 249).

Delia produced a biographical effect at odds with the official histories; she showed the messiness behind the "unified truth" of natural history museums. Delia dwelt on the sickness and injury of early collectors and explorers; she remarked pointedly on insects, weariness, and failure in the past and contrasted that with the experience provided the current (1930) traveler, the tourist, and museum visitor. She foregrounded the devoted and unrewarded wife who kept camp in the jungle and house at home. The wife-manager of Carl's safaris, aware of the material mediations in the quest for manhood and natural truth, showed pique at all the attention given her scientist-husband: "The thrilling story of the accident and his miraculous escape from a frightful death has been told many times by himself from the lecture platform. But a personal account of my equally thrilling night journey to his rescue through one of the densest, elephant-infested forests on the African continent is not nearly so well known" (D. Akeley 1930: 233). This is not the wife who devotes herself to her husband's authorship of wilderness. Indeed, she insisted on "darkest Africa" throughout her book.

Delia foregrounded her glory at the expense of her husband's official nobility. Delia's reader discovers Carl frequently sick in his tent, an invalid dangerously close to death while the courageous wife hunts not only for food for the entire camp, but also for scientific specimens so that he may hasten out of this dangerous continent before it claims him. In the elephant hunt following the mauling, Carl was still searching to restore his endangered "morale." But this time his wife was his companion in what is portrayed as a dangerous hunt terminating in a thrilling kill marked by a dangerous charge. Delia's story demurred on who fired the fatal shot, but "fatigue and a desire to be sure of his shot made Mr. Akeley slow in getting his gun in position" (D. Akeley 1930: 93).

Delia published an extraordinary photograph of a dashing Carl Akeley smoking a pipe and lounging on the body of a large fallen elephant; her caption reads, "Carl Akeley and the first elephant he shot after settling the question of his morale." A reader will not find that particular photograph of Akeley in any other publication than Delia's. Further, my hunt in the Museum's archive for the image of Akeley lounging astride his kill caught Delia in a lie (hoax?) about that elephant. But the lie reveals another truth. The accompanying photos in the archive suggest a version of reality, a biography of Africa, which the Museum and its official representatives did not want displayed in their Halls or educational publications.

The images from the photo archive upstairs haunt the mind's eye as the viewer stands before the elephant group in African Hall. First, the particular elephant with the lounging Carl could not have been killed on the occasion Delia described. The cast of characters evidences a different year; a picture clearly taken on the same occasion shows the white hunter, the Scotsman Richard John Cunninghame, hired by Akeley in 1909 to teach him how to hunt elephants, lounging with Delia on the same carcass. The Museum archive labels the photo "Mrs. Akeley's first elephant." It is hard not to order the separate photos in the folder into a narrative series. The next snapshot shows the separated and still slightly bloody tusks of the elephant held in a gothic arch over a pleased, informal Delia. She is standing confidently under the arch, each arm reaching out to grasp a curve of the elephantine structure. But the real support for the ivory is elsewhere. Cut off at the edge of the picture are four black arms; the hands come from the framing peripheral space to encircle the tusks arching over the triumphant white woman. The museum archive labels this photo "Mrs. Akeley's ivory." The last photograph shows a smiling Cunninghame anointing Mrs. Akeley's forehead with the pulp from the tusk of the deceased elephant. She stands with her head bowed under the ivory arch, now supported by a single, solemn African man. The Museum's spare comment reads, "The Christening." [Figure 3.2]

Here is an image of a sacrament, a mark on the soul signing a spiritual transformation effected by the act of first killing. It is a sacred moment in the life of the hunter, a rebirth in the blood of the sacrifice, of conquered nature. This elephant stands a fixed witness in Akeley African Hall to its dismembered double in the photograph, whose bloody member signed the intersection of race, gender, and nature on the soul of the western hunter. In this garden, the camera captured a retelling of a Christian story of origins, a secularized Christian sacrament in a baptism of blood

Figure 3.2   The Christening. Negative no. 211526. Published with permission of the Department of Library Services, American Museum of Natural History.

from the victim whose death brought spiritual adulthood, i.e., the status of hunter, the status of the fully human being who is reborn in risking life, in killing. Versions of this story proliferate in the history of American approaches to the sciences of life, especially primate life. With Delia, the story is near parody; with Carl it is near epiphany. His was authorized to achieve a fusion of science and art. Delia, the more prolific author, who neither had nor was a ghostwriter, was erased—by divorce and by duplicity.

Safari: A Life of Africa

> Now with few exceptions our Kivu savages, lower in the scale of intelligence than any others I had seen in Equatorial Africa, proved kindly men. . . . How deeply their sympathy affected me! As I think of them, I am reminded of the only playmate and companion of my early childhood, a collie dog. . . . (M. J. Akeley 1929b: 200)

The great halls of the American Museum of Natural History would not exist without the labor of Africans (or South Americans or the Irish and Negroes in North America). The Akeleys would be the first to acknowledge this fact; but they would claim the principle of organization came from the white safari managers, the scientist-collector and his camp-managing wife, the elements of mind overseeing the principle of execution. From the safari of 1895, dependent upon foot travel and the strong backs of "natives," to the motor safaris of the 1920s, the everyday survival of Euro-Americans in the field depended upon the knowledge, good sense, hard work, and enforced subordination of people the white folk insisted on seeing as perpetual children or as wildlife. If a black person accomplished some exceptional feat of intelligence or daring, the explanation was that he (or she?) was inspired, literally moved, by the spirit of the master. As Mary Jobe (1929b: 199) put it in her unself-conscious colonial voice, "It was as if the spirit of his master had descended upon him, activating him to transcendent effort." This explanation was all the more powerful if the body of the master was physically far removed, by death or trans-Atlantic residence. Aristotle was as present in the safari as he was in the taxidermic studios in New York or in the physiological bodies of organisms. Labor was not authorized as action, as mind, or as form. Labor was the marked body.

Carl and Mary Jobe Akeley's books elucidate safari organization over a thirty-year span. The photographs of solemn African people in a semi-circle around the core of white personnel, with the cars, cameras, and abundant baggage in the background, are eloquent about race, gender, and colonialism. The chapters discuss the problems of cooks, the tasks of a headman, the profusion of languages which no white person on the journey spoke, numbers of porters (about thirty for most of the 1926 trip, many more in 1895) and problems in keeping them, the contradictory cooperation of local African leaders (often called "sultans"), the difficulty of providing white people coffee and brandy in an "unspoiled" wilderness, the hierarchy of pay scales and food rations for safari personnel, the behavior of gun bearers, and the punishment for perceived misdeeds. The chapters portray a social organism ordered by the principles of organic form: hierarchical division of labor called cooperation and coordination. The safari was an icon of the whole enterprise in its logic of mind and body, in its scientific marking of the body for functional efficiency

(Sohn-Rethel 1978; Young 1977b; Rose 1983). In western inscriptions of race, Africans were written into the script of the story of life—and written out of authorship.

Few of the black personnel appear with individual biographies in the safari literature, but there are exceptions, object lessons or type life histories. Africans were imagined as either "spoiled" or "unspoiled," like the nature they signified. Spoiled nature could not relieve decadence, the malaise of the imperialist and city dweller, but only presented evidence of decay's contagion, the germ of civilization, the infection which was obliterating the Age of Mammals. And with the end of that time came the end of the essence of manhood, hunting. But unspoiled Africans, like the Kivu forest itself, were solid evidence of the resources for restoring manhood in the healthy activity of sportsmanlike hunting. Hinting at the complexity of the relation of master and servant in the pursuit of science on the safari, the life story is told from the point of view of the white person. Wimbia Gikungu, the Kikuyu known as Bill who joined Carl Akeley in British East Africa in 1905 at thirteen years of age, did not write—or ghost write—my sources. He was not the author of his body, but he was the Akeleys' favorite "native."

Bill began as an assistant to Delia Akeley's "tent boy," but is portrayed as rapidly learning everything there was to know about the safari through his unflagging industry and desire to please. He was said to have extraordinary intelligence and spirit, but suffered chronic difficulty with authority and from inability to save his earnings. "He has an independence that frequently gets him into trouble. He does not like to take orders from any one of his own color" (Akeley 1923: 143). He served with Akeley safaris in 1905, 1909–11, and 1926, increasing in authority and power over the years until there was no African whom Carl Akeley respected more for his trail knowledge and judgment. Bill got into trouble serving on the Roosevelt safari, was dismissed and blacklisted. Nonetheless, Akeley immediately rehired him, assuming he had had some largely innocent (i.e., not directed against a white person) eruption of his distaste for authority (Akeley 1923: 144).

Akeley describes three occasions on which he "punished" Gikungu; these episodes are icons of Akeley's paternal ideology. Once Bill refused to give the keys for Carl's trunk to other white people when they asked, "saying that he must have an order from his own Bwana. It was cheek, and he had to be punished; the punishment was not severe, but coming from me it went hard with him and I had to give him a fatherly talk to prevent his running away" (Akeley 1923: 134). The "father to the game" claimed the highest game of all in the history of colonialism—the submission of man. Later, the Kikuyu shot at an elephant he believed was charging an unsuspecting Akeley. Akeley had seen the animal, but did not know his "gun boy" did not know. Akeley slapped Gikungu "because he had broken one of the first rules of the game, which is that a black boy must never shoot without orders, unless his master is down and at the mercy of a beast." Realizing his mistake, "my apologies were prompt and as humble as the dignity of a white man would permit" (M. J. Akeley 1940: 132). The African could not be permitted to hunt independently with a gun in the presence of a white man. The entire logic of restoring threatened white manhood depended on that rule. Hunting was magic; Bill's well-meaning (and well-placed) shot was pollution, a usurpation of maturity. Finally, Akeley had Gikungu put in jail during the 1909–11 safari when "Bill" actively declined to submit when Carl "found it necessary to take him in hand for mild punishment" for another refusal of a white man's orders about baggage (Akeley 1923: 144). Gikungu spend two weeks in jail; the white man's paternal solicitude could be quite a problem.

Akeley relied on Gikungu's abilities and knowledge. Always, his performance was attributed to his loyalty for the master. Collecting the ivory of a wounded elephant, organizing the rescue after the elephant mauling, assisting Mary Jobe Akeley after Carl's death—these deeds were the manifestations of subordinate love. There is no hint that Gikungu might have had other motives—perhaps a non-subservient pity for a white widow in the rain forest, pleasure in his superb skills, complex political dealings with other African groups, or even a superior hatred for his masters. Attributing intentions to "Bill" is without shadow of doubt; the African played his role in the safari script as the never quite tame, permanently good boy. Bill was believed to be visible; other Africans largely remained invisible. The willed blindness of the white lover of nature remained characteristic of the scientists who went to the Garden to study primates, to study origins, until cracks began to show in this consciousness around 1970.

Institution

Speak to the Earth and It Shall Teach Thee. (Job 12:8)[28]

Every specimen is a permanent fact.[29]

From 1890 to 1930 the "Nature Movement" was at its height in the United States. Conventional western ambivalence about "civilization" was never higher than during the early decades of monopoly capital formation (Marx 1964; Nash 1982). The woes of "civilization" were often blamed on technology—fantasized as "the Machine." Nature is such a potent symbol of innocence partly because "she" is imagined to be without technology. Man is not *in* nature partly because he is not seen, is not the spectacle. A constitutive meaning of masculine gender for us is to be the unseen, the eye (I), the author, to be Linnaeus who fathers the primate order. That is part of the structure of experience in the Museum, one of the reasons one has, willy nilly, the moral status of a young boy undergoing initiation through visual experience. The Museum is a visual technology. It works through desire for communion, not separation, and one of its products is gender. Who needs infancy in the nuclear family when we have rebirth in the ritual spaces of Teddy Bear Patriarchy?

Social relations of domination are built into the hardware and logics of technology, producing the illusion of technological determinism. Nature is, in "fact," constructed as a technology through social praxis. And dioramas are meaning-machines. Machines are maps of power, arrested moments of social relations that in turn threaten to govern the living. The owners of the great machines of monopoly capital were, with excellent reason, at the forefront of nature work—because it was one of the means of production of race, gender, and class. For them, "naked eye science" could give direct vision of social peace and progress despite the appearances of class war and decadence. They required a science "instaurating" jungle peace; and so they bought it.

This scientific discourse on origins was not cheap; and the servants of science, human and animal, were not always docile. But the relations of knowledge and power at the American Museum of Natural History should not be narrated as a tale of evil capitalists in the sky conspiring to obscure the truth. Quite the opposite, the tale must be of committed Progressives struggling to dispel darkness through research, education, and reform. The capitalists were not in the sky; they were in the field, armed with the Gospel of Wealth.[30] They were also often armed with an elephant gun and an Akeley camera. Sciences are woven of social relations throughout their tissues. The concept of social relations must include the entire complex of interactions among people; objects, including books, buildings, and rocks; and animals.[31]

One band in the spectrum of social relations—the philanthropic activities of men in the American Museum of Natural History, which fostered exhibition (including public education and scientific collecting), conservation, and eugenics—is the optic tectum of naked eye science, i.e., the neural organs of integration and interpretation. After the immediacy of experience and the mediations of biography and story telling, we now must attend to the synthetic organs of social construction as they came together in an institution.[32]

Decadence was the threat against which exhibition, conservation, and eugenics were all directed as prophylaxis for an endangered body politic. The Museum was a medical technology, a hygienic intervention, and the pathology was a potentially fatal organic sickness of the individual and collective body. Decadence was a venereal disease proper to the organs of social and personal reproduction: sex, race, and class. From the point of view of Teddy Bear Patriarchy, race suicide was a clinical manifestation whose mechanism was the differential reproductive rates of anglo-saxon vs. "non-white" immigrant women. Class war, a pathological antagonism of functionally related groups in society, seemed imminent. And middle class white women undertaking higher education might imperil their health and reproductive function. Were they unsexed by diverting the limited store of organic energy to their heads at crucial organic moments? Lung disease (remember Teddy Roosevelt's asthma), sexual disease (what was *not* a sexual disease, when leprosy, masturbation, and Charlotte Perkins Gilman's need to write all qualified?), and social disease (like strikes and feminism) all disclosed ontologically and epistemologically similar disorders of the relations of nature and culture. Decadence threatened in two interconnected ways, both related to energy-limited, productive systems—one artificial, one organic. The machine threatened to consume and exhaust man. And the sexual economy of man seemed vulnerable both to exhaustion and to submergence in unruly and primitive excess. The trustees and officers of the Museum were charged with the task of promoting public health in these circumstances.

Three public activities of the Museum were dedicated to preserving a threatened manhood: exhibition, eugenics, and conservation. Exhibition was a practice to produce permanence, to arrest decay. Eugenics was a movement to preserve hereditary stock, to assure racial purity, to prevent race suicide. Conservation was a policy to preserve resources, not only for industry, but also for moral formation, for the achievement of manhood. All three activities were prescriptions against decadence, the dread disease of imperialist, capitalist, white culture. Forms of education and science, they were also very close to religious and medical practice. These three activities were about the transcendence of death, personal and collective. They attempted to insure preservation without fixation and paralysis, in the face of extraordinary change in the relations of sex, race, and class.

Exhibition

The American Museum of Natural History was (and is) a "private" institution, as private could only be defined in the United States. In Europe the natural history museums were organs of the state, intimately connected to the fates of national politics (Holton and Blanpied 1976). The development of U.S. natural history museums was tied to the origins of the great class of capitalists after the Civil War (Kennedy 1968). The social fate of that class was also the fate of the Museum; its rearrangements and weaknesses in the 1930s were reproduced in crises in the Museum, ideologically and organizationally. The American Museum, relatively unbuffered from intimate reliance on the personal beneficence of a few wealthy men, is a peephole for spying on the wealthy in their ideal incarnation. They made dioramas of themselves.

The great scientific collecting expeditions from the American Museum began in 1888 and stretched to the 1930s. By 1910, they had gained the Museum scientific prestige in selected fields, especially paleontology, ornithology, and mammalogy. The Museum in 1910 boasted nine scientific departments and twenty-five scientists. Anthropology also benefited, and the largest collecting expedition ever mounted by the Museum was the 1890s Jesup North Pacific Expedition so important to Franz Boas's career (Kennedy 1968: 141ff). The sponsors of the Museum liked a science that stored facts safely; and they liked the public popularity of the new exhibitions. Many people among the white, protestant, middle and upper classes in the United States were committed to nature, camping, and the outdoor life; Teddy Roosevelt embodied their politics and their ethos. Theodore Roosevelt's father was one of the incorporators of the Museum in 1868. His son, Kermit, was a trustee during the building of African Hall. Others in that cohort of trustees were J. P. Morgan, William K. Vanderbilt, Henry W. Sage, H. F. Osborn, Daniel Pomeroy, E. Roland Harriman, Childs Frick, John D. Rockefeller III, and Madison Grant. Patrons of science, these are leaders of movements for eugenics, conservation, and the rational management of capitalist society.

The first hall of dioramas was Frank Chapman's Hall of North American Birds, opened in 1903. Akeley, hired to prepare African game, especially elephants, conceived the idea for African Hall on his first collecting trip for the American Museum. Osborn hoped for—and got—a North American and Asian Mammal Hall after the African one. The younger trustees in the 1920s formed an African Big Game Club that invited wealthy sportsmen to join in contributing specimens and money to African Hall. The 1920s were prosperous for these men, and they gave generously. There were over one hundred expeditions in the field for the American Museum in the 1920s discovering facts (Kennedy 1968: 192).

There was also a significant expansion of the museum's educational endeavors. Over a million children per year in New York were looking at the Museum's "nature cabinets" and food exhibits circulated through the city public health department. Radio talks, magazine articles, and books covered the Museum's popular activities, which appeared in many ways to be a science for the people, like that of the *National Geographic*, which taught republican Americans their responsibilities in empire after 1888. Both *Natural History*, the Museum's publication, and *National Geographic* relied heavily on photographs. There was a big building program from 1909 to 1929; and the Annual Report of the Museum for 1921 quoted the estimate by its director that

2 1/2 million people were reached by the Museum and its education extension program.

Osborn summarized the fond hopes of educators like himself in his claim that children passing through the Museum's halls "become more reverent, more truthful, and more interested in the simple and natural laws of their being and better citizens of the future through each visit." He maintained that the book of nature, written only in facts, was proof against the failing of other books: "The French and Russian anarchies were based in books and in oratory in defiance of every law of nature."[33] Going beyond pious hopes, Osborn had the power to construct a Hall of the Age of Man to make the moral lessons of racial hierarchy and progress explicit, lest they be missed in gazing at elephants. He countered those who criticized the halls and educational work for requiring too much time and money better spent on science itself. "The exhibits in these Halls have been criticized only by those who speak without knowledge. They all tend to demonstrate the slow upward ascent and the struggle of man from the lower to the higher stages, physically, morally, intellectually, and spiritually. Reverently and carefully examined, they put man upwards towards a higher and better future and away from the purely animal stage of life."[34] This is the Gospel of Wealth, reverently examined.

Prophylaxis

Eugenics and conservation were closely linked in philosophy and in personnel at the Museum, and they tied in closely with exhibition and research. For example, the white-supremacist author of *The Passing of the Great Race*, Madison Grant, was a successful corporation lawyer, a trustee of the American Museum, an organizer of support for the North American Hall, a co-founder of the California Save-the-Redwoods League, activist for making Mt. McKinley and adjacent lands a national park, and the powerful secretary of the New York Zoological Society. His preservation of nature and germ plasm all seemed the same sort of work. Grant was not a quack or an extremist. He represented a band of Progressive opinion terrified of the consequences of unregulated monopoly capitalism, including failure to regulate the importation of non-white (which included Jewish and southern European) working classes, who invariably had more prolific women than the "old American stock." Powerful men in the American scientific establishment were involved in establishing Parc Albert in the Congo, a significant venture in international scientific cooperation: John C. Merriam of the Carnegie Institution of Washington, George Vincent of the Rockefeller Foundation, Osborn at the American Museum. The first significant user of the sanctuary would be sent by the "father" of primatology in America, Robert Yerkes, for a study of the psychobiology of wild gorillas. Yerkes was a leader in the movements for social hygiene, the category in which eugenics and conservation also fit. It was all in the service of science.

The Second International Congress of Eugenics was held at the American Museum of Natural History in 1921 while Akeley was in the field collecting gorillas and initiating plans for Parc Albert. Osborn, an ardent eugenicist, believed that it was "[p]erhaps the most important scientific meeting ever held in the Museum." Leading U.S. universities and state institutions sent representatives, and there were many eminent foreign delegates. The collected proceedings were titled "Eugenics in Family, Race, and State." U.S. lawmakers were one intended audience. "The section of the exhibit bearing on immigration was then sent to Washington by the Committee on Immigration of the Congress, members of which made several visits to the Museum to study the exhibit. The press was at first inclined to treat the work of the Congress [of Eugenics] lightly . . . but the influence of the Congress grew and found its way into news and editorial columns of the entire press of the United States."[35] In 1923 the United States Congress passed immigration restriction laws, to protect the Race, the only race needing a capital letter.

The 1930s were a hiatus for the Museum. The Depression led to reduced contributions, and basic ideologies and politics shifted. The changes were not abrupt; but even the racial doctrines so openly championed by the Museum were publicly criticized in the 1940s, though not until then. Conservation was pursued with different political and spiritual justifications. A different biology was being born, more in the hands of the Rockefeller Foundation and in a different social womb. The issue would be molecular biology and other forms of post-organismic cyborg biology. The threat of decadence gave way to the catastrophes of the obsolescence of man (and of all organic nature) and the disease of stress, realities announced vigorously after World War II. Different forms of capitalist patriarchy and racism would emerge, embodied in a retooled nature. Decadence is a disease of organisms; obsolescence and stress are conditions of technological systems. Hygiene would give way to systems engineering as the basis of medical, religious, political, and scientific story-telling practices.

The early leaders of the American Museum of Natural History would insist that they were trying to know and to save nature, reality. And the real was one. The explicit ontology was holism, organicism. The aesthetic appropriate to exhibition, conservation, and eugenics from 1890 to 1930 was realism. But in the 1920s the surrealists knew that behind the day lay the night of sexual terror, disembodiment, failure of order; in short, castration and impotence of the seminal body which had spoken all the important words for centuries, the great white father, the white hunter in the heart of Africa. The strongest evidence in this chapter for the correctness of their judgment has been a literal reading of the realist, organicist artifacts and practices of the American Museum of Natual History. Their practice and mine have been literal, dead literal.

4

A Pilot Plant for Human Engineering: Robert Yerkes and the Yale Laboratories of Primate Biology, 1924–1942

[I]t is not the activity of the subject of knowledge that produces a corpus of knowledge, useful or resistant to power, but power-knowledge, the processes and struggles that traverse it and of which it is made up, that determines the forms and possible domains of knowledge. (Foucault 1979: 28)

For, Lady, you deserve this state
Nor would I love at lower rate.
(Andrew Marvell, "To His Coy Mistress")

The Servant of Science

Like Carl Akeley, Robert Means Yerkes (1876–1956) loved the great apes. And like Akeley's taxidermy, Yerkes's science was a practice of second birthing and rational fatherhood. But the forms of love and paternity were not the same. If Akeley's ethos, to be father of the game, is iconically represented in the mounted figure of the Giant of Karisimbi, the image haunting this chapter is Yerkes's sorrowing deathwatch for a special chimpanzee child, recorded in "The Light that Failed: A Tribute to Prince Chim" (Yerkes 1925: 253–55). Probably a pygmy chimpanzee, *Pan paniscus*, Chim was one of Yerkes's first ape research subjects. Yerkes bonded closely with this endearing youngster, who seemed to embody the childhood of a future primate Order promised by science. Stressing the need for order and discipline, kind treatment, play, variety, and directed activity as the apes matured, Yerkes guided the development of Chim and of his female comrade, Panzee, at his New Hampshire summer farm in 1923. [Figure 4.1] Photographs from 1925 show Yerkes's later scientific child-wards—Bill (named for William Jennings Bryant), Dwina (in memory of Charles Darwin), Wendy, and Pan—eating at table in a New England pasture. [Figure 4.2] Yerkes suggested that Mme. Rosalia Abreu provide for her young apes at her estate in Havana, Cuba, "a long table and chairs with facilities for use as a playroom or school-room as well as dining room" (Yerkes 1925: 210).

Ill and bothersome from the start, Panzee died little-lamented in the winter of

Figure 4.1   Robert Yerkes holding Chim and Panzee, 1923. Published with permission of the Yerkes Regional Primate Research Center of Emory University.

1924 in Washington, D.C. In June 1924 Yerkes brought Chim with him to Quinta Palatino, where the psychobiologist spent the summer observing Mme. Abreu's extraordinary personal primate collection (Hahn 1988: 25–40). Chim died there of a respiratory illness. Deeply moved by the dying child's behavior, Yerkes wondered at his seeming anticipation of death and his appearance of waiting for something or someone who was absent. Prince Chim had been especially bright and avid to learn; and in that desire, he served Yerkes's quest to assist and guide an epochal unfolding of potential from animal to man, from instinct to consciousness. Chim, a captive, was no longer wild. A surrogate son, his noble epistemological and moral

Figure 4.2 Chimpanzees Bill, Dwina, Pan, and Wendy at meal time, 1925. Robert M. Yerkes Papers, Manuscripts and Archives, Yale University Library. Published with permission.

status inhered not in his closeness to wilderness and to man as hunter, but in his promise as a bright, lively, and docile child in Yerkes's dream of establishing that most modern of institutions—the experimental laboratory.

Yerkes's enduring contribution to science was the founding of a paradigmatic laboratory. That was the theme of *Chimpanzees: A Laboratory Colony*, published just after he retired from direction of the Yale Laboratories of Primate Biology in 1942. The prologue, "The Servant of Science," is a dense tapestry composed of the major threads in this chapter: the fundamental duality of Yerkes as rational, progressive leader and dependent servant of power; the ape laboratory as pilot plant in the service of human engineering; and the structural ambiguity between object and image, servant and master, child and man, model and reality in the scientific relation of ape and human being. Yerkes meant to refer unambiguously to the chimpanzee as the servant of science; but before the prologue was complete, the servant and the animal, the master and the human being were tangled threads in the social fabric. "Man's curiosity and desire to control his world impel him to study living things" (Yerkes 1943: 1). With that banal but crucial assertion about the foundation of human rationality in the will to power, Yerkes opened his book. For him the tap root of science is the aim to control. The full consequences of that teleology become apparent only in the sciences of mind and behavior, where natural object and designed product reflect each other in the infinite regress of face-to-face mirrors, ground by the law of Hegel's master-servant dialectic.

For Yerkes experimental inquiry was the natural telos of naturalistic study, primarily because the laboratory produced the kind of knowledge which enhanced control. Extended to the field, the logic of the laboratory and the clinic was the logic of rationality in all kinds of spaces. Since the first and final object of Yerkes's interest was the human being, the pinnacle of evolutionary progress, where the structure of dominion of brain over body was most complete, greatest curiosity and utility were centered on natural objects yielding greatest self-knowledge and self-control. The laboratory animal in general possessed the highest value for human beings precisely because it was *designed* and standardized, in short, engineered, to answer human queries. But the animal's epistemological status was also as *natural* object yielding objective understandings. This is the circular structure of systematic alienated knowledge—alienated in the technical sense of product (knowledge) postulated to be separate from the producer (scientist). Separate from the producer, the product rapidly becomes prior to its maker, and determines the future by functioning as a fetish.

Because of the success of experimental sciences, "as a direct result, and evidently because of the myriad discoveries which have been made and the natural urge to apply them practically, man is now on the high road to human engineering" (Yerkes 1943: 2). People have been negligent in accepting responsibility for full knowledge of themselves; but the power of science now makes survival depend on self-control based on the human sciences. The paradigmatic human science for Yerkes was psychobiology. Those animals most like people should be used as the most practical producers of knowledge. Availability and social inhibitions prevent most direct use of human beings as experimental objects, but "another consideration is the possibility that study of the other primates may prove the most direct and economical route to profitable knowledge of ourselves, because in them, basic mechanisms are less obscured by cultural influences" (Yerkes 1943: 3). The evolutionary scale of perfection provided the rationale for use of a model system. This conjunction of the reductionistic *desideratum* of simple quantifiable object and the organismic requirement of parts which mirror the whole was the key to the power of evolutionary naturalism.

In Yerkes's ideology *culture* contrasted with *engineering*, a contrast related to primatology's claim to mediate between categories of culture and nature. The apes were natural objects redesigned to produce useful knowledge. They had no culture, (although they did seem to have "social traditions"). They were unobscured mirrors. Culture meant the undesigned, inherited, irrational, fatalistic accretion overlying basic mechanisms; culture was the cloudy, unintended, uncontrolled social state. Engineering remade the natural object, which offered no resistance, in the interest of a rational society. Conceptually, both culture and human engineering assumed a plastic human nature, where underlying biological mechanisms supplied the raw materials for social manipulation. Human engineering assumed rational control of nature so as to manage society for conscious production and reproduction. Engineering properly replaced inefficient culture.

There were several chimpanzee properties that made the ape such a suitable material for human engineering, before scaling up on the Great Chain of Being. At least for Yerkes, the chimp was docile, i.e., responsive to the clinical and educational manipulation of humane, scientific management. While for lesser mortals, the adult chimpanzee returned shit for service (literally), for Yerkes even the adults showed their adaptability in a promising neotenous retention of the plastic cooperativeness of children—those perfect servants of science and engineering in their provision of the raw material for the future. The essence of the evolutionary concept of adaptation for Yerkes was always functional learning, called adaptivity. The gliding confusion of ontogenetic and phylogenetic learning was built into Yerkes's version of human adaptive plasticity. No doctrine of the mechanism of inheritance, no resolution of the nature/nurture dichotomy, affected that central position. Evolution and education were sides of the same coin, and the concepts applied both to animal and human being. Docility (cooperativeness, teachability) must be properly understood.

Like many primatologists, Yerkes loved chimpanzees in a serious and life-changing way. In them nature, medicine, hygiene, and reform converged in a personal and professional satisfaction. This love was intrinsically bound up with power; the colonizing logic of paternalistic domination—a kind of civilizing mission—pervaded every level of Yerkes's science, including the personal relation of human being and animal. This domination stood as the foundation of rational cooperative society, of adult love. Natural authority which produced the benefits of civilization must be accepted. The mark of acceptance was appreciation of the good done from above. The perfect act of appreciation was total response to the lover who remade the beloved in his own image. The beloved was patient. Not accidentally, the primate story echoed primal themes in monotheistic religion, for there sex, power, and fatherhood are not strangers either.

Yerkes wrote moving passages describing the attitude of chimpanzees who received fair treatment, who were given useful work to do (participation in scientific experiments seemed best here), and who received medical attention—from human doctors—when they were sick.[1] It takes little imagination to substitute the words *child, slave, patient,* or *woman* for *chimpanzee* in a rationally managed household to understand the structure of the primate laboratory dedicated to human engineering. In all of this, Yerkes was benignly typical, representative of the best in Progressive reform in America.

Similarity to people was held to be characteristic of apes on all relevant planes— physiological (for the study of glands, nerves, and other mechanisms); medical (for the study of human diseases); psychobiological (for the study of learning and ideational processes, drug addiction, emotions, mental disorders); and social (for the study of tradition and proto-culture).

> I am thinking for example of the drive for social superiority, dominance, and leadership; of the relations of the sexes, with behavioral evidences of "right," "privilege," "conscience"; of the evidence of primitive forms of social service, social dependence and awareness, sympathy and attachment, as are observable in chimpanzees. . . . Even problems of social organization which may be termed governmental may be pressed toward solution, or studied with interest in origins and developments, by the intelligent use of great apes. (Yerkes 1943: 7)

It is no wonder that Yerkes considered that "from the vantage ground of our incomplete knowledge and limited understanding, these animals seem like psychobiological gold mines" (Yerkes 1943: 7). Mining is a common trope for modern science. Gold, the chief fetish of colonial imaginations, here is sought in the tropical primate body. Through the supreme utility of chimpanzees—practically and epistemologically—Yerkes returned to the problem of "human adaptive improvement." He fervently believed "there is no more unprofitable assumption than that of the unalterability of human nature" (Yerkes 1943: 9). For Yerkes such a belief was not only false, it was pernicious. The primate laboratory was meant to be a pilot plant, a demonstration project for rational re-design of human nature.

> It has always been a feature of our plan for the use of the chimpanzee as an experimental animal to shape it intelligently to specification instead of trying to preserve its natural characteristics. We have believed it important to convert the animal into as nearly ideal a subject for biological research as is practicable. And with this intent has been associated the hope that eventual success might serve as an effective demonstration of the possibility of re-creating man himself in the image of a generally acceptable ideal. . . . If as servant of science the chimpanzee should help to make clearer and more attractive to mankind ways for the achievement of greater social-mindedness, dependability, and cooperativeness, how immeasurable our debt to it! The really important things for us at present are recognition and active acceptance of the principles of modifiability, controllability, and consequently improvability, of human nature. (Yerkes 1943: 10–11)

This was the ideal of service built into the laboratory architecture, social relations among animals and people, cage design, committee service, experimental protocols, and publication practices in the entwined paternal practices of love and knowledge in the Yale Laboratories of Primate Biology.

Mirrors and Tools

The themes of modern America have been found mirrored in detail in the bodies and lives of animals. Nature is given our history, even as our history is made to seem natural because we *see* ourselves in these animate, multiform mirrors. Primates have been distinguished from other animals by their eyes and the hypertrophied visual cortex of the brain. Perhaps that is why monkeys, apes, and human beings seem to move about, reflecting each other in elaborate show, in a house of mirrors.

In addition to their symbolic meaning for human beings, other animals, especially primates, have a utilitarian role to play. They are laboratory animals essential to investigation of basic physiology, behavior, and social organization. Their nervous and reproductive systems are the raw material of biomedical science; without this material human beings would know much less about their own bodies and minds. Both the symbolic and utilitarian aspects of people's use of other primates have been bound up in the project of imagining and producing the ideal human being in the ideal society. This is the project of human engineering, closely associated in concrete detail and in metaphoric extension with the history of scientific management of material and human factors of production in industrial, familial, and general social organization in America in the twentieth century. Human engineering in primate studies meant the selection and design of experimental animals and the production of research and of trained people useful in the management of pressing social problems, such as stabilization of the family; structuring of the personality; division of labor by sex, race, and class; promotion of cooperation and stability at work; definition and treatment of social pathology; and the production of acceptable ideologies to motivate efficient functional behavior in war, labor, and reproduction. Yerkes worked to establish the utility of primates for interpreting the place of human beings in scientifically managed, corporate capitalism—called nature. Scientific frameworks for interpreting primate behavior and biology have changed radically since Yerkes's work. Knowledge of primates has responded to general development of biology, psychology, and sociology, as well as to political conflict and debate. But a constant dimension has been the naturalization of human history.

Robert Yerkes founded a major institution for primate studies in America between his arrival at Yale University's Institute of Psychology in 1924 and his retirement from Yale and the Yale Laboratories of Primate Biology in 1942. The themes of cooperation and control, of sex and mind, of other primates and human beings in the project of human engineering permeated his foundational labor. His laboratory fostered cooperative research in the comparative psychology of learning, reproductive physiology, mental evaluation and testing, psychopathology, drug addiction, the relation of behavior to physiological condition, and experimental sociology. Efforts were directed overwhelmingly to a single species, the chimpanzee, to produce an ideal research animal for human beings. Yerkes built his laboratory on the utopian borderlands where fact and fiction commingle in primatology's origin narratives. The story requires attending to contemporary scientific developments, especially in reproductive and nervous physiology, and to contemporary social developments, especially the projects of liberal corporate reform and scientific management of society.

In the first half of the twentieth century, primatology, a particularly revealing science of the body and its forces, existed precisely at the interface of life and human sciences, precisely between biology and social-psychological science. Yerkes called his science psychobiology and defined his primary object of study to be "personality." He considered psychobiology to be a physiology of personality. He spent his life building structures (laboratories, funding, students, ideas, logics of application, myths) for the study of primates as the most revealing objects for a psychobiology of human engineering because he believed nonhuman primates were pure units of personality unobscured by disordered artifice called culture. Therefore, they were particularly plastic to reason, called engineering, and could be models for control of the productive forces of human life. Constructing a primate model was for Yerkes building a powerful technology to remodel persons.

"Personality" as an object of knowledge became a recurrent theme because it was central to the two key levers for psychobiology as a technology of power over labor: the worker and the family. Labor is giving birth and producing to satisfy needs; labor is the great productive force discovered precisely when biology discovered its objects in the reproducing, expanding, fruitful organisms which operated by the laws of productive systems.[2] Expansion, increase, multiplication, diversification, and above all regulated accumulation and control of the forces of production: these are the heart (the gonads?) of life and human sciences. Personality, constituted as an object of knowledge by psychobiology in the twentieth century, was an anchor for the control of expanding bodies through the control of work and sex. To develop a proper personality became a basic primate project, indeed, the very source of control of the Order.

A science of the person required not only a natural model, the nonhuman primate, but also detailed knowledge of production-control relations in expanding bodies.

Therefore, psychobiology was inextricably bound up in studying gonads and brains, emotions and intelligence, motivation and problem-solving strategies. The basic operation of psychobiology was the construction of a comparative science of person-alities enforcing proper relations of mind and body, shown by scales of ascending complexity of the interactions of nervous and reproductive systems. Yerkes's pro-duction of intelligence tests and his efforts in sexual psychobiology were of a piece. He set out to know what was normal; he produced a science of norms in sex and intelligence, in family and work, in body and mind.[3]

The Therapeutics of Labor: Human Engineering

Human engineering was "the movement to control the human element of produc-tion at the individual and group level through the study and manipulation of human behavior."[4] The concept of engineering had its origin in liberal reform of the Progressive Era, roughly 1890 to World War I, especially in science-based industries linked to chemical and electrical engineering. In the language of the emerging literature, the main problems were the inventory, redesign, and maintenance of the human being for incorporation into production and consumption processes of industrial capitalism. That project involved deliberate removal of "irrationalities" in the system, including "maladjustments" and "inefficient" discrimination due to race, sex, or class. By the 1930s personnel management integrated the methods of physical, biological, and social sciences in order to produce harmony, team work, adjustment—in short, cooperation in the face of the imperatives of "modern soci-ety." The structure of cooperation involved the entire complex of division of labor and authority in capitalist production and reproduction. Motivation of cooperation was a management problem. It was also a biomedical problem. It necessitated detailed scientific knowledge of the "irrationalities"—i.e., of instinct, personality, culture. Engineering meant rational placement and modification of human raw material.

The term *human engineering* entered the language about 1910 through the work of "welfare secretaries," social workers attached to industries. When enlightened managers of science-based industry modified the bare system of scientific manage-ment associated with Taylorism, they incorporated the functions of the welfare secretaries in building the modern structure of industrial relations programs and work design. These liberal revisionists were dominant by 1915. Human engineering was built into the nascent social sciences in the service of power (Baritz 1960). The clear purpose of personnel management and industrial relations was "to transform the energy of potential conflict into a constructive, profitable force within a larger cooperative framework" (Noble 1977: 265). Maintenance of stable families and rational management of changing sexual and racial relations were intrinsic to the overall project.

Biological theories of instinct and drive were crucial to social management of production and reproduction. At the same time, bio-psychological theories were a popular element in self-expression and self-improvement ideologies. The irrational self could be managed by the individual as well as by larger social agencies; the popular movements and institutionalized social control worked in tandem ideologi-cally in the 1920s and 1930s (Burnham 1968a and 1968b). The Jazz Age ideal of self-expression, finding embodiment in the new psychologies based on psychoanalysis, on the study of glands of internal secretion, and on behaviorism, relied on techniques of manipulation of the organic, sub-rational self discursively constituted by the bio-medical sciences, including primatology. These popular doctrines were profoundly political; they helped redefine politics as the rational management of the sexually constituted organic household of drives and other irrational elements. The self-improving technologies rooted in sciences of sex and mind, reproductive and nervous systems, were an organon, or tool, of community in which political conflict was profoundly medicalized and referred for treatment to the therapeutic clinic. Practices of planned environment, self-reconditioning, or gland therapy were all connected to a primitive, mechanistic idea of the material, plastic organism structured by technology, in society, in the interests of individualism. It was a short step, in changed historical circumstances of the 1930s, to turn the same doctrines into strong justifications for more directly coercive social control in the interests of stability and recovery of health of the body politic.

The themes of division of labor in the science-management enterprise, of team cooperation, and of individual control permeated the establishment and practice of primatology from 1915 to 1940. The differentiation of knowledge and the social function of this organic division of labor were evident in the intricately elaborated hierarchical relations of pure and applied research, of funding policies within the big foundations which carefully divided fortunes into specialized organs which mediated social control, and of life and human sciences and their indirect connections to social policy and agencies. The division of labor in scientific practice facilitated mediations from the dangerous, alienated internal self to the ordered body politic managed by experts. The movement from organic structure and process, to behavior and personality, to culture and social groups, to agencies treating individuals within a family ideology in the interest of adjustment and integration was a logical ascent from instinct through culture to control.

The connections of Yerkes with human engineering were direct and multi-leveled. He held a key place in liberal corporate reform linked to human engineering institutionalized through the National Research Council, its Division of Psychology and Anthropology, the Engineering Foundation, the Research Information Service, the Personnel Research Federation, Yale's Institute of Psychology and Institute of Human Relations, the Rockefeller and Carnegie philanthropies, and critical people like James Rowland Angell, the president of Yale who helped Yerkes build his primate laboratory. Angell had been a prominent advocate of "functional psychology" at the University of Chicago early in this century.[5] He chaired the Department of Psychology at Chicago from 1905–1919, was dean of the university faculties from 1911–19, and acting president from 1918–19. He chaired the National Research Council from 1919–1920, was president of the Carnegie Corporation from 1920–21, and president of Yale from 1921–37. Angell was a director of the New York Life Insurance Company and the National Broadcasting Company, served on the National Committee for Mental Hygiene, and was a trustee of the Rockefeller Foundation from 1928–36. Angell and Yerkes planned Yale's graduate-oriented Institute of Psychology, funded by the Laura Spellman Rockefeller Memorial in 1924. Angell paradigmatically represented the elaborate interconnections of university, industry, philanthropy, and science policy in the development of the material structures and ideologies of scientific management of society.

The National Research Council was organized as a preparedness move preceding

World War I. It was also a logical development in the coordination of research in support of increasingly science-based, corporate industry; and the council played an important role in rationalization of the social structure of science after the war (Kargon 1977). The NRC was financed by the Engineering Foundation, a philanthropy directed by the leading engineering-reformers who were committed to the promotion of scientific research for industry. It was not necessary, or even desirable, that such research be directly utilitarian; the ideology and practice of the division of pure and applied science was in place. Before the end of the war, NRC leaders actively sought to shift planning to post-war industrial needs in science. One of the pure research activities of the council was administration of the Rockefeller-funded National Research Fellowship program. While the council paid most attention to the physical and material engineering sciences, its organs were suited as well to the biological and human engineering disciplines. Yerkes chaired the Division of Psychology and Anthropology for a year after the war as part of his extensive science management duties while seeking permanent funding for a primate facility. That division was a clearinghouse for projects in educational reform for corporate industrial needs that succeeded military influence in colleges and universities during the war. Testing programs and vocational guidance, as part of human evaluation for rational placement according to capacity and merit, were important foci for the Division and for many psychologists. These aims were explicit in Angell's and Yerkes's plans for Yale's Institute of Psychology. The ambitious Rockefeller-funded Yale Institute of Human Relations that succeeded the Institute of Psychology was a type example of efforts to coordinate research on personality, development, social groups, culture, physiology and medicine, psychiatry, law, and religion under the rubric of social harmony and personal adjustment.

The Research Information Committee of the wartime NRC became the Research Information Service co-directed by Yerkes after the war. The service initiated scientific indexing and abstracting tools, catalogued personnel information, and developed a worldwide science information exchange system. One engineer-advocate called it an "intelligence service," whose explicit purpose was "the corporate comprehension of, and thus power over, the vicissitudes of a 'knowledge-based' industrial society" (Noble 1977: 161). The NRC's Committee on Industrial Personnel Research was headed by Yerkes while he chaired the Division of Anthropology and Psychology. Its tasks included the study of labor unrest, medical and pathological aspects of labor health and productivity, and specification of qualities for employment. The heir of the committee was the Personnel Research Federation, which became the prime agency for coordination of personnel research for industry. Both Angell and Yerkes were leading figures in the Federation.

There was no conspiracy to place science in service of white, male-dominant capital involved in this remarkable concatenation of organizations and people. Operating from the "unmarked category," where the social conditions of one's existence are invisible, these men saw their role to be human service and believed strongly in democracy, individual rights, and scientific freedom. They saw themselves to be self-sacrificing reformers who believed in and practiced science. They helped define what could count as rational, making to this day the act of marking the privileged category with modifiers look merely ideological. Under this banner of rational reform, Yerkes created a chimpanzee community for human service.

Managed Difference: Labor, Race, and Sex

As temporary chairman at the 1920 annual meeting of the Personnel Research Federation, Yerkes developed themes permeating his work for human engineering. He began with a call to "look confidently to disinterested research to guide our race to a wise solution" of the problem of whether "the industrial system and its products [shall] be treated as ends or means to human welfare" (Yerkes 1922: 56). He saw personnel research as the key discipline of the new era. Yerkes believed that industrial systems had evolved from slavery, to the wage system, to the present system based on cooperation. Only now could the value of the person be realized. Equality clearly did not mean organic sameness; therefore it must mean that "in the United States of America, within limits set by age, sex, and race, persons are equal under the law and may claim as their right as citizens like opportunities for human service and responsibility" (58). Equality was limited "only" by the inherent properties of marked bodies, which fitted one to occupy a natural place determined by disinterested science.

Differences were therefore the essential subject matter for the new science. Personnel research would provide reliable information for the employment manager and proper vocational counseling for the "person." The "vocations" themselves were regarded as neutral products of industrial progress. The problem was one of human inventory in a (white, male-dominant, and capitalist) democracy. "Is it desirable, even if possible and practicable, that we should move toward likeness or homogeneity by taking full control of evolutional processes and replacing nature's trend toward heterogeneity? . . . Is it not more desirable that we preserve and use, appreciate and accentuate, individuality, sex, and race instead of trying to reduce, eliminate, or ignore them?" (*Testament*, 225). The unit of analysis was the person, transformed by the scientific concept of personality. Personality then tied physiology, medicine, psychology, anthropology, and sociology into the service of management. The scientific study of instinct (or drive) was like an inventory of raw material in the production of efficient, harmonious society through human engineering. The person and personality had strong anti-materialist meaning (i.e., not socialist) at the same time that the associated ideology permitted scientific reduction by "objective" methods—like intelligence testing, motivational research, and sexual psychobiology. In short, "[i]ndustry now has abundant opportunity to develop suitable methods of measuring persons with respect to qualities of character, mind, and body, and to make this information immediately available in connection with placement, vocational choice, and guidance" (60).

Although the person should be the *object* of scientific management—an essential structure of domination in the science of cooperation—the ideology of self-expression was intrinsic to Yerkes's exposition. Personal differences, democracy, and self expression were inseparable. Satisfaction of basic instincts, themselves known through science, was the essence of self expression in this model. Science, not class, race, or gender conflict, could provide for further human adaptive evolution. Yerkes considered the drive to be socially useful to be a kind of organic instinct compatible with the biological evolution of cooperation which was at last finding adequate industrial development. That theme necessitated his approach to chimpanzees as willing subjects in scientific endeavor because of their mental (and so moral) status.

In his autobiographical *Testament* written after retirement, Yerkes set forth his guiding ideals. "It has been my single and unifying purpose to present and defend the thesis that there are ways of life which are potentially better than the religious. Among them is the way of knowledge and enlightenment, truth seeking and willingness to carry responsibility instead of casting it upon the infinite."[6] He believed science freed human beings from "inherited structure," from outmoded, rigid, maladaptive forms. Yerkes was oriented to the future and his responsibility to prepare for it. His deepest faith was in progress, and he vehemently rejected the idea that he might be judged as a "conservative." "By the results of critical studies of this total picture we are justified, I think, in asserting that man is evolving toward greater altruism, more intelligent cooperativeness, and increasing desire and ability to seek to understand and to control his environment and himself. . . . The story which I read from the records is that of increasing natural morality" (Testament, 401).

Yerkes received a Ph.D. in psychology at Harvard in 1902 and remained in Cambridge until World War I. During those years he published on a range of organisms in the chain of being—from daphnia, earth worms, fiddler crabs, and turtles to dancing mice, monkeys, and people—concentrating on the comparative bio-psychology of sense organs, learning, and instincts.[7] For him, psychobiology was a branch of experimental physiology with logical and historical ties to developmental mechanics (experimental embryology). Meaning was given to the whole by a teleology of intelligence. Nothing made sense without scales of mental perfection. Yerkes was talented in the design of apparatus for comparative mental diagnosis. The institutional context of his vocation stretched from teaching and research at Harvard, to consulting and clinical work at the Boston Psychopathic Hospital, to a decade of scientific administration service in Washington under the auspices of the National Research Council, to the long-sought establishment of primate laboratories at Yale beginning in 1924. These tasks must be viewed as a whole within Yerkes's philosophy of cooperation and control in the promotion of science for human betterment. He had no sympathy for an individualist point of view in scientific work, and yet saw competition as the basis of cooperation.

At Harvard Yerkes had been impressed by an early founder of industrial psychology, Hugo Munsterberg, and his ideas of a natural hierarchy of merit. Placed in context of belief in individual and racial progress, the idea of human engineering was closely related. In his service at the Boston psychopathic Hospital from 1913 to 1917, Yerkes developed a Point Scale Test of mental function which he felt demonstrated the usefulness of psychology in diagnosis and in initiating treatment (Yerkes, Bridges and Hardwick 1915). Logically and practically, medicine and engineering were similar for Yerkes. Boston Psychopathic provided Yerkes with an "apprenticeship in psychopathology" (*Testament*, 140) that would be put to use in the study of nonhuman primates. In its environmental and hereditarian aspects, the medical model prevailed. The lessons learned from the hospital were applied to the army in its (promoted) need to categorize men in modern war efficiently (Kevles 1968). Chief of the Division of Psychology in the Office of the Surgeon General, Yerkes (1969 [1920]) was in charge of psychological examining in the army.

Yerkes did science administration tasks under the NRC in Washington until 1924. His two most important activities for the foundation of the life science of primate studies were his chairmanships of the Committee on Scientific Aspects of Human

Migration (financed by the Laura Spellman Rockefeller Memorial) and the Committee for Research in Problems of Sex (supported by the Rockefeller Foundation through the Bureau of Social Hygiene). These early NRC committees helped establish the model for basic research promotion and funding which partially prefigured the molecular biology program of Warren Weaver and the mechanisms of funding adopted by the National Science Foundation. These committees of public-minded specialists operated among their scientific peers in establishing research projects and centers; supporting "Baconian" ideologies of the relation of basic and applied science; and mediating among the quasi-public philanthropies, social policy, and university science. As important as explicit policy was the web of criss-crossing personnel on the committee, in universities, and on the staff of the big philanthropies. There was simultaneously division of labor and the integration of the scientific organism.

All four proposed original members of the staff of Yale's Institute of Psychology in 1924 were central to the research program of the migration committee in 1923–24.[8] From their research site on Ellis Island, Carl Brigham of Princeton and Yerkes worked on internationalizing mental measurement. Raymond Dodge investigated measurement of primitive forms of human response; his work was at the base of the pyramid of neural mechanisms and drives underlying social behavior. Clark Wissler, who joined the sex committee in 1925 and chaired the migration committee after Yerkes, studied physical inheritance and environmental changes among offspring of mixed marriages. Other work favored by the committee included study of the need for labor in relation to emigration and immigration, investigation of the influence of race upon pathology (directed by Raymond Pearl, who had also been consulted by Yerkes in 1913 on a tropical breeding station for genetic study—a project closely related to plans for primate research), and development of methods of personality and mechanical ability assessment.[9]

The Committee for Research on Problems of Sex intended for Yerkes's position in the Institute of Psychology to provide the needed center for infrahuman sexual psychobiology (Aberle and Corner 1953). The sex committee experience had shifted Yerkes's attention from exclusive concern with mental phenomena to a larger interest in drive and motivation. Those shifts paralleled movements within the community of life sciences in the same years. Race and sex were complementary concerns, in their mental and emotional aspects rooted in the nervous and reproductive systems. Yerkes looked at his work as chairman of the sex committee as a major satisfaction of his life (*Testament*, 228). "Sexual intercourse is of peculiar biological significance because natural reproduction and the continuance of the species depend on it. Celibacy is, in effect, the repudiation of a basic principle of life and therefore a violation of natural law" (404). Morality, biology, and a life of scientific service were inseparable. They were ruled by obligatory reproductive heterosexuality. Sex was a biological function, one among others and therefore not to be overrated, but important in individual and racial adaptation. Natural function was made a moral criterion; adjustment collapsed into adaptation; research provided norms (typological, statistical, ethical) to guide social life. The concept of the political hardly appeared in Yerkes's writing, and never in discussion of treatment for sexual and racial health. "It seems logical to say that, as science of moral values, ethics should be thought of as an extension of biological sciences of function: physiology, psychology, sociology" (407).

In the young institution for the study of primates, Yerkes at last was able to join research laboratory, model clinic, science promotion and cooperation, and pilot plant for human engineering. Housing four young chimpanzees in a refurbished barn made into a research facility, the modest primate laboratory associated with the Institute of Psychology was the start of realizing Yerkes's dream, first recorded in notebooks in 1902, for a psychobiological research station. Possibly his summer experience at Woods Hole led him to desire for psychology what had come to exist for marine biology, especially developmental physiology. A modern research laboratory was the mark of success for a field struggling for social and intellectual space. By 1913 he was making plans to go to the German station in the Canary Islands to study chimpanzees; World War I intervened. Instead, he was invited to come to G.V. Hamilton's estate in Montecito, California, to study an orangutan bought for the purpose, and monkeys (Yerkes 1916).

Yerkes explicitly modified for other primates the tests he had devised at Boston Psychopathic. Looking seriously for a site to establish a primate research station, he searched throughout the tropics and subtropics before settling on Florida for a breeding laboratory.[10] At the Institute of Psychology, Yerkes had a grant for the primate work from Edwin Embree's Division of Studies of the Rockefeller Foundation, and he had somewhat less than $10,000 per year from the Committee for Research in Problems of Sex. But, contrary to his intentions, the sex research aspect of primate studies did not flourish until the mid–1930s and the arrival of Edgar Allen at Yale (Allen 1932; Long 1987a, 1987b; Hall 1973–74; Hall and Glick 1976). From his base in New Haven, Yerkes began active solicitation from the foundation for a permanent ape breeding station. Finally, in 1929, the last year before the Depression, he was successful in obtaining a ten-year grant of $500,000 from the Biological Sciences Division of the Rockefeller Foundation for the breeding facility for chimpanzees. At last, primate research existed in its three necessary parts: a breeding station that would also allow research in reproduction and social life, specialized laboratories in New Haven closely associated with a Medical School, and provision for tropical field work.[11]

The year the breeding station was funded saw the demise of the Institute of Psychology, which was subsumed under the more ambitious Institute of Human Relations. Yerkes did not go into the Psychology Department, and he was only nominally associated with the Institute of Human Relations. Instead, he joined the Physiology Division in the Medical School. That choice re-emphasized the psychobiologist's close connection with physiology and medicine. At first, emphasizing their common vision, Yerkes prepared a memorandum stating his desire to have a separate Department of Comparative Psychobiology within the new institute. "Increasingly satisfactory adjustment of the individual to his physical and social environment is the chief objective of the Institute of Human Relations. On every hand lack of such adjustment is exhibited in physical and mental diseases . . . and the discontent or unhappiness which exists in a large proportion of men and women in every walk of life." Yerkes pointed out that the biological sciences "and their applications in medicine and human engineering, and the social sciences, together with applications in law and social organization"[12] would jointly build the new synthetic humanism, joining the fragments of the human organism separated by advances in specialized knowledge. The theme was a common rallying refrain for the ideology of the Age of Biology.[13]

Psychiatry provided the logic of integration—or domination—for all other organic functions, particularly sex. Sex was fulfilled in family, the social context for the life sciences ruled by neurophysiology and psychiatry.[14] Yerkes felt strongly that the Laboratories of Comparative Psychobiology merited autonomous administrative status on scientific and organizational grounds. To be a department in the Medical School, with loose association with the Human Relations Institute, would be a just arrangement. Yerkes did not succeed in obtaining an autonomous department, but he was an independent director of his laboratories within the department of physiology newly headed by John Fulton. The connection was a happy one in facilitating collaboration between Yerkes's and Fulton's laboratories on cerebral dominance and nervous organization in primates. Experimental psychopathology responded to the crises of civilization: "The family, the nation and the state are all desperate . . . for solutions to overtaxing of the nervous system in complex modern society."[15]

Yerkes was properly wary in 1929–30 of close association with the Institute of Human Relations, not because of its point of view, but because of its overly expansive organizational, administrative scheme, which he feared might result in failure, precipitating funding problems for the primate laboratories. Primatology could stand on its own in the foundations, Yerkes believed. Before the full consequences of the Depression were evident and during the period of joint programming by the Divisions of Medical and Biological Sciences of the Rockefeller Foundation, Yerkes's caution might have seemed excessive. But as the new head of the Division of Biological Sciences, Warren Weaver, began to assert his plans for molecular biology by 1935, and therefore to assess with a hard eye the old responsibilities of his division, things were different. The Institute of Human Relations appeared unwieldy; and true to Yerkes's fears, he was lumped with the Institute by Rockefeller officers who found the nice details of his official location hard to follow. If Yerkes foresaw that possible problem, another and more serious one was entirely unexpected. When as a result of the reorganization of the Rockefeller Foundation after 1928 primate studies were placed in the Division of Biological Sciences instead of Alan Gregg's Medical Sciences, nothing seemed amiss. But by 1935 Alan Gregg was still sympathetic to the basic evolutionary naturalism of Yerkes's approach, and Weaver was taking his division toward biophysics and biochemistry. The resulting shaky position of the Yale Laboratories of Primate Biology in the late 1930s and their subsequent reorganizations go beyond our story.[16]

Behavioral Adaptivity: Organic Function and a Teleology of Mind

By 1905, Yerkes's psychobiological doctrine was firm: he was seeking diagnostic signs of psychic life within the life science problematic of functional organization.[17] His structural signs were sense organs and parts of the nervous system. The invisible essence of psychic life was consciousness, but its study had to be rooted in visible objects. No behaviorist, Yerkes still insisted on operational criteria for mental diagnosis. He was an excellent clinician. The functional meaning of psychic life was found in the concept of levels of "behavioral adaptivity," Yerkes's term for intelligence. Adaptivity for Yerkes meant learning; adaptation meant evolutionary adjustment. The close apposition of ontogenetic and phylogenetic development was typical of the strain of Spencerian thought permeating life sciences deep into the twentieth century. Yerkes was not interested primarily in complex ecological parameters of adaptation, which were essential to a neo-Darwinian, but in scales of neural complexity as markers of increasing behavioral capacity, i.e., of expression of mind which was hidden forever from direct view. Re-constitution of ecology around the "ecosystem" and neo-Darwinian concepts of adaptation occurred after World War II.

Within the pre-war non-Darwinian meanings of ecological adaptation, Yerkes studied scales of mental function as the indicator of increased organizational complexity, the *physiological* meaning of intelligence explaining why that concept was essential to everything Yerkes touched. In this frame the neo-Darwinian breeding *population* was not fundamental; but the *family*, structured like an organism, was properly an object of study for psychobiology. If he could not construct, quantitatively, a scale of the principal function, behavioral adaptivity, Yerkes could not speak scientifically about the psychobiology of an object—be that object a species, sex, race, family, or individual. On the appropriate level, the basic atom was subjected to the basic procedure—ranking in a scale of nature. Scientific method consisted of test design, the technology that made Yerkes's evolutionary naturalism objective.

The implication deriving from his doctrine of the role of intelligence in physiology was that society (i.e., present and future human development or adaptation and adjustment) expressed the evolution of consciousness measured through adaptive, organic, visible acts called behavior. Cooperation was just such a diagnostic category of adaptive, therefore intelligent, acts. The purpose of evolution was mind. The first seat of cooperation was the family. Yerkes's doctrine was technically idealist and vitalist; mind and life were necessary complements. Within a dialectic of cooperation and control—informed by a logic of domination (triumph of mind) and historical explanation based on reproduction (foundation of sexual drives)—psychobiology studied reproductive and nervous physiology in their individual and familial expression. This is the kind of practice underlying the scientific construction of compulsory heterosexuality.[18] Understanding nervous and reproductive physiology based on mind and family produced the fruit of scientific service—a naturalistic ethics of evolutionary adaptation. In that sense, the life sciences of behavioral adaptivity were the foundation of practices of human engineering. Three polar pairs summarized the framework: cooperation/control, comparative method/experimental method, and natural object/designed object. The poles were connected by development (individual and collective) and function (with vital and mental teleology).

Yerkes's psychological doctrine was a functional associationism (Young 1970). The mind was divided into faculties, which in turn were arranged into a scale of increasing powers from sensation to insight. The three principal levels of function were called "monkeying" (fooling with things, getting useful experience by accident); "aping" (imitating another organism, goal direction unclear); and "thinking" (solving problems by insight or foresight). Each human individual was thought to go in order through those three stages of mental evolution, a fact underlying comparative study of primates for human mental science.[19] Each stage involved the inseparable aspects of feeling and thought; the direction of evolution was *not* from emotion to reflection, but toward ever more complex processing of sensations

(Yerkes 1925: 98). The senses provided the elements for fashioning self and environment and the materials for producing thought. The study of sensation was the first step in psychobiology, as the study of instinct came first in the functional associationism of the social organism, and the study of personality was primary in understanding family or culture. Potent with meanings in the history of science, exploration, quest, and progress, *sight* was pre-eminent for the primates.

Processed by the perceptual system, sensations might be combined into images persisting as memory. Freely combined images resulted in imagination. The crucial passage was from representation to ideation. Yerkes looked for behavioral evidence of the detailed nature of these processes in each type of organism he studied. Since his end point was ideational behavior, or insight, his underlying question was, "Do animals in solving unusual problems, or in adapting themselves to the demands made by their environment, act essentially as we do when we use ideas?" (Yerkes 1925: 102). There were two methods for investigating this problem: 1) watching free animals acting spontaneously, and 2) controlling conditions to produce observations. The methods combined the sympathetic attitude of the naturalist with the critical attitude of the experimentalist; both were essential to scientific advance. Both were embodied in a doctrine of representation like that evident in Akeley's taxidermy.

Simple in principle, performance tests demanded creativity in experimental design. No difference existed in principle between tests used for people and for animals; direct comparison should be possible. The real gift was in distinguishing solution of problems by trial and error from solution by insight. In his study of the young female gorilla, Congo, Yerkes constructed a set of diagnostic criteria for the discrimination. The list included the nature of attention shown by the animal, the kind of activity or action pattern (definite and persistent), sudden abandonment of unsuccessful methods, pause between action patterns, quick passage from method to method, and most important, sudden solution of problems. The last criterion resulted in a learning curve with abrupt breaks; a smooth curve tracked the cumulative improvement expected from trial and error strategies. Scrutiny of action patterns required detailed strategies of observation, resulting in evolution of data recording forms to check observer objectivity. Photography also became an important "objective" tool.[20] The purpose of diagnosing problem-solving strategies was to assess the evolutionary position of the organism. The organism engaged in adaptive acts, the signs of a mental stage that showed its capacity for individual and evolutionary learning and its susceptibility to engineering improvement called teaching (Yerkes 1927b: 137).

Yerkes's functionalist associationism organized the table of contents of *The Great Apes*, written to summarize psychobiological knowledge of apes and to announce the birth of a new field in the laboratories at Yale (Yerkes and Yerkes 1929). Each of the major types of ape—gibbon, orangutan, gorilla, and chimpanzee—was discussed in an order moving from structure, to habits in nature and life history, to affectivity, to receptivity (including sensation and perception), to ideation. Field studies had a place in the framework of evolutionary naturalism, but a very different one from that they assumed later under a post-doctoral student in Yerkes's laboratory, C.R. Carpenter. For Yerkes, naturalistic studies were important, just as sympathy was important, just as instinct was important, just as comparative method was important. The culmination of the whole series, however, was in experimental studies, which enlisted the cooperative perspectives in the final movement toward rational control. The field in nature supplied raw materials for the laboratory pilot plant for the production of adjusted human beings.

Explaining the insistence of the logic of domination of mental function over all other organic functions, mental teleology with functional associationism was basic to justification of life and human sciences in human engineering. In discussing the inevitable extinction of the apes, especially gorillas, Yerkes invoked not environmental destruction, but relative anthropoid mental inferiority, in particular, low levels of curiosity sustaining sciences of control.[21] In the summary chapters of *The Great Apes*, Ada and Robert Yerkes were especially clear about the logical structure of their science. Mental faculties were ranked for approximation to the human form. Curiosity, the condition for progress in primates, underlay the other faculties, which advanced from imitation to instrumentation. The last quality commanded great attention because tool use was "indeed the prophecy or threat of steadily increasing control of environment" (Yerkes and Yerkes 1929: 549). Observing apes put sticks together or stack boxes to reach bananas was really watching the origin of adaptation of the environment to the needs of the self versus self-adaptation to environmental determinism.[22] Tool use and construction were at the origin of the highest stage of adaptation, that which led to natural science and technology. The highest form of knowledge was not related to introspection and acceptance, but to rational control. The themes of science over religion as a way of life had many echoes. In giving marks for behavioral adaptivity, *The Great Apes* asked who is quickest (the analogue of "genius")? And who is docile or teachable (the analogue of cooperation, altruism, and the ability to participate in scientific production)? Those two marks defined a natural elite among primates. They determined who could be a servant of science; Prince Chim's royal name signified his noble calling.

Mental quickness and teachability defined the most appropriate laboratory animal. Apes provided the material for "new and daring pedagogical experiments" (Yerkes 1927a: 190). The published descriptions of the actual research plan were like the appeals to the foundations; harmony of scientific philosophy and social space was heard at many levels. Based on the functional associationism culminating in rational control, primatology would illuminate motivation, adaptivity and education, language and symbolism, norms and differences, social relations, and conduct and morality. The laboratories provided the basis for an experimental science which insisted that "between physiology and psychobiology there is no logical boundary" (Yerkes 1927a: 186). And finally,

> A naturalistic ethics is more likely to be achieved through the utilization of the infrahuman primates, beginning with the anthropoid apes, than in any other manner. Suggested by observations already on record are studies in social regulation, for example, by taboo, communistic and individualistic tendencies, sympathetic cooperation, forms of dependence, and altruistic attitude. (Yerkes 1927a: 193)

Cooperation and control became even more central to Yerkes's work in the 1930s. On the one hand, "chimpanzee culture" was strictly for human ascendancy; the high cost of maintaining the apes could only be justified by objective experimental science conducted in well-managed laboratories producing ideal research animals to specification. The cultivation of natural history was meaningful only to tell the experimentalist how best to approach the organic material.[23] On the other hand, for an evolutionary naturalist, the very nature of work with apes demanded attention to cooperation as a method and sympathetic insight as an explanatory category. Cooperation was not just for human scientists—it was essential to engage in *mutual* adaptation and cooperation *with the animals* if science were to be served. Chimpanzee and scientist were like labor and capital in the new era. The methodological require-ment was a consequence of the objective nature of the organisms, human and anthropoid. Yerkes insisted that many experimental procedures, even those painful to the animals, should be done with the animals' intelligent cooperation, not by constraint and compulsion. Callous treatment of animals was not permitted in Yerkes's laboratories for *scientific* as well as humane reasons. Intelligent, acquired cooperation characterized chimpanzees; that trait was a key reason for their utility in human advance in altruism in these times of crisis for civilization. And just as dominance was inextricably bound up with cooperation in chimpanzee behavior, so human beings must dominate the animals by commanding their cooperation to develop the last great area of life science: experimental sociology, complementary to and based on the model of experimental medicine.[24]

The structure of the natural evolutionary and the social scientific enterprises was mirrored repeatedly in animal body and mind; in family as atom of society; in research committee policy for mutually complementary, coordinated centers; and in daily procedure for research and animal care in the laboratory. Team coordina-tion, not competitive individualism, was the theme in large and in detail. The form of rationality implied by cooperation-control was functional integration, not enlightened atomic self-interest. The frame stands in sharp contrast to notions of strategic rationality in post-World War II sociobiology.[25] For Yerkes, the organism and the laboratory were equally bureaucratic organizations necessitating both for-mal and informal, rational and sympathetic, processes integrated in a developmental (production) project. In a sexual context laced with business language, "The Profit Motive in Ape and Man" stressed the drive for superiority and the origin of true cooperation and conscience (*Testament*, 316–35). Yerkes felt the excessive competi-tion of early industrial society would have to yield to a team approach. *Hierarchy* would provide the structure for cooperation. His Depression era, widely shared critique of "unrestrained" capitalism was explicit (Yerkes and Yerkes 1935: 1031). Hierarchical functionalist associationism was not just a common psychological point of view. Informed by the vital principles of life and mind, the rationally organized and internally differentiated objects of life and human sciences reinforced political and economic origin narratives.

Yerkes's commitment to cooperation was constantly visible—perhaps most clearly in the regulations for shared care of animals and recording of life history data. The cumulative records were essential to the manufacture of the ideal research animal. Yerkes's concern for establishing a cumulative *file* as part of the laboratory for human engineering was consistent with his pervasive theme of scientific manage-ment of production.[26] Documentary duplication of the productive process was a mark of the modern bureaucratic enterprise singled out by Max Weber for its role in ensuring a smooth, impersonal functioning. The file functioned in the division of labor to give management control of actual production and to ensure dependence of a deskilled, and therefore docile, work force (Braverman 1974: 126). Science is a writing practice mapping the bodies of the world as resource in a culture driven by a logic of expanding production. Laboratory inscription transfers "skill" from "nature" to "science."

Unsurprisingly, the primary object of attention for the fledgling science of experimental sociology was "the family." Within functionalist associationism the "faculties" of the family were its members, which could be further analyzed into constituent organic drives functionally integrated by the nervous system. The family economy, like the mental one, involved division of labor, (re)productive efficiency, and ultimate unity from an integrating hierarchical principle expressive of higher functional adaptation. Experimental sociology of the infrahuman primates drew from a rich fund of beliefs about natural social units. Argument about whether the apes in nature were promiscuous, monogamous, or polygamous—and what meaning to attach to the answer—abounded in popular and scientific literature. Yerkes, from his predisposition to see the apes as model systems for human engineering because of their pre-existing natural similarity, inclined to the opinion that the chimpanzees approached the state of monogamy (Yerkes and Yerkes 1929: 542). Not only was the individual laboratory chimpanzee an unobscured, uninhibited mirror, but also their groups were naturally revealing. Consequently, the animals were deliberately caged in monogamous family units, basic objects for experimental manipulations of their members within an associationist framework. The social atom—elementary particles bonded through obligatory heterosexuality—was taken for granted. "For the Chimpanzee the unit is undoubtedly the family" (Yerkes and Yerkes 1929: 541).[27]

If organic drives were related to role differentiation, then it was essential to identify and measure the drives. Since the breeding colony's success in producing young chimpanzees depended upon correct understanding of reproductive physiology and behavior, investigations into familial experimental sociology necessarily concerned the sexual drive. The relevance to human population and family questions was explicit. "Yet the understanding of sexual phenomena in their neural, hormonal, and general behavioral relations may be of supreme importance for the profitable conduct and regulation of our lives" (Yerkes 1939a: 80). Other workers, notably at the Carnegie Institution's Department of Embryology in Baltimore, had devised methods of behavioral measurement of sexual drive in rhesus monkeys. Yerkes's intention was to adapt them to his animals. The aim was to predict sexual behavior. His concerns were with female "receptivity" and male and female "dominance" in relation to willingness to mate. The Yale Laboratories had established a period of estrus behavior in chimpanzees. A guiding assumption was that periodicity of female sexual receptivity declined with increasing intelligence and male dominance. How would chimpanzees score relative to rhesus, and what light could be thrown on human sexual adjustment in marriage? Yerkes wished to distinguish and measure female "receptivity," male "desire," and mate "acceptability." The procedure was to observe hundreds of mating trials and record observations on a standardized scoring sheet. The issues of observer expectation determining observations despite clearly honest precautions in data collection deserve detailed discussion. Belief in objectivity endowed the hundreds of standardized data sheets with great powers. Even more important for our present discussion is the complementary organic drive crucial to experimental sociology—dominance.

"Inclusively considered, what I have termed dominance behavior is simply an attempt to satisfy the hunger for social status and all that it implies by way of

privilege, opportunity, and responsibility. All organic drives, all varieties of motivation, vary both in strength and effectiveness of expression. In this, dominance is no
exception" (Yerkes 1943: 46). It would be difficult to overstate the interest in
dominance as a physiological, psychological, and social principle in human and life
sciences in the 1930s and after.[28] Dominance as an organic drive was linked to
competition and cooperation, which generated large numbers of studies in the
1930s. Comparing democracy with other social systems, particularly fascism and
communism, occupied comparative psychologists; and social psychologists developed environmentalist theories of the pathological "authoritarian personality" with
substantial bridges to biological analyses. When Carpenter's howler monkeys start-
ingly showed little dominance behavior, howler society was compared seriously to
human socialism.[29]

Understanding the dominance principle and its limits was central to encouraging
"natural" hierarchies of efficiency compatible with democratic ideologies constrained invisibly by gender, race, and class. For chimps the principle "operates as
a sort of rating system which serves to classify the individual in accordance with
such social values as are suggested to us by the terms initiative, responsibility,
leadership" (Yerkes 1943: 49). There was no myth in ape life of equality of ability;
social position was earned as a function of natural characteristics and developmental
stage. "Sometimes one wonders whether this type of social organization might not
be valuable for man" (Yerkes 1925: 155). Yerkes's interest was in dominance as an
organic element in the dynamic of cooperation and control. Raw dominance had
given way in evolution to higher forms of integration. "[C]ertain facts seem to
indicate rather clearly that the chimpanzee's social system tends to favor the evolution of social service, and even the development of certain sorts of cooperation"
(Yerkes 1943: 49).

Metanarratives at the Food Chute

A choice opportunity presented itself for the experimental investigation of dominance in the context of family-centered experimental sociology. The opportunity
allowed coordination of sexual drive, status hunger, masculine and feminine personality types, and evolutionary transformation to higher forms of social control. Further, the study had noteworthy implications for counseling and human social services. Drive, personality, and social order came together. In the course of tests for
delayed response and representational processes, as part of the study of phylogeny
of language, Yerkes observed that sexual periodicity and dominance-subordination
appeared to influence which animals in cage pairs would come to the food chute to
be examined. Yerkes then conducted competitive food experiments on four kinds
of paired cage companions: heterosexual mates, mature females, mature females
with immature females, and immature females (Yerkes 1939b). Pieces of banana
were presented one at a time in a series of ten, through a chute in the cage. The
observer recorded which animal of the pair took the piece. Results were correlated
with sexual status of the females interpreted in terms of dominance-subordination
and response by "right or privilege." Right or privilege meant that in the period of
maximum genital swelling of the female, the ordinarily dominant male granted her
the privilege to take the banana. Dominance itself was not seen to reverse, yet the
female acted as if by right. Yerkes recognized various problems with the data: for

example, observations were made in only one case for an entire cycle and variation in response pattern overwhelmed the postulated regularities. Tests of statistical significance were not reported. In female pairs, sexual swelling affected performance on the food priority tests, but the effect could be in either direction. Even among "mates," it seemed presence or absence of prior "friendship" greatly affected results. Nonetheless, Yerkes spent most of the paper describing in detail a heterosexual pair, Jack and Josie, who seemed to show substitutions of right and privilege for dominance in exchange for sex. The tone of this father of primatology was tentative, yet expectant that these observations were the beginning of very important studies.

Yerkes's lab wrote the male-dominant economics of power and sex into the food chute exchanges. Here is the origin narrative of prostitution in the market and cooperation in marriage. In a feminist critique of anti-feminist natural history, psychologist Ruth Herschberger, who had corresponded with Yerkes to object to his interpretations,[30] took the part of Josie, the female chimpanzee whose psychosexual adjustment was the subject of such serious analysis. "Josie" complained that experimenters thought that sex made females act dominant although they were naturally subordinate.

> People have been very kind to me at Orange Park, . . . and I don't like to look a gift horse in the mouth. However I would like to register a protest against the attempt to discredit my food chute score in the Thirty-Two Day test. . . . Please note that on March 21, as well as on other occasions, Jack came up to me repeatedly at the chute and gestured in sexual invitation. Doesn't that suggest that he was trying to get me away from the chute by carnal lure? Or was Jack just being . . . an impulsive male? The experimenter took it as the latter. But who knows that Jack wasn't about to exchange sexual accommodation for food?" (Herschberger 1970 [1948]: 7,11)

Dominance as a drive was not sex specific, in Yerkes's opinion. It was a basic organism hunger for social status. "Assuming that dominance is hereditary and that inheritance is independent of sex, men and women might be expected to become creative leaders with approximately equal frequency" (Yerkes 1939b: 33–34). Culture accounted for the observed predominance of male leaders. But the association of "leadership" and biological dominance was natural.

Yerkes was liberal to moderate on the sex role controversies of the day and made clear his opinion that human females should have greater "opportunity" than allowed by tradition. The issue here is not whether Yerkes or other spokespeople for comparative psychobiology were or were not liberals in their own time, but the logic of naturalization of the issues in terms of the hierarchy from instincts to rational control through personality and associated educational and medical therapies. With the passing of religion, the new bedrock for value decisions, the more evolutionarily adaptive ground for judgment, was comparative life science. Sex was fundamental to managed society. Sex was mobilized by science and medicine as it was discursively freed from religion. In the division of labor in the family, the model for division of labor in all of society, naturalization of the bio-politics of reproduction was a cornerstone of historical explanation.

The investigation of sex drive and dominance-subordination was explicitly in the context of contemporary controversies. Yerkes assumed that feminism was equivalent to the proposition that males and female were biologically "equal"; i.e., that the concept of rights in political philosophy was rooted in natural economy. On "scientific grounds," Yerkes rejected the proposition that males were mentally superior, or for that matter, naturally dominant. Males and females had the same psychological (ideation) and drive (motivation) structure. But there were differences in expression of drives as a result, presumably, of hormones. The *result* was personality. The physical marker for the internal state was always required in life science. Relation of psychobiology with the contemporary biology and physiology of sex, the first two legs of the Committee for Research in Problems of Sex's promotional program, was well articulated. If differences in drive expression could be correlated with the social division of labor, then the feminists of Yerkes's time were misguided.[31]

Discussing differentiated techniques of social control adopted by males and females, Yerkes described biologically determined differences in drive expression. Differences among chimps in "techniques of social control" suggested that human modes were also psychobiologically legitimated and inevitable.

> In a word, the masculine behavior is predominantly self-distracting; the feminine, primarily favor-currying and priority-seeking. . . . To the observer the male seems often to be trying hard to blot out awareness of his subordination; the female, by contrast, to be hopefully trying to induce the male to give place to her at the chute. . . . As for the females, wiles, trickery, or deceitful cunning, which are conspicuous by their absence in the male list, are favorite resources. But even more so are sexual allure and varied forms of solicitation . . . . That the female is, cameleon-like, a creature of multiple personality, is clear from our observations. (Yerkes 1943: 83)

The foundation for these "observations" was still the experimental sociology of the food chute test. The lesson for the *limits* of cultural formation of personality, and therefore of possible social change, was not left to the imagination. "I am impressed by the contrasted attitudes and activities revealed by the competitive food situation, and I offer them as evidence that male and female chimpanzees differ as definitely and significantly in behavioral traits as in physique. I am not convinced that by reversal of cultural influences the pictures characteristic of masculinity and femininity can be reversed" (Yerkes 1943: 85). This opinion persisted in the face of Yerkes's belief in human malleability and perfectability through engineering. "Personality differences" should be managed, not foolishly denied.

Personality studies using anthropoid materials were especially favorable because of the absence of social taboos and personal inhibitions. "Therefore, I submit that such observational items as appear in this report, and in related studies of the psychobiology of sex in the anthropoid apes, should have exceptional value for those who concern themselves with problems of social behavior, and, especially at this juncture, for those psychopathologists who are intent on appraising, perfecting, and using psychoanalytical method of observation and interpretation" (Yerkes 1939b: 130). Personality, though less differentiated than in the human species, stood nonetheless in chimpanzees also "very clearly, as the unit of social organization." Personality meant the functional whole, "the product of integration of all the psychobiological traits and capacities of the organism." In a normal personality, inherited characteristics and basic organic drives were integrated with the conscious self. Sex and gender were mediated by the natural-technical object of knowledge called personality. To have a masculine or feminine personality was not a minor matter; on proper development hinged the possibility of adjustment and happiness of the individual and the body politic. Diversity and variability, as for all organic phenomena, were ubiquitous. Comparative science was designed precisely to deal scientifically with variability. For drives as central as sex and dominance and expressions as consequential as masculinity and femininity, nurturance of personality was a matter for responsible scientific service. In fact, the possibility of prescription of social role on rational grounds was at stake. If drives and personality could be measured early, proper treatment could be initiated. Yerkes was cautious, but hopeful.

> If in man dominance as personality trait is highly correlated positively with leadership, as it evidently is in the chimpanzee; if it is a condition of or markedly favorable to individual initiative, inquiringness, inventiveness, and creativeness; and if, further, it should prove to be reliably measurable during childhood, it may very well come to possess conspicuous usefulness and therefore also the basis for differential advice might be affected by it, for congeniality or social fitness may depend appreciably upon similarity or the reverse in dominance as personality trait of mates or companions. (Yerkes 1939b: 133)

The significance of the dependence of the culture concept on personality in the anthropology of the 1930s should be clear. It is not surprising that the controversies surrounding the ideology of the nature-culture split rested on a shared logic of functional integration of life and human science. It was a logic of domination, i.e., of cooperation and control. We have moved with Yerkes from instinct, through personality, to culture, to human engineering. Sex, mind, and society have been interwoven in a vocation of scientific service. There has been a fruitful ambiguity and mutual complementarity of cooperation and control in the establishment of a promising new life science of comparative primate psychobiology, reaching from learning through motivation to experimental sociology. Primatology served as mediator between life and human sciences in a critical period of reformulation of doctrines of nature and culture.

So were man and animal in dual evolutionary partnership as master and servant. The natural child, Prince Chim, and the natural heterosexual couple, Jack and Josie, embodied Yerkes's historical hopes. Perhaps the science of primates was a coy mistress, but the rational management of nature's functions promised a high rate of return. Yerkes produced a science of objects that expand and increase; the result was knowledge of the body as a particular technology of power over personalities in their two key manifestations—sex and mind. Primates, especially as physiological models of persons, became bodies invested with power relations. But power is not possessed; it comes into being through its exercise. Power is exercised as the effect of strategic positions, of a series of highly mediated relationships by which certain forms of domination are maintained through the process of producing knowledge. Within the boundaries of pre-World War II comparative psychobiology, the contours of the primate body became such an object of knowledge. This body was dissected according to endlessly echoing polarities; sex and mind, the reproductive and nervous systems, cooperation and control, reproduction and production, organic drives and problem-solving strategies, culture and engineering, religion and science. The unifying object was the concept of personality, which was developed as the basis of diverse sciences of organic capacity and rational management. Primates became models for the management of human beings as persons with the fruitful capacity to labor, to increase and multiply and fill the earth. This is the power contested as the heart of primatology.

5

A Semiotics of the Naturalistic Field: From C.R. Carpenter to S.A. Altmann, 1930–1955

A *picture* held us captive. And we could not get outside it, for it lay in our language and language seemed to repeat it to us inexorably. (Wittgenstein 1953)

A Tale of Island Colonies

Functionalism is fundamentally a theory of communication. The story of the mutation within functionalism from an organics to a technics of communication in the primate body is a story of semiotic theories and technologies. This is a major chapter in post-World War II constructions of the relations of organic bodies and technical artifacts. Functionalism is a logic for the mediation of control through self-sustaining processes, not a logic of direct visible command. From a functionalist perspective, an organism works by internal principles consequent upon its organization, not by external direction like a puppet. Organic functionalism was transformed into a cybernetic technological functionalism broadly in life and human sciences from the 1930s to the 1950s. There are many threads in the story of this momentous mutation in control theories and practices.

One of the threads spins out from an unexpected place: the primate field. The naturalistic field is a culturally specific social space crafted by scientists in their historically mediated practices of watching free-ranging monkeys and apes. Beginning with Clarence Ray Carpenter in the 1930s and ending with Stuart Altmann in the 1950s, let us follow the construction of the naturalistic field, as an epistemological and material space for producing knowledge about the primate order as a problem in semiotics. Echoing the endlessly recursive structures of primate narratives, the story begins and ends in a colony within a colony, on Cayo Santiago, a 37-acre island off the coast of Puerto Rico. On Cayo Santiago, Carpenter established a rhesus monkey colony in 1938; and Altmann did his doctoral dissertation there in the mid–1950s, at the start of the post-World War II revival of naturalistic primate studies.

Cayo Santiago on the Eve of World War II

On December 2, 1938, approximately 450 captured Indian monkeys arrived in Puerto Rico, after a difficult 6-week ship passage.[1] Carpenter, a young comparative psychologist in the Anatomy Department of Columbia University's College of Physicians and Surgeons, released the animals onto the island and watched them form social groups. He was well prepared to watch this organic process; in the 1930s Carpenter conducted a series of groundbreaking field studies of wild monkeys and apes that established the discourse of naturalistic studies of primate social behavior (Carpenter 1964). He would remain associated with the Cayo Santiago rhesus until he left Columbia in 1940.

Carpenter began his studies of the physiology of primate social groups with a question dictated by his most basic assumptions: Why do animals live in groups? If the animal physiologist needed ultimately to explain the coordination of the whole organism, the student of animal society needed to account for "whole animals in *complete groups* and these in their natural, complex and dynamic environment" (Carpenter 1964: 161). That which needed explanation was the *whole*, i.e., the phenomenon of organic organization. In this context, Carpenter's primatology in the 1930s was a discourse on the organic semiotics of sex and dominance. The first priority in the field notes written on ship with the monkeys was, "Mates must be graded as to sexual potency and their rank order dominance established." The list continued with plans to study castrated animals, maternal behavior, and structures of dominance before release. "Select from the males to be released on the Island a number of individuals. Test them for sex drives and dominance by time sampling procedures and Murchison. Castrate them and then release them. Keep running records and test at intervals of three months." "Produce experimental homosexuality." "Produce intersexes by injections of internal secretions." "Work on sex difference of dominance—Determine hierarchies for both sexes."[2]

Time and money did not permit all of these problems to be pursued. The field notes from the period December 1938 to May 1940 sketched the priorities: (1) a study of dominance as the primary integrating mechanism of primate society, (2) sociometric mapping of dominance relations and other social bonds, and (3) analysis of inter- and intragroup interactions as signs in a functioning system. The basic technical practices were counting numbers and categories of animals in groups, construction of force-vector maps to represent their positions and the probabilities of splits and fusions, and determination of the females' sexual status as a function of their significance in bonding the males to the group. Females were bound to the group by the dominance of males; males were bound by the sexuality of females. Both were bound to each other by a logic of control. The product was the reproduction of primate society.[3]

Carpenter performed one three-week "defect experiment" incorporating the physiological, sociometric, and semiotic principles of his primate story. He operated on the most dominant group that had formed on the island, named Diablo after its "leader," with about 85 animals, including 7 adult males. He determined the dominance order for the males only—they were the presumed axis of greatest activity in primate organization.[4] With these determinations, experimental manipulations could begin. On June 4, the imposing Diablo was removed to a nearby outdoor cage. For Carpenter, Diablo's removal was a surgical intervention into the social body; it was cutting off the head. For a week, Carpenter watched the group's behavior and territorial range. He witnessed confusion and disorganization, marked by increase in fights and decrease in space for the group's activities. He watched to see which of the remaining males would reestablish order, but for days saw only a "persisting lack of integration" and movement that was "amoeboid . . . sluggish and uncertain."

On June 11, he trapped and removed M 174, the next male in the prestige hierarchy. On June 17, he removed the third most dominant male. The observed result was that previously subordinate males of the group occupied central positions among the females and young; but the group as a whole was in physiological decline. He noted female fights during this period with special interest: "It would seem that these female fights relate to the restructuring of a group when a dominant male has been removed or displaced."[5] As in a hydroid polyp, the previously subordinate axes of organization became more visible when the dominant region was cut away. The spatial force-vector diagrams constructed before the trappings disintegrated as stable relations gave way to social flux. On June 26, Carpenter released all the captive males and recorded the reestablishment of social order. The analysis and construction of this defect experiment demanded the integration of tools and concepts from several disciplines: physiology, comparative psychology, social psychology, and neo-positivist linguistics. Life and human sciences came together in the study of the organic "anlagen"—embryological primordia—of human control.

Beginnings: Pigeons, Monogamy, and Primates

Clarence Ray Carpenter began his scientific life during the height of physiological functionalism in the late 1920s and early 1930s, in the Stanford University psychology laboratory of Calvin P. Stone. Continuing work begun in his master's thesis under William McDougal at Duke University, Carpenter studied the effect of gonadectomy on sexual activity of male pigeons in heterosexually mated pairs.[6] That is, he performed the classical defect experiment to investigate the physiology of what he called "monogamic relations." Formed in physiology and comparative psychology, from the start Carpenter viewed sexual activity as the unifying locus of the individual organism and of organic society. He began from the widely held premise that societies of higher animals could be explained in terms of the bionomics of sex: the basic forces of social order—cooperation and competition—must at root be aspects of sexual interaction. Together, sex and mind were not only the principal objects of study for endocrinologists, neurophysiologists, and psychologists, but they were believed to constitute the material foundation of organic social integration and the greatest threat of disintegration. Sex and mind were the keys to scientific control of life, the fundamental natural-technical object of biology.

Behavior was like morphological structure; both were characterized by form and function, and subject to pathology. The defect experiment was the appropriate scientific manipulation to reveal the physiology of individual and society. In studying sex, the procedure was to cut out glands and organs; in studying mind, scientific procedure dictated altering or removing the head. This logic recurred repeatedly in Carpenter's work. Sexual behavior was a privileged handle to the theoretical understanding and therapeutics of natural cooperation ordered by male-female dominance and male-male competition. The physiological structure of coordination was produced by mechanisms of hierarchy and competition among organic parts. The physiological animal-machine generated harmonious wholes from the materials of *complementary* and *antagonistic* difference, the only available logics of difference for ordering the biopolitics of the natural body in colonial and liberal discourses.

Carpenter deepened this physiological point of view in his first work on primates, a field study of wild howler monkeys in the Panama Canal Zone, on Barro Colorado Island. Carpenter spent a total of 12 months in the field from 1931–34, 8 months of that time watching howlers. His next (and last) major research was the study of gibbons in Thailand as part of the Asiatic Primate Expedition of 1937. Here, Carpenter adopted both a neo-positivist linguistic theory of signs; i.e., *semiotics* as practiced at the University of Chicago, and a sociological-psychological field theory of complex small group structures, *sociometry*, to explain the pattern and boundaries of primate social organization. In all these cases, Carpenter sought to understand governance, the physiological structure of control, through studying sex as steersman. The result was a bionomics of organic social groups, a crucial discourse in the management of human life according to medical logic.

Primatology has been pervasively determined by borrowings from human social science. That fact explains the ease with which strategies of biological reductionism could be developed; the biological disciplines were already built like other contemporary functionalist discourses.

The Sexual Bionomics of Primate Society

Completing his doctoral thesis, Carpenter sought help from his Stanford advisers, Stone and Lewis Terman, to obtain a National Research Fellowship in Yerkes's laboratory "investigating some aspect of sex behavior in primates."[7] Yerkes was far from indifferent to the post-doctoral application from the promising Stanford Ph.D. In Carpenter he believed he had found the man who could help realize his program of establishing an experimental sociology of primates, with its heart in scientific study of family relations as a basis for general social therapeutics.[8]

The sketch Carpenter sent the National Fellowship Board of his proposed postdoctoral work reaffirmed that hope. Carpenter included in three possible research problems the essential elements of the unified vision of comparative psychology and reproductive physiology focused on the foundation of social life in animals. First, Carpenter proposed to study social reactions in animals by "not[ing] especially the attachment of one animal for another of the opposite sex." He would study posture, facial expression, and motor action patterns of chimpanzees in response to each other in an effort to measure the strength of social, especially sexual, attachment. The traditional method of comparative psychology, the multiple choice problem, seemed useful for measuring social bonds. Heterosexuality became a scientific natural-technical object of knowledge in just such practices. Second, Carpenter proposed to set up a movable chain with a food reward, requiring the coordinated effort of more than one chimpanzee to procure the goal. The third problem proposed to extend to monkeys work that had been carried out on small laboratory mammals, poultry, and pigeons, i.e., correlation of sexual behavior and physiological state before and after castration. The "action patterns" of behavior must be as carefully scrutinized as the secretions of glands for their characteristic form and function. Sexual physiology was where the relations of dominance and subordination, activity and passivity, initiative and receptivity, and stimulus and response were most evident and crucial.

Carpenter expected to do his research in the laboratory, but he was wrong. Vigilant for opportunities to foster naturalistic field investigations to complement laboratory research, Yerkes had been looking since 1929 for someone to follow up a rich chance.[9] On February 1, 1939 his friend Frank M. Chapman, the founder of the Bird Department at the American Museum of Natural History, who worked extensively on Central American birds from a base on Barro Colorado Island, in the Panama Canal Zone, wrote Yerkes about the wonderful opportunity to study howler monkeys there. "In the trees nearly over my house there are at the moment seventeen howlers, three of which are carrying young—a unique opportunity here to study the individual, the family, the clan, the inter-clan relations under a natural but controlled environment."[10] In his capacity as a first-rate field ornithologist crucial to Carpenter's initiation, Chapman contributed to the development of field methodology and fundamental concepts in primatology. The study of birds was second only to laboratory-based physiology and comparative psychology in the inheritance of primate studies. Carpenter reworked this dual inheritance to produce a field biology of nonhuman primates.[11]

From Birds to Monkeys

Counting

Chapman encouraged Carpenter in a simple operation: making a complete census of the howlers of Barro Colorado Island. Since 1905 American economic ornithologists had conducted breeding bird censuses, and in the middle 1930s the National Audubon Society inaugurated a project of quantitative breeding bird censuses. These procedures produced much more precise awareness of territory size fluctuations, fluctuations of population density from year to year, and the influence of slight changes of vegetation (Mayr 1975). Ornithologists like Chapman were keenly sensitive to the importance of counting as a foundation for basic biological statements, particularly population forecasting and descriptions of the growth form of a population. A census was not just a quantitative description, but was related to the problem of prediction, to the discovery of the laws of population growth, decline, or stability in particular conditions. The population was a biological body that could be in a state of health or pathology like any other organism.[12]

Cooperating

Like territoriality and population biology, animal sociology was not limited to studies of birds in the 1920s and 1930s. But the bird work was a conspicuous thread in the fabric that Carpenter further embroidered. Two figures were central in establishing the dominance concept within the frame of physiological functionalism: Thorlief Schjelderup-Ebbe, a Norwegian who worked in France, and Warder Clyde Allee, at the University of Chicago. Schjelderup-Ebbe (1922) was credited with the discovery that birds were organized into social hierarchies by a strict dominance chain, or pecking order. Studying over 50 species of birds, he thought he had determined that "despotism is one of the major biological principles" (Allee 1951: 136). The dominance hierarchy as a mechanism of social coordination became a major preoccupation of zoologists and comparative psychologists in America in the 1930s. A colleague of Yerkes could correctly write in 1939 after a decade of experimentation on the dominance hierarchy in nearly every conceivable animal that could be induced to move:

> Explorations of the significance of the concept of dominance have hardly begun, since little is known about how dominance may influence all sorts of social interaction. Also, the factors which determine dominance status are yet to be fully elucidated. Among the primates the subject is particularly fascinating, since there the relation between dominance and sex behavior is perhaps most striking. Full exploration of this relationship may suggest solutions of many baffling problems of sex attitudes and perversions as well as indirect means of social control. (Crawford 1939: 418–19)

Allee, a Quaker sympathizer important to the history of ecology based on the community concept, took exception to the Norwegian's narrow sense of the foundation of biological order. Allee emphasized cooperation as the most fundamental biological force. "Disoperation," i.e., disorder, not competition, was the opposite of cooperation for Allee. However, the common ground between the despot theorist and the Quaker is revealing; dominance need not mean a principle of autocratic rule. Dominance and subordination must rather be conceived as forms of social coordination. Like any other physiological object, dominance had to be understood in the full variety of its forms through careful prosecution of comparative biology. At Chicago, Allee studied organisms ranging from bacteria, planaria, crustaceans, goldfish, and protozoa to small mammals and birds as he explored the evolution of mechanisms of coordination from physiological mass action to differentiated nervous systems.[13]

The Community as Organism

Allee defined the community as "a natural assemblage of organisms which, together with its habitat, has reached a survival level such that it is relatively independent of adjacent assemblages of equal rank; to this extent, given radiant energy, it is self-sustaining" (Allee et al 1949: 9). The self-sustaining community had all the other characteristics of an organism: development over time, differentiation of parts by form and function, organization by gradients into sub-wholes called fields, and above all, tendency toward homeostasis, "one of the major inclusive principles of ecology" (1949: 6). The community concept was initially developed by the generation before Allee, especially by the plant succession phytogeographer F. E. Clements and, about 1913, by the animal ecologist under whom Allee earned his doctorate, Victor Shelford.[14]

The arguments of two University of Chicago organicists important to Carpenter, Charles Manning Child and Alfred Earl Emerson, illustrate the depth to which physiological functionalism was structured by dominance.[15] For the developmental physiologist, Child, dominance initially meant the rate of energy expenditure. Differential rates of exchange established dominance. The detailed study of *rates* in biological systems was the principal motor of the transformation from physiological to cybernetic functionalism in developmental biology and ecology. Closely connected with the measure of rates was the measure of pattern maintenance or communication. For Child (1924), the pattern of protoplasm, in all its forms from simple axiate organisms to complex society, was a behavior pattern determined by the "dominant region" of greatest activity in production and consumption of energy: "Apparently all that is necessary for the beginning of orderly integration in protoplasm is a quantitative difference in rate of living and the possibility of communication. Dominance or leadership in its most general physiological form apparently originates in the more rapid liberation of energy" (Child 1928: 32,42).

Organization without dominance had to remain simple. If the dominant or active region of an organism were removed—the classic logic of the physiological defect experiment—a new center of dominant activity must be established or organization would not reappear. Higher behavioral patterns depended on differentation of parts and specialization of function. By no means did the greatest differentiation result from the most autocratic organization; effective control producing greatest freedom was the fruit of the most complex development of communication of materials and processing of energy:

> [The] more completely communistic organism complexes, consisting of many similar zooids, have remained relatively primitive in character, the chief relation between the components being nutritive. . . . As in the organism [in social integration], with the development of means of communication and transport, the effective range of social dominance increases greatly. (Child 1940: 395,397)

Based on his studies of insect societies, Emerson, a co-author of the important 1949 ecology textbook, *Principles of Animal Ecology*, was a principal advocate of the superorganism. The superorganism concept carried the corollary that homeostasis was the correct term to denote social integration. Discovering the laws of dynamic equilibrium, that is, the organic variation and regularity of patterns maintained by dominance, was the task of the biologist in the laboratory and in the field (Emerson 1939, 1954).

The Howlers of Barro Colorado

On December 17, 1931, Carpenter was on a United Fruit Company boat bound for Panama, supported by a grant to Yerkes from the Committee for Research in Problems of Sex and by a National Research Fellowship.[16] Carpenter's first enthusiastic letter back to Yerkes announced the expected themes: he looked immediately for sexual behavior and for social bonds—expressed by "social skin treatment."[17] In close contact with Frank Chapman, he began the census of howlers on the island. Carpenter's early questions and records in the field followed closely the plan Robert and Ada Yerkes used in *The Great Apes* (1929).[18]

Sexuality and its relation to group coordination was central. In January, 1932, Yerkes sent Carpenter the new book by Solly Zuckerman, *The Social Life of Monkeys and Apes*, with a note saying "it will be most valuable." Zuckerman argued that constant female sexual receptivity was the foundation of primate society. He argued further that dominance hierarchies formed by fighting among males and male control of females to amass a docile harem were the mechanisms of social formation and maintenance in all primates.[19] Zuckerman developed his views on the origin of human society in response to Bronislaw Malinowski's ideas on the origin of the family on the basis of unique human female physiology (menstruation) and of the original cultural institution (fatherhood). By March 25, 1932, Yerkes cautioned Carpenter against Zuckerman's extrapolations, especially to apes. But Yerkes still regarded Zuckerman's generalizations about monkeys to be reliable.

Finding Zuckerman's suggestions congenial, Carpenter collected animals "to determine the physiological state of the reproductive system of the females closely associated with males." He had "two very interesting experiments in progress." "1) I have selected a group of nine animals with one male. I plan to remove the male and to observe constantly the group during the period of reorganization . . . [and] 2) The group that we have recently found consists of five animals . . . [I]t will be possible to completely isolate this clan, and since it contains about the average number of females per male, and since the animals can be identified individually, many important questions can be answered . . . ."[20] Following the logic of Child, the plan to remove the luckless male was the first of many such efforts to perform the classical defect experiment on the physiological social group. Collecting female reproductive tracts for anatomical study in order to answer questions about the relation, if any, of physiological state and social association continued throughout Carpenter's early work. Observations of deliberately isolated social groups, especially with reference to detailed study of pairwise relations of identified animals, was a basic technique for studying social bonds microscopically. The "average number of females per male" suggested a basic bionomic law. Later, Carpenter also collected males, either solitary or in all-male groups, to determine their age and sexual state in relation to their peculiar social grouping. It was hard to account for all-male *groups*, made up of males in good physical condition, on the reigning hypothesis of heterosexual attraction as the core of organic society. A corollary hypothesis of saturation of available female sexual drives, thus producing supernumerary males, was predictable and was forthcoming.[21] Carpenter's field notes for the howler study showed from the beginning his personal and theoretical foundations in reproductive physiology informed by dominance, the homeostasis of social groups, and comparative psychology of drives and learning. The howler monkey monograph, published in 1934, put this framework at the textual root of modern field studies of wild primates.

The Monograph

The organization of the paper established a pattern maintained in all Carpenter's further publications. The constant order was: field procedures; postures and vocalizations; feeding; territory and nomadism (progression through space); organization of population; integration of social group; coordination of social group; conclusions. Using field techniques adapted from ornithology to map ranges, Carpenter followed the movements of several distinct groups. Mapping movement allowed him to make judgments about group leadership, composition, and mechanisms of maintaining territory (e.g., vocal battles between adjoining groups). The languages of objective animal psychology, establishment of neural pathways in the brain, and movement in external space were telescoped into a single set of terms and relations. Carpenter mapped the physical-psychological space of 23 groups, one in detail.

Based on the census, Carpenter's fundamental concepts were the bionomic principles of (1) the *central grouping tendency* and (2) the *socionomic sex ratio*. The central grouping tendency or "the mean tendency of gregariousness" was expressed in a formula for the proportion of animals of different categories characteristic of that species. Carpenter considered the tendency to be the result of fundamental physiological and psychobiological forces, not of ecological conditions. Ecological crises could disrupt the central grouping pattern, but they did not account for it. He believed it should be possible to construct a formula for the central grouping tendency of each species. This formula was the foundation for the second basic concept, the socionomic sex ratio, which expressed the number of adult males to females in a stable group. Male and female sexual drives should balance each other. If a sex ratio temporarily existed that left such forces out of balance, the observer could predict group recruitment, fission, fighting, or some other sign of instability that could trigger a homeostatic regulating mechanism. Carpenter called group splitting "apoblastosis," a term for cell-division. Subgroupings of animals were explained within the framework of balanced forces structuring the whole group around the basic organizing axis of heterosexual drives.

Integration referred to processes occurring over the whole life span of individuals and groups, while coordination concerned a time slice of a community in terms of the interrelated forces of its immediate cohesion. The subject of integration was socialization; the subject of coordination was communication. Both were part of the theme of control. In Carpenter's argument, control began with direct contact between infants and mothers and ended as symbolic control across space by posture, vocalization, gesture, and glance. The mechanism accounting for the change was the interaction of inherited tendencies and learning understood by conditioning theory. Carpenter practiced the normal science of the interactionist paradigm that resolved the heredity-environment controversies of the preceding period (Cravens 1978). Interactionism was invoked as much to account for complex animal behavior as for human doings.

Carpenter's basic procedure for studying integration was to give detailed descriptions of elemental relations between selected pairs of organisms.[22] But not all pairs were equally important in equilibrating forces to establish a whole. The important ones had to do with sexually defined interactions ordered by the dominance concept. "The male and female 'vectors' of motivation function reciprocally to bring the animals together and cause them to engage in inter-related patterns of satisfying behavior.... With repetition of the reproductive cycle in the female and with uninterrupted breeding throughout the year, the process of group integration through sexual behavior is repeatedly operative, establishing and reinforcing intersexual social bonds" (Carpenter 1964: 66).

Overall, the field theory of social groups was grounded on mutual dependence analysis like that Henderson used to study stability of blood pH (Parascondola 1971). Carpenter's conclusions for howlers emphasized the low slope of gradients of dominance. The explanatory need was not for particular mechanisms, like a linear dominance hierarchy, but for understanding the precise forms of dominance operative in a given social group. Coordination concerned the mechanisms of social control considered synchronically. Gesture and vocalization were minutely described and recorded in the field on camera and tape. The framework was psychological, with attention to reciprocal conditioning and stimulus-response patterns effective in maintaining a whole social group.

Laws of Language, Laws of Space, 1936 to World War II

Let us follow Carpenter to view the gibbons of Thailand, in order to understand the connections of the physiology of dominance with theories of communication in language and space in the late 1930s. These sciences of communication from both physiology and linguistics were a crucial bridge to the cybernetic functionalism—rooted in post-war systems technologies and theories—adopted in many areas in biology in the 1950s. Harold Coolidge, of Harvard's Museum of Comparative Zoology, assembled the money and people for a major primate research expedition to Asia. Carpenter's object, the gibbon, was an alluring ape. Gibbons had upright posture and a monogamous sexual life.[23]

Carpenter was in the field in Siam from February to June, 1937. Two months and 14,000 miles from the U.S., a discouraged Carpenter recorded in his field notes that he had still not seen a gibbon, despite hearing them constantly. Compounding Carpenter's difficulties was the extensive collecting going on too near his study area. A move of camp produced opportunity for observations underlying the "Field Study in Siam of the Behavior and Social Relations of the Gibbon (*Hylobates lar*)," originally published in 1940 (Carpenter 1964: 144–271). In this monograph, Carpenter clarified his evolutionary framework, as well as the relation of this sort of study to the human sciences.

> [I]t is sufficient to assume that just as there are structural relationships which place man in the same categories with the more complexly developed primates, so there are basic human needs, drives and types of behavior which have elements in common with similar functions of the non-human primate level. For example, many aspects of sexual behavior are similar in man and the apes. Perhaps in these primates one may observe *anlagen* of human motivation and behavior, free from cultural veneers and far enough removed to avoid the well known errors involved in man's study of himself. (Carpenter 1964: 160)

Primate Anlagen and the Culture Concept

Carpenter's program was not unacceptable to many cultural anthropologists in the United States. In particular, it was consistent with the rationale for primate studies developed by Clark Wissler, of Yale's Institute of Psychology and the American Museum of Natural History. Carpenter's claim represented the position that emerged *after* it was unacceptable to study "primitive" people as simple windows into "civilized" behavior. The version of the nature-culture distinction in social science ascendant in the 1930s did not deny the organic nature of human beings. But this version of the nature-culture relation asserted the cultural, language-based control of organic raw materials, especially sex, by means of unique human mechanisms, like kinship, within socially determined systems, called culture.

Wissler was a grantee of the Committee for Research in Problems of Sex from 1928–33 as part of the committee's effort to establish a program in human psychobiology of sex. Sophie Aberle and Beatrice Blackwood actually conducted the research, among American Indians and Solomon Islanders, which was intended to probe "simpler" reproductive relations that might be useful in treating the sexual pathologies found among "more complicated, civilized" people.[24] In 1933 Wissler reflected on this abandoned framework:

> The reasons why a program among primitive peoples was considered promising may be formulated as follows: 1) The assumption has been made that civilization is artificial as opposed to the natural. In keeping with this assumption, it is often said that the artificial settings of civilized society interfere with normal sex life. Accepting this as a working hypothesis, the procedure would be to gather information on sex life as observable among the uncivilized. 2) Experience eloquently preaches the necessity of comparative methods when confronted with problems of behavior and morphology. Hence it was assumed that the contrasts in pattern between uncivilized and civilized would assist in clarifying the patterns of modern sex behavior.

But all humans had "artificial" culture; only primates could show what once was sought among people. The culture concept did not challenge sexualization at the center of explanation of society, but it did displace the position at which one expected to see the operation of drives unobscured by culture and self-consciousness:

> Now it appears that the surviving so-called primitive peoples, also, present artificial settings comparable to those observable in civilization; they exercise social control in forms similar to those exercised by civilized peoples. Hence no important research lead is apparent. Turning next to the comparative approach we note that the intensive study of chimpanzee behavior has been encouraged by the Committee. The preliminary results in hand suggest that the patterns of response for these non-human creatures will lead to a formulation of the human pattern.

Two directions for sex research relevant to people were reasonable: comparative work restricted to primates, not primitives, and direct sexual studies of civilized humans, like those later conducted by Kinsey: "One set of behavior patterns seems to prevail throughout mankind, and so can be better studied among ourselves. . . . In the main, the sex behavior pattern and the integration of the sexes in group life can be clarified by comparing the anthropoids and man. Hence, the intensive study of the great apes is to be favored as the immediately important step."[25]

If Carpenter's point of view is misjudged as old-fashioned biological reductionism, the history of primatology since 1930 becomes inexplicable. Wissler was only one spokesperson, but Carpenter's framework could appear comfortably with those of prominent social scientists of the ascendant Boasian culture school: Alfred L. Kroeber, M. F. Ashley Montagu, and Robert Park. Controversies certainly remained, but primates had a very important role to play in post-Spencerian American social science.[26]

Sociometry

Carpenter imported the sociological techniques of sociometry into his biological study. [Figure 5.1] The direction of borrowing was immaterial; precisely at the period marking the end of creditable biological reductionism in Amerian human science, both biological and social disciplines shared a logic that elaborated function-alist field theories. These field theories were the material directly transformed by cybernetic functionalism during and after World War II, in the profound refigura-tions of biological and social sciences through the physical and technological sci-ences. The important difference distinguishing the gibbon study from the howler monograph was the degree of detailed use of sociometric and semiotic analysis to explain an integrated control system. These theoretical tools, borrowed from human sciences, were keys to primatology's capacity to bridge the natural and social sciences in the mid-twentieth century.

Sociometric analysis was common in social psychology in the 1930s. Its basic tool was the construction of the sociogram. In the words of the man who considered himself the founder of the sociometric movement, J. L. Moreno, "The proper placement of every individual and of all interrelations of individuals can be shown on a sociogram. It is at present the only available scheme which makes the dynamic structure of relationships within a group plain and which permits its concrete structural analysis." (Moreno 1945: 71)[27] The purpose of sociometry was to facilitate

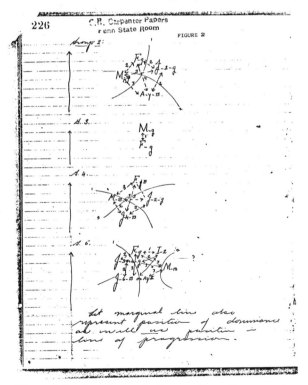

Figure 5.1   A sociometric diagram from Carpenter's Asiatic Primate Expedition field notes. C. R. Carpenter Papers, Penn State University Archives.

the "self-realization" of group goals. "Maximum spontaneous participation" was seen as the only reliable means to achieve human social control. Social control and social structure were intimately linked concepts; the technical expression of this was the integration of the participant-observer into the group studied. In the internalization of control, the technique was maximally invasive. The fully integrated participant-observer, the only person who *knew* the group structure, could help the group to its end. Sociometry was essentially a "microscopic analysis" from inside the group. Thus it was only compatible with seeing sociology as analysis of small-group structures. For Moreno, sociometry revealed *motives*, "the psychological geography of a community" (Moreno 1945: 72). The focus was not the individual, but the atomic social relation. Constructing boundaries of the group as a whole resulted from microscopic knowledge of parts studied in functional-structural relation. "The nucleus of relations is the smallest *social* structure in a community, the social atom" (1945: 72). A geometry of social relations allowed determination of the "tele" of a group—the goal around which it is *actually* organized (no matter how people might *think* it is organized), called the "group criterion." Once the criterion was known, the investigator could predict future group states and develop strategies effective in achieving goals—or in thwarting them. That is, sociometry included essentially "therapeutic and political procedures, aiming to aid individuals or groups to better adjustment" (1945: 72). Carpenter's adoption of sociometric technique was consistent with his approach to building primatology as a science of social therapeutics.

Group criteria differed in complexity; that was the fundamental distinction between animal and human organization. The sociometric expression for the difference was depth in a solid geometric volume. Animal sociometry could be adequately accomplished on surfaces, because relationships were unobscured by depth-producing factors like self-consciousness and language. Therefore the problem of interference by the observer was simpler. A sociogram for an animal society would be a two-dimensional map of physical relations in space, with arrows to indicate vectors of movement. Time was an additional factor pictured by a series of sociograms. Correct valuation of vector forces of attraction or repulsion should allow prediction of physical distances at a given future time. The psychological and physical maps were fully congruent. The geometry of social structures was totally different from chance spatial arrangement; the sociogram showed the actual psychosocial network tending to the achievement of group function. For babies and primates (not for "primitives"), sociometry was simply charting movement in space through time.

At the same symposium in 1945 at which Moreno explained the relation of human and animal sociometry, Ashley Montagu waxed enthusiastic over the technique for anthropology. Montagu, a spokesperson for the human sciences that insisted on a unique human biology, whose product was language and culture, saw sociometry as the realization of Bronislaw Malinowski's "method of cultural analysis." Sociometry was the technique for structural-functionalism. It would play the key role in the "study of man as a functioning unit in the social continuum." The small groups studied by the ethnographer were ideal:

> Such a sociometric mapping of a group would not only yield an invaluable account of the social psychology of the group, it would also provide a more accurate picture of the culture as a whole than is obtainable by the usual means of ethnographic investigation. . . . But whereas functionalism is to anthropology as physiology is to

anatomy, sociometry is to functionalism as histology and biochemistry is [sic] to physiology. (Montagu 1945: 62)

Carpenter, far from reducing human society to primate levels, adapted a well-regarded social science technique to the appropriate level of complexity. He had used sociometry without naming it in the howler studies. In the gibbon work, the approach was all-pervasive. In 1945, he outlined explicitly what he had done in mapping the pattern of a group. First, he made scatter diagrams of the whole group of animals in space; then he made vector diagrams of relations between individuals and reconstructed a map showing the summed positive and negative valences of interaction. Basic data were observations of spacing and duration of interactions. Each relationship took final shape by "organismic summation." The basic concept was a *social relation*, "the reciprocal interaction of the behavior of two or more individuals which stimulate and respond to each other. The behavior, in turn should be considered as expressions of individual motivation or psychological processes." (Carpenter 1945: 59) The maps revealed group social control. Control was always based on some degree of dominance, usually but not always by adult males. *Status* was a measurable control quantity directly derived from dominance. Primates, unlike people, did not have elaborate extragroup controls. The semi-closed primate groups were managed by the socionomic sex ratio, social bonding patterns, and psychological conditioning to territory. Intragroup differentiation was limited by the degree of elaboration of dominance axes (status hierarchies). Intergroup relations were restricted by competitive mechanisms like territory, defense, and the absence of language.

In his field notes, Carpenter drew several sociograms in an effort to predict future group states. His constant questions were: How do groups form? How are they maintained? The corollary of these questions was: How might groups be scientifically managed? Behavior like fighting and competitive aggression, which establishes dominance gradients, was a mechanism of group integration. Appropriate amounts of such behavior were called leadership, control, and initiative. Inappropriate amounts were insufficient to hold a group together or were excessive and pathological expressions, which also disrupted groups. Carpenter believed these sorts of sociometric analyses for nonhuman primate groups were appropriate pilot studies for human sociometry, where control was much more complex. The connections were carefully made between animal dominance hierarchies as regulators of social integration and pathological human personalities, like the much-studied authoritarian character structure with its dire social consequences.

Semiotics

Semiotics was the second borrowing from the human sciences important to the foundational field studies of primatology. Semiotics theorized communication as a problem in control systems. In its roots in the work of Charles Saunders Peirce, William James, James Dewey, and George Herbert Mead, this branch of semiotics was intimately intertwined with American pragmatism and behaviorism; a basic problem was to understand how systems of signs affected behavior patterns. Charles Morris, a philosopher at the University of Chicago, defined semiotics as the science of signs, studying things and properties of things in their functioning as signs.

Morris believed that semiotics was the needed organon or instrument of all the sciences. It would be the tool of the unification of sciences in the twentieth century:

> The significance of semiotics as a science lies in the fact that it supplies the foundation for any special science of signs, such as linguistics, logic, mathematics. . . . The concept of sign may prove to be of importance in the unification of the social, psychological, and humanistic sciences in so far as these are distinguished from the physical and biological sciences. And since it will be shown that *signs are simply the objects studied by the biological and physical sciences related in certain complex functional processes*, any such unification of the formal sciences on the one hand, and the social, psychological, and humanistic sciences on the other, would provide the relevant material for the unification of these two sets of sciences with the physical and biological sciences. (Morris 1938: 2, my emphasis)

Foundations was the second number in the International Encyclopedia of Unified Science, whose editors were Otto Neurath, Rudolf Carnap, and Morris. Carpenter cited and used Morris's formulation as soon as it was published. Carpenter was nothing if not a positivist; semiotics met his need for a unifying theory in a way that the emerging synthetic theory of evolution could not. Carpenter's constant goal was to produce a science of control. The modern evolutionary synthesis produced a form of knowledge of nature further removed from instrumentalism. Carpenter's construction of the primate field remained very close to the laboratory.

Positivism and functionalism connect richly in the history of life and human sciences; these connections are critical to understanding Carpenter's version of the primate story. The founders of the school of urban sociology at the University of Chicago included Charles Horton Cooley, with his sociology of sympathy, and George Herbert Mead and Harry Stack Sullivan, with their psychological role theories. These approaches stressed interpersonal relations and processes of communication in the production of a functional social whole. Morris followed logically in this tradition in building a positivist, organicist theory of communication. When Carpenter systematically adopted these tools to dissect the primate social body, he crafted a tissue of elemental social relations. The pathologies to which these tissues were subject—e.g., excess aggression, sexual malfunction, or disorganization of dominance relations—were understood as diseases of communication. Behavior, a dynamic physiological structure, was fundamentally the process of *signaling*. The organism produced signs, which permitted the interactions constituting or threatening the social whole. Therapeutics and interpretation of signs were never far apart. Life and human sciences have not strayed far from the clinic.

The content of semiotics was semiosis, or the process in which something functions as a sign. It was necessarily a relational study; i.e., ruled by a functionalist approach. Morris named the three parts of semiotics as semantics, syntactics, and pragmatics; he pointed to the three corresponding branches of comparative biology—anatomy, ecology, and physiology. "The properties of being a sign, a designatum, an interpreter, or an interpretant are relational properties which things take on by participating in the functional process of semiosis. Semiotics, then, is not concerned with the study of a particular kind of object, but with ordinary objects in so far (and only in so far) as they participate in semiosis" (Morris 1938: 4). Language was like any other organismic object studied functionally by positivists.

Ordinary human languages, called universal languages because they could potentially represent anything, were the richest systems of signs. But, certain purposes required special and restricted languages. While the most complex focus of semiotics might be human linguistics, all organisms could be considered from the point of view of semiotics, i.e., from the point of view of an organism's response to sign vehicles. No hint of mentalistic language need or should enter such a science.

Since most, if not all, signs have as their interpreters living organisms, it is a sufficiently accurate characterization of pragmatics to say that it deals with the biotic aspects of semiosis, that is, with all the psychological, biological, and sociological phenomena which occur in the functioning of signs . . . . The interpreter of a sign is an organism; the interpretant is the habit of the organism to respond, because of the sign vehicle, to absent objects which are relevant to a present problematic situation as if they were present . . . . The response to things through the intermediary of signs is thus biologically a continuum of the same process in which the distance senses have taken precedence over the contact senses in the control of conduct in higher animal forms . . . . Considered from the point of view of pragmatics, a linguistic structure is a system of behavior. (Morris 1938: 30–32)

In the context of the semiotic project, in practice the primate investigator studied gestures and vocalizations, as well as the development of symbolic social control systems out of the primitive contact-control appropriate to the early infant. Sound recordings, films, taxonomic and functional classifications of calls and movements, charting of the ontogeny of communication: these were the appropriate artifacts Carpenter produced in his gibbon study. Theoretically these artifacts were the data for constructing a picture of primate society as a dynamic, integrated communication-control system. Within the project of the unification of human, life, and physical sciences, a semiotic understanding of primate society helped link the many levels of objects studied by science together without reducing one level to another. Human language was only one kind of control system. The limitation of gibbon integration and coordination was a consequence of the limitations in the power of their systems of signs (Carpenter 1964: 242).

Neither semiotics nor sociometry contradicted a physiological approach to organic wholes. All three were action-oriented; all focused on behavior, on motion, on function. Their respective origins in linguistics, social psychology, or biology did not make them mutually incompatible, but rather different windows onto the same problems of coordinated wholes understood in terms of control and regulation. The organismic field theories from biology; the mapping techniques for showing force equilibria and disequilibria in social systems; and the unification of physical, physiological, psychological, ecological, and social levels of organization in terms of sign systems were all a proper part of Carpenter's primatology. His product was a detailed analytical system for moving from physiology (e.g., endocrines, reflexes), to psychology (behavior, learning, drives), to sociology (integrated, coordinated group and breeding population). The ideological lessons—slow change permits adjustment, frustration produces aggression, dominance is a mechanism of cooperation—were conveyed without resort to conflating nature and culture or their proper sciences. The traffic between nature and culture was mediated by communication.

Celebration of Consensus

The fiftieth anniversary celebration of the founding of the University of Chicago marked these relations of the life and human sciences that enabled Carpenter's practice in the 1940s. The divisions of biological and social sciences came together to produce *Levels of Integration in Biological and Social Systems*; Carpenter's paper appeared physically midway between the contributions of Allee and Emerson, animal sociologists who based their work primarily on social insects and birds, and those of Kroeber and Park, an anthropologist and a sociologist identified with the triumph of Boasian social science in America (Redfield 1942). The September 1942 symposium had two origins. First, the social science division desired to have a conference exploring the newly opened, promising research areas on the "borderland" (Redfield 1942: 1) of the study of human society: (1) rapprochement of anthropology and sociology; (2) recent investigations of social behavior of monkeys and apes relevant to the origin of human society; and (3) work in mammalian and bird society that had caught the interest of sociologists and anthropologists. The borderlands could be explored now that the *boundaries* of life and human science seemed secure. Second, the biologists had planned papers exploring the problem of parts organized into wholes, from multicellularity to society. Clearly, the two planned sessions should be incorporated. The result was the celebratory volume signifying the relations of these sciences on the eve of America's entrance into World War II. The problems of integration were transformed by that war, in theory and in practice.

In August 1940 Carpenter accepted a position in the Department of Education and Psychology at Pennsylvania State College. His relations with Smith at Columbia were strained, partly over the disposition of gibbons from Cayo Santiago, but largely due to different priorities over use of the rhesus.[28] Carpenter's behavioral emphases were not central for Smith. Never holding a stable position at Columbia, Carpenter was not a powerful figure in the central research areas of reproductive and neural physiology. His behavioral studies were informed by reproductive physiology, but the interest was not entirely mutual. Most laboratory physiologists tended to look at primates in terms of supply of individual animals for experiments. Because most funding was directed to laboratories, this bias plagued students of social behavior who did not have an independent research base. Carpenter's new location in a department of education greatly affected his role in primate studies. While not absent from the postwar rebirth of field studies, he was not a major figure. His efforts in education echoed broad postwar themes in a global order in which new communications technologies played leading symbolic and material roles.[29] Partly due to the war, primatology was at a low ebb in the 1940s. Its international rebirth in the 1950s must be accounted for in terms of processes outside itself.

A Sociobiology of Communications Systems

A bio-technological picture of organisms and animal societies as command-control-communication systems began to take shape in evolutionary biology in the decade after World War II. This picture is significant for our understanding of the relations of the life and human sciences and of the level at which biological theory is simultaneously and necessarily political theory; i.e., a statement about power.

Coding and copying, communication and replication are the key concepts. The emergence of this picture is part of the pre-history of sociobiology and part of the deep transformation of central areas of biology since World War II, from a discourse on physiological organisms, ordered by the hierarchical division of labor and the principle of homeostasis, to a discourse on cybernetic technological systems, ordered by communications engineering principles and a tightly associated principle of natural selection. The kinds of natural-technical objects embedded in communications theories emerged complexly from practices of war-related operations research, positivist theories of language and information, telephone industry research, and wartime labor management, as well as in the extensive promotion of these points of view through foundation policies, conferences, new technical opportunities, and social networks after the war. A significant thread of the primate story ties into this complex fabric through the doctoral field research of Stuart Altmann on Cayo Santiago in the 1950s.

Born in St. Louis in 1930, Stuart A. Altmann earned his Ph.D. at Harvard in 1960; he was one of E. O. Wilson's first graduate students and also consulted with Ernst Mayr. Altmann attained his B.A. (1953) and M.A. (1954) at the University of California at Los Angeles, where George Bartholomew was an important influence.[30] Altmann did his M.A. thesis on the mobbing behavior of birds, another link between ornithology and primatology. From 1960–65, Altmann moved from assistant to associate professor of zoology at Emory University, Atlanta; and from 1965 to 1970, he held the title of "sociobiologist" at the Yerkes Regional Primate Research Center. From 1970, he has been professor of biology and anatomy at the University of Chicago. Altmann has always worked closely with his mathematician-wife Jeanne Altmann (Altmann and Altmann 1970).

Like his dissertation superviser, Altmann helped turn the widely felt promises of cybernetics and communications theories generally into technical achievements in evolutionary animal behavior science in the 1950s. Wilson has become famous and infamous for his versions of sociobiological theory, while Stuart Altmann has moved more in the direction of socioecology and away from the communications theory-oriented approaches that characterized his first field study. But, Altmann's early debts to 1940s and 1950s communications technologies illustrate an important thread in the weave of evolutionary biology as a discourse on technology and the organism as a natural communication system.[31]

Operations Research, Ergonomics, And Cybernetics

We must begin in the midst of World War II, while Wilson and Altmann are still in high school. The extraordinary organization of scientists in the war effort in Britain and the United States threw biologists together with engineers, linguists, physicists, mathematicians, and administrators in intense activity that had profound consequences for the conceptual structure of biology after the war.[32] The geneticist, developmental biologist, and systems theorist, Conrad Hal Waddington, illustrates the experience of large numbers of life scientists whose professional lives were redirected by their scientific war work. He articulated the meanings for biology of operations research, from roots in his work against U-boats in the Royal Air Force Operations Research Sections. Waddington learned to produce models for decision-making to optimize probabilities of meeting goals for any kind of problem. Goals in systems control were not formalized in terms of micro-control of individual components, but in terms of probabilities for controlling error rates at key points in a system. Identifying boundaries and constriction points for determining rates of information flow became crucial operations. Boundaries were constituted by differential flow rates of information and energy. Control of boundaries constituted system control.[33]

Wartime science provided biologists with a second systems theoretic tool for reconceptualizing organisms and societies: *ergonomics*, the discourse about optimizing the energy-information relations of all components in the organization of labor.[34] Ergonomics is known in the United States as human factors research. Human error rates were the crucial bottleneck in sophisticated technical systems. Integration of human operators as factors in a total system allowed solution of the design problem of optimizing defense performance. Ergonomics includes all aspects of the organism considered as part of a machine carrying out tasks. A cybernetics of the hierarchical division of labor, ergonomics began as the study of human beings in terms of the technical laws of work.

Ergonomics is specifically not an aspect of psychological-sociological human relations research; it is, rather, rigorously directed to studying labor in terms of technical systems design, especially attending to the operational breakdown of any factor under stress. Ergonomics seeks answers to questions like: What information does an operator need? What are the most efficient channels for getting information to the receiver-operator? What communication loads are tolerable for each component? Stress, a psychiatric and medical concept crucial to post-war ideology and practice, is intimately linked to these communications theoretic questions about system potential and design limits.[35] Associated with the notions of breakdown and obsolescence, stress is also fundamentally part of the conceptual apparatus of cybernetic evolutionary biology, like ethology and sociobiology. Stress limits and machine communication conceptually imply each other. Communication in ergonomics refers to flows of information considered in terms of altering error rates at crucial points in the system. Communication design *is* system design.

Converging in a view of a cybernetic evolutionary theory of animal behavior, operations research and ergonomics were joined by the related linguistic theories, called by neo-positivist Charles Morris "semiotics" and by anthropologist Thomas Sebeok, in reference to the post-war biological context, "zoosemiotics." As earlier, therapeutics and interpretation of signs were not far apart in the post-war cybernetic form of semiotics. This approach to psychiatry, in its cybernetic communication theory garb, drew heavily from semiotics; zoosemiotics from the beginning bore a close relation to the therapeutics of communication disorders and overstressed communication systems.

A technological relocation of the principles of semiotics has been important in the transition from physiological to cybernetic logics in many biologies, including the biology of social behavior. In the transition, the organism as living responder to the sign vehicle lost its privileged position. The more powerful analysis of sign systems, cybernetics, dispensed with the need for a biological organism, in the same way that ergonomics considered a human worker as a technical system component whose status as a living organism was interpreted in strict communication engineering terms. Organisms appear in both ergonomics and machine theories of communication. What has gone definitively is the *privileged* status attaching to life or consciousness. Organisms become biotic components, highly interesting, but not ontologically special, in cybernetic systems sciences.[36]

Within these frameworks, the dichotomy between animal and human weakens. The boundaries and connections of living and non-living similarly shift fundamentally. In post-World War II biology within this frame, the social is theorized strictly in terms of the exchange of information. Receivers and senders of signs need to be known in terms of channels, capacities, frequencies, error rates, and so forth. Social theory can thereby be constituted without reference to hidden, mentalistic, or willful parameters. Cybernetic theorists claim a uniquely satisfying union of social, organismic, and mechanistic because of their perception of the power of their communications theories and machines. Cybernetics in the 1950s was a parent to cognitive science in the 1970s and 1980s (Edwards 1988).

If pragmatics was the aspect of semiotics emphasized by Morris in pre-cybernetic versions of semiotics, syntactics became the key to zoosemiotics as a branch of communication machine theory. Syntactics was understood in the sense meant by Claude Shannon and Warren Weaver; i.e., as an analysis of messages in terms of the frequency of occurrence of a signal.[37] Semantics—broadly, meanings—was not the key in cybernetic zoosemiotics to understanding animal society as a communication system. To treat society as the result of stochastic communication processes, where units could be quantified as information and effectiveness measured in terms of altered probabilities of uniquely characterized motions in space (an exhaustive catalogue of behaviors), promised major advances in evolutionary behavioral biology. Thomas Sebeok understood the essence of these ideas in discussing zoosemiotics in terms of "coding of information in cybernetic control processes and the consequences that are imposed by this categorization where living animals function as input-output linking devices in a biological version of the traditional information-theory circuit with a transcoder added" (1967: 363). Natural selection would come progressively to be characterized as a principle for redesigning communication systems at all levels of biological organization, assessed, naturally, in terms of reproductive efficiency.

The Macy Conferences

The hoped-for unity of the sciences through the new communications theories was actively promoted by an important social mechanism immediately after World War II: the series of ten Macy Foundation conferences on cybernetics, beginning in 1946 with a meeting entitled "Teleological Mechanisms in Society" and ending in 1953 with the last of the conferences, entitled "Cybernetics: Circular Causal and Feedback Mechanisms in Biological and Social Thought."[38] These meetings had profound consequences for cybernetic evolutionary theories of animal behavior developed in the 1950s. The Macy Foundation's sponsorship grew out of its commitment from its origin in 1930 to a holistic approach to health through the integration of biological, medical, and social sciences. Macy's prior interests in psychosomatic illnesses were quickly translated with the beginning of war into work on traumatic shock and war neuroses, which was fundamental to post-war development of stress biology. The Foundation's special technique was the conference series, in which about fifteen scientists were invited to form an original nucleus to meet for two days at least once a year for three to five years. Between 1940 and 1950, the Foundation sponsored 132 conferences with over 800 participating scientists. The interdisciplinary emphases of Macy, coupled with its long-term interest in mental health, human relations, and the physiology of integrative, homeostatic processes, led the Department of State to ask the Foundation in 1946 to conduct sessions on human relations in the Department, drawing on its resources in clinical psychology and cultural anthropology. In 1948, Macy facilitated the London International Congress on Mental Health, which preceded the founding of the World Federation of Mental Health, charged with applying mental health principles to human relations within and among nations and serving as a consultative organization to UNESCO and the World Health Organization. Thus, McCulloch's and Bateson's efforts to organize a conference to explore the integrative possibilities of the new conceptions emerging from mathematics, engineering, physiology, behavioral science, and linguistics were well received by Macy.

Communication was a favored theme of the Foundation both as a technical and mythical concept; communication was to be realized and idealized. Its connection to the rhetoric of humanism has remained central to cybernetic biologies, including sociobiology. The conferences "centered around functional analogies between the central nervous system and a class of machines in which information storage, communication, control mechanisms, negative feedback, and the ability to carry out mathematical and logical operations were of central importance" (Heims 1975: 369). The social mechanism of the Macy conferences made available for social-biological theory the ideas and techniques of systems and communications theories, the most potent fruits of wartime work on integrating biological components into comprehensive technical control systems. Neither the war nor the conference was a unique cause of subsequent scientific developments, but they were part of a cluster of social mechanisms that informed important figures across the life and human sciences about the promise of the new communications approaches. These people were linked in many ways: for example, location in Boston-Cambridge at M.I.T. and Harvard; ties between Harvard and the University of Chicago; connections between home institutions and federal science bodies with concerns spanning from defense to mental health; and formal and informal research interests across usual disciplinary boundaries. These scientists published widely, taught at major institutions, and were publicly identified with the promise of cybernetics. Wilson and Altmann inherited their approaches to animal behavior through these kinds of multiple connections. How did the general promise of cybernetic social-biological theory take shape in Altmann's early field work?[39]

Semiotics and the Unexpectedness of Communication

Stuart Altmann was probably the first person to apply the 1948 Shannon-Weaver equation for analyzing sequences of messages to animal behavior, producing the explanation of animal society as a problem in the syntactic aspect of zoosemiotics. Altmann has been a central figure in the social organization of post-World War II primate studies. The work of the group associated with him in research at Amboseli, Kenya, on East African baboon social groups and populations is one of the longest continuous field studies and has yielded data and theory of major importance

(Altmann and Altmann 1970). Altmann was an early facilitator for the communication of Japanese primatologists with their western counterparts. His work in the 1950s on Cayo Santiago rhesus monkeys was at the beginning of post-war western field studies of primates. Stuart Altmann's primate biology in the 1950s and 1960s illuminates the meanings of zoosemiotics in the reconstitution of natural-technical objects in biology as command-control-communication systems ordered by a teleology of reproduction. Altmann used a rich brew of semiotic, information, and structural linguistic theories to approach animal communication. These approaches have complex social, intellectual, and technical histories not reducible to the intense social crucibles of war and industry, but the deep retheorizations of organisms and societies as technological communications systems have been charged with meanings in those furnaces.[40]

In the Army Medical Service from 1954–1956 working on parasites, Altmann was aided in setting up his first research on wild and free-ranging primates by the civilian director of the Walter Reed Army Institute of Research, Division of Neuropsychiatry, David McKenzie Rioch. Rioch had long acquaintance with primate biology through reproductive and neural physiology, the major traditional connections between medicine and primatology. In physiology at the Johns Hopkins Medical School from 1931–1937, Rioch had associated with the outstanding figures in primate physiology, e.g., G. B. Wislocki. Both Wislocki and Rioch had supported organized primate research expeditions. Rioch facilitated Altmann's 1955 howler monkey population census work on Barro Colorado Island in the Panama Canal Zone and put the younger person in touch with C.R. Carpenter. Giving Altmann the data from his pre-war census work on howlers, Carpenter worked with Rioch to put Altmann on Cayo Santiago in 1956 under more favorable conditions than the colony had experienced since just before World War II. Both were in touch with Robert Morrison, director of the Division of Medical Sciences of the Rockefeller Foundation, about permanent funding for the island rhesus colony. Both Carpenter and Rioch were also part of the committees organizing under the auspices of the National Research Council in the 1950s to ensure the laboratory supply of primates and generally support burgeoning primate research. In short, Altmann began his research career in the midst of an important social network that put him in touch with the major funding and theoretical resources of a field that expanded dramatically after 1955.[41]

Although Altmann had not worked in Rioch's unit at Walter Reed, their theoretical dispositions were similar in the 1950s and 1960s. Rioch, an invited guest at the Eighth Macy Conference, constantly cited philosopher-physicist Donald MacKay on cybernetics, language, and communication theories. Rioch's special interest was in the promise of cybernetics, or control theory in general, for psychiatry, especially the psychiatry of aggression. Rioch expected experimental field studies of primates, a kind of comparative social psychology, to be valuable "in view of increasing interest in clinical social psychiatry and preventive psychiatry."[42] Rioch and his associates emphasized the ideas of information overload and stress in relation to behavioral psychology. Deviant behavior was a kind of communication system breakdown calling for informational therapies. Rioch considered the study of stress, data processing in the organism, and problem-solving capacities to constitute "sociobiology." This framework unified work from neurophysiology of the brain with comparative psychology and sociology. Social behavior was constituted by and controlled by the exchange of messages; social therapeutics was constituted by application of effective information. Since communication was here measured in terms of altered probabilities of behavior of the receiving organism, cybernetics provided simultaneously a theory of communication and a promising technology of control through statistical procedures (Rioch and Weinstein 1964).

Altmann expected to spend about two years on Cayo Santiago 1) determining population size, sex ratios, age distribution, disease problems, and individual identification; 2) conducting an observational study of behavior and social structure to obtain baseline data; and 3) performing intensive observational and experimental studies on facets of behavior, population ecology, and social relations. The first priority after studying population structure was an investigation of communication. We meet here with the two critical objects of the biology of cybernetic systems: population and communication—the principle of expansion-replication and of social exchange-control.

Altmann's organizing questions were an appropriate translation of Carpenter's semiotic-sociometric analysis to a framework of cybernetic functionalism (Altmann 1962).[43] These were the questions underlying the budding science of sociobiology.

> What are the roles of the various sensory modalities in communication? What is the function of the communicative signals in the integration of the society? For every signal: what are the necessary, sufficient and contributory stimuli; what members of the society respond; what is their response? What is the relation between communicative feedback and homeostasis? Are there any social communicative networks that are "self-damping"? Does metacommunication exist? Are there (a) signals whose only function would be "acknowledgement" of a signal emitted by another, (b) signals "asking" for a signal to be repeated, or (c) signals "indicating" a failure to receive a signal?

The section on "Reproduction," which contained the largest number of queries (thirty), began with "What are the behavioral signs of sexual receptivity?" The logic of the section on "Social Dominance" was that dominance structure was part of the communication system. Questions under "Territoriality" were about boundaries and their maintenance. Under "Leadership," Altmann asked about the frequency with which the behavior of one individual determines behavior of others and about changes in social structure tied to changes in leadership or patterns of aggression. Under "Gregariousness," Altmann asked two sociometric questions, now fitted into the probabilistic communication framework: "To what extent are the spatial relations between pairs of individuals an index of the amount of interaction taking place? Can changes in spatial relations be used to predict changes in social structure?"[44]

The degree to which Altmann constructed his discourse out of the material of social and technological sciences like anthropology and social psychology, structural linguistics, and communications engineering is striking. Claude Shannon's development of information theory at the Bell Telephone Laboratories, aimed at putting the maximum number of signals on a transmission line, yielded a way to study the stochastics of social signalling in monkeys (Altmann 1965a). George Miller at Harvard taught Altmann the fundamentals of mathematical communication theory, and Miller owed much of his insight to work growing from the wartime psychoacoustics lab.[45]

Sex, dominance and aggression, and cybernetic communication emerged as the chief themes in the publication of his thesis results in 1962. Like Wilson in his study of social insects, Altmann strove to provide an unambiguous taxonomy of signals or signs to which a quantitative cybernetic analysis could be applied. Definitions, taxonomies for organizing data, and the structure of arguments were mutually determining in this constitution of a natural-technical object. " 'Society' will be defined as an aggregate of socially intercommunicating, conspecific individuals that is bounded by frontiers of far less frequent communications" (Altmann 1962: 373–4).

The two major sections of Altmann's dissertation paper were sexual behavior and social status. Since Solly Zuckerman's influential hypothesis in 1932 that mammalian society was constituted by sexual attraction, and primate society uniquely character- ized by constant female receptivity, research attention was necessarily directed to evaluating that idea. So Altmann established copulation and birth periodicity for the Cayo Santiago rhesus and concluded that, at the least, primate society needed more than sexual interaction for its explanation. He also developed a mathematical model for sexual selection of mates, predicting "the probability that the male of any dominance rank in a group of any size will have access to a sexually receptive female" (1962: 430). The model, dividing males into haves and have nots, assigned selective meaning to male dominance rank. In addition to pointing out that ascribing evolu- tionary meaning to dominance rank *needed* testing, Altmann cast the question in statistical terms consistent with his framework.

The second section, social status, was an extended discussion of several related topics: agonistic and aggressive behavior patterns, stability and functions of domi- nance hierarchies, various agonistic strategies open to rhesus monkeys, optimization strategies for animals at various positions in a hierarchy, dominance behavior in the formation of male-male coalitions, and the relation of such male behavior to the evolution of social cooperation. The last topic was the use of status indicators as communication about communication, i.e., metacommunication. The search for the structuring of "metacommunication" has been a privileged quest in the life and human sciences in the late twentieth century. Metanarrative sets the limits for the field of possible stories; metacommunication sets the limits for the field of possible semiosis. In the darwinian frame of reproductive advantage, the general organizing idea in Altmann's discussion of status was a familiar evolutionary, cost-benefit calculation: "Rhesus monkeys tend to act in such a way as to maximize their gains while minimizing their losses" (1962: 399).

The threads in Altmann's early work combining to reconstitute primate society in the frame of cybernetic communication theory converged into a finished fabric in the conference on primate social communication, which Altmann organized in 1964 in Montreal (Altmann 1967). Typically for primatology, many scientific disciplines were represented at the conference. Rioch was the discussant for two symposium sections—causal mechanisms and agonistic behavior, in which he ex- plained the psychiatric meanings of stress and communication for sociobiology. The anthropologist-linguist Thomas Sebeok linked zoosemiotics, the synthetic interpre- tative framework for the conference, with the communications engineering and information science tools. Study of primate systems was an investigation of commu- nication with the help of a "typology of the sign systems used by the animals in the different sensory modalities at their disposal." Providing the research program for the future, the relative informational and energetic properties of chemical, optical, tactile, and acoustic systems remained to be worked. Sebeok called this work a study of languages as the basic tool for construction of society; zoosemiotics was a study of the technical design features of society, a system of communication operating through information exchange that controlled the behavior of the sign vehicle-elements. Animals were reconstructed as a new kind of natural-technical object— i.e., biotic components in technological communications systems.

Altmann's own paper, "The Structure of Social Communication," made the striking claim that all the design features developed by the structural linguist Charles Hockett and others to describe human language apply to anthropoid social communication systems.[46] Two points are important in Altmann's claim: 1) his recognition of the direction of borrowing in primatology from human sciences to the biological discourse of primatology, and 2) the definition of the social in both biological and human sciences as an information exchange and control system. Sociobiology is not necessarily a discourse that reduces human social organization and behavior to biological levels of explanation. Engineering, labor sociology, linguistics, philosophy, operations research: these were contributing discourses for early versions of sociobiology. Sociobiology consistently theorizes social events (the basic structural units of a communication system) in the same cybernetic engineering terms which pervade the human sciences since World War II. Exchange, selection, and control are the principal processes which constitute an information system. Altmann stressed that communication, by definition, occurs when the probability distribution of behaviors, categorized in taxonomies like that of his 1962 paper, is altered. Social communication is the process of controlling the behavior of social actors. This theory of social communication lends itself to a predictive technology for social control. With a common tool kit of cybernetic machines and theories at all levels of analysis, the consultant evaluating social design is the communications engineer specialized to assess the system under consideration, whether biological, economic, or familial.

Hall's Island: Defect Experiments Telemetrically Reconsidered

C.R. Carpenter's 1938 field notes from Cayo Santiago had contained another idea for an experiment: "Problem: Given animals with bi-lateral and uni-lateral frontal lobe-ectomies. To learn, what adaptations are made to a free-ranging environment and competitive social conditions." But he did not get the chance to participate in defect experiments using brain lesions until many years later, on a different island colony—in 1971 on brain-implanted, telemetry-controlled gibbons at the Bermuda Primate Center, on Hall's Island. These experiments highlight the transition from an organics to a technics of control of the social body. This Atlantic island and these experiments are a tragic, ugly image for Carpenter's last field work. The man whose considerable skill and commitment had established the practices enabling scientists to watch gibbons living freely in the 1930s ended his field work surveying a brain-damaged colony in a high-tech narrative of remote control.

This time Carpenter's field work was made possible by the psychiatrist and specialist in technological control of human aggression, José Delgado, who like Yerkes, was based in physiology at the Yale University Medical School. Hall's Island was a 1 1/2-acre piece of land near the Bermuda Biological Station, a marine and oceanographic research laboratory established in 1910. Rental of the island was arranged for the gibbon research proposed by Delgado of Yale; A. H. Esser, the director of the Social Biology Laboratory of the New York Rockland State Hospital; and N. S. Kline, director of research at Rockland. Delgado, Esser, and Kline were associated with the Psychiatric Research Foundation in New York, an organization founded to promote controversial investigations in psycho-pharmacology.[47] Carpenter, principal investigator on a separate but related proposal, observed the five juvenile and one adult gibbons on the island from mid–June to September 1971 and made a series of recommendations for the improvement of Delgado's behavioral study. Esser's grant proposal clarified the context of the research:

> Life in a group provides protection for the individual. However, the group member has to pay a price for this protection; he has to submit to the social order. In today's society there are increasing numbers of people unwilling or unable to pay this price; for such people the guidance and restraint traditionally supplied by family and classroom is no longer effective—we call these people alienated. Social planning cannot hope to reduce alienation if it unwittingly flaunts the deep-rooted biological laws which underlie social cohesion. Utopias are notoriously short-lived; in our troubled cities today the traditional forces of governmental action give little relief, yet radical proposals are justifiably suspect. We lack the necessary knowledge of man as a social animal to evaluate our social plans in advance.[48]

Carpenter's career-long belief that human and animal aggression were homologous at the level of brain and endocrine mechanisms justified his part in this research plan. In the 1960s he and Delgado had both participated in the plethora of conferences to consider war, aggression, stress, and territoriality.[49] Gibbons had a social structure that seemed especially interesting for human beings theorized within a western universalizing discourse: they defended territories and lived in monogamous family groups. The Delgado-Esser application proposed to study naturalistic populations, believed to be more relevant to modern human "stress" and "alienation" than the prison and mental hospital patients and laboratory monkeys to which they had access ordinarily. Recognizing constraints on human experimentation, they sought as an experimental system a free-ranging primate with important social analogies to "man." The proposed research was a straightforward extension of work Delgado had done for over twenty years. He had been instrumental in developing the multichannel radiostimulator, the programed stimulator, the stimoceiver, the transdermal brain stimulator, a mobility recorder, chemitrodes, external dialtrodes, and subcutaneous dialtrodes. These were cyborg organs within cybernetic functionalism. Behavioral observations would be automated as far as possible.

> The main purpose of the project is to investigate the possibility to induce lasting modifications of free-ranging behavior by means of longterm stimulation of the brain. . . . We hope that our project will increase the knowledge of the cerebral bases of anti-social behavior, and thereby contribute to a better understanding of the many factors involved in the study of this important part of our social life. . . . The methodology and technical expertise for the behavioral aspects of the project have evolved over the past nine years in continuous studies of the behavior and biochemical functions of a group of schizophrenic patients in the Behavioral Research ward of the Research Center at Rockland State Hospital.[50]

The situation on Hall's Island upset Carpenter because he felt insufficient care had been exercised in keeping baseline behavioral data on social interactions before operations. He did not object to the basic research goals, but directed his recommendations to improving the research to meet them. Noting that Delgado and Esser wished to develop an "early warning system" for potential suicides and "depressive psychotics," Carpenter reported, "They hope, through the Hall's Island project and parallel experiments in laboratories in the States, to get to the stage where a patient in a hospital can be given a battery of psychological tests—including a period of MTS (Mobile Telemetry System) and results would then be fed into a computer which would spell out the correct medication."[51] This is a vision contested by many who have followed Carpenter into the special historical place called the naturalistic primate field. It is a vision made possible by learning to see that field as the primordium for a semiotics of control; it is a vision that theorizes communication—especially language—as a remote control system.

Maps of Meaning, Maps of Action

The structure of a command-control-communication system pervades the discourse of Delgado and his community, whether or not explicit military metaphors or social ties appear. This constitution of the social is not militarized as a function of funding, direct control by the Defense Department, or evil intentions. The metaphorical and conceptual structure of the world "for us," late industrial people, has the form of a problem of strategic control at deep levels in post-World War II discourse, including the sciences of animal behavior. These are not uncontested structural constitutions of what may count as nature, nor do they exclude other meanings and further translations. But neither does the constitution of the world for us as a problem in communications engineering and strategic reasoning have *no* effects. These constitutions are fundamental to cultural maps for relationship and action. That is what cultural maps of nature have always been about. If the natural *versus* the social sciences is a false issue in the discourses of sociobiology, so is the opposition of humanism and science. If anything, sociobiology is a hyper-humanism, in some of its spokespersons' claims to provide an effective way to achieve goals of "human fulfillment" through accurate knowledge of the requirements of human design and redesign. There are other humanisms, but they all rest on the rhetoric of commitment to the historical fulfillment of "man." Nothing in sociobiology opposes such a commitment. From one point of view, sociobiology is simply a communications humanism seeking the computer capacity to simulate human semiosis.

Stuart Altmann has been generally professionally uninterested in human primates. He and most of his colleagues are zoologists in the important sense that they write and talk about animals, which they observe carefully for years; and they are not ideologues for various kinds of social control. These people use funding sources like the National Institute of Mental Health because the money is available. And in any case, the intention of the scientist is neither knowable nor relevant. The multiple connections of sociobiology to psychiatry are important, but even without such connections, the political meaning of scientific knowledge pertains. Science produces meanings and possibilities.

I have been calling attention to the kinds of objects of knowledge which historically can exist and are made to exist by the mundane practice of science in the world really structured by war, capitalist economic organization, and male-dominant social life. Such structures enable and constrain meanings; they do not directly produce them. There are other determinants of scientific knowledge as well, obviously including such beings as monkeys. But these objects are radically mediated for us. There are also competing explanatory possibilities. In Altmann's work, animal societies were not "nothing but" command-control-communication systems. But, the technological and biological systems discussed in this chapter are formally similar; they are constituted by the same practical procedures. It would be absurd to label all cybernetic-related science and technology "bad science" or ideology. It is also absurd to claim that the only political dimensions of "good science" concern institutionalized application. Meanings are applications; how meanings are constituted is the essence of politics. No one can constitute meanings by wishing them into existence; discourse is a material practice. The meanings of cybernetic communication systems include *particular* structurings of objects of knowledge—not as ideology, but as that which can be known in a particular time and place, called Nature.

Natural-technical objects of knowledge are *contested*; they are the product of social engagment, in and out of the perimeters of science. People like Altmann and Carpenter have been quite aware that their work rested on historically new languages, machines, and patterns of daily interaction in the production of discourse. Science fiction writers also stake their craft on the premise that knowledge is fundamentally political; i.e., dialectically constituted by and constitutive of social possibility. Knowledge about the world is historically contingent. The historical contingencies of capitalist patriarchy have meant, most starkly, that biology can be a war baby and like to play with guns. Knowledge can be a militarized zone. It can also be demilitarized, but not automatically. Semiosis is politics by other means. [Figure 5.2]

Figure 5.2   Stuart Altmann in the field in Amboseli National Park. Courtesy of Stuart Altmann.

PART TWO

DECOLONIZATION AND MULTINATIONAL PRIMATOLOGY

6

RE-INSTITUTING
WESTERN PRIMATOLOGY
AFTER WORLD WAR II

Their political ability is invested in the heart of doing science.
The better politicians and strategists they are, the better the
science they produce. (Latour and Woolgar 1979)

New Actors and New Authors in
Post-World War II Primate Studies

Post-war primatology and its natural-technical objects of knowledge have been highly heterogeneous. The laboratory-based scientist has dreamed of standardizing "the monkey" into a design catalogue of primate material for given research purposes (Clarke 1985). Monkey and ape variation would then represent the basis and product of a kind of materials engineering, working from a pre-existing taxonomic catalogue. For "the monkey" to exist, natural variation had to have an enforceable social and technical status of raw material. Historically, this status has depended on the social relations of extractive colonialism and neo-colonialism. The tight epistemological-political constraints of medicine, aimed at the direct control of hostile disease phenomena, result in the construction of a natural-technical object of knowledge whose variations must be tamed for use. There are not many degrees of freedom in this field; the singular outnumbers plurals.

But the questions asked by the field worker are inherently hostile to the existence of "the monkey." And the conditions of access to animals internationally have become hostile to ongoing constructions of primates as natural-technical objects of knowledge exclusively within western agendas. Even in the earliest behavioral field studies by Carpenter, initiated from within a medical ideology under colonial rule, "the monkey" dispersed into different species social types and appropriate roles for age-sex classes. Further studies broke the species type apart, into an array of variants

115

responding to ecological, historical, demographic, and genetic influences, as well as to the heterogeneous ways primatologists wrote about what they learned to see. The epistemological constraints of field work, aimed at "ultimate" evolutionary explanation, demand ways of knowing that increase observed variation of many kinds. Variation is, of course, still tamed and used, but within fields with many more degrees of freedom. The convergence of materials engineering and evolutionary theory, centrally represented in late twentieth-century molecular genetics and bio-technology, remains practically and epistemologically distant from most field prima-tology. Field primatology can be characterized as constructed around its dispersed, interacting centers, i.e., its field sites, each full of animals and people with multiple strategies and histories. "The field" is more than a term for the physical-social space where people watch uncaged animals; "the field" is what makes "the monkey," a kind of univocally defined instrument, disappear in favor of the story-laden animals embedded in the interacting "narrative fields" which this book explores. The term suggests the metaphor: primatology exists within and is constructed by social-epistemological force fields, each of which alters the structure of the others, while ensuring that primatology is polyvocal and polysemous. "The monkey" of medicine becomes only one impoverished, albeit highly powerful, generative center of mon-keys and apes as natural-technical objects of knowledge.

Multiple dimensions determine the deceptively simple word *primatology*: More than 240 species, primate habitats from snow areas of Japan to tropical rain forests and from deserts of Ethiopia to swamps of South America, disparate national traditions among those who study and write about them, conflicting institutional priorities, con-tending conservation agendas, struggles over the terms of decolonization in primate-habitat countries, incompatible disciplinary and explanatory approaches, struggles over field and lab methods, appropriations by various ideological tendencies, unset-tling hierarchies in the scientific division of labor in each social setting where primates are described, painful displacements of founding generations of field workers by recent Ph.D.s. Throughout the coming chapters another dimension of the heteroge-neity of primatology, the different university- and field site-linked networks and lin-eages of primate scientists, will be prominent. Complicating these lineages and net-works is the thread of gender difference, explored especially in Part III.

There are also the homogeneities that permeate the field of primatology: the strong western bias in setting the research agenda, the numerical predominance of European and Euro-American researchers among the western scientists, the common threat of primate extinction through habitat destruction, linked to war and "devel-opment." But part of the reason behind this book's avoidance of the convention of capitalizing "the west" and "western" is to disrupt the ideological stance that the West is One, even while sometimes indulging in that fiction in order to characterize lines of force in powerful story fields. Within "itself" western primatology has been cacophonous and contested. An even more important reason for the small letter is the international history of the field. In science and in other story-telling practices, Europeans and Euro-Americans are not alone in their fascination with monkeys and apes. Post-war primate studies recommenced in Japan. Further, the growing involvement of women and men from Latin America, Asia, and Africa in research and conservation is the promise of a future for the Primate Order. But let us begin with a problematic founding gesture: a filmmaker's construction of the Laboratory as a mythic origin point for the primate story.

Primate: An Unauthorized Picture

The seven regional Primate Research Centers of the National Institutes of Health have a product: primate bodies and written texts. Frederick Wiseman, the radical documentary filmmaker who has explored several major institutions in the United States—meat packing, fashion modeling, high schools, a prison mental hospital, military basic training centers—turned his camera on the apparatus for the production of the textualized primate body. Wiseman's view of the Yerkes Regional Primate Research Center in Atlanta produced *Primate* (1974), a picture of love and knowledge. The camera recorded the laboratory as a cybernetic organism, a cyborg internally ordered to give birth to fully controlled, rational, reproducing, textualized systems. Wiseman ground a distorting mirror to tell an origin story. The scientists at the Yerkes center deplored Wiseman's film. My purpose is not to defend its accuracy or fairness, but to examine its own symbolic story, its vision of the production of nature as cultural artifact with a phallocratic birth, as an image for introducing the complexities of race, sex, class, and species in post-war primatology. In *Primate*, the jungle and extraterrestrial space are very close together; this is the condition of the Primate Order in the last decades of the Second Millenium.

Through Wiseman's lens, primatology is about first and last things. Hidden in *Primate*'s elusive narrative, the *cinema verité* documentary is about metanarrative in the primate order. Structured as an origin story, *Primate* should be read as science fiction, as well as social commentary on the production of scientific fact. *Primate* is a myth of modern self-birthing, i.e., of the achievement of "man's" humanist goals of self-knowledge in science and technology. The film opens with a hall of portraits of human bearded male fathers of primatology, from Linnaeus to the present. The film closes with another sort of masculinist birth, in which animal, technology, and human being are co-engineered to gestate in space. In between, Wiseman envisions a system of production of scientific knowledge in which animal, machine, and human are integrated in a self-regulated, techno-organic whole. The dominance hierarchies of species, sex, class, and race are all pictured as components of a system whose product is *Primate*. The film begins with animal birth and ends with technological birth, and the two stories are one reproductive whole. Reproduction and communication are both about the reconstitution of the self in the primate field.

Leaving the portrait gallery, the camera takes the viewer up the drive into a complex of scientific buildings, to a hall of cages, where filmmakers (bearded) and scientist (bearded) discuss gorilla sex in front of a male-female pair of that ape species. The sound track gives us a fragment of a discussion on data collection in ape sex research. The scientist explains the observation and caging system: "We don't want them doing things when we can't see it." In the scientist's office we see photographs of copulating gorillas and get a discussion of sexual behavior observed in the field by George Schaller.

But the film will not explain these matters externally; we must learn to see from within the text. *Primate* is a film about the structure of observation, about how to produce knowledge, about daily practice in a scientific laboratory, about objectivity, and about alienation. Each scene dissects the means of production of what will be allowed to be seen. A major theme is the distancing of observations, the structuring of vision. From the opening sequence of photographs and portraits, of observers in front of cages with tape recorders and standardized data sheets, and of pictures of gorillas discussed in front of a visible filmmaker, to the ending scenes of a technological birth watched on radar, there is no *immediate* vision in *Primate*. There can be no illusion of immediate nature in science. Vision is mediated by writing technologies. The body becomes an inscription device.[1]

Next, the camera surveys a birth scene; a newborn orangutan crawls clumsily over its mother's bloody head. She sits placidly sucking the placenta and stroking her infant. Outside the cage a white-coated, white female technician takes structured notes while a white male scientist speaks into a tape recorder, translating the physicality of the birth into frequencies of motion in space. We are then taken into the infant reception room of the laboratory. A black woman feeds a baby ape with a bottle; a white woman rocks an ape baby in a chair and croons "mama's baby." But this is no scene of unleashed maternal nature across the barriers of race and species. All the responses of apes and humans are integrated into a data collection system, where hugs and kisses are ways of relaxing infants for insertion of thermometers and translations into marks on paper. At the end of the scene a white male human enters to check charts and rapidly scan the room; he does not interact with other humans or animals. The following sequence shows young apes returned to their cages after a period in the exercise yard. Several of them cling onto the body of a black man. There is no interaction.

The saga of birth and growth appears to be interrupted in the next sequences; but the fragments will come together later. A rhesus monkey has a blood sample taken; we see an automated blood analysis system; a black woman does hemoglobin counts in a microscope. Two white men discuss a stimulus device to get semen samples from the chimpanzee John. At one frequency, the technician can generate an erection, at another, an ejaculation. Shortly after, we see a male chimpanzee taken from his cage, anaesthetized, and stimulated. Five men hover over to assess the amount of semen, while another cleans the teeth of the insensate animal.

Wiseman offers a multiplicity of visions, taking us into the language experiments with the chimpanzee Lana; into a trailer to observe locomotion in apes to study the evolution of bipedalism; into cerebral localization laboratories; and into chambers to study the physiology of weightlessness. Throughout, we walk down hallways where black women empty garbage and white women receive orders from white men on preparation of samples. The laboratory is a workplace with a social division of labor, for which race, sex, and class are fundamental. We do not know the product of this workplace yet; we see only fragments. But purposefulness is everywhere.

Three sets of sequences structure *Primate*. They appear in bits, but assemble into a system of meanings, a myth about bio-technical systems of production. The three sequences concern experiments on brain localization and sexual and aggressive behavior, on gravity effects, and on artificial insemination. The first two studies use rhesus monkeys, the last, chimpanzees. All use human beings and elaborate technology.

In the experiments on cerebral localization and behavior, *Primate* shows the viewer two white women using a banana reward to insert electrodes into a box installed on the head of a rhesus. The women must be new at their task; they and the monkey emit frequent fear grimaces in repeated unsuccessful attempts to connect the hardware. Finally, the preparations are complete. The male rhesus with his box is released into a test cage, and two females are added, as if they were chemical reagents in a reaction vessel. Scientists in another room full of stimulus and recording devices turn the monkey on and off to elicit and damp out aggressive behavior. The visible attacks and pauses are translated into data on unit behavior frequencies and stimulus strengths. This sequence set is an eloquent picture of a communication system animated by a teleology of control. Human, animal, and machine work smoothly in mutual communication.

Only gradually could a viewer conclude that another sequence set is about gravity variations and their physiological effects—and even more gradually this work emerges as part of the U. S. space program. The Cold War enables this primate text. The viewer sees a room crammed with equipment—viewing screens, automatic recorders, dials. Nothing is explained; all the other inhabitants of the room seem to the viewer to know what is going on—except, we see, one other. A lay white woman in street clothes, perhaps a reporter of some kind, seems to be visiting. Her squeamish, mystified face witnesses the insertion of a rhesus strapped in a standard lab restraining chair into a large black box. The woman seems in the way. She asks dumb questions; and a white male scientist talks about baseline data before the flight, about mimicking lunar or martian gravity. The door of the box is closed; the monkey spins faster and faster inside. Human beings watch the monkey intently— on a TV screen showing the animal's contorted face. Various recording channels have been activated; streams of data pour from several inscription devices. The multiple layers of observation, the conventional scientific metaphor of a black box, the relations of lay woman and male cleric-scientist, the monitored womb for generating data: these are the components for Wiseman's self-assembling primate system.

Returning to the chimpanzee John and the careful efforts to get a semen sample for artificial insemination experiments, Wiseman leads us into another kind of scientific space, the seminar-conference room, for one of the funniest glimpses in recorded history of scientific intercourse. The room, devoid of machinery, plainly furnished with a long table, is full of white men of all ages conversing about schedules for obtaining John's samples for inseminating Flora, Cherry, and Banana. The viewer hears that they have frozen sperm for a backup, hears cost-benefit analyses of sampling on Tuesday or Wednesday, hears all the details of experimental logistics. No omniscient narrator tells why these preparations make sense. The conference ends with the overheard phrase summing up the deliberations: "Let nature take its course." An acne-faced young man exits. He reemerges later, a lab bottle full of grape juice in one hand, a tubular stimulus device in another, to entice John with squirts of juice, while skillfully executing a scientific masturbation. This is a very curious technology of compulsory reproductive heterosexuality.

What is to be born in this film about communication and reproduction technologies? Why this elaborate labor? The laboratory is not sterile; that, anyway, is the lesson preached by its director, Geoffrey Bourne, in the penultimate sequence, the second scientific conference Wiseman allows the scientists for explaining their rites. The inheritor of Robert Yerkes's scientific legacy, Bourne appears a kind of elder or priest sermonizing the workers in the lab in a period of funding crises to motivate them to fight for the material foundation of their efforts.[2] Bourne addresses an all-white male audience on the relations of pure and applied science, the frontiers of research, and the threat of displacement of America's lead in biomedical science by European competitors. He calls on his ancestors, Abraham Flexner and Sir Alexander Fleming, a man of vision who modernized the structure of medical-scientific research and a winner of a Nobel Prize for the serendipitous discovery of penicillin.

The lesson Bourne preaches from this text is the absolute necessity of basic research, the stock on which we draw for applications. The stock can dry up and wither. We must as a culture again learn the usefulness of useless knowledge. It is as nature to culture—the resource for triumph over sickness, over others; the ground of our birth as men.

Wiseman finally leads us to the payoff of *Primate*. A van draws up to the rear of the laboratory to remove its product, a rhesus in a plastic restraining chair. The monkey is driven to a military airfield and placed with a mechanical lift into the belly of a large plane. The animal shares the space with several high-altitude-suited men. But the monkey is placed in a black box and watched by the men on a TV screen for the rest of the film. The channels of the monkey's brain are recorded by other equipment. The plane carrying this cargo is watched on a radar screen at the airfield. We, in turn, see all the screens: the airplane diving and rising; the radar monitors; the humans floating in the belly of the plane, connected by thick cables to life-sustaining equipment in the placental periphery; the monkey staring out at us from its box; the film credits. The documentary is over. Primates are in space, connected to earth by the technology of communication, birthing streams of information, transforming nature into culture according to an everyday logic of domination in a scientific division of labor. It is only a myth, an origin story among others. A payload.

Post-War Primate Science in the United States

After World War II, largely as a result of the extraordinary wartime mobilization of science in the Office of Scientific Research and Development (the OSRD), the organization of natural and social sciences in the United States was permanently altered and greatly expanded. With the establishment of the National Science Foundation, for the first time in the United States regular government funding was available for civilian science. The National Science Foundation began modestly, but shortly was the major funder for individual reseachers and university-based projects in biology and somewhat later in anthropology. "Behavioral science" came under the rubric of natural science, where it was suspect if its practitioners had too much to say in the way of social criticism. "Social sciences" seemed to have an onomatopoetic relationship to socialism, and they got less of the largess and under more strictly policed ideological terms.[3] In the immediate post-war years, the Office of Naval Research funded a wide variety of scientific research tenuously related to military needs. Ironically, scientists' experience of permissive military scientific management honed their sense of entitlement to pursue publicly financed "pure research," i.e., without much social accountability or democratic process for setting scientific and medical priorities.

"Pure research" funded by NSF after the 1950s required some paragraphs of social justification, perhaps in the broad terms of contributing to a cure for cancer for the developmental biologist, or in the terms of contributions to human child and maternal health for the primatologist, who likely believed that understanding some of the 240 species of simians in zoological terms was more important to a humane multi-culturalism and "multi-speciesism" than proposing yet another theory of mother-infant bonding for psychologists' and advice writers' tool kits. But in the unprecedentedly lush times in a U.S. entering into a post-war global hegemony, money was available on the slimmest of arguments, because the material and ideo-logical bases for "disinterested" and "objective" natural science had been laid during and right after the war. Because that money was there, it was not necessary to win the argument for interest in monkeys and apes themselves, not just as surrogates for industrial people. In addition to NSF funding, mission-oriented, federally funded, biomedical civilian science greatly expanded, with the establishment of the National Institutes of Health in their several disease-specific branches, such as cardiac disease, neurological disease and blindness, and mental health.

All of primatology, but especially the field studies of wild monkeys and apes, rode the coattails of these much larger financial and institutional developments. If early post-war primate field observers George Schaller, Alison Jolly, and Irven DeVore had NSF grants, it was not because the ruling elites seriously considered gorilla, lemur, and baboon behavior of crucial national importance or because of a ground swell of democratic demand to know how gorilla families stay together. Young graduate and post-doctoral students were funded because the scientific system had more niches and links in the food chain than it did in Yerkes's and Carpenter's pre-war worlds. To be sure, the field primatologists' niches and money-foraging behaviors were justified in their mentors' grant proposals by references to antici-pated insight into such issues as conservation, social management, etc. But none of those arguments for the importance of naturalistic and most behavioral laboratory primate studies would have mattered had there not been a new and hefty federal commitment to funding civilian science.

The question of monkey supply for biomedical research, however, did involve matters of broad specialist and popular concern. The scientific prevention of infec-tious disease has been a world-changing social force for a century, and polio in the 1950s was not an obscure disease (Latour 1988). Hope had been aroused that a vaccine was near, and the scientific and political stakes were high. India briefly cut off monkey exports in 1955, in response to popular outrage about cruel treatment of the sacred Hindu monkeys in trapping, shipment, and experiments. The trade was restored on the basis of formal U.S. assurances to use the monkeys exclusively in research for the "benefit of all humanity," provide humane treatment, and—especially—exclude the monkeys from all military and space research. The 1955 embargo precipitated U.S. National Research Council action to arrange a secure domestic breeding supply of research primates, one event leading to establishing the Primate Research Centers in the early 1960s.

By the mid-1950s, the demand for a sure supply of research material was vastly larger than in 1938, when Carpenter released rhesus monkeys on Cayo Santiago to forestall the effects of another Indian embargo. On the supply rested the campaign against polio, but also many other biomedical and psychological projects and the cosmetics testing required by the Food and Drug Administration. Monkey supply became an issue affecting the highest levels of the U.S. scientific establishment, operating through the National Academy of Sciences and the National Institutes of Health. Field studies and most behavioral research with monkeys and apes in the U.S. were enabled by the budgets and primate colonies of the primate research centers. But the rhetoric surrounding establishment of the centers made the under-lying priorities unmistakable. At the 1955 NRC conferences called to deal with the supply crisis, the Chairperson of the Division of Biology and Agriculture of the NRC-National Academy of Sciences, Paul Weiss, summed up the matter: "It occurred to me right from the start that conservation, that is, merely the conservative aspect is not really the only answer, but what we have to have is a balance sheet between production and consumption. . . . Irrespective of how large the global supply may be, what counts here is merely what we can get our hands on."[4]

There are seven NIH-sponsored regional Primate Research Centers, the first opening in Oregon in 1960 and the last in 1965. The Yerkes Laboratories of Primate Biology moved to Atlanta, affiliated with Emory University, and became one of the regional centers. Others located in Davis, California; near Boston (affiliated with Harvard); New Orleans; Madison, Wisconsin; and Seattle. Their mission is to use nonhuman primates to solve human health problems. By 1970, about 8,000 primates (12,000 by 1980) representing 45 species, studied by about 400 scientific investigators and 700 supporting staff, at a cost of approximately $15 million per year, were busily engaged in this mission. They helped produce about 500 papers and 20 books annually at that time. Until 1962, the regional centers were under the administration of the National Heart Institute, and since then under the Division of Research Resources. The initial cardiac oversight was partly due to hypertension in the Cold War. Producing a kind of "hypertension model" gap, the Soviet Union seemed to be making more rapid research progress in developing animal models for high blood pressure than the United States (Bowden 1966). In 1956, Dr. James Watt, director of the NIH National Heart Institute, visited the Soviet primate research facility on the Black Sea, the Sukhumi Center established in 1927. Watt returned to recommend to the NIH Council that a primate colony be established in the U.S. to study cardiovascular diseases. The final result was a much larger undertaking, with the first Congressional funds appropriated in 1960.[5]

But for at least the first twenty years of their existence, the NIH primate centers were not self-sufficient in primate production. In the early 1970s, the U.S. as a whole still imported 90% of its lab primates. To address the continuing high demand, in the 1970s the Charles River Breeding Laboratory of Wilmington, Massachusetts, undertook a major rhesus breeding project on an island in the Florida Keys. The description of hygienic precautions in their trapping and shipment from India—in order to provide less-diseased animals for research—reads like science fiction (Bermant and Lindberg 1975: 165–66). Charles River annual sales of laboratory animals (90% rodent species) in 1979 was $30 million, leading to favorable Wall Street attitudes to shares in the company.[6] The commodity aspect of primate research frames the narrative of the history of primate story-telling. By 1982, the centers were about 75% self-sufficient in primate breeding, and officials expected to achieve full self-sufficiency in the following decade.

The Science Information Exchange reported that use of primates in research swelled from 666 projects of all types in 1965 to 1183 projects in 1971. Total primate references listed by the Primate Information Center of the Washington Primate Research Center grew from 5000 in 1960 to 35,000 in 1971.[7] Medical uses continued to predominate; but Stuart Altmann's (1967: xi) graph documenting the explosion of primate field studies in the early 1960s showed that between 1962 and 1965, the number of "man-months" devoted to field studies of ecology and behavior of nonhuman primates more than exceeded the total of all previous research. Field workers point out that more has been learned about wild primates since 1975 than in the entire history of the field before then (Smuts et al 1987).

The U.S. tradition of private philanthropy also mattered to field primatology.

The Ford, Wilkie, and Louis Leakey Foundations, National Geographic Society, New York Zoological Society, World Wildlife Fund, the Grant Foundation of New York, and others have all made important grants. But the Wenner-Gren Foundation for Anthropological Research was the most important in the first 15 years of postwar primate studies. Their conferences paint a miniature portrait of the ideological and theoretical debates in primate studies and paleoanthropology from the late 1950s through the early 1980s. The foundation's dollar figure for research has been dwarfed by federal sources, but its encouragement of approaches and international networks has been formative.

International Conferences: Rebirth of a Field

Primate field studies seemed to ignite simultaneously and spontaneously from several foci. There were three independently planned international conferences in the spring of 1962: one sponsored by the London Zoological Society; a conference on "The Relatives of Man: Modern Studies of the Relation of Evolution of Nonhuman Primates to Human Evolution," organized by John Buettner-Janusch and sponsored by the New York Academy of Sciences; and a meeting in Giessen, Germany.[8] A list of participants in these events provides a nearly complete account of the research centers and personnel, both senior and junior, who reinitiated primate studies in the west.[9] After scientists realized what the three independent 1962 conferences implied, the Wenner-Gren Foundation sponsored an international meeting in Frankfurt that led in 1964 to the International Primatological Society. Although initiating primate field studies first, in 1948, the Japanese did not appear in the international meetings until after the three April 1962 affairs. There was a Japanese participant, Mizuhara Hiroki, at the Stanford Primate Year in 1962–63, organized by Sherwood Washburn's network, and funded by a NIH grant to David Hamburg and Washburn. Those participants complete the snapshot of the international primate networks that developed rapidly from the mid-1950s.[10] All four early primate meetings were strongly structured by anthropological concerns, especially versions of the belief that early "man" became "human" through hunting.[11] A fifth early conference, through which the strands of 1950s and early 1960s primatology may be traced, was organized in 1964 in Montreal by Stuart Altmann (1967), where the emphasis was on interpreting primate society with the help of new communications theories.

Strong zoological interests represented at the early conferences by people like Peter Marler, Stuart Altmann, Alison Jolly, and George Schaller were never entirely at ease with the hunting preoccupations of the anthropologists. For example, commenting a bit later on the evolution of the notorious human love of starchy foods and the reasonable suggestion that hominid manipulative skill evolved from our insect-grubbing history, Jolly wondered if heroic hunting stories might not be "a superstructure of speculation [that] has been based on the weekend hunters' delight in bagging a buck and the delight of nearly everyone but the Brahmin in attacking a juicy steak" (1972: 67). She pointed out prominently what everyone knew but did not always want to remember in those heady (and well-funded) days of the 1960s, when all the world seemed once to have been an australopithicine hunting reserve: "We can hardly expect to reconstruct the behavior of protohominids from the behavior of primates alone when we cannot extrapolate from a North Indian langur to a South Indian langur" (1972: 4). The history of field studies of monkeys and apes has richly confirmed Jolly's insistence.[12] Thelma Rowell established that insight for baboons, an important "model" species for protohominids in the practice of anthropologists.[13]

Third World institutions have been crucial to primate field studies. For example, Makerere University, founded in Kampala the year after Ugandan independence of 1962, appeared often in primatologists' accounts. Niels Bolwig pursued behavioral observations in South Africa while on the staff of Makerere, and A.J. Haddow was at the Virus Research Institute at Entebbe (the Ugandan capitol from 1894–1962), making field observations of African monkeys, who were vectors in human diseases like Yellow Fever.[14] The bridge from medical tropical research to field studies was facilitated at Makerere. Visiting from Bristol University in anthropology, Vernon and Frances Reynolds were aided by the Makerere staff in their 1962 chimpanzee study in Uganda's Budango Forest.[15] With her husband, Richard, Alison Jolly spent 1963–64 at Makerere University. Robert Hinde's former doctoral students, Thelma Rowell and her neuro-ethologist husband, C.H. Fraser Rowell, were at Makerere from 1962–67.[16] The chairperson of the Zoology Department at Makerere during Thelma Rowell's period there was Professor David Wasawo. Both Hugh and Thelma Rowell were actively committed to scientific education for black Ugandan citizens. Both Rowells helped Alison Jolly. Those were the years of Thelma Rowell's important studies of Ishasha baboons. Stephen Gartlan passed through Makerere in 1963–64, where he and Rowell discussed their shared doubts about dominance hierarchy interpretations of baboon and macaque society, resulting in Gartlan's early critique of the dominance hypothesis. Gartlan met Rowell through her friendship with his mentor, K.R.L. Hall, who was in Uganda to study patas monkeys.[17] Also in 1963, Jeanne and Stuart Altmann, beginning their work on baboons, and Thomas Struhsaker, studying vervets, were "nearby" at Amboseli in Kenya. Goodall was at the Gombe in Tanzania. Jolly characterized an exciting Nairobi conference of that period that brought most of these people together "in Leakey's East Africa . . . [confirming] primatology as a field of study in its own right."[18] Jolly recounted a stimulating network of influences in the 1960s from these people, who intersected in the ex-colonial universities and game parks.

This introduction has been a crude snapshot of intersecting people and historical threads that make up one swatch of the fabric of the international re-emergence of primate studies from the late 1950s. Perhaps the snapshot can serve to remind us of the dense, concrete particulars against which grand cinematic overviews of the primate story must be viewed.

A Perverse Periodization: A Cinematic Sweep from U.S. lenses

There are many ways to periodize a science; each implies a politics of the history and philosophy of science linked to social commitments and beliefs about the world. It would be possible to tell the history of post-World War II primatology by mentioning only developments conventionally believed to be "inside" the sciences, in this case scientific areas of medicine, zoology, anthropology, psychology, demography, etc. Such a story would emphasize the growth of knowledge in field and laboratory primatology, the convergence of anthropological and zoological perspectives on nonhuman primates, and the improvement of theory as quantitative, problem-oriented studies accumulated. Decade by decade, primatology would seem less and less permeable to noise from the "outside" society, so that the embarrassing presence of ideological disputes about male dominance in the 1960s, for example, would be seen to pale beside the mature production of more objective studies by the 1980s. In this tale, the hegemonic position in the 1980s of sociobiological and socioecological explanations of the lives of monkeys and apes over the 1960s and 1970s anthropological approaches called structural functionalism would appear to be the result of better training of the scientists, who learned to ask better questions addressed by better methods. Past bias, such as a rather monomaniacal attention to one sex or the other, would be seen to be happily corrected by the double move of admitting to the practice of science the excluded category (race, sex, etc.) and learning how positive science, dominated by problem-oriented, quantitative studies, inherently reduces the element of interpretative bias.

The above abbreviated narrative, common among practicing scientists, is not groundless; it is merely interested—committed to the ideology of progress modeled by the natural sciences, and so committed to the rational superiority of positivism over hermeneutics and other critical representational practices. This narrative about progress is a method of tidying up politics by making some things exist inside and others outside a kind of "nature reserve" called science. The ideology of progress makes the sciences seem like wilderness preservation areas of the mind, free from the ravages of human culture and history, even if the commodification of the border areas is a bit unrestrained, for example in the rapid translation of pure research in molecular biology into products for sale. But unlike the vanishing wilderness areas of the real planet earth, the natural sciences seem to be spreading around the mapped globe and extending their rational domain. The old unlamented natural histories, progressive in their time, as well as the old-fashioned ethnographies of simians, rooted in the field diaries of neophyte observers, are replaced by the unreadable, mature texts of the hard-minded, who know how to police politics and subjectivity from the documented lives of simians.

In the spirit of subversion of this particular story of primate progress, I will briefly chart the flow of cultural and ideological concerns evident in the work of both laboratory and field scientists from various primate-related disciplines since the 1950s, primarily in the United States, but also in other western primatologies. That is, I will periodize primatology in order to confuse the inside/outside boundary. My argument is not that "outside" influences have continued to determine primatology into the period of problem-oriented, quantitative studies, but that the boundary itself gives a misleading map of the field, leading to political commitments and beliefs about the sciences that I wish to contest. My periodization takes pleasure in confusing boundaries.

My tale breaks up the story field into two general periods in post-war primate studies: before and after the early 1970s. The brief characterizations should be read as caricatures emphasizing particular features to make a point. The sketch provides an interested map for reading the following chapters. Instead of the line "inside/outside of science" related to a periodization privileging scientific progress, my privileged time boundary suggests different moments in the history of post-war social history, in dialectical play with explanatory possibilities in primate studies. I do not wish to tell a causal tale, but to multiply associations and clusters, so that the chronology of discoveries and explanatory transitions in primatology becomes provocatively problematic. The main point is not that the following associations are indisputable, but on the contrary, to illustrate how a chronology is necessarily a political history.

My first period dates roughly from the mid-1950s, with the resurgence of primate field studies, growing medical demand for primates as laboratory material, and the developing interest in primates as models for psychobiological questions about humans. This first phase closes about the mid-1970s, with the restriction of the growth of funding for field primatology, but also the beginnings of results from long-term sites making a serious difference in what can count as a reasonable primate story. The first period ends also with the budding prominence of sociobiological and behavioral ecological explanatory strategies.

In the face of deep post-war anxieties about the fate of modern civilization, the image of man the hunter dominates the first period, coupled to the equally compelling image of the mother-infant pair. Social stability was the burning question, along with a growing list of practical concerns about "achieving" female monkeys, adequate infant care, and stress in urban environments for men adapted to low-density hunter-gatherer conditions. The stability and evolution of "the family" informed both laboratory and field imaginations. Depression became a popular research topic, although it proved difficult to get a reliable primate model, even in the imaginative conditions of confinement that had worked with people. Gibbons on Hall's Island were implanted with telemetric devices in a high-technology effort to model monogamy and pathological aggression in the interests of public social hygiene and patient management in state psychiatric institutions. Female monkeys turned out to have orgasms, but for the most part it was middle-class gay men's organizations who were pleased with the news, while women's liberation groups were leery of the message of sexual liberation. Communication vied with social stability as the most important theoretical concept, and communication was redefined in the hard science terms enabled by the machines and information-processing theories of military and communications industry research.

Debates raged about whether the peaceful chimpanzees and gorillas or the tight-assed baboons were the better model for "man's" evolutionary past and so future hopes. Books were published on the connections of primate studies to man's putative propensities toward xenophobia and territoriality (Holloway 1974). E.O. Wilson named "the xenophobic principle" in *Sociobiology* (1975: 286). Man's natural aggressiveness versus his inherent peace-loving capacities were argued on college campuses while the Vietnam War killed the Vietnamese and the darker skinned and white working-class contemporaries of the young people in the audiences (Carthy and Ebbing 1964). Questions about the origin of man seemed critical to the prophylaxis of stressed twentieth-century "First World" populations. The key hominid fossils of this era of Louis and Mary Leakey in East Africa were the famous australopithecines from both East and South Africa, generally interpreted in terms of man's hunting past. The *Harvard Lampoon* renamed Leakey's famous Olduvai Gorge in East Africa the "Oh Boy, Oh Boy Gorge" for the excitement that the seeming avalanche of fossils and artifacts brought to a distant culture consumed with the problem of origins.

The popular books of the period included Robert Ardrey's dramas, fit for the silver screen for which he also wrote, *African Genesis* (1961), *The Territorial Imperative: A Personal Inquiry into the Animal Origins of Property and Nations* (1966), *The Social*

Contract: A Personal Inquiry into the Evolutionary Sources of Order and Disorder (1970), and his swan song, *The Hunting Hypothesis* (1976); Desmond Morris's lively tales, *The Naked Ape* (1967) and the *The Human Zoo* (1970), sandwiching his edited volume of technical papers, *Primate Ethology* (1967); and Lionel Tiger and Robin Fox's inflamed *The Imperial Animal* (1971). Konrad Lorenz's *On Aggression* (1966) was widely read and appreciated. Less read, but remarkable for its lack of doubts, Steven Goldberg's *The Inevitability of Patriarchy* (1973) appeared just in time to slay dragons released by the women's liberation movement. Every library in the U.S. continues to teach new generations of young people the lessons of male agency through "Man the Hunter" in their Life Nature Library book, *The Primates* (Eimerl and DeVore 1965). The hunting hypothesis included the corollary of woman as the mediator of culture, in male exchange of women as the origin of human marriage transcending natural sex. Time-Life Books highlighted both messages.

Other popular books opposed the messages of aggression, territoriality, biological determinism, male hunting as the motor of humanization, and male dominance as the structure of human cooperation. Jane Goodall's *In the Shadow of Man* (1971) conveyed a message of natural peace, and it was authored by the only scientist-woman of the popular list. Elaine Morgan's *The Descent of Woman* (1972) offered an auto-didact housewife's smart reply to Desmond Morris, and Evelyn Reed argued from Engels in her *Woman's Evolution* (1975). Neither Reed nor Morgan was a match for their credentialed opponents. The women who would challenge the messages of the 1960s were getting ready to start graduate school or just beginning to write. Sally Linton, later Sally Slocum (1975), a charter member of the American Primatological Society, presented "Woman the Gatherer" at the 69th American Anthropology Association meetings in 1970; but it was a decade before a standard academic publisher got the word out, with Nancy Tanner's *On Becoming Human* (1981) and Frances Dahlberg's collection, *Woman the Gatherer* (1981). Ashley Montagu repeated (1968) his message on the primacy of culture and the possibility of peace, but he too was no match for the stunning prose of Lorenz, Morris, and Tiger.

Few other voices were in print for large audiences, although Thelma Rowell (1972) and Alison Jolly (1972) sounded discordant, careful notes in primate behavior and evolution textbooks. Many practicing primatologists probably shared more perspectives with Jolly and Rowell than with Ardrey or Tiger about biological determinism and human nature. But the dominant popular literature of the 1960s and early 1970s was ideologically conservative to reactionary, emphasizing the difficulty for the stressed hunter in the city in creating social stability and happy families in the face of modern perverse refusals to follow nature's laws or in the face of eruptions of his violent natural propensities in a crowded world.

In the second primate period, from roughly 1975, the ideological and technical agendas both changed; both agendas were publicly announced by E.O. Wilson's *Sociobiology* (1975). Perhaps the chief difference from the previous literature was de-emphasis on the structure of social stability, replaced by intimate and loving attention to strategic possibilities and the cost-benefit analysis of everything. In Wilson's adaptationist version, no development was without a function and most likely a gene; but not all players conducted the game that way. Many of the same fictive actors were on the natural and cultural stages, including Wilson's version of man the hunter. But his logic was different from that of a decade before, and his investment strategies were hotly contested by others in the market. The key fossil was no longer the hunter australopithecine confidently striding out into history, but the diminutive bipedal Lucy facing a reproductive crisis as her body failed her in difficult ecological times, requiring that she tie herself to a husband at all costs. Our ancestors were re-imaged to appear not as hunters, but as small-minded gatherers and scavengers.

The mother-infant pair took an ontological backseat to maternal investment strategies and the problems of dual career mothering. Woman the gatherer contested for Act One in the evolutionary play with her former hunter husband. The peaceful chimpanzees were found murdering, conducting warfare, and killing and eating their babies, while the previously macho male baboons were given their due as good providers of paternal care and documented to need the friendship of lots of females to be able to get into a new troop. To be a good social strategist counted for more than other fictive forms of social power, like dominance hierarchies. Everything turned on a strategy: energy budgets, foraging patterns, genetic investment possibilities, sexual deceit payoffs, social manuevers. Sociobiology emphasized the genetic aspects of social strategies; socioecology emphasized the intricate current market conditions of making a primate living given a particular body size, sex, species, habitat, age, and so on. Primatologists began to theorize and document complex mental and emotional capacities of simians whose lives evidenced extensive, likely conscious, strategic behavior. Theorizing nonhuman and human primates as social strategists is another form of traffic between the western categories of nature and culture, expressed in the philosophical categories of liberal individualism in late industrial society. Nature, like culture, is replete with strategic moves and beings capable of making them, consciously or not. Genes and minds are the key "strategic" players in late capitalist bio-politics. No one sex, community, or species could model "the human"; to be animal, including to be human, was reconceived as a shifting strategic game. The human sciences made the same moves as the natural sciences in their fascination with all things strategic. No wonder a major 1980s weapons system, rooted in the deadly dialectic of fact and fiction in technoscience, was christened the Strategic Defense Initiative. The name expressed a broadly shared metaphysics in late capitalist culture.

The popular primate literature continued, as always, to be a hybrid genre in which credibility drew from the scientific credentials of the author and in which lots of perfectly good science may or may not be found. Sarah Blaffer Hrdy's *The Woman that Never Evolved* (1981) and Frans de Waal's *Chimpanzee Politics: Power and Sex among Apes* (1982) endowed all sexes and species of primates with the crucial abilities to conduct their business in the chancy natural economy, which no longer gave much respect to natural rights, but rewarded a clever strategist handsomely. The stakes were survival and reproduction in a world where everyone was ultimately for oneself. Primates became model yuppies.

Crossing the line into the technical literature category, books appeared inscribing primate bodies with meanings about paternal infant care, female hierarchies, the wide range of what it means to be a female mammal, and the complex sex role possibilities in primate life.[19] Curiously, in this post–1975 period, in which the progressing science ought to be muting explicitly political tracts coming from *inside* the discipline, these recent professional books seem more conscious than ever that basic cultural and political meanings are at stake in the sober science of animal life. If anything, it is harder to categorize a book as either popular or technical than it was in the 1960s. But the sense of more actors (people and animals) contesting for the meanings, as well as a consciousness of scarcity and the chanciness of social and natural ties of all kinds, pervades the literature.

Rehabilitants and Surrogates: Modeling Social Problems after World War II

> Then hunger closed the gap between them, and the son of an English lord and an English lady nursed at the breast of Kala, the great ape. (Burroughs 1976 [1912]: 30)

Using monkeys and apes as stand-ins for people was not new, but the array of western-defined human problems in the last half of the twentieth century which called for pliable simian modeling clay is impressive. Population regulation, mother-infant bonding in white families with middle-class mothers working outside the home, formation of heterosexual affectional systems, depression, pathological aggression, child abuse, infanticide, and male-male corporate social cooperation were only some of the foci.[20] As nonhuman primates became widespread surrogates for late twentieth-century western people or for western definitions of other people's problems, the simians themselves were simultaneously rehabilitated back into nature. Some monkeys and apes entered the laboratory or were constructed in the field to model "culture," while others were rescued from pet owners or illegal smugglers and restored to "nature" under the well-advertised care of scientific white women. Both moves were crucial to mapping the dilemmas of late twentieth century social reality in the west.

In 1977, the Mount Asserik Chimpanzee Rehabilitation Center in Senegal's Niokolo Koba National Park was the only program for rehabilitating confiscated orphan chimps to live independently.[21] Stella Brewer and Raffaella Savirelli taught wilderness survival to the young apes. Meredith Rucks started a similar program in Ghana based on the Senegal center; and in 1977 Janis Carter did the same in The Gambia, beginning with the Temerlin's chimp daughter Lucy and her companions, who came both from home-rearing and confiscation from the illegal pet trade. Chimpanzees had disappeared from The Gambia by the early 1900s, and Carter reported that Gambian officials were eager for the rehabilitation work to succeed.[22]

The women who ran these camps often have given years of their lives to the projects in a practice very different from the usual university-based field studies by academics. But their practice also shared the other researchers' many-layered relationships to gender, decolonization, class, race, and other large historical constructions. Compounded by physical hardship, economic uncertainty, severe loneliness, and cultural dislocation, the emotional complexity of the relationships among women and chimpanzees in these stories of exile and rehabilitation is stunning. The forms of love and knowledge in these narratives are rich, often wrenchingly painful, and far from innocent extensions of the possibilities of animal-human contact imagined and enacted within western culture in the Third World. The narratives tell of profound loss and also of major achievements—by people and by animals. The people and the animals in these stories are *actors* enmeshed in history, not simply objects of knowledge, observers, or victims. The 1986 photograph of Janis Carter in an embrace with Lucy, then an adult rehabilitant in The Gambia after nine years, captures part of the tone of these narratives of relationship. [Figure 6.1]

Figure 6.1   Janis Carter and Lucy embrace on Baboon Island in the River Gambia in 1986, a year before Lucy's death. Copyright Janis Carter. Published with permission.

At The River Gambia National Park, Carter's camp was enclosed by wire; the chimps had the rest of the island. "From the beginning, the chimps would have to accept that I alone lived in a cage" (Carter 1982: 100). But being outside the cage was no picnic. Hardship and intra- and inter-specific separation were at the heart of Carter's understanding of Lucy's historical situation.

Lucy made the emotional transition to "nature" via an "adopted son"—bonding with a chimp infant at the rehabilitation station (1982: 103). In the story, Carter and Lucy communicated partly with AMESLAN about re-entering a constructed wilderness. But Lucy's sign vocabulary was larger than Carter's, and the human woman reported that she and Lucy communicated more fully in other ways (personal communication, 7 Jan. 1989). However, for neither animal nor human was this to be a narrative of a wild child returning to the forest after a period in civilization. In this narrative, both Carter and Lucy *began* as adults in western civilization, marked by personal and global histories full of substitutions, bonds, and breaks, which endowed each being with her strengths and vulnerabilities.[23] For Carter, Lucy did not have immediate access to both original animal and acquired human nature. In a tragic story of exile, rehabilitation, accomplishment, and death, the chimpanzee had free access to *neither*, and yet was constructed by both. Carter portrayed Lucy as a strong adult being, with emotional strengths from a childhood in a human home; but Carter did not see Lucy—or herself—as a romantic innocent. The emotional drama between Lucy and Carter was a complex narrative of two displaced beings tied inexorably together in an unequal, power-charged process of separation. Neither had access to an innocent biology or nature, either as animal, female, or woman. Neither had ever lived in "the wild," but its culturally constructed meanings were at the heart of what they had to do in The Gambia. Symbolically and materially, they invented a border space neither original, wild, nor free, but also not a space of victimization and reification. Fragile and requiring survival skills from both humans and animals that ruled out sentimentality, it was a space of complex semiotic traffic and allegory. Lucy's and Carter's cross-species contact may be read as an allegory of reinventing nature in a world where the cost and the work of the construction can no longer be invisible.

After a long struggle, Lucy made the transition from a Euro-American middle-class home to a very different life with other chimpanzees on Baboon Island in The Gambia. Unhappily, Lucy appears to have been killed by poachers looking for other prey (Carter 1988). Confident of her abilities with human beings, the chimpanzee may have aggressively approached the hominid intruders in order to defend her domain. Carter speculates that the people may have shot Lucy in self-defense, and then routinely skinned her. Yet, marked by the risks of living in the profoundly historical borderlands between nature and culture, Lucy's fate was happier than most former ape language subjects, many of whom were abandoned to make the transition from "partners in communication to surplus lab equipment."[24]

Reversals fill the western reports from the potent zone at the boundary between nature and culture. Brewer described her teaching young chimps rescued from human homes how wild chimpanzees communicate by gesture and calls; the chimps also learned from older rehabilitants. The babies learned from their human rehab teachers what to fear and how to react appropriately, what to eat (the humans make exaggerated food grunts and place the appropriate wild foods in the young chimps' mouths), how to open hard fruits, and how to build sleeping nests. The aim was material and emotional independence for the rehabilitant youngsters, so they could "return" to the wild they never knew themselves. Brewer stressed their roles as foster mothers, as surrogates for the natural chimpanzee mothers described by Jane Goodall. So, Goodall's mother chimps became models for the human teachers in West Africa. The dramas highlighted the wrenching process of surrogate mother and maturing child separating, as the apes accepted their expulsion from emotional dependence on humanity and entered society with each other. Integration with existing wild groups seemed unlikely, so the hope was that the rehabilitants would form a self-sufficient new procreative group. Some of the rehabilitants came to Brewer from Simon and Peggy Templer of the Chimpanzee Rescue Center in Breda, Spain. The Templars restored the health of chimps confiscated from resort beach photographers and animal dealers and then sent them to rehab stations in the Ivory Coast and Senegambia. Otherwise, the beach chimps would be killed when they became too old to drag around bars and beaches.

Nonhuman primates' status both as surrogates and as rehabilitants rested on their semiotic residence on the borderland between western constructions of nature and culture. Many heading for laboratory colonies, some for forests, simians were literally in a busy two-way traffic between these two domains because they lived in an epistemological buffer zone; and like the European Balkan peoples, those in a buffer area between contending forces rarely control their own political status. Similarly, the border status of primates in an age consumed with questions of communication across barriers of culture and history structures the fascinating study of language capabilities of apes, especially gorillas and chimpanzees. Historically, the language studies were tied to attempts to cross the border between ape and human by raising young apes with human children in the homes of white, middle-class scientists.[25]

Underlying the language and home-reared ape studies is the simple, enduring western question: what would it be like not to be barred from nature? Is touch possible? Since traditionally language had been imagined to be the source of the barrier, perhaps if a language could be shared, contact with apes, almost as extraterrestrials, could be made, and "man" would not be alone. So while researchers in homes and laboratories worked to teach apes a human language, either computer-based or the gestural American Sign Language, researchers engaged in field and caged-animal studies labored to learn how to recognize and use the simians' own vocal and gestural modes of communication. The rehabilitant stories make clear how humans, having learned ape language, teach it to culturally deprived apes rescued from ignorant humans. Roger Fouts, a teacher of chimp Washoe (who has in turn taught chimp infants to sign), advised the makers of the movie *Greystoke: The Legend of Tarzan, Lord of the Apes* (1984) on ape gestural language. Again, as in the rehabilitant/surrogate pair, the traffic was two-way. Teaching apes human language and learning from them how they communicate with their own kind were both efforts to open the border inherited from the separation of nature and culture. Surrogates, rehabilitants, language students, and adopted children: apes modeled a solution to a deep cultural anxiety sharpened by the real possibility in the late twentieth century of western people's destruction of the earth.

7

APES IN EDEN, APES IN SPACE: MOTHERING AS A SCIENTIST FOR NATIONAL GEOGRAPHIC

I turn to Koko: "Are you an animal or a person?"
Koko's instant response: "Fine animal gorilla."
(Patterson 1978: 465)

Communication

The Nature of Gender for the Seven Sisters

In 1984, to mark nine years of underwriting the National Geographic Society's television specials, the Gulf Oil Corporation ran an advertisement entitled "Understanding Is Everything." Winning several prestigious Emmy Awards, the National Geographic specials have been among the most-watched programs in public television history. Immensely popular sources of narrative and visual pleasure, the educational documentaries also promise accuracy and improvement. Five programs have been about nonhuman primates: *Miss Goodall and the Wild Chimpanzees* (1965), *Monkeys, Apes, and Man* (1971), *Search for the Great Apes* (1975), *Gorilla* (1981), and *Among the Wild Chimpanzees* (1984). The advertisement photograph, occupying half of the full-page color ad, shows two wrists and hands reaching in from opposite sides of the page to meet in gentle embrace. The hands are richly sensuous; they fill the space in which they intertwine. One is white, young, with well-trimmed nails; the other, about the same size, is brown, hairy, showing signs of a harder life. Both hands are open and vulnerable. "In a spontaneous gesture of trust, a chimpanzee in the wilds of Tanzania folds his leathery hand around that of Jane Goodall—sufficient reward for Dr. Goodall's years of patience." [Figure 7.1]

Understanding, touch, communication, spontaneity—these are the unmistakable themes. Carl Akeley's African Hall dioramas brought the viewer into immediate

Figure 7.1   "Understanding is everything." Advertisement from *Natural History Magazine*.
Copyright Gulf Oil Corporation, 1984. Published with permission of Chevron U.S.A., Inc.

visual communion with nature. But it was a different nature, one whose purity and strength could stay the threat of decadence. The distance was as important as the touch. The viewer of the dioramas confronted no barrier, and yet was held not in manual, but in visual contact. The fantasy of the hunter and inventor of cameras was about freezing a moment in time, assuring a timeless presence. Here, in the

d, the creature reaching across the white page from the
l Jane Goodall promises communication, the specific
Understanding is everything." The post-World War II
failure of communication, the malfunction of stressed
it language, about the immediate sharing of meanings.
ins, "Our goal is to provoke curiosity about the world
of the natural order; to satisfy that curiosity through
create an understanding of man's place in the ecological
ility to it—on the simple theory that no thinking person
of anything whose value he understands." This is the
science, by the slight move in the advertising text from
l.

embrace from chimpanzee to human, as in the experi-
can Hall, the question is inescapable: what mediates
tory of this touch? If Akeley measured his manhood by
e could achieve of a male gorilla before shooting with
it is the measure of distance in the National Geographic
films, and what is being measured? From whose point of view is distance calibrated
and closeness negotiated? Another way to put these questions suggests itself: when
may (white) woman best represent (species) man? How do the race, species, gender,
and science codes work to reinvent nature in the Third World for First World
audiences within post-colonial, multinational capitalism?

Gulf took on sponsorship of the series the same year the giant company, ranked
number 8 among the Forbes 500 in 1980 and traditionally one of the Seven Sisters
of international oil companies, was in the news for major slush fund scandals to
U.S. and overseas politicians, resulting in the forced resignation of the Chairman
of the Board. The 1970s were not only scandal-ridden for the oil giant; they were
also economically disastrous because of the formation of OPEC (Organization of
Petroleum Exporting Countries) and Kuwait's subsequent takeover of its oil fields.
Gulf moved from the unbeatable ally of British Petroleum, ruling the rich oil fields
of a small Middle Eastern country, from which Gulf's profits amounted to more
than a $1 million per day in the mid-1970s, to a supplicant for crude oil from an
assertive Kuwaiti government. The years of intense media focus on the system of
international oil profits and politics and on the "energy crisis" brought a new form
of advertising by the energy multinationals. Ads were full of scenes of natural
integrity and environmental restoration and preservation through the enlightened,
scientifically guided practices of the giant companies. It was a good time for Gulf
to associate itself with the thesis that "no thinking person can share in the destruction
of anything whose value he understands. . . . Association with the National Geographic Society and the television specials is only one aspect of Gulf's lively concern
for the environment. But it is an especially proud one."[1]

The association of the leading corporations of the industrial world with science
is not new. Nor is the interest of the great oil fortunes in underwriting research,
exploration, and teaching new—the Rockefeller Foundation almost single-handedly
paid for the establishment of modern biology and medicine in the United States
before World War II. Well before industrial capitalism, science and the commodification of the world grew up together; "progress is our most important product"
(Westinghouse). But the images of nature that began to be emphasized in early

modern Europe and continued to intensify in brutality and cogency were coded quite differently from Gulf's deeply reassuring touch. Gender is the explicit key to the code. The latent key is the ambiguity between the human races, marked by colonialism to be living in nature, and the nonhuman species, whose touch is so desired, that pervades post-World War II National Geographic television, photographic, and prose texts.

In *The Death of Nature*, Carolyn Merchant (1980) showed the centrality of female representations of nature at the origin of western capitalism. In a reconstruction of the western masculine imaginary with vast consequences, joining masculinized human ambition to the discourses of science and technology, earth was recoded in order to legitimate deeper penetration and relentless unveiling by a virile lover/actor/knower. The deeper penetration imagined and practiced by an expanding Europe at the "birth" of modern science and capitalism violated a closed cosmos forever, generating simultaneously the technological and imaginative projects, fears, and desires both for reclosing the broken globe in a recreation of the garden and for escaping finally from her gravity into a purely abstract space. Science from the "beginning" was coded as sensuously erotic, with the knower engaged in making love with all the force at his powerful command. The kind of lovemaking and the object of love have not remained constant.[2] But there was no question until very recently that the knower had to be socially male in his relations to the earth and to the natural-technical objects of knowledge which his fertile mind and hands generated from her raw materials. Akeley had to shoot before nature got too close, or the manhood he sought to assure would have been hopelessly compromised. He was engaged in a delicate negotiation between masterful touch and decadent embrace.

But with the Gulf Oil ad and the National Geographic specials, Jane Goodall is the scientist, not the beloved or the veiled object of knowledge and desire. Prominently female and white, Dr. Goodall guides the television viewer in the *ars erotica* of science, in which "man's place in the ecological structure, and his responsibility to it" can be assured (advertising copy). What she discovers, in an art of observation coded as requiring years of patience and yielding quiet triumphs, is a means of shared, earthy touch. Jane Goodall's touch was redemptive; its power saved others, who could not be allowed to repeat her original actions if the wildness of the animals and the safety of the humans were to be preserved. Dr. Jane Goodall inhabits one half of the system of desire mediated by modern science and technology, the half dreaming of reclosing the broken cosmos, known in its natural-technical form as an ecosystem.[3]

Primates in Space

The other half of the system of desire has a different locus for its inhabitants—not the ecosystem, but the extraterrestrial, i.e., "space." An ecosystem is always of a particular type, for example, a temperate deciduous forest. In the iconography of late capitalism, Jane Goodall did not go to that kind of ecosystem. She went to the "wilds of Tanzania," a mythic "ecosystem" reminiscent of the original garden from which her kind had been expelled and to which she returns to commune with the wilderness's present inhabitants to learn how to survive. This wilderness is close in its dream quality to "space," but the wilderness of Africa is coded as dense, damp, bodily, full of sensuous creatures who touch intimately and intensely. In contrast, the extraterrestrial is coded to be fully general; it is about escape from the bounded globe into an anti-ecosystem called, simply, space. Space is not about "man's" *origins* on earth but about "his" *future*, the two key allochronic times of salvation history. "Space" has formal properties; it can be warped, for example, like a topological mathematical figure. Space and the tropics are both utopian topical figures in western imaginations, and their opposed properties dialectically signify origins and ends for the creature whose mundane life is outside both: civilized man. Space and the tropics are "allotopic"; i.e., they are "elsewhere," the place to which the traveler goes to find something dangerous and sacred.

The first primates to approach that abstract place called "space" were monkeys and apes. A rhesus monkey survived an 83 mile-high flight in 1949. Jane Goodall arrived in "the wilds of Tanzania" in 1960 to encounter and name David Graybeard, Flo, and the rest of the famous Gombe Stream chimpanzees introduced to the National Geographic television audience in 1965.[4] However, other chimpanzees were vying for the spotlight in the early 1960s. On January 31, 1961, as part of the United States man-in-space program, the chimpanzee HAM, trained for his task at Holloman Air Force Base, 20 minutes by car from Alamogordo, New Mexico, near the site of the first atom bomb explosion in July, 1945, was shot into suborbital flight. [Figure 7.2] HAM's name inevitably recalls Noah's youngest and only black son. But this chimpanzee's name was from a different kind of text. His name was

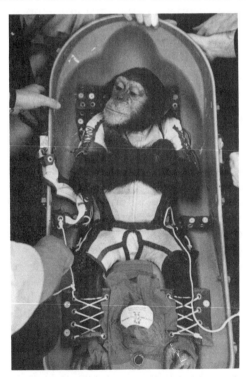

Figure 7.2   Ham awaits release in his couch aboard the recovery vessel LSD *Donner* after his successful Mercury Project launch. Photograph by Henry Burroughs, 1961. Published with permission of Wide World Pictures.

an acronym for the scientific-military institution that launched him, *Holloman Aero-Medical*; and he rode an arc that traced the birth path of modern science—the parabola, the conic section. HAM's parabolic path is rich with evocations of the history of western science. The path of a projectile that does not escape gravity, the parabola is the shape considered so deeply by Galileo, at the mythic moment of origins of modernity, when the unquantifiable sensuous and countable mathematical properties of bodies were separated from each other in scientific knowledge. It describes the path of ballistic weapons, and it is the trope for "man's" doomed projects in the writings of the existentialists in the 1950s. The parabola traces the path of Rocket Man at the end of World War II in Thomas Pynchon's *Gravity's Rainbow* (1973). An understudy for man, HAM went only to the boundary of space, in suborbital flight. On his return to earth, he was named. He had been known only as #65 before his successful flight. If, in the official birth-mocking language of the Cold War, the mission had to be "aborted," the authorities did not want the public worrying about the death of a famous and named, even if not quite human, astronaut. In fact, #65 did have a name among his handlers, Chop Chop Chang, recalling the stunning racism in which the other primates have been made to participate.[5] The space race's surrogate child was, as the popular Time-Life nature series book, *Primates*, put it, an "understudy for man in the conquest of space" (Eimerl and DeVore 1965: 173). His hominid cousins would transcend that closed parabolic figure, first in the ellipse of orbital flight, then in the open trajectories of escape from earth's gravity.

HAM, his human cousins and simian colleagues, and their englobing and interfacing technology were implicated in a reconstitution of masculinity in Cold War and space race idioms. The movie *The Right Stuff* (1985) shows the first crop of human astrau(gh)ts struggling with their affronted pride when they realize their tasks were competently performed by their simian cousins. They and the chimps were caught in the same theater of the Cold War, where the masculinist, death-defying, and skill-requiring heroics of the old jet aircraft test pilots became obsolete, to be replaced by the media-hype routines of projects Mercury, Apollo, and their sequelae. After chimpanzee Enos completed a fully automated orbital flight on November 29, 1961, John Glenn, who would be the first human American astronaut to orbit earth, defensively "looked toward the future by affirming his belief in the superiority of astronauts over chimponauts." *Newsweek* announced Glenn's orbital flight of February 20, 1962, with the headline, "John Glenn: One Machine That Worked Without Flaw."[6] Soviet primates on both sides of the line of hominization raced their U.S. siblings into extraterrestrial orbit. The space ships, the recording and tracking technologies, animals, and human beings were joined to form a new kind of historical entity—cyborgs in a postmodern theater of war, science, and popular culture.

Henry Burroughs's famous photograph of an interested and intelligent, actively participating HAM, watching the hands of a white, laboratory-coated, human man release him from his contour couch, illuminates the system of meanings that binds humans and apes together in the late twentieth century (Weaver 1961). "Under his pressure suit are telemetric devices to monitor his body functions" (Eimerl and DeVore 1965: 172–73). HAM is a cyborg, the perfect child of space. *Time* described chimponaut Enos in his "fitted contour couch that looked like a cradle trimmed with electronics" (8 Dec. 1961: 50). Enos and HAM were cyborg neonates. Linguistically and materially a hybrid of cybernetic device and organism, a cyborg is a science fiction chimera from the 1950s and after; but a cyborg is also a powerful social and scientific reality in the same historical period (Haraway 1985). Like any important technology, a cyborg is simultaneously a myth and a tool, a representation and an instrument, a frozen moment and a motor of social and imaginative reality. A cyborg exists when two kinds of boundaries are simultaneously problematic: 1) that between animals (or other organisms) and humans, and 2) that between self-controlled, self-governing machines (automatons) and organisms, especially humans (models of autonomy). The cyborg is the figure born of the interface of automaton and autonomy. There could be no more iconic cyborg than a telemetrically implanted chimpanzee, understudy for "man," launched from earth in the space program, while his conspecific in the jungle, "in a spontaneous gesture of trust," embraces the hand of a woman scientist named Jane in a Gulf Oil ad showing "man's place in the ecological structure." On one end of time and space, the chimpanzee in the wilderness models communication for the stressed, ecologically threatened and threatening, civilized human. On the other end, the ET chimpanzee models social and technical cybernetic communication systems, which permit "man" to escape both the jungle and the city, in a thrust into the future made possible by the social-technical systems of the "information age" in a global context of threatened nuclear war. The closing image of a human fetus hurtling through space in Stanley Kubrick's *2001: A Space Odyssey* (1968) completes the voyage of discovery begun by the weapon-wielding apes at the film's gripping opening. It is the project(ile) of "man."[7] The Cold War has been, so far, simulated ultimate war; the media and advertising industries of nuclear culture produce in the bodies of animals—paradigmatic natives and aliens—the reassuring images appropriate to this state of pure war.[8]

A Primate Ethogram

From the primate cyborg birthed in filmic, fictional, and scientific-technical translations of war research and the space program, let us turn back to "nature" and its ecosystems to look at communication. One of the first tasks facing Jane Goodall as a field worker was the defining and cataloging of the chimpanzee's gestural, vocal, and facial expressions for intra-specific communication. Her rhetoric in this task was a rigorously scientific prose, not the adventure travelogue of seemingly participatory, popular, democratic science characterizing the *National Geographic*.[9] Making the signal lexicon was part of constituting the natural-technical object called an "ethogram." A graduate student of Robert Hinde at Cambridge, Goodall was within the ethological tradition that emphasized the evolution and form of social signals among animals.

The early comparative ethologists worked primarily from fish and birds; their students and their students' students extended the analysis to primates, including humans (Morris 1967b). Chimpanzee facial expression had been examined within that tradition, in captivity, by J.A.R. van Hooff, graduate student of Desmond Morris, who in turn had studied with one of the founders of ethology, Niko Tinbergen, at Oxford. Tinbergen was an important influence for Robert Hinde at the origins of the field station that became the Sub-Department of Animal Behaviour that gave Goodall an academic home for her Gombe studies, initiated independently of academic authorization. Peter Marler, whose second Ph.D. was also from the Sub-Department of Animal Behaviour, was one of the first visiting scholars at

Gombe; he came to record the vocal repertoire of the chimpanzees for ethological analysis. His students and post-docs have contributed extensively to primate social studies. N.G. Blurton Jones, at the Institute of Child Health of the University of London, had also been a Tinbergen student. After his 1964 Oxford thesis on birds, Blurton Jones constructed British nursery schools as an ethological object of knowledge and practice. Apes could be "humanized"; humans could also be "naturalized." The issue has little to do with the ideological object called "biological determinism" or the philosophical object called "reductionism." Like each of the modes of communication examined in this chapter, ethological tools were crafted to infiltrate the academically and politically fortified border between nature and culture. Ethological analysis was "naturalistic," i.e., focused on the form, function, and evolution of social signals meaningful to organisms conceptually located in their evolutionary and ecological homes. The privileged place to conduct an ethological analysis was "the field," although "the laboratory" was the scene of its refinement.

Ethology was philosophically and practically distant from the construction of primates as human understudies in space. Yet, ethology had multiple interfaces with the cyborg HAM and his kin. In a famous series of papers in the 1930s, Konrad Lorenz's conceptual foundations of ethology elaborated an organismic hydraulic machinery, where the "specific action energy" of "fixed action patterns" was released by various signal stimuli.[10] Ethologists emphasized the schematization and modification of movement patterns occurring first in some other context to serve as social signals. Focus was on chains of action patterns, culminating in a "consummatory act." Energy could be dammed up, held back from release, and so built up and released in intense behavioral explosions, or diverted into displacement activities. The matching of motivation and behavior in an evolutionary theory of drives was the goal. These kinds of explanatory concepts were translated from hydraulic metaphors into cybernetic ones progressively from the end of World War II, when those metaphors acquired a wide social and technical cogency. The webs and chains drawn by ethologists were full of feedback loops and other cybernetic control techniques. The principal natural-technical object of analysis was, as before, communication, but now understood in relation to a different, and far more general, information technology.

Several different kinds of ethological analysis were pursued with the aid of the information sciences' social, metaphoric, and technical resources, but they have one crucial similarity. The line between natural organisms and constructed technical systems was redrawn in a radical way, so as to produce the cyborg as the central natural-technical object of knowledge in the last half of the twentieth century. Both organism and technology were theorized and encountered in practice as communications engineering problems, where the ontological distinction between the natural and the artificial lost meaning.[11] The naturalistic primate studies in the ethologically constructed field intersected the extraterrestrial primate studies of the space program in the electronically recorded and telemetrically implanted simians beaming information to listening scientists in the field, laboratory, and command center. In the universe of information, the antipodes of the earthly ecosystem and extraterrestrial space meet in a shared code.

Dream of a Common Language

In addition to the "spontaneous" manual gesture towards the white female man, the ET cyborg communications machine-organism interfaces, and the "natural" ethological signaling systems, the chimpanzee is also a model in a fourth way in a post-war western world obsessed with communication: in the use of American Sign Language (AMESLAN) to converse with human beings and other apes. Ironically, in this search for connection, each of the four communications systems rests on a fundamental transgression or boundary crossing: man represented by woman, masculinist reproductive "organism-machine" chimeras birthed in space, natural communication theorized as command-communications cybernetic signaling, and finally, animal speech.

With funding from the National Geographic Society, in 1966 Allan and Beatrice Gardner began to teach a young female chimpanzee, Washoe, to sign. The Gardners reasoned that past failures of apes to learn human language, for example in Cathy and Keith Hayes's efforts from 1947–53 with chimp Vicki, were caused by trying to use vocal instead of gestural media. Roger Fouts, working first with Washoe and then several other chimps, developed the AMESLAN research with chimpanzees, who signed with humans and with each other and taught chimp babies to sign. Beginning at the San Francisco Zoo in 1972 and then in a private trailer in Woodside, California, Stanford University graduate student Penny Patterson, inspired by the Gardners, taught the gorillas, Koko and Michael, to use AMESLAN (Patterson and Linden 1981). Deaf native human users of AMESLAN have been participants in the AMESLAN projects. Since Koko apparently understood much spoken English, a logician and head of Stanford's Institute for Mathematical Studies in the Social Sciences, Patrick Suppes, with his colleagues, "designed a keyboard-computer linkage that enables Koko to talk by pressing buttons linked to a voice synthesizer" (Patterson and Linden 1981: 109). Koko, too, is a species of cyborg, whose communication modalities can be translated and re-synthesized to cross species and machine-organism barriers. A *Reader's Digest* (Oct. 1985: 202) account of Koko's and Patterson's conversations links the coupling to the hopes of children, the status of scientists, and the achievements of saints: "[Koko] was one year old when she met Penny Patterson, a Stanford University Graduate Student, who as a child had idolized St. Francis because he could converse with the beasts of the earth."

Many scientific reasons have been given for studying the apes' abilities to learn and use human languages, but perhaps the deepest motivation was expressed by Biruté Galdikas in a *National Geographic* cover story on her Indonesian orangutan rehabilitation and field work. Galdikas invited Gary Shapiro, involved in language teaching with captive apes, including Washoe, to teach AMESLAN to rehabilitant Sugito. Galdikas explained, "I had often regretted that I would never be able to talk to Sugito, so that I could examine how he perceived and interpreted the world. By teaching orangutans sign language in their native habitat, we might find out what was important to them, rather than to us" (Galdikas 1980: 845). This is the ethnographer's dream of knowing the world from the other's point of view. That language does not grant this grace even among people of the "same" culture has not deterred the imagination of communion across species. However, Galdikas's *National Geographic* narrative noted pedagogic problems. Galdikas's first and favorite orange rehabilitant child, Sugito, had experienced a humanized childhood. Confused about his species identity, but not his (hetero)sexual, identity, Sugito was going through adolescence and saw Gary Shapiro as a male rival; he failed to learn anything. Shapiro found a more apt pupil in an adult female, a former captive named Rinnie. Shapiro swam across the river and taught her sign language in her home, the forest, and at her will, since she was not confined. She learned fast, but restricted her conversation to requests for food. Ethnography has often been disillusioning (Galdikas 1980: 845–8).

Sugito, called by Galdikas's husband, "one of your orange children," proved himself a murderer of other rehabilitant orangutans, of whom he was presumably jealous. That plus his troublesome male adolescence resulted in his banishment to a distant section of the forest (Galdikas 1980: 846). "Sugito was something different. Perhaps the biblical analogy was apt: Raised by a human mother and exposed to human culture, he had eaten of the 'tree of knowledge' and lost his orangutan innocence" (Galdikas 1980: 832). Ironically, Sugito's species confusion, his "hominisation," became the cause of his being removed as far as possible from the scene of his "rehabilitation," which had become the theater of his interconnected crimes of speciecide and sexual disorder. The sexual disorder was the result, in the narrative, not of sexual, but of species, transgressions. Likewise, his murder of his rehabilitant siblings was built into a story of a transgression of his natural community, or rather lack of community. Male orangutans in nature are mostly solitary, likely a feeding adaptation for a large-bodied fruit eater. Sugito had to deal with a lot of "stress" as he matured, including a cross-specific sexual rivalry. Similar sexual identity problems have beset home-reared apes, who develop a taste for reproductively inappropriate sexual partners—people. Narratives of these problems are rich sources for exploring the scientists' preoccupation with reproductive heterosexual "normality" at the dangerous edge of the garden of nature, where confusions abound.

Shapiro also worked with a bright young rehabilitant, Princess, who in turn taught signs to Galdikas's toddler son, Binti (Galdikas 1980: 846–49). Galdikas reported being worried about her son's constant imitation of Princess, the boy's inseparable companion. The *National Geographic* cover photograph pictured the orange and pink babies, Princess and Binti, in a plastic bathtub together. "It would not have been any cause for concern except that, with no other children in camp, orangutans were becoming his role models . . . He didn't talk to Princess; he signed to her as he did with other non-signing orangutans" (848). Although Galdikas mentioned Binti's biting and other "animal" behaviors among her worries, the account of Binti's learning *language* from the ape child, rather than from his parents, inflected the anxiety in a curious way. It seemed that the ape was an inappropriate *human* model for the *orang* surrogate, who was really a boy. Galdikas reported that Binti became fully human in his behavior later, when he had human playmates, but Galdikas herself closed that same paragraph by updating the reader on the fates of her orange rehabilitant children who had gone back into the forest. "I saw Akmad and Sobiarso, my two original female monster babies, traveling across the river with wild subadult males. I know that it is only a matter of time before I again become a 'grandmother'" (852). The thin quotation marks that barely contain the powerful sign, *grandmother*, cannot close the border between the scientist-mother and her multi-specific children.

Both science and gender have combined in a narrative of ambiguously innocent and transgressive reproductive polity. The "human" quest for the origin of "man" led late twentieth-century Euro-American women into the vanishing forest gardens at the dangerous historical moment of decolonization. The explicit scientific frame for their heroic actions was conservation and evolution, the disciplines of preserving the past as augurs for a future. There, the scientific women inscribed in *National*

Geographic's text took on the task of restoring violated creatures, which western scientific and popular stories conceived to be on the border between nature and culture, to their natural wildness. In the crossing, they and their children spoke with the animals, an act impossible since the original sin at precisely the time of origins they sought to restore and to know. Linguistic, scientific, and sexual conversation—open and productive systems of signs—constitute the commerce between human and ape. Interrupting the communication were the noisy sounds of lumber companies felling the forest all around the fragile parks.

But we must return from the traffic between nature and culture in endangered Indonesian forests to a similar transgressive commerce in Koko's world in California's Silicon Valley. The main issue is less a linguist's view of the niceties of her patterns of word order, length of utterances, invention of jokes and insults, prompting answers in tests for her younger companion gorilla, reports of past events in her life, or her vulgar sense of humor, than her photographic appearance on the *National Geographic* covers with a camera (Oct. 1978) and a kitten (Jan. 1985). It is not only that Koko speaks; the female gorilla also bears the other defining stigmata of "man." This chapter opened with Koko's signed response to Patterson's question about her identity as an animal or a person: "Fine animal gorilla." She names herself. The theme of self-awareness is even sharper in National Geographic's photograph of Koko photographing herself in a mirror with a 35 mm poloroid flash camera, while the accompanying text reads: "Viewing a copy of this picture, she identified herself and signed, 'Love camera' " (Patterson 1978: 464–65). [Figure 7.3] Akeley's capture of the gorilla through the practices of naked eye science pales beside the gorilla self-portrait made with the light personal camera. In the 1978 article, Koko

Figure 7.3   Koko taking her picture in a mirror. Copyright Ronald Cohn, the Gorilla Foundation, Box 620–530, Woodside, CA 94062. Published with permission.

signs "naughty" in response to the pictures of the crying three little kittens who lost their mittens; this human surrogate is capable of moral discourse. [Figure 7.4] She holds a cup up to her chimpanzee doll at a play tea party with Penny; Koko creates imaginary worlds. In response to photographs and drawings, Koko signs in front of her mirror, not only imitating the creatures in picture books, but watching herself imitating and signing. National Geographic builds a strong bridge across the divide of nature and culture. It is no wonder that the ethics of placing a language-using gorilla back in the zoo, to be watched as an object and not conversed with as a subject, aroused the public around 1976–77, when it looked like Patterson might be forced to return Koko to the San Francisco Zoo, so that she might form bonds with gorillas in order to reproduce, as well as be displayed to the public that owned her. The intervention of the mayor of San Francisco saved Koko from the zoo, allowing Patterson to buy the "humanized" gorilla (Patterson and Linden 1981: 64). The stories about Koko at this point reassure the reader that she is not isolated from her species; she has a proper signing gorilla companion in the younger Michael, acquired in a kind of pre-arranged marriage for the gorilla whose language propelled her across species barriers. As on Noah's Ark, the animals crossed in heterosexual twos on *National Geographic*'s journey.

Koko's status in relation to zoo animals was affirmed by a quite specific act: she kept pets herself. With a history of befriending a rabbit that frequented her trailer home, Koko asked Penny for a kitten, pouting when Patterson produced a stuffed version. Then the gorilla chose a tailess little male from a litter, naming him Ball in AMESLAN. [Figure 7.5] Ball had been suckled by a dog, who had adopted the

Figure 7.4   Koko signing in response to the storybook, *The Three Little Kittens Who Lost Their Mittens*. Copyright Ronald Cohn, the Gorilla Foundation. Published with permission.

Figure 7.5   Koko with her kitten. Copyright Ronald Cohn, the Gorilla Foundation. Published with permission.

orphaned kitten litter.[12] Koko's keeping a pet was not placed in a cultural and historical context that distinguished specific forms for people's and other animals' interactions outside the time of origins, in actual times and places. That Koko, the AMESLAN-speaking gorilla wanted a kitten and befriended a rabbit rhetorically located her in a shared world with "man," not in a shared world where rabbits and kittens have particular resonances in European-derived industrial cultures. It was precisely to renaturalize "man" in the context of decolonization and the Cold War's nuclear culture—both of which banished "man" once again from "nature"—that apes and white people, especially women represented as surrogate mothers and scientists, were placed together *simultaneously* in the "natural" world of the forest, sharing adolescence and child-rearing practices, and in the "social" world of language studies, sharing a love for cuddly pets. The natural and the social were both mythic spaces of origin in this structure; neither led to history and specificity; in each Man was coded through white women and anthropoid apes.

The *National Geographic* narrates that Penny was pleased by Koko's serving as a second surrogate mother to the kitten, allowing him to bite her without retaliating, treating him gently, and carrying him about in her rehearsal for the desired reproductive event planned for her with her signing companion, Michael. Koko's rehearsal was enhanced by other efforts to prepare her for the big event: "Nursing human mothers visit Koko to provide role models, and Penny talks to her about bearing and caring for young" (Vessels 1985: 113). But Koko has had trouble conceiving; still childless in the late 1980s, she seems destined for the symbolic and practical world of late twentieth-century reproductive heroics for those who wait too long to start their families. The Gorilla Foundation placed an ad in *The American Scholar* (Winter 1988: 160) inadvertently representing Koko as a potent sign simultaneously for the struggles over homelessness, reproductive politics, the constant extinction discourses in nuclear culture, and the continuing myth of stripped-down languages at the boundary between nature and culture. Appealing for funds the ad quotes Koko: "Koko need home hurry . . . want gorilla baby." Her signs are transcribed into the syntax of babies and "primitives" in racist discourse. The text continues, "Koko's childbearing years are few in number. [Your membership can help provide a home] where we can watch and record Koko as she communicates with her baby in sign language. Please join today and help Koko realize her dream for a baby of her own." Perhaps a surrogate pregnancy can be arranged.

The story of compulsory reproductive sexuality is never far in the background in primate visions. The multiplicity of surrogates confuses the question of alliances and the nature of progeny, but not for a moment does all the boundary crossing—of species barriers, machine-organism barriers, language barriers, earth-space barriers—relax the injunction to be fruitful and multiply, heterosexually. Communication is the foundation and goal of the whole innocent-transgressive enterprise. The progeny are cyborgs, creatures with ambiguous and permeable boundaries: monkeys, apes, and humans, all entwined in a compulsory reproductive politics.

Reading Out History

The Traffic between Nature and Culture

In the placement of apes and people in this curious world composed of space, the "wilds of Tanzania," a trailer near the Silicon Valley, and a jungle in Indonesia, species, gender, and symbolic location are explicit variables in the stories. Another narrative possibility is conspicuous by its absence: history. How do tales of primate lives narrate either nature or culture, or both, but narrate them so as to exclude a reader's consciousness of mediation, history, and construction?

Koko seems to be placed in culture, the realm of the human, by her keeping a pet, and above all by naming him. The lesson is that she is like "humans." But "humans" do not keep pets; members of particular societies do. Koko is given the attributes of the camera, the mirror, and the book—all to endow her with the "human" property of self-consciousness. But a "self " is a complex historical construction that emerged in the forms characterizing Koko during the modern development of class, gender, and race (Lowe 1982). Koko's self seems to have been crafted in the political theory and political economy, not to mention the consumer culture, of the modern west. The gorilla has been brought into culture without bringing her, and her kind, into history. Language narratively marks her off from the zoo animal in relation not to specific constructions of animal and human, but in relation to ontological statuses in an implicit framing myth about origins. The organizing story axis of the duality, nature-culture, forbids an account of historical mediation.

The naming of nonhuman primate subjects is a key rhetorical device bestowing a particular kind of individuality in the form of an apparently timeless, universal selfhood. The combination of naming and photographic portraiture can be quite extraordinary. Two recent examples are Frans de Waal's *Chimpanzee Politics: Power and Sex among Apes* (1982) and Dian Fossey's *Gorillas in the Mist* (1983). The physical appearance of the books conveys the point.

The endpiece of Fossey's book contains drawings from photographs of 12 gorillas' heads, each markedly individual. The faces capture the reader immediately. The names of the gorillas appear to the side: Uncle Bert, Flossie, Digit, Effie, Icarus, Beethoven. The book is dedicated to some of the murdered gorillas: Uncle Bert, Digit, Macho, Kweli. Inside, the individualizing themes are elaborated into prose family portraits across the generations. The frontspiece is an evocative drawing, described on the copyright page as a "portrait of Uncle Bert, Flossie, and their two-week-old son Titus, made after field observations. Jay Matternes has captured the essential nature of the gorilla—the protectiveness of the silverback, the maternal concern of the mother, the vulnerability and dependence of the infant."

Perhaps field work was necessary for this portrait, but it bears comparison with Alfred Brehm's nineteenth-century "Gorilla Family" engraving (Reynolds 1967: plate 60). Although imagined, Brehm's nuclear holy family is nearly identical with Matternes's version. The combination of detailed individuation, life stories, portraiture, and the timeless harmony of the organic whole group, iconically rendered in the ideal patriarchal, heterosexual, reproductive family constructs the gorillas as the embodiment of both individuation and community, bypassing the conflict-laden realm of history. The gorillas have personality and nuclear family, the two key elements of the bourgeois self represented simply as "man." History enters Fossey's book only as a disrupting force in the Garden, through murderous poachers, selfish graduate students, and mendacious politicians.

The cover jacket of De Waal's *Chimpanzee Politics* features a patently scheming chimp sitting alone in a grassy area. The agent of politics, the scheming individual, dominates this saga, subtitled "power and sex among the apes." The epigram, from *The Imperial Animal*, which Tiger and Fox meant to refer to *Homo sapiens*, is appropriated for the chimpanzee state of nature: "The political system is a breeding system. When we apply the word 'lust' to both power and sex, we are nearer the truth than we imagine." The reader is willy-nilly implicated in the authorial "we." "We" are given no choice but identification, unless "we" choose to be contentious readers from the start, losing all the pleasure of complicity with the author and identification with the engaging chimpanzee political actors we are about to meet. The disengagement from the "we" is made marginally easier for a female reader, who will learn that female doings among the chimps do not get to count as politics. For example, a female rescuing a stolen child from another female, returning the child to its mother, and thereby restoring social peace, all done with every appearance of conscious understanding of the negotiation, is explicitly labeled "non-political," i.e., merely "social," the realm of the female (182–83).

The female hierarchy, first discussed on p. 185, after fulsome description of strategic reasoning among the males, is labeled a "subordination hierarchy." Females just don't seek power; they avoid trouble and help out their friends and family. The females in this book are explicitly described as intelligent, but not really "rational," a quality which means "strategic reasoning." The male is rational and status oriented; the female is protective and personally oriented (197). De Waal asks disingenuously if that conclusion is only another example of his prejudices. Through De Waal's disarming insistence on the inescapability of interpretation in ethology and his disavowal of positivism, the reader is constantly reminded *by the author* of the interpretive frame, and thereby caught all the tighter in the web.

Chimpanzee Politics is endowed with quotes from *The Prince*, an enthusiastic Malinowskian functionalism, and elaborately detailed confirmation of the book's major premise that "chimpanzees never make an uncalculated move" (42). The absolutely essential property of liberal man is abundantly evidenced: strategic rationality, the defining property of the individual. Females come out with an ambiguous status, ideologically analogous to that of their human sisters. They are plainly capable of strategic reasoning, i.e., in principle of participating in "chimpanzee nature"—politics—but ordinarily they live in a special realm called the "social." Their cooperation, or lack of it, was crucial to any male who wished to play for high stakes, but they were better off staying out of the game themselves. The one female, Mama, who had been in the top position before the adult males joined the Dutch zoo colony, was deliberately dethroned by the scientists, who were merely hastening "this more natural course of events" (59). Mama had proved unworthy of her political power because she was too violent. After her deposition, Mama slowly developed into an arbiter of social harmony: "Her central position is comparable to that of a grandmother in a Spanish or Chinese family" (56). Why not Dutch? Female nature makes the theorists a bit nervous, but can be accommodated in all its nonrational intelligence in the convenient category of the "social," leaving the drive for power over others, and the schemes to achieve it among potential equals, in the rational zoo polis—the scene of fulfillment of primate nature.

The complement of strategic rationality for the bourgeois individual is personality. Personality and strategic rationality together make a "self." De Waal's book has the researchers dreaming of the chimps as persons, as strong individuals (54); and the rich descriptions of each animal lead the reader into their social circle. The chimps are as vividly full of specific personality as any human group. The prose narrative is emphasized in the portraiture (52–53), which would do credit to a bourgeois household equipped with a personal camera for the informal snapshot, the icon of selfhood for mass bourgeois culture. All of these rhetorics convey large pleasure to the culturally prepared reader. Pleasure in these stories is absolutely essential to "our" understanding their grip in scientific as well as popular explanation.

All these devices place the animals just at or over the line into "culture." People are also placed at or over the line into "nature" in the complete origin story. In the post-World War II *National Geographic* versions, in the jungles of Africa and Indonesia, white women undergo the trials and rigors of the questing hero to receive a particular grace: the spontaneous touch of the other, the bridge between "animal" and "man" built *by the animal* as a spontaneous and supremely meaningful gift. Running a halfway house on the edge of the forest, Galdikas restores orangutans from the civilized status of pets to the state of nature. In that guise, her body becomes a generalized primate maternal body; she is covered in clinging rehabilitants and orphans. The cover photograph from the October 1975, *National Geographic* shows her accompanied by an orange toddler holding her hand and an orange infant clinging to her torso. The photograph emphasizes her breasts. The cover of *National Geographic* for January 1970, features a similar photo of Dian Fossey with two gorilla orphans walking in a mountain meadow.

But Galdikas, like Fossey, is simultaneously studying wild apes. In that persona, she, again like Dian Fossey and Jane Goodall, must seek the desired touch, yet seem less to *achieve*, than to *accept* it. The women's work is to create receptivity, to produce the conditions in which the animal *can* approach. In the 1975 Gulf Oil-National Geographic special, *Search for the Great Apes*, Galdikas is shown beginning her arduous task, given only the sounds of food dropping in the forest and the abandoned shit samples to analyze as signs of the presence of her quarry. More than 5000 hours of observation and three years later, and after immense efforts to habituate a single adult male, whom Galdikas and Brindamour named Nick, "on the 41st day comes the moment I never dared hope for. Nick comes to us" (soundtrack). To create permissive conditions is no mean task. The words are followed by a long still shot of Nick approaching the viewer, who sees from the vantage point of Galdikas and Brindamour. The immense orange male utterly fills the screen, covering the viewer with his physical reality. *Nick* succeeds in bringing the viewer into nature; Galdikas, as scientist and woman, made ready to receive the gift, while Brindamour was ready to film it. This is the holy family of post-World War II natural science.

Part II opens with Dian Fossey's drama of acceptance by the gorillas, who originally fled from the pale strange intruder, but later accepted her into their "hidden world" (soundtrack). The film is full of Dian and a young male gorilla, Digit, whose first touch seemed to convey his need to communicate. It "span[ned] a chasm of immeasurable time." His touch ended Fossey's solitude, bringing her into the community of nature. In the story of another *National Geographic* author's first touch from the primate subject she came to Africa to observe, Shirley Strum (1975: 673) opens her account with: "I was watching Naomi, one of my favorite baboons, sitting with her friend Queenie. Entering notes about Naomi's behavior on my clipboard data sheet, I suddenly felt small hands touch my back, so softly at first I could not identify the sensation. I slowly turned and saw that it was Robin, Naomi's two-year-old daughter, grooming the thin cotton of my shirt. It was a gesture that thrilled me."

In all of these dramas of touch, nature approaches man through white woman. Man is placed in nature through his emissary, woman, just as the apes were placed in culture through a human female emissary in the Koko story. The stories are about modes of *communication*, not about history. "Communication" in all four modes examined in this chapter (gesture from animal to human, extraterrestrial cyborg, ethological signaling, human-animal conversation) is about boundary crossing, about the drama of touch across Difference, but not about the finite, difference-laden worlds of history. The duality nature/culture works well within this frame of communication because both of its terms are allochronic; i.e., existing in a time outside the contentious, coeval time of history, with all its differences and uncertainty of understanding.

A Triple Code

Gender

Woman in these narratives fulfills her communicating, mediating function because of a triple code, only one part of which is gender. Gender in the western narrative works simply here: Woman is closer to nature than Man and so mediates more readily (Ortner 1972). Positioned by the symbology, real women are put into the service of culturally reproducing Woman as Man's channel. That modality of gender is required to heal man's expulsion from the garden after the bomb and in the ultimately threatening world that followed.[13] The National Geographic's late twentieth-century story is not about transcendence, but immanence, the possibility of survival on earth. Man's mediation of the touch with nature in Akeley's sacred status as "father of the game" required a greater distance to be permanently maintained, marked by the camera and gun, as well as by the craft of taxidermy. In the National Geographic narrative, the camera remains firmly in the hands of men. Hugo van Lawick, husband of Jane Goodall and photographer; Ron Brindamour, husband of Biruté Galdikas and photographer; Timothy Ransom, "consort" (Shirley Strum's term) and photographer; Ronald Cohn, companion to Penny Patterson and photographer. Dian Fossey remains virginal in the National Geographic account, but her photographer, Robert Campbell, is true to gender type.[14] The woman-scientist of National Geographic is married to the camera's eye, or she is a virgin wise woman, married only to nature in the touch of a male ape. The attribute of the camera is like strategic reason, a difficult technology for Woman to master, the stutter that keeps her safely in the marked category.

Science

The National Geographic woman is multiply typed as a scientist. The above story about Strum prominently included her "clipboard data sheet." All of the articles and films emphasize the activity of "research" and the status of the women reseachers as either graduate students in science or as already holding a Ph.D. Various phrases function as signs to stress the scientific theater of both the forest excursions and the AMESLAN conversations, e.g., the "5000 hours" of Galdikas's study or the lists of questions for future research at the ends of articles. The *National Geographic* articles and films are not structured as professional scientific reports; their genre is quite different. But they are replete with signs pointing toward Science. To be woman and to be scientist in this world requires patience and receptivity, but also a spirit of adventure and an ability to endure hardship for an important goal. The woman/ scientist presented to an urban TV audience is fully rational, but not in the sense of the "unmarked," type-defining chimpanzee nature narrated by de Waal. Instead, she is like the intelligent but deposed chimp, Mama, mediatrix of peace and knowledge, the one without whose cooperation nature will not be fulfilled. Such a scientist does not hold the camera; she is still the one photographed for millions to view. Koko gets a dispensation to take a picture of herself in the mirror of her trailer in Woodside, in order to endow her with the nature of man. She gets to be both photographer and photographed, in an icon of wholeness.

The feminine gender of the woman scientist is crucial to the National Geographic's account of the transmission of scientific knowledge. There is here no picture of the modern laboratory akin to those in Wiseman's *Primate*, where "nature" has been fully transcribed into linear data streams. Rather, as in *Search for the Great Apes* (1975), at the end of the film after Digit's reach to Fossey that "span[ned] a chasm of immeasurable time," Fossey is shown teaching a young woman scientist (Kelly Stewart) how to watch gorillas. The soundtrack announces that now Fossey is drawing students and visiting scientists from around the world. "Dian passes on a legacy of understanding only she possesses. . . . [She] leads a handful of others into a world that until now only she has known." The film ends with a long shot of a gorilla's eyes. Fossey passes on a secret lore to another woman. The links with Cambridge University's Sub-Department of Animal Behaviour, where Kelly Stewart was a graduate student, are not mentioned, nor the difficult relations between Fossey and Stewart in "real" life. The picture is about a wise woman and her initiate. It would be hard to misrepresent more systematically the social organization of scientific work in the late twentieth century. The misrepresentation is made plausible by the insistent double code of gender and science. *Search for the Great Apes* pictures a feminine secret world as the space where science is perfected.

However, all of the accounts of the National Geographic women scientists locate the origin moment of that female space in a patriline. The code is about "women's" science, but not about "feminist" science. The scientist-women of the National Geographic films had a common thread in their histories; Goodall, Fossey, and Galdikas were all facilitated in their work by the Anglo-Kenyan paleontologist-anthropologist, Louis Leakey. Leakey considered that investigations into "man's" origins, as well as reconstructions of the behavior of the Miocene apes whose fossils he had found, required study of living primates. He set up one young man in 1946 to study wild chimpanzees, but he quit after six months (Cole 1975: 326). In the 1950s, Leakey began to support young women in the study of living primates. As his biographer put it, "Even when Louis was in his sixties girls in their twenties found him attractive; he clung to his immortality, and naturally he was flattered and pleased. He collected satellites like the planet Jupiter, using them to run errands and help with his writings; but he also encouraged some of them to pursue their own studies under his direct supervision" (Cole 1975: 325).

Many women worked under Leakey's aegis over several years, mostly at the Tigoni Primate Research Center near Nairobi that he established in 1958, with Cynthia Booth as the first scientific director. Leakey's efforts culminated in the successful, ultimately independent, famous trio of Goodall, Galdikas, and Fossey. Whatever his personal needs and whatever the scientific merit of particular women students, Leakey rationalized his sponsorship of women in traditional gender terms, which were convenient to him in his renegade resistance to the new organization of scientific knowledge in archaeology and paleoanthropology that was leaving him behind—including behind his wife, Mary Leakey, and his son, Richard Leakey. He imagined the young women to be without the professional scientific prejudices of the masculine world of modern research. None of the famous "ape women" went to the field with a Ph.D., and Leakey helped set up connections with Cambridge University's Robert Hinde to credential Goodall and Fossey. Leakey felt the "un-biased" women would make better observers than men because they would be more patient, not be threatened by the male apes in masculine rivalry, and be sensitive to the mother-young interactions. He is said to have worried about the young women's emotional involvements with the apes, seeing in them a romantic sexual thread. The story is more important than its shaky truth value.

Leakey worked hard to fund the projects of these women scientists, and much of the energy of his last years was given to fund-raising for primate behavior studies. He turned often to his friend, Leighton Wilkie, senior partner of Wilkie Brothers, inventors, manufacturers and distributors of machine tools in the U.S. and Canada. Wilkie gave money for Jane Goodall because he was fascinated by the prospect of her finding evidence of chimpanzee tool using. At the Pan African Congress on

Prehistory in Rhodesia in 1955, Wilkie had been much taken with Raymond Dart's blood-splattered version of a hunting, tool- and weapon-using human origin, which Dart had read into the famous early australopithecine fossils from South Africa (Cole 1975: 213). Goodall did not fail Leakey's sponsor, promptly finding both tool-making and meat-eating at Gombe. In a tour of the U.S. in 1959, after Mary Leakey's discovery of the jaw of "nutcracker man," *Zinjanthropus*, Leakey arranged to meet the president of the National Geographic Society. Leakey inspired the men of the National Geographic and drew on them frequently for his projects. The NGS made Leakey and his protegees household figures. In 1968, Leakey's friends founded the L.S.B. Leakey Foundation for Research Related to Man's Origins, which often funded Leakey's renegade projects. Lecture series sponsored throughout the 1970s and early 1980s to raise money and popularize the ape research were extremely successful. This ideological, philanthropic, and sexual nexus made the gender coding of the ape science of Goodall, Fossey, and Galdikas more than a semiotic game. Grounded by that nexus, and by a global "ecosystem" crafted by Gulf Oil and the other Seven Sisters, the narratives' gender and science coding became the key to the primate order for millions of western readers and viewers.

Race

"Race"—the prominent *whiteness* of the women, refusal of the signs of full humanity to people of color, and the "Third World" status of the animals—works as the third essential sign in the National Geographic system.[15] Repeatedly in the stories, ritual phrases like "the strange pale ape intruder" announce that it is not simply man who has entered the forest in the body of woman, but white man. "Wild chimps flee the pale-skinned stranger invading their domain." "It means that not yet can the blond stranger draw near" (soundtrack, *Miss Goodall and the Wild Chimpanzees*, 1965). It is the strange pale intruder who enacts the drama of touch. The claim is made that, thereby, "Man" is brought into touch with his origin and nature. The marked quality of the pale ape trailing the gorillas, chimpanzees, or orangutans, or of the blond woman washing her hair in the pure stream (Goodall 1963b: 284) disappears in the final achievement. If whiteness were too much emphasized, the universal nature of the saving act would be compromised. The race marking is *sotto voce*, but essential. *White* cannot be said quite out loud, or it loses its crucial position as a precondition of vision and becomes the object of scrutiny. Yet, whiteness must be attended to, if the re-entry of the west into Africa at the moment of decolonization is to be narrated. It is western, scientific, European, and Euro-American "generic" man that is really at issue here, not as he incorporates (white) woman, but as he is represented by her. She is his surrogate. It is he who has been excluded from "nature" by both history and a Greek-Judeo Christian myth system; and more immediately, he is being thrown out of the garden by decolonization and perhaps off the planet by its destruction in ecological devastation and nuclear holocaust. It is time to call in the blond and female mediator to negotiate the discourses of exterminism and extinction in space and the jungle. The animals are (colored) surrogates for all who have been colonized in the name of nature and whose judgment can no longer be repressed.

It would hardly do to address this specific "white" dilemma, which has global consequences, with a woman of color, scientist or not.[16] It would have seemed strange to dramatize the exclusion of Third World people of color from touch with the animals when they had just succeeded in gaining the political right to control their own territories, within which these very primates lived. Third World people had quite different origin stories to tell in the 1960s: nationalist stories. A western woman of color would have made the cultural particularity of the story of the woman in the forest too obvious because she could not have as easily carried the ambiguity of marked and unmarked categories required for the narrative resolution of the dilemma in universal terms. Both the specific dis-ease of white westerners in the post-war world and that race's continuing practice of reading its history as the story of the family of man made the pale ape's color significant. The color was sufficiently critical to bring the shading of the representatives of "man" to the edge of explicit visibility, an extraordinary condition for members of an unmarked category, a condition that ran the risk of leading to questions about what maintains the "normal" invisibility of the race of "whites," but not of people of "color." Just what about race is visible to whom in the primate story? How was the race of people of color displaced onto animals and the generic status of man displaced onto white women in the decades after World War II, when race as a scientific object was crumbling? There was plainly more than the idiosyncracies of young white women's personal histories that made several into heroes for National Geographic in the 1960s and 1970s.

There is another aspect to the working of the color code in National Geographic's narratives: in the history of the sciences of man, including anthropology and prima-tology, people of color were constructed as objects of knowledge as "primitives," more closely connected to the apes than the white "race." The concept of race itself was inextricably woven out of the history of the conjunction of knowledge and power in European and Euro-American expansion and economic and sexual exploi-tation of "marked" or "colored" peoples. Race as a natural-technical object of knowl-edge is fundamentally a category marking political power through location in "na-ture." People of color could not mediate the required touch with nature that could reassure "man" within the myth and science system of National Geographic and the Seven Sisters because they were still implicitly (if no longer officially) assigned to a lower rung on the chain of being, insufficiently differentiated from the nonhuman primates. In addition, their oppositional political and economic activities in and around the "garden" forest (or in the urban "centers" of civilization) made it difficult in hegemonic white media discourse to code a person of color as a lonely hero traveling across a great temporal and spiritual gulf to receive the animal's gift of touch. Black people *were* the beast; it is written into the history of lynching and into the history of biology. It is their accepting touch, coded back onto the animals, that was so ardently sought in the ongoing western narrative of threatened apocalypse.

In a broad range of popular and official discourses in the west, perhaps especially in the former white settler colonies like the U.S., *white* is a color code for bodies ascribed the attribute of *mind*, and thus symbolic power, not to mention other forms of power, in social practices like "intelligence" tests. The *body* is coded darker, denser, less warm and light, less constituted by number. Who may be seen as scientist, as mathematician, as any "genius"? Without its own principle of order, the body is properly the subject of control and the object of appropriation. Women and animals are set up *as* body with depressing regularity in the working of the mind/body binarism in story fields, including scientific ones. The man/animal binarism is crosscut by two others which structure the narrative possibilities: mind/body and light/dark. White women mediate between "man" and "animal" in power-charged historical fields. Colored women are often so closely held by the category *animal* that they can barely function as mediators in texts produced within white culture. In those cultural fields, colored women densely code sex, animal, dark, dangerous, fecund, pathological.[17] The "body" in revolt is often accused of irrational "terrorism." In the United States, political imaginations of white people color terrorists as dark and dark people as dangerous. The *body* in western political theory is not capable of citizenship (rational speech and action); the body is merely particular, not general, not mindful, not light. It is sex (woman) to mind (man); dark (colored) to light (white). The body is nature to the mind of culture; in primate narratives, white women negotiate the chasm.

The period of the early post-war field studies of monkeys and apes was a time of instability and reformulation of the categories of race in biology and anthropology. But the touch across the "immeasurable gulf of time" from gorilla to scientist-woman implicitly still required one of the hands to be white if it were to represent humanity for the National Geographic audience. White people could be considered to be "alone in nature" when they were prominently in the immediate company of black people. The failure of people of color to count as human could hardly be more plain. In *The Search for the Great Apes* (1975), Dian Fossey is introduced as a woman who has lived alone for nearly a decade. She is then shown listening to an African man playing a wind instrument for her and his wife and child. The sound-track announces immediately after that "at long last the mountain gorilla sanctuary stands open to one human being." The nearby Rwandan village is announced as the "only real link with the outside world." African people are links to "human," i.e., European, human community; and they are irrelevant to the community sought with the animals. Even visually present, they are continuously absent. The same point is made by the elaborate naming of nonhuman primate subjects in the field studies, accompanied by the precise naming of all white scientific personnel with first and last names. Third World people are either not named at all, referred to as field staff without differentiation, or given first names. If naming and portraiture are the signs of rational selfhood for the great apes, those signs were not affixed to people of color working in the research projects until the late 1970s.

European culture for centuries questioned the humanity of peoples of color and assimilated them to the monkeys and apes in jokes, medicine, religious art, sexual beliefs, and zoology.[18] [Figure 7.6] That a person of color would seek a healing touch from these animals borders on the absurd. The romance of the women of National Geographic was also a romance of race, with its plot mapped out in the history of western expansion. Woman was closer to nature than man, and that was required in the story, but woman had to be white, or the closeness would have underlined a different and disturbing construction of nature. The scandal of racism would have erupted within the film's narrative, breaking its harmonious surface. Contemporary history would have erupted instead of timeless origins.

Finally, the color coding of the primate story also reflects the difficulty students of color have had in obtaining scientific credentials in the countries that sent out their scientists to study wild animals. Neither Goodall nor Fossey was a science student when she began. But their stories were the exception, and Leakey set them both up in a prestigious Cambridge doctoral program. Most of the field workers

Figure 7.6 Tarzan (Johnny Weissmuller), Jane (Maureen O'Sullivan), Boy, and Cheetah.

from the late 1950s were graduate students or post-Ph.D. scientists with philanthropic or government grants. In the United States, black students who could go to college and who had an interest in science were channeled into medicine, not whole organism, ecological, and evolutionary biology. For example, the first black Ph.D. in developmental biology, E. E. Just at Howard University, in the early decades of the twentieth century was actively forced by the big foundations and leading white scientists to devote his energies to training black medical students, an activity that greatly hindered his own research (Manning 1983). Shirley Malcolm, head of the AAAS Office of Opportunities in Science and herself one of the few black doctoral behavioral biologists in the U.S. in the mid-1980s, noted that a high proportion of black bachelors degrees are granted by the traditionally black colleges. Catalogues from those colleges list few courses in whole organism or ecological-evolutionary biologies. It has been very difficult for a black student to be exposed to the whole area (Malcolm, telephone interview, Sept. 1985; the specific factors for students from other U.S. communities of color reflect other histories). In addition, for many students with religious backgrounds common among Afro-Americans, courses in evolution are more threatening than other aspects of biology. The degree to which students choose not to enter fields that had been historically most racist, such as physical anthropology and primatology until mid-century, is difficult to measure, but not an unlikely factor. What is clear is that people of color have not authored the primate story.

In summary, in all of these stories humans from scientific cultures are placed in "nature" in gestures that absolve the reader and viewer of unspoken transgressions, that relieve anxieties of separation and solitary isolation on a threatened planet and for a culture threatened by the consequences of its own history. But the films and articles rigorously exclude the contextualizing politics of decolonization and exploitation of the emergent Third World, obligatory and normative heterosexuality, masculine dominance of a progressively war-based scientific enterprise in industrial civilization, and the racial symbolic and institutional organization of scientific research. Instead, the dramas of communication, origins, extinction, and reproduction are played out in a nature that seems innocent of history. If history is what hurts, nature is what heals. It is precisely to renaturalize "man" after the calamities of advanced industrialism, and especially after the bomb, that apes and (white) people are placed together in *both* the "natural" world of the forest and the "cultural" world of language-users and pet-keepers. This union is a crucial part of the ideology of relief of "stress," a failure of communication that has been constructed since World War II. The nature-culture myth is restored by the lush, filmic mediation of tropical animals and white women. The myth appears universal, not the very expensive story of one people addressed to members of that same group.

And finally, history is also excluded from the location that is neither nature nor culture, but beyond both frontiers, in space. The cyborg chimps and monkeys, products of communications engineering, are not presented to the mass popular audience for their activities as understudies of Cold Warriors, but of "mankind" venturing bravely into a new world, where new freedom is promised. That new freedom, new frontier, is premised precisely on the escape from history. In this scenario, it is not the forest, nor sign language and kittens, that mark the history-free locus. Rather, it is marked by technology itself—in the frozen and fetishized form of the artifacts of the space race, concealing the particular human agency, craft, and social relations that both made the machines and are built back into them. Thus mystified, technology decontextualizes a captured and orphaned chimpanzee child and designates him by an acronym, HAM, if he is not first aborted in a mission failure, and launches the future in the space fetuses of *2001* and *Primate*. Here the gender code is rigorously masculine; the story is about man's escape from the body. Space flight, and the primates engaged in it, is narrated not as a chapter in contested and partial history, but as cosmic project. "The choice is the universe—or nothing" (Bryan 1987: 352). The line between science and science fiction blurs, reading out an oppositional system of meanings and practices called history.

The National Geographic: Readers and Writers

Readers

Robin's gentle grooming touch of Shirley Strum's cotton shirt not only thrilled the anthropologist taking her data; it thrilled millions of magazine readers; National

Geographic claimed "six million member-families" in the instructions for "Writing for *National Geographic*" that the Society sent Strum, a prospective author, in 1972. All of the *National Geographic*'s women scientists have been mediators of deep pleasure for millions of readers. To understand how that pleasure is constructed, let us look briefly at the institution of the National Geographic Society and at one writer's effort to control the editorial process.

The crucial years of policy making for the *National Geographic* magazine began after 1898, the year of the Spanish-American War, when Alexander Graham Bell became president of the NGS.[19] The war enhanced the importance of geography for United States citizens, and articles in the magazine after the war stressed the benefits of colonialism, geography, and the commercial possibilities of America's new possessions. Committed to forging a national society from the scattered local branches and creating a broadly popular magazine, Bell brought in his future son-in-law, Gilbert Grosvenor, as assistant editor. With little sympathy for professionalism, Grosvenor, who set policy until 1954, had the task of increasing circulation. Under his son Meville Bell Grosvenor, his policies have essentially remained in force into the 1980s.

He was wildly successful, with a formula that melded science and adventure and brought in, not mere subscribers to a magazine reading passively to cheer on elite science, but member-families receiving a card assuring them of election to a scientific society. Grosvenor's formula was not *popularization* of science, which leaves the reader with little to do but acknowledge the legitimacy of distant activities watered down for the "lay person." Rather Grosvenor stressed *participation* in doing science, in a nineteenth century sense. Generations of U.S. newlyweds have received as wedding gifts from their families and friends certificates affirming their membership in a scientific society, receiving monthly a "vividly illustrated" document with "authoritative reports on the many fields of geography, science, and natural history . . . to be read, reread, and carefully preserved as a work of reference" ("Writing for *National Geographic*," 1972). The professional geographers initially fought the changes, but they lost, retiring to form another scientific geographic society.

But participation did not mean setting policy, which remained securely with the editor and the board. Participation meant a particular reader self-image and a style of sponsoring and reporting scientific research. In 1908, the NGS sponsored the Peary expedition to the North Pole, arbitrated disputes about the authenticity and priority of discovery, and became a "potent force in the fields of exploration and scientific research."[20] It was an anomalous force in the twentieth century, when scientific professionalism swept away most popular natural history societies and created the category of "lay people" with the status of voyeurs and supplicants. Readers of the *National Geographic* are certainly voyeurs; its supremely lush photography is probably the single greatest factor in the magazine's popularity. But they are not supplicants at the temple of science. People got—and wanted—a sense of "immediate experience of the world, not the systematic lessons professionals sought to provide" (Pauly 1979: 527). The annual membership fee pays for the magazine, but also funds the NGS's many research expeditions. Grosvenor claimed that the reader gets "first-hand accounts in his [sic] own magazine" (530). The reader is asked to identify, to have an adventure, not to acquiesce in a distant elite undertaking. Never has the genre of science as travel literature been more evident and popular.

Scientific primatology's continuing anomalous status, never quite demarcated from popular and contentious debates about human primates, made it the perfect subject for the NGS's brand of democratic, participatory science. Primatology's popularity was enhanced, while removing the politics from view. Dedicated to eschewing controversy, the NGS in sponsoring primate studies provided what millions of people, whether scientists publishing only in professional journals like it or not, understand to be the latest word from the sciences of monkeys and apes. It was this complex and popular research tradition that provided part of the funding for the primate studies of Jane Goodall, Dian Fossey, Biruté Galdikas, Francine Patterson, Allen and Beatrice Gardner, Jane Teas, Shirley Strum, and others.[21] The magazine and films made Fossey, Galdikas, and above all, Jane Goodall, into household figures.

As Pauly notes, the magazine was, and is, a first cousin to the natural history diorama, a species of "naked eye science." But the post-World War II still photography and motion picture film make a world of difference in the power of the proferred participation. The NGS is a master narrator, steeped in the techniques of giving visual pleasure and satisfying the desires for pleasure through the filmic rhetorical strategy of identification. The viewer of the diorama was spiritually a young man learning citizenship in the state of nature. However, the reader of NGS films, photos, and prose may be explicitly or spiritually female. Generations of young women have identified with Jane Goodall and the other women, who have vividly portrayed how to be a scientist, an adventurer, and a woman, all at the same time. Coded as participation in the "life of the mind," the vocation of scientist has been particularly marked as masculine in European and Euro-American culture; to be a scientist woman has been a contradiction in terms. The National Geographic stories have recoded the vocation of science for a mass audience. This recoding of science and gender was part of a rearrangement of the terms of citizenship in both nature and culture in western myth and political systems in the last half of the twentieth century.

Writers

All of the National Geographic women scientists play a gender-coded rhetorical role in the magazine and films, irrespective of their personae in other publications or their own agenda for the NGS media. But Strum's article is particularly interesting, since she successfully managed *not* to be coded as a mother, and still her article is a powerful vehicle for gender-marked scientific rhetorics common to the post-World War II period. Strum, whose attribute is the data sheet, continues to embody nature's acceptance of scientific man, represented by scientific woman. Strum fought hard with the editors of *National Geographic* to avoid the image of the primate mother for herself and to present the animals with the dignity and complexity she felt they deserved.

> I was so involved and committed to the baboons that I didn't want that [presented first] with this dramatic woman in the wild . . . . I wanted women to have a model of something to do that was legitimate. . . .The majority of women who work in the field are like me. Serious about it, they may have attachments to the animals, but I felt it was unfair . . . not to show that this was a serious endeavor that a woman could do, and do it as science, with all the rigor. I wanted to counter the existing

stereotypes because I thought here were opportunities for women that were much more satisfying than [the romantic image].[22]

Strum did manage to keep herself as an icon covered with the attributes of primate motherhood off the cover, which instead featured a baboon infant clinging to its mother, resisting the curious overtures of a juvenile. "The thing I stood firm on was romanticized pictures of me. . . . It's doing work, not posed" (interview, 29 June 1983). The only picture of Strum in the article was a side view of her walking quietly within the baboon troop, with a clipboard and camera. The caption was, "Not a boring moment in 16 months" (Strum 1975: 681). [Figure 7.7]

The appearance of easy immediacy in the NGS article is belied by Strum's account of extensive negotiations about the content and rhetorics in the presentation of the baboons in the Pumphouse Gang.[23] She recounted that the idea for the article came from her "consort," Timothy Ransom, who went to the National Geographic to interest them in his taking photographs of the baboons. Ransom (1971) had just finished a Ph.D. based on research on Gombe baboons, but wanted out of academics. The NGS gave him film with the promise that if the material looked interesting, he could do the pictures and Strum, still a graduate student, could do the prose article. Strum stressed her naivete in the first stages of relation to the magazine. An editor "wrote back and said, 'I didn't know she was that attractive. Take lots of pictures of

Figure 7.7   Shirley Strum watching baboons in Gilgil in the 1970s. Photograph by Timothy Ransom. Published with permission of Shirley Strum. Unreadable as an icon of the first family in Eden, this image shot from behind the observer with her clipboard was published as the photograph of the author on the back of the dust jacket of Strum's popular book, *Almost Human* (1987).

her with the baboons.' That should have warned me!" (interview). Strum wrote her first version, about the saga of the male baboon, Ray, joining a troop by first forming "special relationships" with the resident females. An editor wrote her on personal stationery to say she loved the article, but Strum should be wary about what would happen to it. Strum recounted that the magazine "tried to focus on me [and] made up anecdotes about the animals. They just completely made them up" (interview). The guidelines for writers suggested an "apt quote" every 250 words. Strum countered that if they wanted to hear what she had to say to the animals, they could have their quotes. The rewriting was underway. Strum said an editor later told her she was the most difficult writer he had ever worked with.

Several drafts later,[24] the article was at the typesetter, and Strum was on the phone at midnight with Washington, D.C., angry enough to insist her name be removed from the piece. So they negotiated some more. "I gave them 'ivory daggers' and they took out 'feminine mystique.' " In the published version, the threat-displaying baboon on p. 685 is introduced with, "Ivory daggers make formidable adversaries of adult males." But on p. 680, Strum's midnight battle resulted in the section headline, "Sexuality Leads to Fray." Strum commented wryly, "Not a whole lot better" (interview). Comparison of photographs before and after touch-up shows another telling production detail. On pp. 682–83, there is a photo of juvenile baboons peering in the window of the author's house at a Siamese cat. On the untouched photo, there is a heavy baboon urine stain on the windowsill and wall below; on the touched up print, the stain is gone. Male aggression was given a dramatic double-page display, headed "Bare-fanged fury" and highlighted by striking red flowers in mid-field. Aggression and excretion both were aestheticized, but the lesson from each is different. One is dramatized, the other erased. All body functions are not equally beautiful and instructive in nature! However, overall, Strum felt she won in the negotiation with National Geographic, leading to an article that emphasized the baboon's social agency and complexity.

Primate Pictures: Monkeys and Apes in the Movies

The Virgin and the Beast

In 1933, movie-goers were treated to the scene of the monster gorilla, King Kong, tragically falling in love with the lovely blond woman, Anne Darrow (Fay Wray). Kong had regularly extracted the tribute of a beautiful maiden from the savage brown people he ruled on a lost island recalling darkest Africa. But he had never seen anything as ravishing as the pure white woman. After she was rescued from him by her daring white admirer, the captured giant beast was brought back to New York City, to be displayed in shackles to the pleasurably horrified urbanites. But the love-besotted Kong broke his bonds and retook his prize. In one of the most famous movie scenes of all time, Kong, atop the Empire State Building, was killed by airplane attacks—but only after he put his beloved gently out of harm's way.

In 1982, I received a Hallmark greeting card, designed by Nancy Carlson, which pictures a dimmutive silverback male gorilla in bed, covered by a pink blanket trimmed in violet ribbon.[25] [Figure 7.8] A framed picture of pink flowers graces the blue-papered wall above the head of the bed. The embarrassed and frightened dark

Figure 7.8 "Getting Even," by Nancy Carlson. Copyright Hallmark Cards. Published with permission.

ape clutches the blanket up to his chin, trying to hold his cover against the tug from the bottom of the bed. There a red-nail-polished white hand, its wrist sporting a broken shackle, pulls at his blanket. Filling the broken window is a huge, blond, blue-eyed face. New York's skyscrapers fill the night sky visible beyond the giant woman. The bright red lips are rounded in a gesture simultaneously suggesting a kiss and curious astonishment. The eyes are wide open and round, peering intently at the recumbent victim. The card is titled "Getting Even."

The two images of the woman and the gorilla are at the two poles of the history of popular images of apes in urban America. Both images must be read with the triple code of gender, science, and race. Together, the images form an ironic diptych summarizing arguments in this book. The 1933 Kong is the image of the gorilla Akeley struggled all his life to replace with that of the Giant of Karisimbi. But the noble love and virile sexuality of Kong, turned to the true worship of the white woman, demanded that the intended Euro-American audience see Kong as a tragic hero too. Kong crossed the boundaries of his species in a tragic overreach, but his sexuality was admirable. Kong sought consummation in the protective possession of his innocent female prize, and thereby established his essential—if masked by the form of the beast—humanity; he could be a father of a new and better race. But his bestial over-reach also had the unmistakable tone of racial crossing; for the white inflamed imagination he was the icon of the captive black man's love for the white woman. Beast and "primitive," Kong was lynched. He was the object of discovery, the prize brought back by an adventure safari, on which the chief white male hero is a film producer. The technology of the camera at the beginning of this self-reflexive movie, constantly commenting on the terms of filmic vision, was complemented by the technology of the airplane and the skyscraper at the end. The entire story of King Kong and the virgin was embedded in science-based urban culture, where the dark male beast in chains threatened escape and destruction. That beast could only be conquered by abstract, technical force, i.e., by the instruments of modern war. He was more powerful than naked white man without his cameras and planes.

Hallmark's Kong looks like a stereotype of a gay gorilla about to be exposed by a Marilyn Monroe sex goddess. Her gaze and implied kiss do not evoke terror and rapture, the proper response of Anne Darrow, and of Jane to Tarzan, but rather blanket-clutching embarrassment. When the lady is so much bigger than her beloved, the gender code does not allow the noble sexual theme of tragic species overreach. Marrying "down" for women seldom does. Galdikas's "first-born" orange-colored son, Sugito, has more tragic stature for his sexual confusion than does the overreaching blonde goddess. Her gaze unmans the beast, just as the scientific gaze of the female man recodes the gender of both nature and science. The usually hidden axes of homophobia and gynophobia in the ubiquitous "normal" theme of heterosexual union become prominent when the sexual roles are confused. The greeting card maintains the implicit trope of heterosexual union for the gaze of the knower upon the object of desire, but the Hallmark Kong story is comical, not tragic. The broken shackle on the woman's wrist is a sign of women's liberation. She, like Kong in 1933, broke free of a captivity where she was held to be seen by the "public." Like Kong, in her freedom she pursued her object of desire. But her achievement takes place at the level of getting even, a petty drama. The racial coding is also obviously reversed, but the meanings become comic when the black beast is in the urban bed and the blonde virgin mounts the skyscraper. The 1980s card draws unconsciously on the traditional binary pair structuring white prceptions— the feminized black man and the savage beast (Stepan 1982). Civilization and white patriarchy are not at stake in this private revenge and private shame—in a hotel bedroom, not on a public monument. The Cruise missile will not enter this domestic scene to save a black homosexual. The reversal of the positions of King Kong and the blonde virgin in the 1980s card cannot simply reverse the semiotic values of the tragic 1933 film and its classic myths of race, gender, and knowledge; it turns them into farce, the risk run even by white scientific women in public. This is the risk run by National Geographic's Jane Goodall.

From Entertainment to Edification

From King Kong and his mutants, let us turn to the context of the history of scientific, popular, educational primate films. With *Monkey into Man*, a 1938 film produced by Solly Zuckerman and Julian Huxley under the sponsorship of the Zoological Society of London and the Paris Musée de l'Histoire Naturelle, the entertaining primate educational film made its debut. *Monkey into Man* opened with the organ grinder monkey, a laughing human audience, and the narrator's deep voice cautioning that the tendency to laugh at monkeys ought to be replaced by learning the important evolutionary lessons that nonhuman primates have a definite social order "arranged like our own, [with] a family with a male head as the basic unit of all society." The camera pans the Hamadryas baboon colony of the London Zoo, a group infamous for having killed most of its babies and wounded just about all the females and males in vicious attacks provoked by conditions of captivity. But that is not the description given by the narrator, who sets the familial theme for the primate order. From the baboons the film rises to the apes, where, with its ability to substitute a human parent figure for the putative dominant male in the wild, the chimpanzee models an evolutionary transference. The result of this transference is pedagogical success; the chimpanzee can learn to act like a human in captivity, showing great intelligence. The filmic scene has chimps at table eating with more or less proper western manners. We then move to the famous ape sense of rhythm, with long shots of the humanoid male chimpanzees "dancing" in their cages to the "primitive" beat of a drum on the soundtrack. Cut to the "larger, more terrifying, man-like gorilla" and intimidating shots of two males fighting. Race, gender, science.

Quickly, speech is added on, and we get Man, in the person, naturally, of a naked black savage. Cut to Greek temple ruins and then to the modern city with an airplane. "The physical world becomes [man's] servant." But the "family remains just as it was among the apes and monkeys." The reassuring conservatism is illustrated by a baboon mother grooming her child, while the narrator continues, "In the family circle we learn pride in our outward appearance." The airplane notwithstanding, the narrator stresses that the family circle is "our" (that slippery first-person plural, forced complicity again) highest achievement, remaining constant year after year, providing "certainty amidst chaos." It would be hard to find a clearer, more edifying statement of the best scientific opinion of the day about the great chain of being, the family, race, and technology—all in less than 30 minutes. This was not exactly anti-fascist science in the late 1930s. Absent from the film as from Zuckerman's books, was any call for sustained field work to see what wild primates might do with their lives. There were no questions in this film, only reassurances.

Yerkes's post-doctoral student, Clarence Ray Carpenter, produced the films appropriately compared to Zuckerman and Huxley's. But Carpenter's films were made with a sober pedagogical purpose, not for a wider popular audience. Carpenter's films, begun in 1940s and based initially on his pre-war field work and footage, initiated a complex tradition of primate research and teaching films.[26] Carpenter's several films were organized like his monographs: first a context-setting ecological overview, a survey of postures and locomotion, and a sociometric and socionomic analysis of each age-sex category within the whole group, beginning with the adult males and proceeding methodically through the social membership roster, ideally looking at every possible pair-wise type of age-sex class interaction. Carpenter believed film could not lie, that it gave a kind of flat truth about the animals; his editing technique came as close as possible to convincing any viewer of the flattest of truths, fact by fact. The virtue of these films was their explicit search for species differences, for the lives led by the simians, and an overt silence about the lessons for human family life and technical progress. These films show what a believer meant by objective description.

Beginning with the new generations of field workers in the 1950s, primate research and teaching films flourished. Sherwood Washburn and Irven DeVore produced *Baboon Behavior* (1966), *Baboon Ecology* (1963), *Baboon Social Organization* (1963), and a summary version called *The Baboon Troop* for social studies curricula. The National Geographic films quoted directly from these films, e.g., in *Monkeys, Apes, and Man* (1971). First of their type, they set a standard and canonized images of baboon troop life, including the social glue of the attractive black infant and the dominant male; the mother-infant bond; the aggression, protective function, and cooperative dominance hierarchy of adult males; and the image of the ecological scene of hominid evolution on the savannah. Social group functionalism predominated, in harmony with the research program of the Washburn lineage in the late 1950s. The theme of the peripheral juvenile and adolescent male rang through the films, sounding a familiar chord for those student viewers in the 1960s and 1970s in the U.S. who responded to the themes of the long lonesome road.

In the same film generation, George Schaller and C.R. Carpenter collaborated on *Mountain Gorilla* (1959), an utterly pedagogic film in Carpenter's inimitable style. The gorillas have no names, but are referred to as "it" and "the male" or "the female." The organizing theme of male dominance dramatizes the generalization that dominance is to primatologists what kinship has been to anthropologists—at once the most mythical, most technical, and discipline-grounding of a field's conceptual tools.[27] The film should be compared to National Geographic's nearly contemporary television specials. There, the viewer left the classroom, headed for the living room, and settled in for the masterpieces of primate drama, where the generic lines between research, edification, and spectacle were sutured in the mode of participatory science and adventure. But between the television set and the avid viewer is the history of a primate research site, the Gombe National Park of Tanzania. Here we find not apes in Eden or apes in space, but apes—and people—on earth.

Intermission at Gombe: History of a Research Site[28]

In 1957, Jane Goodall met Louis Leakey. She began as an assistant secretary for him at the Coryndon Museum of Natural History (later the National Museum of Kenya) in Nairobi, where Leakey was curator. After her stint in the archaeological-paleontological digs at Olduvai, Leakey suggested that the 23-year-old Goodall undertake the study of wild chimpanzees at the Gombe Stream Reserve. Leakey expected that the lakeshore animals would be particularly relevant to reconstructing the hominid drama of origins, since he had found many early human remains on lake shores. Leakey did not share the vision of hominid origin on the dry plain, exiled from the garden. In 1960, with the help of the Game Ranger, David Anstey, and others, Goodall and her mother, Vanne Goodall, arrived at Gombe, 12 miles from the town of Kigoma. Goodall was traveling with her mother because Tanganyika officials had refused permission to a young British woman to work "alone" (without European company) at the reserve.

En route to Gombe from Nairobi, the two women were delayed in Kigoma by an event that silently marked the post-World War II primate literature: the successful revolution against colonial rule in the Belgian Congo. Goodall's account of the events reflected the view of her culture: "Violence and bloodshed had erupted in the Congo"; and "First, it was necessary to wait and find out how the local Kigoma district Africans would react to the tales of rioting and disorder in the Congo" (Goodall 1971: 26–27). Boatloads of Belgian refugees filled the town of Kigoma, 25 miles across Lake Tanganyika from the Congo. There, while George Schaller was forced to abandon his study of mountain gorillas in the Virunga Volcanoes of the Congo, Jane and Vanne Goodall spent their enforced delay in traveling to Gombe by fixing 2000 spam sandwiches for the fleeing Belgians (Goodall 1971: 27).

A decade after the events, *In the Shadow of Man*'s chapter on "Beginnings" for the Gombe chimpanzee research is silent on the fact that 15 African nations achieved independence in 1960 alone. Those episodes of "rioting and disorder" had fundamental consequences for the primate story.[29] But Goodall's "beginnings" were for a different narrative.

In the beginning of Goodall's work at Gombe, the reserve with its lake shore was home to two game scouts and "the few Africans who had permission to live permanently at the reserve so that the scouts should not be completely isolated" (Goodall 1971: 31–32). Just to the north was a substantial fishing village. Incredulous that a young woman would travel from England merely to look at apes, the villagers suspected that Goodall was a government spy. David Anstey arranged that Goodall be accompanied on her field observations by a game scout and by the son of the chief of the fishing village. Anstey also recommended Goodall employ an African to carry her haversack, "for the sake of my prestige" (33). Convinced the chimpanzees would never make contact in such a crowd, the British woman was dismayed at the prospect of all this company. The chief's son's company was intended to ensure that she did not write she had seen ten chimpanzees, when she had seen one. "Later I realized that the Africans were still hoping to reclaim the thirty square miles of the reserve for themselves" (Goodall 1971: 33).

Searching for "man's" origins and motivated by a love of animals that characterized her since childhood, Goodall entered the political world of Africa at the moment of decolonization. Her description of the greeting she received as she set foot on the shore of the reserve is a classic of colonial discourse: "We stepped ashore, splashing into the sparkling wavelets, and were greeted . . . , with great ceremony, by the honorary headman of Kasekela village, old Iddi Matata. He was a colorful figure with his red turban, red European-style coat over flowing white robes, and white beard. He made a long speech of welcome to us in Swahili, of which I understood only fragments, and we presented him with a small gift that David had advised us to buy for him" (32). Her innocence and the political complexities of the situation, mediated for her by Anstey, are marked in the juxtaposition of fragments of prose from different genres—"first contact" rhetoric of the colonial explorer and phrases of political analysis appropriate at the inflection point, where colonial discourse is forced into the syntax of a post-colonial Africa.

This syntax is underlined in the acknowledgment structure of Goodall's book, as well as in the acknowledgments of research permission in professional primate literature since various primate habitat countries gained independence. In Goodall's 1971 book, the first acknowledgment is to the father of the research, Louis Leakey, then to the husband, Hugo van Lawick, and then to the father of newly independent Tanzania, President Mwalimu Julius Nyerere. Nyerere and Iddi Matata stand at the two poles of western contact literature in Africa: the nationalist president and the village headman. Anstey, the mediator of the human social network laced between these poles, is the fourth person acknowledged. The syntax of patriarchy was stable through the many narrative genres present in Goodall's widely read book, aptly titled *In the Shadow of Man*. The chimpanzee David Graybeard would take his place in this patrilineal network. He and the model primate mother, Flo, were the last individuals acknowledged: "To David Graybeard and Flo in particular we owe much" (Goodall 1971: 12).

Goodall discouraged the son of the chief of Mwamgongo from accompanying her, so her work was initially conducted with the assistance of the game scouts, identified in her publications as Adolf, Saulo David, and Marcel. In addition to herself and Vanne, a cook identified as Dominic, along with his unnamed wife and daughter, camped at Gombe. Hassan, Leakey's boatman, soon joined the project and stayed over 10 years. Her mother remained about 5 1/2 months. Her sister visited early in the research. Goodall's 1971 acknowledgments listed about 20 Africans who aided the research in some fashion, as well as an extensive list of white colonial, European and U.S. officials, observers, relatives, or consultants. Out of this dense social world, National Geographic fashioned its story of the young woman alone in the wilderness.

Recommended by Louis Leakey, the Baron Hugo van Lawick arrived at Gombe in 1961, after Goodall agreed to permit a professional photographer to record the study. "Louis wrote to tell me about Hugo and his abilities. At the same time he wrote to Vanne telling her that he has found someone just right as a husband for Jane" (Goodall 1971: 82). Born in Indonesia, educated in England and Holland, and experienced in wildlife photography from two years on the television program "On Safari," Hugo van Lawick became central to the work at Gombe, returning for a nine-month filming stint in 1963. Goodall and van Lawick married in Europe at Easter time in 1964. In 1961–62 Goodall began graduate work at Cambridge under Robert Hinde, visited England for a six-month period of study, and gave papers in New York and London.[30]

In 1962, the adult male chimpanzee, David Graybeard, already a special figure in Goodall's narrative of first contact, began visiting the camp itself, soon bringing other chimps with him. Growing out of those visits, in 1963, in order to observe the shy chimpanzees better, since they had proved very difficult to follow through the dense and steep forest country, Goodall initiated a banana feeding station, where more systematic records were begun on infant development and social interactions, interpreted in terms of status hierarchies. Banana feeding became the focus of chimpanzee observations, and in late 1964 about 45 chimpanzees visited camp regularly for the treats. The regular feedings were discontinued in 1968, when the disorder caused by the frequent feedings and dense social aggregations of chimpanzees and baboons got badly out of hand (Wrangham 1974). Permanent camp buildings were under construction in 1964, and the Dutch woman, Edna Konig, joined the project as the first assistant. Sonia Ivy was soon employed for secretarial help. These two women made the study observations when Goodall and van Lawick left Gombe in March 1965 for two terms at Cambridge University. That year inquiries from international students began to be made about coming to Gombe to observe wild chimpanzees, and the first National Geographic film on Gombe was completed. In 1967, Gombe became a National Park, with all the long-range implications of that status for park visitors and wildlife conservation politics.

Goodall's Ph.D. was granted in 1966, and she and van Lawick returned to a Gombe about to be transformed into an international research site avidly sought by large numbers of students. Between June 1960, and March 1965, Jane Goodall had spent 45 months at Gombe, with only a very few other chimp observers. Their status was unambiguously as her assistants, her family, or game reserve personnel. In 1966, Caroline Coleman, a graduate student from Bristol; Sally Avery, who had helped operate a dairy farm in Kenya; and Alice Sorem (Ford) and Patrick Mc-Ginnis, graduates from the University of California at San Diego, arrived at Gombe.

They were soon followed by Geza Teleki, C.R. Carpenter's first primate graduate student, and Patricia Moehlman, recent graduate of the University of Texas, who went on to do a Ph.D. with John Emlen at Wisconsin on wild asses. Their arrival marked the beginning of an international network of students, who have become professional colleagues in primate field studies and beyond. Their impact on primatology has been strong in both theory and field practice, and their social networks mediate the most exciting intellectual currents in the discipline years after Gombe has been essentially closed as a long-term research site for foreigners. Former Gombe students became prominent in behavioral ecology, socio-ecology, and socio-biological approaches to primates and other social mammals.

Between 1966 and the closing of Gombe in May, 1975, as a result of the kidnapping of four students by guerrillas from Zaire, about 60 European and Euro-American undergraduates, graduate students, and doctoral scientists came to Gombe. They came primarily to study chimpanzees, although a few came to observe the Gombe baboons or red colobus or to do other ecologically oriented projects. Initially, students came from several places, but two institutions became dominant because of formal connections: Stanford and Cambridge. [Figure 7.9]

Beginning about 1972, Goodall and David Hamburg, an advocate of experimental animal model research in the Psychiatry Department of the Stanford Medical School and a founder of the undergraduate program in Human Biology, began a collaboration between the Gombe Stream Research Center and the university. A captive outdoor colony of chimpanzees was established at Stanford, called Gombe West, for comparative observations. Goodall spent one academic quarter a year at Stanford, and Human Biology majors who took the appropriate preparatory course work were able to go to Gombe as research assistants for academic credit. They contributed to the pooled site data.[31] Shortly after Goodall's own 1966 Ph.D., other Cambridge University graduate students came to Gombe. Robert Hinde, head

Figure 7.9 The research team at Gombe in 1973. Published with permission of Harvard University Press.

of Cambridge's Sub-Department of Animal Behaviour at Madingley, and David Hamburg were both scientific advisors for the Gombe Research Center. Patrick McGinnis, who finished his Ph.D. at Cambridge in 1973, early became central to Gombe research, taking charge of the research program and records regularly during his four years there.[32]

Cambridge graduate student Richard Wrangham and Anne Pusey, Hamburg's graduate student who married Craig Packer, had been undergraduates together at Oxford and were influenced by the ethological tradition there. Hetty Plooij's husband, Franz, had done his Ph.D. at Gronigen on Gombe chimps, and he came to Hinde's Sub-Department of Animal Behaviour as a post-doc. Hamburg's student, Barbara Smuts, who was one of the kidnapped students and so was unable to complete her doctoral study of the chimps, visited briefly at Cambridge, and the network of Madingley students (including primate people Dorothy Cheney, Robert Seyfarth, Kelly Stewart, Sandy Harcourt, Carol Berman, and others) has been important to her. After her field work was ended at Gombe, she went to Harvard, where she participated in Irven DeVore's primate seminar. Smuts had been a Radcliffe undergraduate and had taken Irven DeVore's primate behavior course, in which Peter Rodman was her teaching assistant. Rodman did his doctoral work on orangutans, assisted by his wife. Later at U.C. Davis, he sponsored several primate doctoral dissertations, including Joan Silk's thesis, and organized the appointment of Sarah Blaffer Hrdy as a Target of Opportunity Professor at Davis in anthropology in 1984. At Harvard, when Smuts was a Radcliffe undergraduate, Robert Trivers was her friend and intellectual mentor. Trivers taught sociobiological evolutionary theory to DeVore and was an important member of DeVore's primate seminar. From Harvard, still technically a Stanford graduate student, Smuts did doctoral work on baboons at Shirley Strum's research site at Gilgil, Kenya. She was in the field at Gilgil with Nancy Nicholson, DeVore's student, who had also been at Gombe as a Stanford undergraduate in 1973. Smuts introduced Richard Wrangham to E.O. Wilson's sociobiology in 1975. Julie Johnson, former Gombe Stanford undergraduate, also did doctoral work on Gilgil baboons, as an Edinburgh graduate student in 1979. To borrow a concept developed by Bonnie and Timothy Ransom, Shirley Strum, and Barbara Smuts at Gombe and Gilgil, these human webs resemble the "special relationships" structuring much of nonhuman primate society.

In 1972, Nancy Nicolson, at Hamburg's invitation, studied the future Gombe West chimps, after Stanford had purchased them but while they were still at the Delta Regional Primate Research Center in Louisiana. There she worked with former Gombe researchers, Caroline Tutin (1975) and William McGrew (1979). Nicolson had been part of primate seminars organized by Stanford, Davis, and Berkeley primate people around 1973–74. At that point sociobiologically inclined students like Nicolson began to see that the Washburn-Dolhinow students from Berkeley had a fundamentally different approach to evolution and primate behavior. The middle 1970s marks a kind of theoretical turning point in primate studies, and after that point the Berkeley anthropology networks declined in their relative prominence in the discipline. The Gombe-interconnected people were instrumental in that change of direction.[33]

After May 1975, the kidnapped Gombe students' release was negotiated over several months, substantially as the result of David Hamburg's passionate commitment to the students' welfare and strenuous organizing work. Barbara Smuts, the only graduate student kidnapped, was released after a week to communicate the guerrillas' demands to the Tanzanians and Americans. The guerrillas had split from Patrice Lumumba's group in Zaire. They spoke no English, and the students' Swahili, in Smuts's words, extended to, "Where are the chimpanzees and what are they eating?" (interview, 18 March 1982). Smuts and the leader of the guerrillas spoke French, which became the language of their interaction. The students were ransomed. Goodall's and Hamburg's collaboration ended after the stress of the kidnapping. Hamburg left Stanford in 1975 to become the President of the National Institute of Medicine of the National Academy of Sciences, and the Gombe West chimps soon left Stanford.

Former Gombe students were hired to work at Stanford after the kidnapping, microfilming and organizing the massive Gombe data accumulated from the beginning until 1975. (The microfilmed records were deposited at both Stanford and Cambridge.) After his post-doctoral plans to study gelada baboons in Ethiopia were dashed by political turmoil there, Richard Wrangham got a post-doc from David Hamburg to arrange long-term procedures for dealing with the Gombe data. Having been at Gombe between undergraduate and graduate school, Joan Silk worked on the mother-infant data at Stanford (Silk 1978). When Wrangham taught primate behavior the next year at Harvard, he found himself in a network that included Barbara Smuts and DeVore's sociobiologically oriented primate behavior students. Between 1977 and 1980, he was a fellow at King's College, Cambridge, in Timothy Clutton-Brock's behavioral ecology mammal unit, along with Robin Dunbar, who had done a Bristol primate behavior Ph.D., and other major people associated with socioecology and sociobiology (Clutton-Brock 1977). Both of those explanatory strands in Britain showed major influence from the ethological tradition rooted in Oxford and Cambridge.

These brief sketches of connections only begin to indicate the dense web of social and intellectual ties among former Gombe workers. This entire structure is systematically invisible in the National Geographic version of the secret life of the chimpanzees. It is hinted at in the acknowledgment and citation apparatus of Goodall's book and professional papers, but its significance has remained implicit. The Gombe network reaches beyond people who actually worked there to tie together major study sites, graduate institutions, and explanatory strategies. The importance of the network was underlined by the collaboration of key people in the Gombe network, with their multiply interconnected colleagues from Cambridge and Rockefeller University webs, in 1983–84 at the Palo Alto Center for Advanced Study in the Behavioral Sciences, in a kind of second Primate Year, after the first organized by Sherwood Washburn and David Hamburg in 1962–63 to codify the results of the first wave of new field studies. Including Barbara Smuts, Richard Wrangham, Dorothy Cheney, Robert Seyfarth, and Thomas Struhsaker, the group was writing, organizing authors for other papers for the volume, and editing a book that would inscribe their network's theoretical approaches and extensive field studies as a new standard and pedagogical tool of the discipline (Smuts et al 1987). This broad field of workers is hardly monolithic or in agreement on every issue—indeed, it is a network defined by its implicit agreements with each other about what the relevant arguments are, not by answers to those arguments. However, to be seen as working within explanatory traditions regarded as fruitless or outmoded by this powerful disciplinary infrastructure became professionally costly. National Geographic's representation of Jane Goodall defined what would count as primate society for the popular audience; the networks reaching out from the much less publicly visible Gombe students have determined what would count as leading-edge science for key audiences, audiences controlling access to critical resources—grants and jobs— for science. Goodall had solid scientific authority, but it was an authority embedded in a contentious collective discourse outside the popular eye.

So Gombe, popularly represented as the solitary world of National Geographic's and Gulf Oil's Jane Goodall, was for nine years a densely social, collective, international research site—perhaps more so than any other primate research site established by western observers. In 1972, about 50 scientific personnel—Tanzanian, European, Euro-American—lived and worked at Gombe. With their families and other staff, the population of the field station was about 100 souls, mostly living on the lake shore, with up to ten people living in individual huts in the forest. Between 1972–73, the study population of habituated, named chimpanzees numbered about 14 adult males, 15 adult females, 19 dependent young, 2 adolescent males, and 4 adolescent females (Wrangham 1975). People from the community called "scientists" considerably outnumbered chimpanzees at Gombe during the most intense years of research activity.

Crafting Data

A sketch of the history of data collection at Gombe can provide perspective on debates about field method, consciousness of social history, sex and gender in primatology, and the changing roles of western visiting observers and African field staff.[34] Goodall's first mode of collecting data was to enter in a daily field diary whatever she observed and to transcribe the field notes nightly by typewriter. That approach prevailed until 1964. When the banana feeding system was initiated in 1962–63, notes were spoken onto tape and transcribed nightly. This highly personal style was progressively broken apart, quantified, and standardized. From 1964, daily charts were kept on 1) group structure and activities, 2) individual activities and gestures, and 3) contents of feces (giving information about chimp foraging). Population data, photographic archives, and lists of gestures and vocalizations were assembled. The feeding station permitted standardized records of attendance, behavioral interactions, and estrus state of females. From 1963, at least one observer was present at the station throughout the daylight hours, charged with collecting the data onto standard sheets or onto tape for transcription. Between 1963–68, observations were largely restricted to the feeding area. The peak of provisioning was 1966–67, with irregular provisioning from 1967.

Beginning in 1968, some individuals were followed from the camp into the forest. Typically, a pair of observers would follow a "target animal" for up to 13 daylight hours. The Tanzanian field assistant would collect standardized data, with categories for behavior of an individual animal, feeding, travel routes, and party composition, while an expatriate research worker recorded on a field check sheet and/or tape recorder and Nikon camera special data relevant to that person's assigned or independent project. Pooled data were the foundation of later papers. For example, Richard Wrangham's 1979 paper in *The Great Apes* utilized data from 6000 hours of observation by more than 30 observers, as well as 1000 hours of data collected by himself during 12 months, plus the daily records of attendance at the feeding station. He had learned how to see and record during the 14 months he spent in 1970–71 at Gombe as Goodall's research assistant. Observation times were recorded for each individual chimpanzee in a particular sampling session. Various schedules of sampling were used by different people as a function of their own projects. Special events, like predation, were in unstandardized collected records, and such descriptions of events were contributed by whoever saw them. After January 1972, about 400 observation hours per month were recorded on Group Travel Charts (Wrangham 1975), giving a critical group resource unobtainable by any single individual. Beginning in 1972, individual chimps were sometimes followed for several days.

This description gives an unrealistically simple view of what should fit into any category. For example, definition of such basic issues as what should count as a "party" might not be achieved until the end of the study (Wrangham 1975). Data collection was constantly negotiated in the practical interactions of observers. Students with varying degrees of experience and understanding of what they were doing could be responsible for significant periods for making judgments about what they saw and what to record. None of the negotiation and inexperience would be visible in the numbers and charts of pooled data, but it was present in the culture of the Gombe people. Method was *embodied* in an interacting international group with dynamically evolving questions about primate lives. The popularity of new questions could make entire practices of recording problematic. Much of the continuity in embodied experience about data came from permanent Tanzanian field staff. Comparability of data taken at Gombe with other chimpanzee sites, for example the Japanese studies sixty miles south of Gombe or in the Budongo Forest in Uganda, could not be assumed. Methods of getting the data, especially provisioning, provoked debates about the "naturalness" of the observed behavior. A behavioral ecologist might be distressed by behavioral data obtained from a feeding station, but obtaining anything like the same density of behavioral observations by following dispersed animals would be very difficult. It was far from simple to determine what foods were available when and to whom and to assess the disturbance introduced into the system by changing patterns of artificial feeding. These matters illustrate the micro-issues that make up the evidence appealed to on all sides in the contests for narrating the primate story. Contested meanings do not exist only at the level of finished accounts; they are the basic stuff of every step in the discipline.

We have seen that initially Jane Goodall regretted the assigned company of the Gombe game scouts on her travels through the Gombe forest. Much of Hugo van Lawick's film record showed her alone, without the game scouts. Whatever Goodall's early personal practice was, by the time animals were followed from camp after 1968, Gombe regulations mandated pairs, one of whom was a Tanzanian permanent member of the field staff. That structure began to show in publication practice (Goodall, Bandora, Bergmann, Busse, Matama, Mpongo, Pierce, and Riss 1979). The kidnapping of 1975 and Goodall's and Hamburg's break made the role of the field staff even more crucial. Without large numbers of western students, collective records at Gombe depended entirely on a growing Tanzanian field staff, numbering about 10–15 in the 1980s. Western observers still came to Gombe for days or weeks, but the station depended as never before on the expertise of Tanzanians. The first paper prepared after the kidnapping made the significance clear; it was authored by Goodall, do Fisoo, Mpongo, and Matama. The days of anonymity for nationals in Third World primate habitat countries were over.

However, the dehistoricizing gestures in literature from Gombe continue through the most recent work, Jane Goodall's monumental *The Chimpanzees of Gombe: Patterns of Behavior* (1986). This is a wonderful book that achieves sustained accounts of the individuality, intelligence, and complexity of the animals; it is a book that never fails to respect the chimpanzees' specific material-social locations in reference to each other. The chimpanzees are firmly subjects, not objects, in this publication; respect for the animals has been enforced as a criterion for access to them. But the specificity of a human history that enabled the accounts is muted in favor of another rediscovery of the time of origins and great discoveries. Desire to know the chimpanzees is described as simply a species of man's desire for knowledge, best exemplified in the history of science and the love of adventure that "led to Christopher Columbus' discovery of America [and] in our generation has landed people on the moon" (3–4). The New World, the Final Frontier, and Gombe. Goodall summarizes studies at the dozen field sites for watching wild chimpanzees since Harry Nissen's (1931) at Neribili, under Robert Yerkes's sponsorship; and she discusses interactions of researchers and animals as they affect the study at Gombe. What is too dim is a dimension problematizing (not erasing) the mythic, scientific, and individual axes; i.e., the historical.

By history I mean a corrosive sense of the contradictions and multiple material-semiotic processes at the heart of scientific knowledge. History is not a completed past simply waiting to be applied to deepen a time probe or to give perspective. It is a discipline reworked by postmodern insights about always split, fragmented, and multiple subjects, identities, and collectivities. All units and actors cohere partially and provisionally, held together by complex material-semiotic-social practices. In the space opened up by such contradictions and multiplicities lies the possiblity for reflexive responsibility for the shape of narrative fields.

Such an historical dimension is not missing altogether in *The Chinpanzees of Gombe*, and it is sharper in the 1986 book than it was before. For example, the kidnapping of 1975 is briefly discussed, with its consequences for the greater role of the Tanzanian field assistants and for a different structure of research. The collective and constructed nature of work from Gombe is clear in an extensive appendix on the history of data collection (Goodall 1986: 597–608). [Figure 7.10] That discussion required acknowledgment of some of the complexities of different kinds of literacy in the history of primate studies, including practices of writing and tape recording by the Tanzanian men after 1975, whose own language (Kiswahili) has no written primate literature (required reading for European and American field workers), but whose mnemonic practices enable almost total recall of field observations (605–6). History implicitly structures Goodall's accounts when she discusses the different rates of production and quality of data collected by people with scientific careers operating on competitive professional clocks *versus* people with jobs that include the usual forms of resistance to time and work discipline in a hierarchy (608).

Sex and gender have been salient variables at Gombe in every aspect of the work, including the processes of crafting data. We saw how the signs of gender, science, and race formed a triple code for reading the National Geographic version of the primate story, but what about the complex Gombe network and research site absent from National Geographic's account? Plainly, both men and women were at Gombe

Figure 7.10   The research team and chimpanzee Goblin at Gombe in 1985. Copyright Christopher Boehm, Department of Anthropology, Northern Kentucky University. Published with permission.

in large numbers and formed interacting social groups not easily characterized by sex or gender, but persistently suggesting that *something* about sex and gender was very much at issue in the network, just as for the chimpanzees these people studied. The story of sex and gender at Gombe cannot be simple, but a hint of its patterning shows, first, in the research projects of men and women doctoral students[35] and, second, in the curious question of whether female chimpanzees should be dependent or independent variables in accounts of the character of a chimpanzee social unit. It turned out that sex, and probably gender, had everything to do with what would count as a chimpanzee "community."

The student assistants who came to Gombe were first assigned to Goodall's projects, so that they contributed to the data pool and learned how and what to see. Many stayed or returned to do their own projects, and later students came originally for their doctoral research, although everybody still contributed to the pooled data. Some students came to Gombe first and then arranged to be accepted into graduate programs for their work, facilitated by Hinde's department. From 1963, Goodall focused on social behavior and infant development. As a crude generalization, women assistants were the ones primarily assigned to watch and later follow mother-infant pairs in Goodall's study of infant development.[36] The men tended to be the ones watching and following males and pairs of males (interviews). Subsequent choices of projects were probably influenced by this early gendered research experience. However, the women and men doctoral students designed their later work for themselves in ways they could not at first.

Although there is no absolute separation, these men and women differed markedly in the foci of their work. Among male students who did Ph.D.s based on Gombe
chimp field work, Bauer, Bygott, Wrangham, Teleki, and Halperin collected data
almost exclusively on male chimps, while McGinnis and Plooij worked on "bisexual"
topics. Only Plooij worked on a topic that crossed stereotypic gender expectations.
Among women who earned Ph.D.s from Gombe field work, none focused exclusively on male chimp data, and all showed marked interest in the content of and
explanations for female lives. Whether they consciously identified with same sex
subjects or not, both men and women students at Gombe focused on "gender-
appropriate" subjects, while making use of similar explanatory frameworks not
differentiated by gender.

There is an interesting absence in the Gombe field work: there has been no
completed study of adult female interactions or female behavioral ecology. Mother-
infant studies abounded; male interactions were examined from several points of
view; sexual behavior was scrutinized; predation ("hunting") got much attention;
tool use was examined; infant development preoccupied many observers. Three
students set out to look at adult female interactions or ecology: Lori Baldwin,
Barbara Smuts, and Richard Barnes. Smuts and Barnes were stopped after only a
few months by the 1975 kidnapping. Baldwin's study was announced by Goodall,
but never appeared (Goodall 1971: 11). The impression is left that the subject
became a personal and scientific priority late in Gombe's history. The consequence
is that sustained field work on the subject at Gombe is likely never to be done. Let
us look at the deepening unhappiness that fact held for researchers' standards of
good scientific explanation by the late 1970s and early 1980s.

The interweaving of personal motivations for studying females, political and
cultural contexts legitimating the interest, and scientific rationales for focusing on
the topic may constitute a type example of the complexity, relevance, and synergism
of all three of these determinants of the narratives that make up primate science.
The three-strand interweaving may be untangled by examining the changing definitions of what constitutes a "unit" of chimpanzee social organization.

Goodall's early descriptions of chimpanzee social organization identified only
one stable social grouping, the mother and her dependent offspring.[37] Otherwise,
chimpanzees were described to associate fluidly and mostly peacefully in nomadic
bands, without defended social or territorial boundaries among bands or parties.
The first phase of research, 1960–66, when Goodall got her doctorate and more
observers began arriving, seemed to reveal a kind of primate utopia—mother-
centered, but with outstanding male personalities engaged in status competitions,
which did not seem to be the organizing axis of chimpanzee "society." The ideological unit of the mother-infant pair was deep in Goodall's culture and evident in many
contemporary and past scientific constructions of objects of knowledge, including
Robert Hinde's. Symbolically, in Goodall's writing, the chimpanzee mother and
infant constituted a perfect model, after which she portrayed her own relationship
with her infant son, Grub. Her personal motivations are unknowable, but the textual
narrative of the personal emphasizes the congruence of her own mothering, the
utopian model, and the scientific inquiry. The forest's peaceable, open chimpanzee
society, full of strong personalities, was a counterpoint to the dominance-organized
and relatively closed baboon unit on the dangerous dry savannah. Culturally and
politically, the early Gombe accounts fit into contemporary European and Euro-

American concerns. The accounts offered a peaceable kingdom, one part of the dual code of a culture obsessed with psychological explanations and therapies for all kinds of historical conflict and pain. Male aggression concerned Goodall, but it did not define what counted as chimpanzee society.

Subsequently, in Japanese accounts of their chimpanzee study population in the Mahale Mountains, observed from 1965, the concept of a "unit-group," a multi-male, bisexual group of 20–100 animals, was emphasized (Itani and Suzuki 1967; Nishida 1979). The group was described as fluid, not very cohesive, breaking into different sub-groups, with exchange of members among neighboring unit-groups. Resulting from their search to identify the social unit as the first task of a proper study, the Japanese emphasis on a unit-group was consistent with their general methods and explanatory patterns. For the Japanese, the rational starting point of an explanation was not the autonomous individual, and they did not begin by asking what could explain the always slightly scandalous (to a westerner) fact that many animals live in groups whose members, beyond the sacred mother-infant primal One, seem to like being with each other.[38]

After the Japanese reports, Gombe workers began to describe chimpanzee groupings in terms of the concept of a "community."[39] The community concept at Gombe was constructed from observations of male associations and interactions.[40] Females were assigned to communities as a function of the frequency of their interaction with males, whose own interactions were thus the independent variables. Bygott described females as living in the male community, more or less as valuables within the shared male ranges (Bygott 1979: 407). This account of community initially seemed unperturbed by the casual assignment of females because it did not seem critical to think about natural selection very seriously, especially in relation to females. The focus of attention among the early men who followed male chimpanzees at Gombe seemed to be the problem of "human" aggression, as that essentialized attribute of "human nature" was ideologically and scientifically constructed in psychological, evolutionary, and mental health terms. Goodall shared this framework with the students. The issue was basic to David Hamburg's interest in the chimpanzees. The chimpanzee community was the ahistorical natural-technical object for examining "male" violence and cooperation. These kinds of studies were part of the means for constituting what it meant to be male in western scientific societies. Chimpanzee males engaged in both more violent and more affiliative behaviors with each other, and the patterns of each established the core and boundaries of a social unit.

Questions about how primate female behavior might lie at the heart of anything, except motherhood in functionalist and psychological terms, were not prominent. They were very hard to formulate, even when the logic of natural selection by 1980 seemed to demand it. Although it came to be seen as a logical scandal, female behavior was not at the center of early sociobiological formulations of natural selection and inclusive fitness, as they began to seep into Gombe accounts. The culturally easy, gender-stereotypic interest of male observers in chimpanzee male behavior of certain types, leading to a natural-technical object of knowledge called a community defined in terms of male associations, was initially unchallenged by the emerging "new" explanatory frameworks.

Meanwhile, women observers were giving detailed accounts of aspects of female lives, like Anne Pusey's study of female transfers between the male-defined communities. She noted the similarity of her picture to Japanese descriptions of female movements. The absence of data on female-female interactions and female behavioral ecology began to be remarked in the literature, and beginning graduate students began to plan their field studies to explore the topics.[41] In addition, primate workers began to understand that sociobiological explanatory strategies destabilized the centrality of male behavior for defining social organization. Female reproductive strategies began to look critical, unknown, and complicated, rather than like dependent (or entirely silent and unformulated) variables in a male drama. Female observers pressed these points with their male associates in the field and in informal networks. In general, since the men were not taking many data on females, they were not in a position to see the new possibilities first. In general, the women had higher motivation to rethink what it meant to be female. Several of the women reported personal and cultural affirmation and legitimation for focusing scientifically on females from the atmosphere of feminism in their own societies. Men also reported the same sense of legitimation for taking females more seriously, coming from the emerging scientific explanatory framework, from the data and arguments of women scientific peers, from the prominence of feminist ideas in their culture, and from their experience of friendships with women influenced by feminism. I do not think that it is possible in principle to build a causal argument from these reports, even if everyone testified to the same thing, but I do think the narrative of the social construction of scientific knowledge is implausible without these dimensions.

Barbara Smuts and Richard Wrangham became friends and colleagues just before Smuts went to Gombe in 1975. Through their collegial relationship, the third metamorphosis in the narratives of sex and gender salient to the scientific construction of the chimpanzee unit of social organization occurred, displacing the sexual biopolitics of the mother-infant pair and the male-defined community. Retrospectively, Smuts's story began with National Geographic's model of Jane Goodall and chimpanzee life.[42] After Goodall's first *National Geographic* article, Smuts was fascinated by chimpanzees. She read the article for a junior high school term paper that she wrote for a woman teacher, whom she remembered as a strong mentor. Her parents encouraged her to follow a scientific career. "From the time I finished [the Goodall article] until the time I went to Gombe ten years later, that was my goal in life. I never wavered! Basically, it was finding out you could make a living studying animals; as soon as I found that out, then it was obvious what I would do" (interview, 18 March 1982). Influenced by Robert Trivers, while still an undergraduate Smuts studied adult female rhesus monkey behavior on the island of La Cueva, Puerto Rico, in the summer of 1970. Her thesis, "Natural Selection and Macaque Behavior," was the first systematic attempt to apply what became known as sociobiological theory to primates. In that sense, sociobiology in primate studies *began* female-centered and woman-authored.[43]

Smuts told the story of her work in terms of an intellectual spark from Trivers and her love of animals. The two joined in her decision to study adult females at Gombe. She went to Stanford for graduate school, after meeting David Hamburg in 1972 at a Congress of the International Primatological Society, shortly after Hamburg and Goodall began collaborating. Smuts took courses in neuroanatomy and neuroendocrinology for a year and a half in the Neuro and Behavioral Sciences program in which the primate students were inserted, and then headed for Gombe in March 1975. She had met Goodall at Stanford and learned much from her about the chimpanzees, but not about "theory"—i.e., fitness strategies. At the beginning of her Gombe work, she was intrigued by Richard Wrangham's questions at the end of his thesis about what the females were doing and how that fit into the picture he had drawn of the males in his study.

Wrangham and Smuts were close friends after the close of Gombe and while Smuts (1982) did her doctoral research on baboons at Gilgil. Wrangham visited the first baboon site Smuts worked at, Masai Mara, and he visited her while she was working at Gilgil. They also interacted briefly at Cambridge and again at Harvard when Smuts was writing up her thesis. For several years, their intellectual worlds were closely related and mutually constructed. In the face of the missing data on Gombe female behavioral ecology, they collaborated on a paper detailing "Sex Differences in the Behavioral Ecology of Chimpanzees in Gombe National Park" (Wrangham and Smuts 1980). Smuts and Wrangham both recalled a rich brew of conversation about females, selection theory, Trivers's ideas about females as limiting resources for males, and missing data on female behavioral ecology. Published during this period of intense interaction, Wrangham's papers developed the theoretical perspective of behavioral ecology to redraw the picture of primate society. His explanations centered female foraging and social strategies as the chief independent variables, in relation to which male patterns would have to be explained. Simultaneously, similar ideas were being developed for evolutionary theory of vertebrate society generally.[44]

Wrangham grew up intellectually in the streams of research practice and theory that converged into sociobiology and behavioral ecology by the late 1970s in Britain and the United States. A zoology undergraduate student at Oxford, he was embued with the Tinbergen approach, having begun reading his books at about age 16. After his B.A. he wrote to about 25 people to arrange an opportunity to study wild animals, but he failed to get funding. Finally, he wrote Jane Goodall, who invited the young ethologist to Gombe. His tutor's daughter, Anne Pusey, was already planning to go to Gombe. As Goodall's research assistant, he studied sibling relationships among pairs of males. Wrangham had not studied the primate literature thoroughly, and he was not primate oriented. He knew the bird literature best. He arranged with Robert Hinde at Cambridge to do a doctoral project on Gombe chimps similar to the one he had originally wanted to do on banded mongooses, namely a study of the relationship of food availability to social organization. After ten years of data at Gombe, very little was known—or had been asked—about that kind of relationship. Wrangham designed his doctoral study so he could finish getting data in a "reasonable" length of time.[45] That meant eliminating confounding variables. "I chose to study males because that meant fewer independent variables, [coming from] the whole reproductive economy of females."[46] "The behavior of females changes according to their reproductive state: in order to focus attention on behavioural responses to environmental change, only adult males, in two communities, were selected as target individuals" (Wrangham 1975: iii). The logical confusion of the decision, aside from the difficulties it caused for *later* explanatory goals, came from the way the marked category could ambiguously stand both for *general* principles of behavioral response to environmental change and for *male* responses to such change.

Wrangham came to regret his choices. Four texts show his progression toward a female-centered theory of primate society generally, during the course of which the notion of a male-defined chimpanzee community, with females assigned to a community on the basis of their interactions with males, disintegrated. In his dissertation (1975), he found nice relationships between food availability and size and range of social groups. "Apparently when food is scarce the behaviour of individuals is directed to maximising feeding efficiency; when food is abundant males may forego maximisation of feeding efficiency for the sake of increased reproductive effort" (iii). Because females did not seem to conform to male patterns of association, Wrangham proposed a model to account for chimpanzee dispersion, where "the form of chimpanzee social structure is suggested to depend on the feeding strategies of mothers" (iv). Wrangham referred to Trivers (1972), Hamilton (1964), and Lack (1966) [1954]; but the framework could not be argued systematically with the data he had collected. The functional explanation that females are distributed by food and males by females was in the bird literature that Hinde urged Wrangham to use.

In his contribution to the Wenner-Gren volume on the great apes, Wrangham argued the same case, going further in his dis-ease with the community concept for chimpanzees, complaining that the assignment of females as a function of their pattern of association with the most frequently observed companion got the relationships logically backwards. He explicitly argued against the idea of the community as a bisexual unit (Wrangham 1979b: 488). Then in a paper that grew out of DeVore's request to contribute a piece developing sociobiological theory for ape societies in general for a symposium at the American Anthropological Association meetings in 1976, Wrangham attempted (1979a) to reconstruct social theory for four groups of apes (orangutans, chimpanzees, gorillas, and gibbons and siamangs). The sociobiological paradigm in that paper was economically defined as the thesis that individuals have been selected to maximize their inclusive fitness. "The best competitor is therefore the individual who competes for whatever resource is most cost-effective in increasing fitness" (338). For females, that would be food, for males, females. This general explanatory strategy pervades sociobiology. The consequence for field practice in primatology was the imperative to get data on females that had not seemed essential to Wrangham as late as 1971–73. The particular form of the problem was to explain female social feeding, since the most "logical" foraging strategy, on the axiomatic assumptions of methodological individualism built into fitness theory, was solitary and dispersed foraging. In a 1980 paper, Wrangham extended the argument to female-bonded monkey societies, like baboons and macaques. The 1979 and 1980 papers prominently acknowledged Smuts.

Plainly, sociobiological theory can be, really must be, "female centered" in ways not true for previous paradigms, where the "mother-infant" unit substituted for females. The "mother-infant" unit had not been theorized as a rational autonomous individual; its ideological-scientific functions were different, located in the space called "personal" or "private" in western dualities. The sociobiological kind of female-centering remains firmly within western economic and liberal theoretical frames and succeeds in reconstructing what it means to be female by a complex elimination of this special female sphere. The female becomes the fully calculating, maximizing machine that had defined males already. The "private" collapses into the "public." The female is no longer assigned to male-defined "community" when she is restructured ontologically as a fully "rational" creature, i.e., recoded as "male" in the traditional explanatory systems of the culture. The female ceases to be a dependent variable when males and females both are defined as liberal man, i.e., "rational" calculators. The practical effect of constructing this "female male" was to legitimate data collection practices that made both men and women watch females more and differently. The phenomenological picture that emerged of female lives has been full of rich contradictions for the logical model of stripped down individualism that legitimated the investigation.

It seems impossible to account for these developments in the micro-space of a particular area of primatology, rooted in a particular research site, without appealing to personal friendship, colleagueship, interacting webs of people planning books and conferences, disciplinary developments in several fields at once, the history of western capitalism and political theory, and recent feminism among particular race and class groups. It is this kind of webbed system of explanation that makes me argue for the idea of a contested narrative field, rather than for other models of construction of scientific knowledge.

Jane Goodall was one of the nodes of intersection for people like Wrangham and Smuts. Her story was part of the field of possibility for their stories. Let us leave the metamorphoses of social order among chimpanzees and scientific chimpanzee watchers at Gombe and return to the popular Jane Goodall tale. Whoever the "real" Jane Goodall has been, this is the set of tales which has *meant* Gombe and chimpanzees for millions of people between 1963 and 1984, the date of the first *National Geographic* article and the last National Geographic TV special. The passage to this Gombe may be eased by an article in the travel section of the 7 April 1985 San Francisco *Examiner*, entitled "Swinging into Jane Goodall Territory." There, travel writer Sally Leville informs the potential visitor about Gombe, the tourist attraction. It tells which train and boat to take, what kind of camera to bring, and how to dress. "The park ranger who greeted us on the beach registered us and collected the park fee, making certain to check our currency declaration vouchers, verifying that we had not changed our money on the black market. . . . We drank and washed with water brought from the same stream or lake as Jane Goodall has done for the past 25 years" (Travel Section, 1). "Strangely enough, my biggest thrill came when the chimpanzees left us and walked off into the trees. . . . These were indeed wild creatures . . . We were the ones, sweating in the "caged" observation hut, trying to catch a glimpse of them living a free and natural life. . . . [T]he hardships of visiting animals' natural habitats [are] often intimidating. But if first-hand experience of the world's wonders is what you seek, perhaps you should consider the chance to stalk the wild chimpanzees at Gombe Stream National Park" (Travel Section, 17). If the hunt was ugly when it was an aristocratic and white colonial practice, it has not become more beautiful as a democratic one-day excursion.

Jane Goodall: Girl Guide to Wise Woman

Miss Goodall and the Wild Chimpanzees (1965) is a film about the negotiation of the terms of observation and an orgiastic, but still innocent, celebration of the grace of touch across crucial cultural boundaries. It is a first contact narrative, recognizable within science fiction conventions. The film is also an argument about the nature of science. And finally, it is a story of the self-sufficiency and complete happiness of a young single white woman in nature.[47] The film opens with a scene of contact between human and chimpanzee, but a corrupt scene in which the chimpanzee has been completely bent into the image of "man" and then re-presented as spectacle, not as healing vision. A costumed chimpanzee, Sam, ice skates with his trainer, while the narrator, Orson Welles, intones the scriptural text ruling relations between human and ape: "Among all the animals, the chimpanzee is closest to man"; and yet all we know is the chimpanzee in captivity, performing human tricks. Potential source of self-knowledge, the chimpanzee has been distorted into self-parody. The film cuts to scenes of chimpanzee youngsters playing in a lush green space, "wild and free," while Welles tells us that an investigation of these chimpanzees could lead to a better understanding of human behavior. Dangers haunt such a study, emphasized by shots of a big male chimpanzee traveling fast through the forest. But "determined to uncover the secret of the chimpanzees," Jane Goodall, shown in a boat on Lake Tanganyika en route to Gombe, will conduct a study that "will lead to a redefinition of the word *man*." The film credits follow.

Welles read H.G. Wells's *The War of the Worlds* for the 1938 Halloween radio performance of a newscast that frightened thousands into believing that a Martian invasion was in progress. That Welles, with his dramatic and authoritatively deep male voice, was chosen by National Geographic to narrate Goodall's story adds a nice touch to reading the film within the genre of science fiction. The play between Jane Goodall's voice and Orson Welles's highlights the gender codings of science in the film. When a "discovery" is to be authenticated, Goodall speaks a few sentences about the find and its significance, to be echoed by Welles's deep tones.

The narrative of first contacts proceeds in several stages, via a rhetoric of distance and touch. Warned she would never get close to her goal, Goodall, constructed as rigorously alone and undergoing hardships and dangers, first is shown spotting the elusive chimpanzees only by signs of their passage—a tuft of hair on a bush. The soundtrack announces she is getting closer. Then she zeroes in on them visually by binocular from a mountain peak. "Jane finds she has made contact at last." She descends toward where she spotted the animals, but "the wild chimpanzees flee the pale-skinned stranger invading their domain." No cameras are visible; no clue has been given so far how Goodall herself has been made visible.

But "Jane" is even more determined. Goodall is called by the familiar first name constantly, marking her status as girl, even while she is engaged on a quest that will change the definition of man. The passage of the night is marked by scenes of Africans fishing on the lake. Morning comes, and Goodall begins another day, filled with long distance observations. Day ends, with Goodall on the mountain top. "Here Jane will spend the night, high above the African forest." Goodall's voice confirms, "There was a special fascination. . . . [I] enjoyed those nights in the mountains with no human companionship." Tomorrow, Welles intones, she must get closer. There is only one jarring note in the scene of the female representative of man alone in the Garden—she eats a spare dinner of pork and beans from a tinned can. The odd sign evokes the history of the transformation of systems of production and of daily habits in the mid-nineteenth century, when large-scale canning in the U. S. got a huge boost from demand created during the Civil War (Boorstin 1974: 309–22). The tin can on Jane's mountain top preserves pork, beans, and the social relations of industrial capitalism enabling the colonial "penetration" and division of Africa.

The next step is Goodall's building a leaf screen, behind which she hopes she can make observations at 100 feet of distance. While Goodall watches, chimps in small groups come to their feeding area. Her first act after this observation, which allows her to study the animals as individuals, is to name them. Here we meet for the first time the Flo family, with daughter Fifi, sons Figan and Faben, and baby Flint. Other individuals are introduced, and "Jane" sees more and more. After two years, the chimpanzees are in the open. Part I ends. Part II opens at a new stage of contact: the chimpanzees are visiting camp. They have initiated the contact that transcends vision and moves into the sensory field of physical touch. They came first when Goodall was out of camp, looking for them. At a pre-arranged signal from the camp cook, Dominic, she runs back to see the guests. The next shots are wonderful scenes of toddler Flint playing on top of her tent. Another chimp is even secure enough to threaten Goodall mildly, while another accepts a banana from her hand.

This first film does not greatly individualize the particular chimps who touched "Jane," but *In the Shadow of Man* elaborates the narrative of touch into the theme of special friendship with Flo's family, but above all with David Graybeard. It was he who first permitted Goodall to watch him. The first paragraphs of "Beginnings" dramatize the moment. "Less than twenty yards away from me two male chimpanzees were on the ground staring at me intently. Scarcely breathing, I waited for the sudden panic-striken flight that normally followed a surprise encounter between myself and the chimpanzees at close quarters. But nothing of the sort happened. . . . Without any doubt whatsoever, this was the proudest moment I had known. I had been accepted by the two magnificent creatures grooming each other in front of me. I knew them both—David Graybeard, who had always been the least afraid of me, . . . and Goliath" (20). Graybeard initiated the camp visits, first took a banana from Goodall, and became above all a friend. In the context of later worry that the habituated chimps who had learned to touch humans might be dangerous to observers because of the animals' great strength and loss of fear, Goodall held onto the legitimacy of her touch with David Graybeard. "I do not regret my early contact with David Graybeard; David, with his gentle disposition, who permitted a strange white ape to touch him. To me it represented a triumph of the sort of relationship man can establish with a wild creature, a creature who has never known captivity. In those early days, I spent many hours alone with David. . . . Sometimes, I am sure he waited for me" (270). Then one day, Goodall offered David a ripe red pine nut. "When I moved my hand closer he looked at it, and then at me, and then he took the fruit, and at the same time held my hand firmly and gently with his own. . . . At that moment there was no need of any scientific knowledge to understand his communication of reassurance. . . . It was a reward far beyond my greatest hopes" (271).

The next stage of the drama of first contacts in the film is dated to 1964, when the barriers between "Jane" and the chimps are virtually all down. The film becomes an exuberant orgy of touching among chimpanzees and between Goodall and the animals. After a scene between toddler Flint and Goodall, the narrator intones that after five years in the African forest, the "struggle for equality with the chimpanzees is won." This is not a story of man's stewardship, but of his homecoming, not just in peace, but in equality, represented by a white girl scientist. It is a narrative not of civil rights, but of natural rights. The film concludes with the drama of Melissa, who has just given birth to Goblin, returning to the group, pleading for touch and reassurance. After a short delay, the other animals accept her outstretched hand.

She is accepted into the group once again. Only at that moment does Orson Welles tell the viewer that "Jane" is now headed to the University of Cambridge to earn a Ph.D., to further the "goal of man's better understanding of himself." Part II ends with a shot of the sunset.

Goodall's coming formal scientific credentials are carefully framed in this film's narrative of good science. In the first scenes of the film, the viewer learned that Louis Leakey sent "Jane" to the field to explore the origins of man because she was a "girl with no special training, but a natural aptitude, . . . with no preconceived ideas." In the course of the film, Goodall "overlooks no detail" that might shed light on man's mystery and the chimpanzee's secret. She is less a twentieth-century scientist than a girl guide. The discoveries she makes—chimpanzees chewing leaves and using them to sop up water like a sponge, modifying grass to probe into termite mounds for tasty insects, predation and meat eating, male animals in a magnificent "rain dance"—all were coded into the narrative of western origin stories. Welles: "Til this day man alone was the tool maker." Leakey: "Now we must redefine man or redefine tools." Man is no longer alone in his possession of the frightening attribute of mastery of nature. These discoveries were not like engineering new macromolecules for biotechnology-based corporations or cracking the atom for nuclear energy and weapons; they were rather at the level of unlocking the secret of the "human nature" that brought these other ambiguous objects into the world. A second Eve, "Jane" names the animals in the garden after the original peace has been reinstituted. Because "Jane" made "friends with the animals, she could make discoveries no one else made" and "reap the rewards of total acceptance." Goodall stands for the promise of science in the service of restoration of peace and of the potent moment of possibility, called the "beginning." Protected by the experience of that moment, she can safely seek a Ph.D., which would otherwise be merely the outer shell of scientific knowledge.

Commenting on this reading of the Goodall story, Karin Knorr-Cetina, a pioneer in ethnographic approaches to the social studies of science, observed that her scientists would be utterly scornful of this approach to their task (personal communication). Engaged in a study of a recombinant DNA lab in northern Germany, Knorr-Cetina insisted "nature" was a kind of annoyance for her informants, something to be cleaned away so science could proceed. The contrast suggests the multiple simultaneous codings of science and nature in industrial society. All of these codes are available for contestations for the practices and narratives of scientific work and "discovery." "Yet this 'world' is itself the outcome of a process of inquiry constructed generatively and ontologically, rather than descriptively and epistemologically. More concretely, inquiry is 'about' ever new procedures in terms of which 'something' can be practically reliably encountered and recognized as an object which displays identifiable characteristics, and which can thereby become incorporated in and constitutive of our future world" (Knorr-Cetina 1983: 136). In Knorr-Cetina's sense, Goodall's chimpanzees, like a DNA probe, were the outcome of such a generative process of inquiry that made them able to be reliably encountered and incorporated in and constitutive of *our* future world.

"Jane's" status almost as virgin-priestess in the temple of science and nature is underlined in *Miss Goodall and the Wild Chimpanzees* by her unmarried state. The issue was not left implicit, any more than her skin color was. Even the chimpanzees at a reunion scene greet each other more "as friends and lovers," than as the family members Zuckerman and Huxley insisted on in 1938 and National Geographic would produce only a few years after the 1965 film. In the 1965 script Goodall says, "I am completely happy here in the forest. . . . This is the life I had always wished for." "Jane" is not subject to a husband, nor is she yet a mother seeking guidance from Flo. Her singleness and her priestess-scientist status complement each other in the overwhelmingly innocent narrative of this first National Geographic primate special. The complexities of power and domination are far from this Gombe. The word "dominance" does not even occur in the film, in the very years the dominance hierarchy concept was supposed to be the ruling axis for organizing knowledge of primate society.

Goodall returns to National Geographic's Gombe with a husband and a Ph.D. The double change in status to married woman and credentialed scientist was first announced in the *National Geographic* magazine, in "New Discoveries among Africa's Chimpanzees." The author line no longer presents Miss Jane Goodall, but "Baroness Jane van Lawick-Goodall," while the first sentence has as its subject the pronoun "we" getting stiff-jointed off an airplane and speeding to Gombe. The end of the second paragraph has "my husband Hugo grip[ping] my arm and point[ing]. 'It's the Flo family,' I breathed, hardly able to believe what I saw" (Goodall 1965b: 802). A few lines later, the reader learns what precipitated the married pair's hurried return to Flo's family circle: "It was only a glimpse, but we could clearly see the tiny black infant clinging to its mother's warm dry belly" (802). A short letter from Dominic to Goodall, finishing up her Ph.D. at Cambridge, had announced the event: "Flo amekwisha kuzaa. . .Flo has had her baby" (802). It would be hard to construct a denser portrait of "the family" than that in these spare lines of the magazine. The story of Flo's motherhood, and her daughter Fifi's rapt attention to its lessons, follows. A full page photograph shows Goodall, with notebook and pencil, and van Lawick, standing behind his camera on a tripod, watching three grooming chimpanzees, while the caption tells of the "husband and wife team. . . . The Dutch nobleman and the British scientist were married in 1964" (821). The days of maidenhood and solitary research ended together. The article includes the announcement of new research buildings. Jane Goodall, now Dr. Goodall, directs a mature scientific project.

Also highlighted by the double message of marriage and scientific credentials, the second appearance of Jane Goodall in the National Geographic TV specials, in *Monkeys, Apes, and Man* (1971), deepens the picture of her as a director of a project, including brief footage of the crowd of scientific observers of chimpanzees at the banana feeding station. But no longer is the viewer watching the restoration of origins; instead the scene at the banana station is of "a tranquil society transformed." The screen is filled by the bickering and comic greed of chimpanzees each trying to garner the whole banana bonanza. In the story, natural scarcity did not exist for these animals, but the politics of scarcity were introduced by the provision of plenty. "A wealth of bananas triggered outright aggression." The scientists were forced to regulate the feeding, in order to continue their observations.

The disturbing note of deliberately constructed scientific observation and intervention in a world far less innocent than the one six years before is softened by the "act of faith" Goodall displays in applying "natural methods of child rearing" to her infant son, Grub. He was "never in a playpen, never alone," in a tribute to ape systems of rearing children. We do not get to see the cage in which Grub was kept to keep him from becoming tasty meat to the hunting chimps. That scene was shown on TV a few years later. This middle chapter of the Goodall television narrative closes with the assertion that "narrowing the chasm between man and apes cannot be from science alone, but by understanding and compassion." Goodall's ability to touch the animals and to model her mothering on theirs is not quite coded as science here; hers is rather a feminine gender-coded practice of identification and compassion. The banana-mad scene recedes. The film goes from Gombe to Desmond Morris walking among the church spires of Oxford, talking of the human zoo and its dire plight. Moving from "tribal hunters to urban masters of the world," man "can become extinct just as easily as any other species." Morris asks rhetorically, "How far can we push man before we push him too far?" This film that began with scenes of Dian Fossey alone with gorillas in the Garden ends with urban crowds and the threat of extinction. Goodall's and Fossey's efforts to learn from the animals are urgent. Obsolescent and threatened man is at stake.

The third National Geographic film of Goodall, *Among the Wild Chimpanzees* (1984), is much more somber. Goodall is neither virgin nor married woman and young mother. She appears as a wise woman returning to the scene of her discoveries and teaching the crowds in a busy airport. Hugo is gone, but we are not told of the divorce. Neither are we told of the kidnapping of 1975. But amidst the continuingly pleasurable and reassuring footage from earlier films, this program is laced with terrible shots of the polio epidemic, which occurred while Goodall was pregnant with Grub, striking 15 chimps in the main study group, and with the later scene of Flo's death, closely followed by that of her now nearly adolescent son, Flint. Flo had failed as a mother; Flint had remained much too dependent on her and was unable to accept weaning and the birth of his sibling, Flame. Flint died of grief and loneliness after his mother. Flame died early; Flo was just too old to cope. Worse than these events, the news of warfare, infanticide, and cannibalism among the chimpanzees emerges. This was not just the comic competition over bananas, but the systematic attacks of one band of chimpanzees by members of another "community." The infanticides were the result of the awful systematic behavior by one mother and her daughter, Passion and Pom, who killed and even ate the infants of other mothers in the community. Goodall is at a loss for explanation, ascribing the horror to a unique psychotic individual. On television, she did not appeal to the popular resources of sociobiology, in which the killing pair might have been increasing their own inclusive fitness at the expense of their conspecifics. But over the "warfare," the message was more pointed: It made the chimpanzees even more like "man."

But the somber world of the 1980s holds messages of hope as well for National Geographic's distant audience, but a hope mediated not by a girl, but by a woman with gray hair. Understanding the daily life of the animals, witnessing the succession of their generations, the pleasure of remembering work done and imagining a successor: these are the lasting goods from Gombe for National Geographic's Goodall. The last scenes show Goodall, her mother, Vanne, and adolescent son, Grub, returning in 1982 to Gombe for a celebration feast, full of speeches in Swahili, replete with the telling of stories, the codification of memory. Past and present, the Tanzanian field staff are prominent, hinting at the identity of the keepers of Gombe's future. "The pioneer who dared to be accepted by wild animals and won," Goodall is filmed speaking in Swahili and then recalling in English her most precious memories—"to have a chimp just sit there, know I was there and not mind." The narrator assures us that Goodall "remembers them all" and that the "future is still uncharted." Gombe "no longer reveals the gentle noble savage." That dream was lost not only at Gombe, but in the entire post-colonial western imaginary and its bestiaries and ethnographies. "Now we see unexplained violence." This world holds the seeds of "terrorism" and guerrilla warfare. The dream of beginnings has been devastated. But there is still a future.

Once again, the old western dream of perfect representation surfaces; for that future, we are told, is "to be written by the animals themselves." The Gombe students, who transcribed the lives of the chimpanzees into the languages of sociobiology and behavioral ecology, believed their approaches were fruitful because they "converged on an empirical reality" (interviews). They described reality as an imperturbable asymptote, not itself deeply constituted by their own multiple practices. In National Geographic's version of Goodall's science, the chimpanzees wrote, and will write again, directly, at least for those who learn to read. In this ideology, the book of nature is not really co-authored. Scientists cede their authorship to "nature"; that is the key to their social authority. Francis Bacon would have enjoyed the National Geographic specials. Both the native scientific folk-epistemology about a finally validating "empirical reality" and the popular ideology of scientific representation in the TV specials rely on the myth of the faithful copy, where interpretation or reinvention disappear, where history and its complexities can be finally repressed. Neither the international network of scientists nor the popular television heroine suggested another theory of representation—that of power-charged, negotiated, embodied, contested, and partial practices and codings that make up the scientific world, as other worlds. From a constructivist and interactionist perspective, the chimpanzees must be active participants in negotiating those practices and codings. Part of the embodiment of the world, the animals are actors, not resources. But ultimately the lesson offered in all the versions of the narratives about Gombe's chimpanzees has been different: The world is given, not made. It is no wonder Gulf Oil was proud to sponsor that message.

8

REMODELING THE HUMAN WAY OF LIFE: SHERWOOD WASHBURN AND THE NEW PHYSICAL ANTHROPOLOGY, 1950–1980

Man meets the problems of the atomic age with the biology of hunter-gatherers. (Washburn et al 1974: 7)

Strong rhetoric functions like the skull-and-crossbones on a poison bottle. (Huizinga 1973 [1934]: 294)

With the tools of narrative history, a research program developed over many uncertain years by a heterogeneous collection of people with problematic ties to each other, may look like a plan, with noble beginnings and tragic endings, masterminded by a founding figure with sure access to unbounded resources. The bones of old papers can be reanimated in the bodies of another generation's professional and political publications, as the bones of fossil hominids can be reanimated in late twentieth century U.S. sexual politics or international anti-racist organizations. It is an old story that evolutionary functional comparative anatomy and historical narrative share principles of composition—reconstruction of the family of man. Allegorical narratives seem to order themselves easily in analogous series: the humanizing way of life posited for the ever older fossils, destined for only two hominid genera, *Australopithecus* and *Homo*[1]; the human way of life of universal man insisted upon in the documents drawn up by the victors of a world war; the primate way of life of monkeys trying to make a living on land constructed as nature in game parks established by colonial practice at the eve of decolonization; and the scientific way of life enacted in a research program in the post-World War II United States science establishment. All these can be reconstructed as elements of a unifying narrative about origins and ends, that turns out to be about the fruitful and always densely particular ambiguity of fiction and fact in story-laden sciences about what it means to be human. What it meant to be universal man and to be human generically turns out to look very much like what it meant to be western scientific men, especially in the United States, in the 1950s.

The following reconstruction of the academic practices of Sherwood Washburn and his associates employs the same rehabilitative narrative technology for yielding a plausible account of scientific life that my subjects needed for their own constructions of the human and hominizing ways of life in their evolutionary physical anthropology. This narrative of humanism grounded a major chapter of the post-World War II textualization of the lives of nonhuman primates in the program for field studies initiated by Sherwood Washburn and his many students.

The main thesis of this account of a research plan is that Early Man in Africa—the focus of the Wenner-Gren Foundation for Anthropological Research's first effort to stimulate hominid paleontology—was conceived as the prototype of the United Nations' post-World War II universal man, in the ecological conditions of Cold War, global nuclear and urban proliferation, and struggles over decolonization. In that context, Early Man in Africa and UNESCO Man became Man the Hunter, the guarantor of a future for nuclear man. In a twenty-year system of research and teaching, Man the Hunter embodied a socially positioned code for deciphering what it meant to be human—in the western sense of unmarked, universal, species being—after World War II. In a sense, this Man the Hunter was liberal democracy's substitute for socialism's version of natural human cooperation. Man the Hunter would found liberal democracy's human family in the Cold War's "Free World." His technology and urge to travel would enable the exchange systems so critical to free world ideology. His aggressiveness would be liberal democracy's mechanism of cooperation, established at the first moment of the hominizing adaptation called hunting. Above all, Washburn and his peers made the hunting hypothesis, and the "new physical anthropology" from which it emerged, part of the modern evolutionary synthesis. Into the modern synthesis was built the doctrine of nature and culture that would establish the basis for human universals in biology and human differences in liberal social sciences, especially social-cultural anthropology, with a broad bridge between biology and social science formed by the "behavioral sciences." Washburn wanted the "new physical anthropology" to be a behavioral science supporting that bridge, as well as to be part of the structure of the account of both human universals and human differences, i.e., nature and culture, biology and society. This was the context for the comparative behavioral studies of primates pursued by many Washburn students from the late 1950s.

Man the Hunter's biological and social existence as universal man was never untroubled. He and his family found themselves in industrial, nuclear, urban society, where stress management offered a far from ideal prophylaxis; and his very existence as a stable natural-technical object of knowledge was undermined by internal disciplinary and other political countercurrents. From its inception, the story of Man the Hunter and of one of his chief academic progenitors must include the corrosive skepticism of many in the anthropological world, especially social-cultural anthropologists. But even more decisively, UNESCO's and Early Man in Africa's promising universal man was unraveled by those who failed to see him as the bearer of their experiences of what it meant to be human. The deconstruction would come from the academic and political right and left, often inextricably intertwined. By the late 1970s, Early Man in Africa in his 1950s and 1960s incarnations had to contend with the pretenders to humanity cast onto the surface of the earth by post-colonialism, feminism, and late capitalism. In his humanist guises, he was assailed by postmodernism in the critical disciplines that used to speak for man.

Finally, he was challenged by the related biological doctrines, especially sociobiology, that disrupted the early post-war biological humanism on which Man the Hunter relied. Legitimate sons and pretenders have been bound together in a contentious discourse on technology, often staged in the high-technology media that embody the dream of perfect communication promised by international science and global organization.

Evolutionary discourse generally, and paleoanthropology and primatology in particular, are highly narrative; story-telling is central to their scientific project.[2] The narratives are complex and protean. They are propagated in oral and written practices, although the oral discourse is premised on a vast and hegemonic technology of writing. The paleoanthropological body is a written body; paleoanthropology is an inscription practice, evident in the profusion of metrical devices for tracing meanings. The tools for producing the body as narrative and as text are themselves complex and multiple, requiring dense description and critical interrogation. Narrative physical anthropological discourse has contentious constituencies, which are both audience and author. There are many claims made on the narratives, and each claim can become part of a new version of the stories. The stories taken together constitute a story field, with axes of organization and rules for producing transformations, distortions, and highlightings. These axes and rules derive both from deep structural cultural patterns and from people's continuing daily struggles over meanings and other means of existence. The stories can only be generated, told, and read in relation to each other. But not any story can be accommodated in the field, and some stories can no longer be told. Readers of the discourse of physical anthropology do not find single tales; but they do find plot typologies, constant narrative closures, a discourse adapted to engage in particular kinds of social struggles, a field mapping certain kinds of meanings again and again. Evolutionary physical anthropological discourse gives material for more than one reading of experience. But, as Gillian Beer points out for Darwin's widely read narratives in the nineteenth century, many people in the twentieth century Euro-centric west pay evolutionary physical anthropology the homage of their assumptions. What is read from fossils and simians becomes common sense, becomes the foundation of other stories in other story fields constituting what can count as experience. Evolutionary theory is a form of imaginary history (Beer 1983: 8), where the requirement to produce narrative reconstructions is the chief rule of the game. And imaginary history is the stuff out of which experience becomes possible.

In the late twentieth century "imaginary," history in paleoanthropology as elsewhere is produced by a vast culture industry. To prepare for a return to Early Man in Africa and his forum in UNESCO at the end of World War II, let us look at some compelling 1980s images embedded in popular evolutionary anthropological discourse. These are scientific cultural images for producing beliefs, engaging desires, and populating imaginations—teaching films, photography for mass circulation magazines, BBC/Time-Life television specials, traveling museum expositions, and commercial feature films. Their product is literally the imagination, but their means of production are the entire system of technologies for inscribing the physical anthropological body in the last thirty-five years.

Reconstructions

From the point of view of adults, all children have large, beckoning eyes; but the deep, empty sockets of the 1 to 2 million-year-old fossil, a kindergarten-age youngster called "Africa's Taung Child," in the holographic portrait on the cover of the November 1985, *National Geographic* seem especially significant.[3] Discovered in 1924 in a South African cave and named by the head of the Anatomy Department at the University of Witwaterstrand, Raymond Dart, this child announced Africa as the birthplace of "man."[4] The holographic portrait of Dart's Taung child is a perfect icon for an uneasy high technology discourse on embryos and ancestors. Holography permits the perception of depth on a two-dimensional surface. The gray, flat, metallic plane on which the dim image of a skull is barely discernible on *National Geographic*'s cover ripples with glossy rainbow colors and fascinating depths when the portrait is placed in bright light. The viewer sees deep into the eye sockets, peering beyond the prominent zygomatic arches to the side of the skull. As the viewer's eyes investigate from different perspectives, the colors change constantly, but always within their metallic rainbow spectrum. There is no illusion of organic life, but the fossil takes on a promising technological life, through which the existence of a child two million years ago can be imagined. The photograph was produced for the *National Geographic* by the American subsidiary of the International Banknote Company. The Taung child was the first to be found of the genus *Australopithecus*, the southern ape, which has become so famous in reconstructions of hominid evolution in Africa. The bones of *Australopithecus* are actors in contending tales of human unity and diversity told so often in terms of "the human family." The contentiousness is iconically represented in the commemoration in South Africa in 1984 of the discovery of Dart's baby; the colorful octogenarian vied with the famous fossil for attention. But many scientists outside the country refuse to travel there to celebrate human origins in South Africa until after apartheid is ended. Until that time, the Taung skull threatens to be more a mascot for late-capitalist white supremacy than the child-witness of polyphonous and collective human ends. The hologram's metallic rainbow child stares at us as from within the stony matrix of a high-tech security state in which South Africa's buried strategic minerals—gold, fossil, and nuclear—fuel global imaginations and technologies for constructing apocalyptic origins and ends.

About the same time that the National Geographic Society published its 1985 cover story on "The Search for Early Man," the *Museums Bulletin* described a new, ambitious Commonwealth Institution exhibit in London, called The Human Story. This exhibit was sponsored by IBM and directed by David Pilbeam and Richard Leakey, who worked to challenge racism by presenting a vision of a long, shared human past, in which recent differences of race and color have little significance. The Human Story was planned as part of the Commonwealth Institute's 1984 Focus on Africa. Africa is again at the center of a discourse on what it means to be human. But contrasting sharply with the 1920s and 1930s African Hall dioramas, called peep holes into the jungle, The Human Story of the 1980s is organized by the presentation of the development of modern media, the narrative of the global communications industries. The western categories of nature and culture are again bridged by a fable of communication, but there the similarity between the exhibits of the 1930s and 1980s ends. The difference is between a discourse on organisms threatened with decadence and one on bio-technical communication systems under stress.

In London in 1985, human unity is embodied in the technologies of newsprint, instant newsflashes, computers, and television. The exhibit is made up of seven modules describing particular segments of humankind's development. The modules are linked by three-dimensional "time boxes," which transport the viewer across great gaps in species history. In the boxes, voiceovers and special sound effects explain the journey. Throughout the exhibit, the viewer might be stopped by Newsflashes and Stop Press presentations of that precious commodity linking humanity into a whole—information. The information is packaged as headlines and news stories. Unconsciously, the exhibit underlines the construction of modern global unity in the form of a commodity: The human story is produced as news. In the later modules, as the time approaches the present and future, the information technology changes. The penultimate module, The Present, "is entered through a tunnel of screens broadcasting live television (history in the making), and key environmental, health, and social issues are presented for discussion" (*Museums Bulletin* 1985). In the final module, The Future, the viewer is asked to vote on how the future should be planned. A computer tabulates the votes of all the visitors, and from these "democratic" signals, a publication is planned. Human history is fully textualized in late twentieth-century form; the means of production of the story are displayed proudly: media technology in a "global village." We are all high-tech primitives, worthy heirs of the holographic Taung child.

Despite the exhibit's explicit story of anti-racism and human unity, the layout people at *Museums Bulletin* could not quite shake the ever-present images of white racial hegemony and dis-ease in the story of Africa. Two illustrations accompany the story. One is a dark ape looking full at the reader, holding a mask in his hand. The mask is of the face of a young white man; beneath the veneer of the white race lies the universal truth of the dark humanoid animal. The eyes that would see out of the eyeholes in the mask are those of the upright apeman, the guarantor of human unity. The story of Tarzan has been evoked by the merest fragments of that important colonial saga. In the second illustration two adult men are bent over, picking up and examining bones. Both wear typical field clothes and fancy field watches. They seem to inhabit the same modern scientific time frame, neither primitive to the other. One man is black , the other white ; they seem in perfect colleagueship, peering at the remains of a shared past to establish the hope of a shared future. But the caption shatters that message: "Richard Leakey and assistant in the field in Kenya." Aristotle could have written the phrase; the master and his tool are in perfectly harmonious relation, the one with a name, the other indicated by a function. It feels like a mere question of syntax, surely not the stuff of global history? But syntax like this is precisely the stuff of the semiotics of master and slave, of the other who labors in the name of the one, the linguistic structure of the human story.

> Lucy. She was disguised by millions of years of fossilization—changed to stone, broken, then exposed to the hot Afar sun. How old was she? What was she? In 1974 she was found in Ethiopia by Dr. Donald Johanson and a team of scientists— the International Afar Research Expedition. . . . But it took a team of experts, and a multidisciplinary approach to paleoanthropology to give her an age and a context. . . . She was 3.5 million years old, fully bipedal, with a smallish, essentially human frame. But her head was primitive, shaped more like an ape than a modern human being. What should they call her? The scientists had to name her a new species— *Australopithecus afarensis*.

At first hinting at bizarre and ancient torture rites, these are lines from the front page of an advertising flier for the educational documentary film, *Lucy in Disguise* (1982), a remarkable narrative of reconstruction, jointly sponsored by the National Endowment for the Humanities and the Marsh and McLennan Companies. [Figure 8.1] The title of the pedagogical film is, of course, a pun on Lucy in the Sky; and well into the film, after an identity has been inscribed on the fossil bones by an extraordinary array of talented machines, Walt Disney-like cartoon animations, computer-assisted special effects graphics, and grave scientific conversation in the living rooms of famous paleoanthropologists, the soundtrack explodes with the Beatles' famous song while the narration gives Lucy her name, stressing her unique, individual personality.

If ever an ancestor were given birth by the adamic scientific inscription technologies and mass communications industries of the late twentieth century, Lucy is she. Eve should have been a fossil, so she could become the Barbie doll of a high-tech culture, which would clothe her in the latest fashions of flesh and behavior. The film is confident, authoritative, dense with the signs of science. The Council on International Non-theatrical Events selected *Lucy in Disguise* to represent the United States in international motion picture events. In the advertising copy, her face and head, which we learn in the film have been almost completely constructed from fragments of other related fossils found in another site, are repeated in series and filled in with maps and an artist's drawing of what she might have looked like in the flesh. The maps give a high resolution location of Hadar and two kinds of geological representations of deposits and stratigraphy. Her body on the front cover of the flier is broken into jigsaw puzzle pieces. In the film, we hear the operator of the

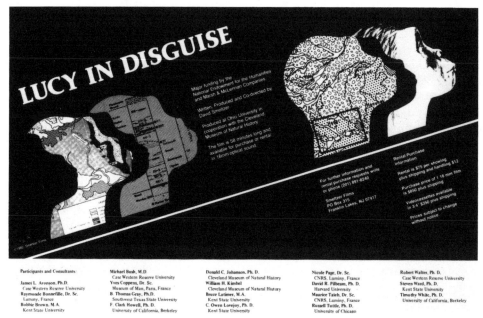

Figure 8.1   Detail from *Lucy in Disguise* advertisement flier. Copyright Smeltzer Films. Published with permission.

CATSCAN machine call his tool an X-ray buzz saw, with which he optically slices into Lucy's teeth from any desired angle, without ever breaking any "real" surface.

Lucy is rebroken and reformed at will, and she is then animated by the cartoonist's art in sequences interspersed among the scenes of the production of science.[5] The animated Lucy is always alone. There is no sign of any companions, children, anyone else, as she is brought to life and then killed in the final cartoon scene that reconstructs the fantasy of her being eaten by a large crocodile (crocodile!) and then preserved in the mud of the water's shore until her technical reanimation more than 3 million years later. (Lucy was stealing the crocodile's eggs, showing the ability of her species to gather and carry food, so her end was merited.) The line between science documentary and science fiction is thin, as special effects and a common narrative, indebted to Frankenstein's quest for the secret of life, provide the dominant experience of both. Both Frankenstein and the prize-winning educational film's scientists and technicians animate their products whose parts were unearthed from the grave in a quest for knowledge of the origins and nature of "man." The film has the structure of a quest narrative. Lucy's disguise peeled away by the special work of science so carefully praised in repeated passages in the film, she comes to us as the most whole hominid yet offered by paleontology. In the penultimate scene of the film, this achieved Lucy is presented to a scientific meeting. Applause from an audience of peers opens the cut. An all male panel of eminent paleoanthropologists sits, ready to answer questions from the floor. At the podium, the father of the multidisciplinary team approach at Omo, Clark Howell, confirms the film's staged consensus about the meaning of Lucy.[6] He intones, "Any experienced vertebrate paleontologist would reach the same conclusions." But about what, precisely? A consensus can be produced quite powerfully by film editing.

Near the scene of the virtual sectioning of Lucy's teeth with the CATSCAN, Owen Lovejoy tells us that from the teeth, we can learn about diet; that point quickly becomes allegory to permit the next statement: from the part, the whole can be reconstructed, the faith enlivening comparative functional anatomy since Cuvier. Even social behavior can be reconstructed from fragments of bone, and later in the film we see Lovejoy present his hypothesis about the evolution of "the human family" based on this procedure. Lovejoy, an expert on locomotion at Kent State University, did the analyses of Lucy's pelvis on which his discussions of the significance of hominid bipedalism have been based. The fact that the australopithecines walked upright, while having small brain cases, has riveted the attention of physical anthropologists. Why did they walk upright; what were the consequences of their adopting this mode of locomotion? Later, this chapter will focus on the relation of these questions to the famous, white, urban, middle class, male citizen of the 1960s in Chicago, Berkeley, and Cambridge—Man the Hunter. Lovejoy did not appeal to hunting, but to that other staple of the colonizer's versions of the human story, sex. The enticements of the hunt and sex both found their justification in domestic revery—the origin of the family. In Lovejoy's account, the line between ape and man is the transformation of the animal matrifocal group to the human bifocal "primitive nuclear family" (Lovejoy 1981: 348). This putative humanizing change was the fruit of enlisting male provisioning and care of the young in the face of a crisis in reproduction of species-making proportions. Females had reached their maximum of ergonomic efficiency and just couldn't make babies fast enough in new conditions without some outside help. Always imagined to be in excess of mere functional requirements, male energy, this time, at least, directed toward helping out at home, propelled the species over the line between nature and culture and into that marvelous adaptation called ever after, "the family."

Reconstruction of the family has been an absolutely central plot of the craft of physical anthropology from Early Man in Africa to Lucy in the Sky—it has been reconstructed quite simply as the human story, the key to the unity of "man." Indeed, family discourse was a chief ingredient of nineteenth-century anthropology's recipe for human unity too (Schneider 1984). The whole organism (Lucy), the nuclear and therefore complete family, the family of man: this tripod supports much of modern discourse in physical anthropology. Family discourse is intimately intertwined with gender, race, and colonial politics; and it has been a major issue of social struggle throughout the century. Various reconstructions of the family in scientific discourses follow particular social debates and movements in intimate, professionally embarrassing detail.

The final scene of *Lucy in Disguise* returns the viewer to a set used several times in the film: the scientist lovingly assembling fragments of bone like jewels on red velvet. Lucy's bones are a treasure, displayed on bright, plush fabric and hoarded by experts who tell us their value. They have been mined from stone, their matrix, and displayed to the world. Early in the documentary, two white women, one French-speaking and the other English-speaking, are shown instructing children in museums in front of a case with Lucy's bones: Lucy is an international figure. Another message is dully familiar: the men create science; the women teach it. At various points in the film, bones have been moved like chess pieces on the cloth, the whole pattern not yet clear. But at the end, Johanson's voice caresses the bones as he completes their array and praises their completeness. However, scientific completeness is of a particular kind; it must open up, not close down, the next questions. Narrative closure in science must prepare for the next repetition of the story, narrative closure is not the same thing as actually resolving the longings that animate paleoanthropology. The human story is never finished, neither in the direction of the future nor of the past. The quest must be permanent, and so the film ends by naming the fondest desire of the paleontologist seeking to push back the boundaries of the craft—the hope for yet older fossils. The origin in physical anthropological discourse is ever receding, not only because new fossils are found and reconstructed, but also because the origin is precisely what can never really be found; it must remain a virtual point, ever reanimating the desire for the whole. Lucy in the Sky with Diamonds. The other Lucy, the one the crocodile ate, must be surpassed.

The holograph of the Taung fossil, the Commonwealth Institute's exhibit of The Human Story, *Lucy in Disguise*: all of these are 1980s, high-technology-mediated images of human unity and diversity produced and consumed by particular, socially positioned constituencies. Each of these images works to promote not just its version of the "human story," but the story of science. And each stimulates a repressed uneasiness about the narrative apparatus of late twentieth-century technologies, in their power to reconstruct "human" unity. Another contemporary image dependent on modern communications media and related technologies, as well as on the configuration of social struggles in militarized multinational capitalism, has a much more ambiguous relation to science, drawing on its powers gingerly, in order to locate discourse on origins, ancestors, and primitives in the much balder propaganda mechanisms of a genocidal, repressive regime in South Africa. Here, in a popular commercial feature film, the images are not of bones reanimated by science, but of living people petrified into media mascots testifying to the opposite of actual social policy towards indigenous peoples.

The gathering and hunting people of the Kalihari desert, the San or !Kung, have been at the center of anthropological attention for many years; their lives have been enlisted in western discourse about aggression, child care, sex roles, women's shared experience across cultural difference, the first tools, diet, the human condition for 99 % of its bio-social history, the nature of the human family, and practically everything else. The people themselves have changed dramatically over the period during which they have participated in and been subjected to this intense scrutiny. Richard Lee, a principal investigator on major grants to study the ecology and social organization of the !Kung San for about twenty years, has stressed that the !Kung are not living fossils, but are modern people, subjected to forced conscription in the South African army, dealing with their own long histories of change and interaction with neighbors, in every way inhabitants of the same time zone as those who wish to interrogate them for answers to arcane questions emanating from other peoples' stories. Dozens of North American graduate students have earned Ph.D.s on these projects, and the desert people are famous in anthropological films.[7] The !Kung have been so deeply embedded in arguments about what it means to be human in some original and authentic sense that it seems impossible to extricate them from the web of universalizing, primitivizing discourse. This struggle matters to current conflicts over race and international politics.

"Welcome to Apartheid funland, where white and black mingle easily, where the savages are noble and civilization is in question, and where the humor is nonstop."[8] These lines open Lee's review of the international hit by South African filmmaker Jamie Uys, *The Gods Must Be Crazy* (1984). In the comedy feature film, the icons of science and commercial capitalism are called into question by the presentation of the klutzy white game biologist and disruptive coke bottle that descend into the lives of fully naturalized, edenic Bushmen somewhere in "Botswana." Lee points out that the film was shot in Namibia, not quite so remote from direct, illegal South African control of the San and much else. Uys's film works like Frontierland in a Disney World. The signs of civilization connote discord, and nature is resurrected as a dream of time before the fall. In fact, as an historical rule, people who are condemned to live in media Edens have first been victims of real genocide—as the San were through the eighteenth and nineteenth centuries, when they were shot as vermin for sport. In the twentieth century, they are used as trackers on borders with the frontline African states in the war against the South West African liberation fighters and the African National Congress.

To act in a film that claims to represent a real and timeless present—the impossible compound tense of colonizing discourse—the !Kung had to hide their transistor radios and beer cans and get out clothing made from game skins that had not been worn for years. "All of the actors in the film had themselves spent time or had relatives on South African army bases" (Lee 1985: 20). Perhaps the final irony of this film industry image of the natural savage in 1984 is that it remains precisely this imagination of nature before coke bottles that legitimates constructing the remaining lands of the living people into a new game reserve in Namibia, on which the !Kung will not be permitted to engage in modern economic activities. "On land that cannot sustain them as foragers they will be asked to hunt and gather with bows and arrows and digging sticks, recreating images of the past for the pleasure of the tourists" (1985: 20). This is the "human way of life" as living diorama, in a world where tourism is perhaps late capitalism's largest single industry. With digging sticks and bows and arrows for props, the 1980s !Kung are indeed techno-savages, twins to the urban high-tech primitives of the Commonwealth exhibition's electronic global village. The !Kung are fed by the industrial food of a world system of agribusiness, while modeling the timeless and innocent origins of technology for subsistence hunting and gathering, under the guardianship of suppliers of strategic metals for a nuclear human order. In a film that explicitly ridicules the commodity form and its attendant scientific civilization, the latent logic is perfected for transforming living people into commodities in the use-value form of natural history spectacles. This repressed logic can and does defeat anti-colonial conservation efforts, with tragic consequences for people, the land, and its animals and plants.

The imaginary quality of both scientific and popular reconstructions of a "timeless" human past is in consequence of the technical-scientific achievements of physical anthropology, not in opposition to them. The timelessness comes from the very depth of time required by modern anthropological clocks. If biblical calendars flattened time by allowing too little depth to see difference, the extreme depths of modern evolutionary calendars make a mockery of mere historical experience; humanity's real time is measured without reference to ordinary experience. It should also be clear that these imagined, timeless theaters of origin could only be produced through the lived social experience and advanced technologies of late industrial peoples. The machines that peeled the disguises from Lucy were birthed by science after the bomb—the radioisotope dating methods coming into paleoanthropology in the 1960s, allowing an accurate geological clock; the scanning electron microscope, permitting depth views of microscopic surfaces important to reading diet, and therefore behavior, from tooth wear patterns; the CATSCAN technique of sectioning fossil parts without destroying them; the molecular biological approach to taxonomy by comparing macromolecules of living primates, another way to establish a clock for evolution; methods of nearly instantaneous transportation of people and data across the planet; satellite pictures of interesting areas of earth surface, made possible by the strategic and commercial importance of continually re-mapping the earth from space-based weather, communications, and spy satellites. The icons explored in the last pages, from the holograph of the Taung child to the !Kung satisfying an international mass market for media images of primitives, are not nineteenth-century artifacts of a past western colonialism, which can now be nostalgically and safely invoked in funny movies. They are post-World War II productions critical to current struggles over "the human way of life." It is hardly surprising that "technology" remains, along with "the family," a central preoccupation of physical anthropological discourse. Lucy and her sisters and brothers live in the late twentieth century. The socio-technical, necessary conditions of existence are very recent for these millions-year-old beings, who are so like other extraterrestrials, whose natures seem both alluringly different from and perfect mirrors for those contemporary people whose cultures support their making reconstructions for a living. Fossils and extraterrestrials, apes and spacemen, hunter-gatherers and Cold Warriors: these are the actors in the current run of the human story, the end of the primate story.

Two final media images in a TV series narrated by Richard Leakey set the stage—the 3.7 million-year-old hominid footprints found in the volcanic ash at Laetoli, near the Olduvai Gorge, by Mary Leakey and others from 1977–79 and the putative 3 million-year-old fossil relatives of Lucy found in the Afar triangle at Hadar, Ethiopia, by Donald Johanson and his colleagues in 1975 and 1976 and presented to the world as "The First Family" (*The Making of Mankind* 1982). "The First Family" included fragments of adult individuals who fall into two distinct size classes, paternally—and controversially—judged by Johanson and his allies to be evidence of large sex differences in a single species with the best credentials yet for human ancestry. Two kinds of founding hearth are implicit: the White House and Eden, politics and religion, both in the trope of the original, primal family.

Ideologies common in science claim that politics and religion have been transcended by the rational discourse of modern science. But rather than having been transcended, the images, narratives, and functions of religion and politics luxuriate in science. Denial and repression are not transcendence. The act of naming the australopithecine "first family" is a totemic claim to the material means (the fossils and the expertise) to name the original conditions of hominid existence. "The First Family," with all its echoes, including the Oedipal ones, is a serious joking name—a stake on the right to produce reconstructions of human nature in the image of that imaginary primal family. Race and gender are the major contentious products of "human family" discourse.

The Laetoli footprints in the volcanic ash suggested the name for Part II of the BBC/Time-Life series. *One Small Step* . . . highlights the always implicit connection between spacemen and fossils, as the first words of the U.S. moon walkers are invoked for the australopithecine stroll in the time of the childhood of the hominids. In the videotape, Mary Leakey discusses upright walking as possibly the first development toward humanization; the famous footprints in the ash on earth and in the dust on the moon echo each other in the founding western myth of man's endless, onanistic journey toward making himself. They also echo expulsion from the Garden, as man's path takes him beyond himself, beyond his origins. Mary Leakey sees, in the footprints of two individuals, one larger than the other, with the smaller feet's impressions indicating a rather heavy person for their size, "perhaps a family party" (sound track). The anatomically expert scientific illustrator, Jay Matternes, whose reconstructions of musculature and whole-body appearance of fossil material appear widely in paleoanthropological literature (Johanson and Edey 1981: 354–57), has painted the family walk at Laetoli. His version traveled with the once-in-a-lifetime exhibit of paleontological remains that made it to the American Museum of Natural History in 1984.[9] The viewer sees the scene from behind the travelers. The painting shows a man walking ahead, face forward, tool/weapon in hand in the form of a large stick. A bit behind follows a woman carrying a large infant in her arms (with no baby sling, which Richard Lee [1968], Adrienne Zihlman [1978b], and Nancy Tanner [1981] imagined as perhaps the first humanizing tool). She glances to the side, not looking steadfastly ahead. A plume of ash is spewing from the active volcano; the family of hominids is walking away from a scene of earthly origins in a world on fire, the sign of destruction and banishment. The first family is on the move, armed with the prophetic humanizing tool in the hands of man.

But Richard Leakey's *Making of Mankind* is about the hope of avoiding apocalypse in a world after expulsion from the Garden. The TV series emphasized the diversity of humankind, its ancient shared past, and the common problems of modern civilization (nuclear weapons, settled communities and their organized aggression, misuse of technology for weapons, etc.). Leakey's Kenyan identity strengthened a kind of Third World perspective in the series, especially in its sustained critique of Cold War world views and big power manipulation of developing countries. The ideological axis of the series is the image of the human way of life as a sharing way of life, threatened in the bellicose conditions of the modern world system. Leakey is intent to highlight the view that hunting did not make us human; sharing did. The nuclear family in small groups, illustrated by Richard Lee's views of the !Kung, is used to emphasize that message. "The family," sharing, flexibility and intelligence, technology for mutual aid and progress, a natural tendency to cooperation and not to war (which is the result of relatively recent settled communities): these are the realities of being human for Leakey's TV series. In the "family of man," technology is tame and safe; only when sharing breaks down is technology a weapon of mass destruction. The fruitful ambiguity between the timeless, reproductive, nuclear family and the equally timeless "family" of the whole species, both carefully dated by the most advanced machines, runs through the fabric of Leakey's story. Exchange consequent on the sexual division of labor, modeled as an egalitarian affair among the non-patriarchal !Kung, functions as the trope for a possible modern social reconstruction of "the human way of life," promising salvation from nuclear war and extinction. Sexual politics and nuclear politics structure a common narrative field. Meanwhile, the !Kung live on army bases, conscripted to help maintain a system of apartheid.

Leakey stresses that race hatred is learned; science demonstrates a shared human past contained in a single surviving hominid species, whose dynamic, shifting diversity does not order itself into natural hierarchy. Exploitation and war produce unnatural hierarchies. The task of human beings after Hiroshima and in a period of decolonization is to rediscover our humanity, i.e., the sharing way of life once lived by all of us, in the heterosexual idyll of the hunter-gatherers. This is a 1980s version of the progressively uneasy narrative of scientific humanism that UNESCO attempted to build into the founding documents of a post-World War II international order. Physical anthropology had a full agenda producing the natural-technical objects of knowledge capable of sustaining the discourse of United Nations humanism inaugurated in the 1948 Universal Declaration of Human Rights and the 1950 and 1951 UNESCO statements on race. For this epic task, physical anthropology turned to Early Man in Africa, where the reanimation of newly discovered fossils seemed to promise hope for recent nuclear warriors, who believed they had won the war against fascism, with its natural hierarchies epitomized by racism. This plot is, of course, like all history of science, a fictional reconstruction of the twinned productions of science and politics.

Nature, Culture, and the United Nations Family of Man

The 1950 and 1951 statements on the nature of race and racial differences published by the United Nations Educational, Scientific, and Cultural Organization stand poised on the boundary between fascism and colonization, on the one hand, and multinationalism, Cold War, and decolonization, on the other. The war against fascism had been won; its perceived roots in racism had to be addressed by the victors, if the united family of man were to be achieved through the mediation of the new international organization, the United Nations. Just before that problem was addressed, the united family of man had been given a portentous definition. When the Universal Declaration of Human Rights was adopted by the General Assembly in 1948, René Cassin succeeded in getting an amendment substituting "universal" for "international," and as Cassin himself later pointed out, "universal" man is not the same creature as "international" man (Cassin 1968). "Universal" man is more easily abstracted from the complications of history—such as one of the nations of some "international" group deciding that a particular human right does not apply. "Universal" man became part of the machinery for bridging the application of the Declaration from the defeat of the Nazis, in which all the victors could share, to the more divisive realities of the Cold War and the dawning struggles against colonialism and neo-colonialism. Abstracted from the political realm of international relations, by the same abstraction "universal" man removed the unity of mankind from the discourse of politics to that of science. Although the importance of the constitutional achievements of the Enlightenment tradition were not denied, the permanent importance of another claimant for grounding non-partisan, universal discourse was strongly asserted. Natural science would be needed to get post-World War II universal man off the ground, launched into the future and unearthed from the past.

At the time the Declaration was discussed, the Cold War had already erupted, and eight of the fifty-six United Nations abstained from the vote. Six of these abstentions were by the Soviet Union and other socialist countries, on the grounds that the Declaration was fatally flawed by giving the new post-war man only political rights, and not economic, social, and cultural rights by which to live at all. The Soviet Bloc representatives thought "science" took universal man a bit farther than to the polls and the supermarket. To them science meant historical and dialectical materialism, and by implication socialism, pointing up a nice detail of Cold War linguistic politics. Applied to humankind, life and social science in the "Free World" came most "naturally" to mean behavioral science: psychology, sociology, physical and cultural anthropology, behavioral and population genetics, paleontology, and the modern synthesis of evolutionary theory. As the Cold War mushroomed in the 1950s, "behavioral" analysis came to be viewed ideologically in institutions like the United States National Science Foundation as the true scientific substitute for Soviet-inspired social(ist) versions of human life (Senn 1966).

By 1950 various currents of social science, especially Boasian approaches to the relation of race and culture, provided sharp tools for the UNESCO construction of man; but for practical, scientific, and ideological reasons, tools provided directly from the biological sciences were also required. The very disjunction between race and culture in Boasian anthropology, along with the overwhelming emphasis on the second term, left Boasian physical anthropology at best ambiguously authorized to speak about the biological dimensions of "man."[10] Boas himself had conducted the classical craniological work demonstrating that head measurements were responsive to social conditions and did not represent a stable mark on a typological racial scale. His follower in that tradition, M.F. Ashley Montagu, played a central role in the UNESCO task; but convincing biological authority on race rested on a different scientific voice unambiguously categorized within the natural sciences. In this context, the authority of the architects of the modern evolutionary synthesis was crucial to the birth of post-World War II universal man, biologically certified for equality and rights to full citizenship. Before World War II, versions of Darwinism, as well as other doctrines in evolutionary biology, had been deeply implicated in producing racist science as normal, authoritative practice. It was therefore not sufficient for social science, set across an ideological and disciplinary border from nature and natural science, to produce anti- or non-racist doctrines of human equality and environmental causation. The body itself had to be reinscribed, re-authorized, by the chief discipline historically empowered to produce the potent marks of race—Darwinian evolutionary biology. For this task, "behavior" would be the mediating instrument.

The constitution of UNESCO stated that the last war had been made possible by the "doctrine of the inequality of men and races." To fulfill its commission to end racial prejudice, UNESCO had to have the scientific "facts." In December 1949, a group of anthropologists, psychologists, and sociologists from Brazil, France, India, Mexico, the United Kingdom, and the United States met in Paris to draw up a document, released in July 1950, as the first UNESCO "Statement on Race." Although the absence of biologists was later cited to explain the need to call for a second statement, the first UNESCO statement on race bore the unmistakable mark of two of the biological humanists who brought the modern evolutionary synthesis into public consciousness: Julian Huxley and Theodosius Dobzhansky.[11] Neither was an author of the statement on race, but both, along with eleven others, had been consulted on revisions before the physical and social anthropologist, M.F. Ashley Montagu, wrote the version released in July, 1950. Montagu was a forceful apologist for anti-racist scientific ideologies available within the modern synthesis' treatments of race, and the pivotal concepts of the 1950 document were those of the architects of the modern evolutionary synthesis. Their doctrine of natural selection and population biology was about complexity, biological efficiency, and adaptive flexibility. As authors of the sacred texts of mid-century biological humanism (called by John Greene "the Bridgewater Treatises of the twentieth century"), they had strong commitments to a version of the human place in nature that emphasized cooperation, human dignity, the control of aggression (war), and progress.[12]

The second sentence of the 1950 document notes the critical "further general agreement" which differentiates this scientific statement of unity from others reaching back to the Enlightenment: "that such differences as exist between different groups of mankind are due to the operation of evolutionary factors of differentiation such as isolation, the drift and random fixation of the material particles which control heredity (the genes), changes in the structure of these particles, hybridization, and natural selection" (UNESCO 1952: 98). This affirmation of human unity will not be a discourse about the developmental stages of a teleological natural type, as they are arrayed on the hierarchical great chain of being. Rather, it will be a unifying discourse about a more recent kind of natural-technical object of knowledge, one with antecedents in seventeenth- to nineteenth-century natural history and political economy, and then economics and biology, but one which did not displace the system of human unity and differences based on developmental types until the mid-twentieth century, in the face of urgent historical reasons. The new object would be the *population*.[13] "From the biological standpoint, the species *Homo sapiens* is made up of a number of populations, each one of which differs from the others in the frequency of one or more genes. . . . A race, from the biological standpoint, may therefore be defined as one of the group of populations constituting the species *Homo sapiens*" (1952: 98). Such innocent statements of the "facts" transformed whole logics of research programs in physical anthropology, model systems for narrating the evolutionary play and ecological theater, allowable analogies and allegories, field practice, measuring techniques, and the literal visible structure of human and animal bodies, as they were mentally and physically dissected into new pieces.

The 1950 document went beyond negative statements that science provided no proof of inherited racial inequality of intelligence; it stated that "scientific evidence indicates that the range of mental capacities in all ethnic groups is much the same" (1952: 102). The double point of mental equality of races and the species trait of plasticity was to be the keystone of the post-war doctrine of the relation of nature and culture, with the associated disciplinary division of labor between biology and anthropology, bridged by the behavioral sciences, including physical anthropology. "The one trait which above all others has been at a premium in the evolution of men's mental characters has been educability, plasticity . . . It is indeed a species character of *Homo sapiens*" (1952: 100). The argument about plasticity had at its core the logic that made the analysis of bones, muscles, and primate social groups into a psychological science. This research logic illuminated the productive collaborations between anatomists and primatologists working within the new physical anthropology and psychiatrists worried about stress and obsolescence. Through two concepts derivative from the synthetic theory of evolution—the adaptational complex and mosaic evolution (as opposed to the typological trait ranked from primitive to advanced)—Man the Hunter would be enlisted in the 1950s and 1960s to provide arguments on the early origin of plasticity and equal human mental capacities. Anti-racist human unity would turn on these very early, shared adaptational complexes. In the mid-century doctrine of nature and culture, human universals would be the fruit of genetics, biology, and the key humanizing adaptational complexes, like bipedalism and hunting, that shaped the capacity for mental productions called culture. From the point of view of the emerging "new" physical anthropology, a psychological adaptationism was built into biology, and a psychological idealism permeated the science of culture. Through the primacy of psychological explanation, both ends, nature/unity/biology and culture/differences/anthropology, were safe from social(ist) versions of the human story.

The conclusion to the 1950 statement compounded the inevitable controversy over its assertions of mental equality of races with the further claim that "biological studies lend support to the ethic of universal brotherhood; for man is born with drives toward co-operation, and unless these drives are satisfied, men and nations alike fall ill."[14] Although the modern evolutionary synthesis would prove highly extendable, its constructions of natural selection and populations proved unable to sustain such an unqualified statement, which disappeared from the 1951 revision. Even so, rooting social cooperation in physical anthropology's incorporation of the synthetic theory of evolution remained a viable part of several research programs through the 1960s. Such steadfastness would eventually result in the damning accusation of group selectionism against Washburn's tradition of primate studies; and by 1975 a claim for a genetic predisposition to cooperation, unless it was figured in strict investment terms, called inclusive fitness, would be the kiss of death.[15] Any late twentieth-century universal brotherhood of man in the last quarter of the second millennium would have to make do with a rational economic calculus, based on strict exchange equity, which was not quite the same thing as the 1950 statement's affirmation that "every man is his brother's keeper" (UNESCO 1952: 103). Last quarter revisions would read more like "everyone is his/her sibling's banker."

The debate about race and intelligence and about the existence of a natural tendency to cooperation in and between human groups was implicitly gendered. Mental excellence and male dynamism have been closely linked notions, and in turn closely tied to beliefs about scientific rationality, the touchstone for post-Enlightenment versions of human "intelligence." Man the Hunter's and UNESCO man's unmarked (and unremarkable) gender were part of the solution to one kind of racism at the inherited cost of unexaminable, unintentional, and therefore particularly powerful, scientific sexism. Since weapons were the preeminent product of the "adaptational complex" of hominizing intelligence linked to bipedalism and hunting, the question of war at the origin of "man" remained inescapable in these scientific myths that were simultaneously organizing hypotheses. Hunting and war were joined twin brothers, and they had to be separated. The tie of science, the great mythic "human" hunt to reveal secrets, and war had never been more evident than in World War II (Kevles 1978). Weapons, technology, war, mind, and language were all linked to the implicit unifying structure of gender, at the same time as the unifying structure of scientific race doctrines was eroded in the modern synthesis. Therefore, the potential for male cooperation was uniquely at issue in the debates about "human" nature. Female conflict, frequently remarked and studied, was bickering, tied to sex, not war; and putative natural tendencies among females to cooperation came and went with fashions in functionalist explanations of the consequences of mothering. Race could be—and politically had to be—reconstituted after World War II. The same was not true for gender until the late 1960s. Though bound into the "human family" together, race and gender have different social and scientific calendars.

Although the reaction to the 1950 UNESCO statement among American social scientists was on the whole quite favorable, elsewhere the document ran onto serious opposition. In the letter columns of the British journal *Man*, physical anthropologists in England and France expressed concern about the "sociological" composition of the first committee, as well as about the tenor of certain of its assertions. Two contentious issues emerged: the inheritance of mental traits, especially "intelligence," and the question of a natural human predisposition to cooperation over competition (UNESCO 1952: 7). As a result, a second UNESCO committee was called together in June of 1951 to issue a revised statement.

Ashley Montagu, as the primary author of the earlier statement, was carried over to the second committee; but the other members were all geneticists or physical anthropologists—those who were licensed to produce scientific knowledge of the biological-technical object of knowledge called "race." The statement they prepared was then submitted to the ninety-six "experts" before the document was released. Within this group, however, the architects of the modern evolutionary synthesis (including Huxley, Dobzhansky, and J.B.S. Haldane) played an even greater role than had been the case in 1950. Several of the leading figures—including Haldane, Huxley, and Lancelot Hogben—had been prominent opponents of mainline eugenics in the prewar period (Kevles 1985: 112–28); among the authors and commenta-

tors were British scientists from the prewar political left, including Haldane, Hogben, and C.H. Waddington (Wersky 1979). By no means all of the experts were
predisposed to racial egalitarianism: C.D. Darlington and Ronald Fisher, neither
minor figures, opposed the statement largely on the ground of their belief that
races do differ in innate mental capacities. Significantly, at this historical moment
poised at the edge of decolonization and civil rights movements, people of color,
including prominently the anti-racist physical anthropologist at Howard University,
W. Montagu Cobb, were hardly to be found among authors or commentators. Even
with appropriate professional degrees people of color did not qualify as objective
"experts." They would *by definition* be partisan; the marked category does not have
the insignia of objectivity.

Although it did not categorically state that there were no inherited racial differences in intelligence and other mental aspects, the 1951 statement stressed that
mental characteristics did not form part of a reputable scientist's way of categorizing
race. And while its wording was much more cautious, it affirmed the same interactionist paradigm between genetics and environment and the same relation between
nature and culture as the earlier document; the fundamentals of biological humanism were unaltered. Rather than phylogenies and types, it was processes and populations—constructed out of gene flow, migration, isolation, mutation, and selection—
which were to be the privileged scientific objects of knowledge. Although nineteenth-century typologizing has remained a strong tendency even in late twentieth-
century racial classification (Brace 1982), the romantic theorist's mode of scientizing
racial becoming through a teleology of racial/moral/spiritual/intellectual development was seriously challenged. The 1951 statement contained one generalization,
not explicitly in the first statement, which became critical to post-World War II
scientific/political struggles over the meanings of race—and later, sex—group differences: "With respect to most, if not all, measurable characters, the differences
among individuals belonging to the same race are greater than the differences that
occur between the observed averages for two or more races within the same major
group" (UNESCO 1952: 12–13).

Precisely these kinds of anti-typological approaches in the UNESCO statement
made the act of deciding what will count as a group in the first place a major site of
political/scientific struggle, first in the area of race and then of sex. What constitutes a
group must be understood to be a function of the question asked, not of an essential
property of an innocently observed nature. This move de-naturalizes the objects of
race and sex and makes their historical construction recognizable. Evolutionary
populational thinking makes the previously obvious become problematic. The "plain
evidence" of the eyes, so relied upon in typological approaches to race and sex, has
been forced to give way, at least in important part, to a biology constructed from
dynamic fields of difference, where cuts into the field come to be understood as the
historical responsibility of the holder of the analytical knife. The most fundamental
concepts of the modern synthesis have been indispensable resources in post-World
War II anti-racism and feminism.

Typological raciological science was not finally defeated—i.e., stably relegated to
the cautionary discard category of "pseudo-science," always defined so as to show
how current science escapes any such taint. But romantic typological thinking could
now be contested more effectively; and the ability to contest racial taxonomies within
physical anthropology, challenging both the theoretical and technical processes of

producing the taxonomies and the particular classifications produced, has been strengthened by the power and status of the modern synthesis and international anti-racist struggle. With all its warts, this was not a minor achievement. World war, fragile international organization, and newly ascendant scientific paradigms were all required to produce such remarkable statements. Scientific and political struggle would make or break the "general agreement" so hopefully written in 1950. Sherwood Washburn's "new physical anthropology"—with its whole apparatus of publishing, student careers, conferences, collaborations and funding—through its unreserved endorsement of the synthetic theory of evolution was one arena of the necessary struggle to birth the scientific reality of the UNESCO prefigurative proclamation.

The Ambiguous Legacy of Earnest A. Hooton

Of the United States physical anthropologists who participated in the formulation of the second UNESCO statement, six had earned their Ph.D.s at Harvard under Earnest A. Hooton, who himself was among those "experts" consulted.[16] Starting with Harry Shapiro, who received his doctorate in 1926 (and who was one of the twelve drafters of the 1951 statement), Hooton produced about forty doctorates in physical anthropology, including all but two of those awarded degrees before World War II. Hooton's students, or their students, headed six of the eight new post-war programs in physical anthropology that had been started by 1960. During the next decade, when doctoral programs in physical anthropology expanded to about fifty, Hooton's academic progeny were everywhere.[17] But participation in a particular dense network of institutional power is not the same thing as sharing a theoretical and methodological community, even in graduate school, much less beyond it.

Hooton's own background was in classics (the discipline of his Ph.D. in 1911), the archaeology of the pre-Romans, and then in craniology. As a physical anthropologist, he favored the statistical methods of the British biometricians, publishing on the correlations of morphological types and criminal behavior (Hooton 1939) and conducting a multi-year project on the characteristics of the Irish race. Never slow to adopt advanced measurement techniques, he may have been the first to use computer aids in the physical anthropology lab, with his IBM-aided work in the 1950s. However, statistics or other methods of quantitative analysis are not equivalent to, nor do they force movement toward, the populational analyses advocated within the modern synthesis. The functional relation of one feature to another, extent of intra-populational variation, possible genetic mechanisms, and patterns of gene flow and reproductive and other demographic patterns are essential components of the modern synthesis' notion of population. None of these basic kinds of questions determined post-1950 racial analysis fundamentally. Instead, "the paradox of typological analysis undertaken with complex procedures, such as multivariate analysis," characterized much post-World War II racial study.[18]

Hooton's popularizations of his field, widely read in classrooms, were still considered for required graduate student reading as late as 1959. In that year, Washburn was consulted by Julian Steward, whose newly autonomous department at the University of Illinois was setting up a graduate program in anthropology. Steward wrote Washburn for a reading list of works in physical anthropology to be required for all candidates for the Ph.D. in anthropology. A colleague had suggested Hooton's

Up from the Ape (1946), which Steward considered "just a bit on the racist side." Washburn replied that race was a difficult field for the teacher because good reading for non-specialists was unavailable and that Steward was right about Hooton's book. Washburn headed his reading list with Simpson's *Meaning of Evolution* and Dobzhansky's *Genetics, Evolution, and Man.*[19] In Washburn's view, behavior was the link between physical anthropology and the rest of the field, and measurement techniques were justified in terms of their utility in producing "interpretations of the lives of populations." Neither genetics nor statistics produced evolutionary population biology; questions about ways of life did.

In practice, there was considerable ambiguity, confusion, disagreement, and continuity with earlier explanatory strategies in physical anthropology and other historical-political agendas in the family of man. Thus Carleton Coon, a 1928 Hooton doctoral student, who edited a textbook *Reader in General Anthropology* (1948), showed confusion, common in physical anthropology until much later, about how populational analysis might undermine romantic racial and cultural teleologies. Coon's explicit message was the need to understand anthropology in the light of the American Anthropological Association's *Statement on Human Rights*, which had been sent to the United Nations during the period of debate on the Universal Declaration of Human Rights. Recognizing that "qualitative" judgments of cultures would be forbidden by the new scientific order, Coon argued that "quantitative" systematic science alone could be a step in the "formation of those codes of international behavior with which the United Nations Commission is concerned" (1948: vii). In Coon's hands, the "quantitative method" supported the thesis that human cultural streams moved from simple to complex, resulting in his traditional cultural typology reaching from "level zero" (Carpenter's version of monogamous and closed gibbon societies) to "level six" (complex political institutions). Coon's great chain of being remained bound to a romantic typological approach inherited from Broca, Topinard, Deniker, Keith, and Hooton. Coon's later racial taxonomies changed little from his 1939 account. Falling out of analytical contact with much of the text, they persisted as powerful, if amputated, appendages.[20]

The example of Coon demonstrates that a Spencerian doctrine of cultural evolution, coupled with romantic racial taxonomies, could and did inform the physical anthropology textbooks by a major United States writer on race for decades after World War II. The modern evolutionary synthesis' approach to unity and diversity entered physical anthropology much more thoroughly in primate studies and paleoanthropology than in explicit race discourse. In that sense, Washburn's work, only sporadically concerned directly with race (mostly because he had to teach on the subject and was concerned with pedagogical reform in his science) is a better place than Coon's to look for the weavings of U.N. humanism with the new physical anthropology.

The principal theme informing Washburn's life work was functional comparative anatomy: the mutually determining relation between structure and function, organ and behavior, interpreted in a broad comparative, evolutionary, and developmental framework. As a teenager, during the summers he went to the Museum of Comparative Zoology in Cambridge; and in Hooton's program in graduate school at Harvard, Washburn's approach to human evolution was through a study of the functional comparative anatomy of the other primates, culminating in a thesis in 1942 on skeletal proportions of adult langurs and macaques. Besides the biometrician crafting numbers in the service of racial and racist scientific discourse, the other persona of Hooton was the behavioral evolutionist. Broadly among physical anthropologists in the 1930s, the human place in primate phylogeny was a major issue; and W.E. Le Gros Clark's *Early Forerunners of Man* (1934), which reconstructed the history of primate evolution on the basis of the anatomy of living species and the fossil record, influenced opinion about human evolution for the next 30 years. Although he greatly appreciated Le Gros Clark's lectures at Oxford in 1936, Washburn departed from him in arguing a much later divergence of apes and humans. Throughout his career, Washburn followed W.K. Gregory (1934) in regarding the human frame as a made-over brachiator, a matter of considerable importance in the interpretation of anatomy, fossils, primate behavior, and finally molecular taxonomy.[21] Against the grain of Hooton emphases, in graduate school Washburn was also reading Malinowski (1927), whose views applied better to animals than to people, in Washburn's opinion, and Radcliffe-Brown (1937). However, the notions of function and social system in Washburn's physical anthropology owed more to comparative evolutionary biology than to the analyses of either of these social theorists. For example, in later years, Washburn's own graduate students in primate studies were required to take medical anatomy, a discipline Washburn continued to regard as essential to any comparison of human behaviors to those of other animals.[22]

From graduate school in the late 1930s to organizing and participating in meetings sponsored by the Wenner-Gren Foundation from the mid-1940s on, the Swiss primate comparative anatomist, Adolph Schultz, was in many ways more important for Washburn than Hooton. After training in anthropology at Zurich, Schultz came to the Department of Embryology at the Carnegie Institution in Baltimore in 1916, where his studies on the comparative fetal ontogeny of primates and his important collections of primate material laid foundations for the new field of primatology.[23] Continuing this emphasis after moving to the Anatomy Department of the Johns Hopkins University in 1925, Schultz, along with Harold Coolidge, Jr., of the Museum of Comparative Zoology, organized the 1937 Asiatic Primate Expedition. From Boston Brahmin society and a friend of the Washburn family, Coolidge was a force in the development of modern primatology. Schultz's social connections may also have been a factor in the field; he was a Frick on his mother's side, and the Frick townhouse in New York occupied the same Fifth Avenue address as the early Wenner-Gren Foundation (Erikson 1981). In 1937, Schultz was after an anatomical series of gibbons; he was aided by Washburn, a young graduate student getting his first field experience on an expedition that was one of the last of the nineteenth-century-style colonial collecting ventures and the first of the new primate behavior field trips conducted in the frame of incipient international conservation politics (Carpenter 1940). Serving as a role model, Schultz was a major figure in the establishment of the institutional instruments of modern primate studies. But intellectually, Schultz, steeped in biometry, had less impact on Washburn than W.T. Demster, his mentor in human anatomy at Michigan, where the Harvard graduate student spent a semester in 1936 en route to a term at Oxford and then to the Asiatic Primate Expedition. Demster emphasized functional systems and the significance of joints, which biometry ignored.[24]

Washburn's first job, from 1939–47, was teaching anatomy at Columbia University Medical School, where, inspired by S.R. Detwiler, he added a developmental and experimental approach to his functional comparative anatomy.[25] Head of the Anatomy Department, Detwiler was an experimental embryologist. While at Columbia, Washburn formed a close professional friendship with the first director of the new Wenner-Gren Foundation, Paul Fejos. Also present at Columbia then were the human and population geneticists important to the modern evolutionary synthesis, L.C. Dunn and Theodosius Dobzhansky. Like Washburn, Dunn was a staunch political liberal. Dunn chaired the committee whose 1935 report contributed to closing down the Cold Spring Harbor Eugenics Record Office in 1940. Both geneticists were active in reconstructing racial discourse and genetics in an anti-racist direction.[26]

In 1947, Washburn moved to the University of Chicago to replace the departing physical anthropologist, W. M. Krogman. Sol Tax, who was instrumental in bringing him, recalled that Robert Redfield, then dean as well as professor of anthropology, had been very excited about Washburn. Redfield (1942) had edited the fiftieth anniversary commemoration volume for the University of Chicago that celebrated the notion of levels of functional integration linking biology and social science; that framework was completely congenial to Washburn's evolutionary functionalism. The move to Chicago meant Washburn began to have many graduate students, as well as significant responsibility for Chicago's general graduate anthropology course on "Human Origins," the first segment of a three-part (Anthropology 220–230–240) course series developed several years before by Tax, Krogman, and the archaeologist Robert Braidwood to give graduate students an integrated training in the traditional "sacred bundle" of anthropological subdisciplines. Tax himself had early interests in bio-anthropology, having done a 1931 Wisconsin B.A. honors paper on the idea of culture in relation to animal behavior.[27] He and Washburn worked together closely in the "Human Origins" course, which became a significant part of the institutionalizing mechanism of their vision of the discipline, not because it was accepted by all students, but because it formed a map of contestation for large numbers of students from one of the most productive departments in the United States. Assuming the equal potentiality of all human groups, it advanced the biological and evolutionary component of the post-war relativist scientific humanism for which anthropology as a discipline had large responsibility. At Chicago, when social anthropology was incorporated into a program with the study of "human diversification" and the evolution of culture, the ideological structure of the first courses initiating new graduate students into the discipline reflected closely the post-war anxieties and hopes about nuclear civilization that were the social matrix of the founding documents of United Nations humanism (Stocking 1979: 37).

The Emergence of a New Physical Anthropology

In 1948, with Viking Funb support for a project on cranial form, Washburn traveled to East and South Africa. After meeting Raymond Dart and Robert Broom, he was convinced of the revolutionary significance of the South African fossils. He was already deeply committed to the modern synthesis and to experimental approaches in physical anthropology. And he was not alone in his discipline. Beginning about 1950 a spate of publishing on human evolution showed an international and nearly simultaneous interest in the synthetic theory of evolution and populational thinking. A 1947 symposium at Princeton had resulted in an influential, easily available volume. With a change of editors in 1943, the *American Journal of Physical*

Anthropology showed a change in patterns of publication from racial classification to focus on processes of change, underlined by Dobzhansky's 1944 paper "On Species and Races of Living and Fossil Man." In principle, in post–1950 physical anthropology, fossils and races shared the distinction of exhibiting the processes of formation of post-World War II universal man.

But by this time, it was more difficult for a fossil to be represented as an ancient representative of living animals. Viewed from the perspective of the evolutionary synthesis, a fossil was a member of its own dynamic population. And there were more fossils to interpret, lots of them. While phylogenies changed rather erratically, high status debate over fossils came to be more in the language of processes of evolution and patterns of selection. Supposed non-adaptive characters were given less weight, while the search for probable adaptive meanings gained salience.[28] Functional morphology was hardly new, but programs to make its practice part of the modern synthesis gained momentum about 1950. To measure a fossil, one needed an hypothesis about ways of life—i.e., behavior. And though Washburn was not alone, he was instrumental in codifying and institutionalizing the modern synthesis in physical anthropology and in behavioral life science, especially comparative psychology and experimental bio-psychiatry, which deeply influenced the directions of primate social studies in the United States. He was one of the organizers of the important Cold Spring Harbor Symposium in Quantitative Biology attended by 120 geneticists and anthropologists. He and Dobzhansky worked out the program, which was designed to promote the doctrines of the modern synthesis in physical anthropology and human evolution.

The argument about evolutionary physical anthropology and the human way of life developed by Washburn at Cold Spring Harbor and in succeeding papers is a lens through which to view the emergence of an organizing hypothesis of a "master behavior pattern of the human species" (Laughlin 1968: 304), as it was formulated in the post-war period of relative United States hegemony and disciplinary expansion. The synergistic technical and political question animating Washburn's research program can be reconstructed teleologically: What evolutionary account of a human way of life could ground the particular post-war constructions of universal human nature that seemed essential both to hope for survival and to anti-racism? Washburn's particular account of Man the Hunter was written into the texts in the 1940s to 1960s Bridgewater Treatises of the evolutionary synthesis. From the vantage point of post-colonial, anti-western, and multicultural feminist discourses from the late 1970s on, the fatal move in Washburn's approach was precisely the requirement to produce universal man, i.e., a finally authorizing and totalizing account of human unity which submerged the marked category of gender and relegated cultural difference to the thin time layer of the last few thousand years, a kind of icing on the cake of the human way of life, in order to bind "man" into one in the face of the history of biological racism. Scientific humanism, United Nations humanism, and United States hegemony have stumbled on the same dilemma of western doctrines of Man in science and politics: One is too few, and two are too many. These weighty issues left traces in bones, flesh, and behavior.

For Sherwood Washburn in 1950, what distinguished hominids from apes was a decisive difference in the shape of the ilium and its gluteal muscles. The recent discoveries and analyses of the pelves of the South African man-apes had established that modern humans and the small-brained australopithecines both walked upright;

for Washburn, that meant that bipedalism was a behavioral complex fundamental to the meaning of being human (Washburn 1951a). Examining the foundation of the human way of life meant deriving the consequences of a great primary adaptation. Washburn was part of a zoological tradition that saw the modification of the hands and feet for grasping as the main primate adaptation from the mammals, followed by the rich diversity of secondary adaptations like stereoscopic color vision and reduced olfactory sense. Apes diverged from monkeys on the basis of another primary adaptational complex—the pectoral modifications for brachiation. Each primary adaptational complex was embedded in a behavioral transition.

Similarly, the primary adaptation of the pelvic complex was followed by the further adaptations to living on the ground, particularly by the adoption of a tool-using way of life, i.e., culture. With the primary humanizing adaptational complex, the animal adopted the behavior to remake itself. This is the key mythic element of evolutionary scientific humanism. In this framework, culture means first of all tools. Culture remakes the animal; this is the universal foundation of human unity and the structure of the persistent western dualism of nature and culture, resolved through a self-making productionist dialectic. Man is his own product; that is the meaning of a human way of life. Both mind and body become the consequence of a primary adaptive shift registered in the bones and muscles, which are signs, literally, of a special way of life (Oakley 1972 [1954]). Technology has, therefore, an extraordinary scientific-mythic significance, built into the comparative anatomy of the bones and muscles of the pelvis. Stories about technology and skill are here the key guarantors of human unity, where the typological genius of race or nation is made to give way to the universal story of plasticity, culture as tool-using skill, language as instrumental naming, and self-making humanity. Meanings about technology for a post-nuclear scientific humanism permeated this specific physical anthropological discourse.

The division of human from animal by upright posture was not new, mythically or scientifically (Cartmill et al 1986). But Washburn's version emphasized two elements crucial to the wedding of comparative functional anatomy and the modern synthesis: experimental analysis and behavioral hypothesis. Finished form, often all one had to work with in fossil collections, could not be interpreted without experimental intervention to understand how physical forces and functional uses during growth determined adult structure. If the pelves of South African ape-men indicated the primary humanizing behavioral adaptation, the task of the physical anthropologist was the experimental examination of locomotion of living forms, to give meaning to metrical analyses of the fossils. In harmony with the architects of the modern synthesis, Washburn argued that selection is for function. The principal problem of the anthropologist is to understand behavior. Behavior, in this case bipedalism, set the stage for the secondary adaptations, such as tool-using culture, which in turn rebuilt the body.

In the Cold Spring Harbor paper, Washburn suggested a defect experiment that has been a kind of recurring icon of mythic-scientific reasoning in primate studies about what it means to be male. Like Carpenter, Washburn worked from the persistent Aristotelian logic, whereby an implicitly male, dominant center of activity animates the whole. In 1951 Washburn suggested removing the canines of dominant males in rhesus groups on Cayo Santiago to test the importance of these fearsome teeth in social organization. The reasoning was, of course, that males have the social role of defense, requiring the appropriate weaponry. Does society fall apart if males lose their equipment? Since the man-apes did not have large canines, argument by analogy with other ground-living animals leads to the conclusion that at the dawn of hominization, culturally made weapons supplied the body's new biological lacks. (The original hominid male body was feminized, i.e., deprived of built-in weaponry.) It was not just technology in general that made man, but the always already gendered technology of defense, posited as the key to sociality itself.

Fossils, living forms, and the classical functional defect experiments on generalized experimental model systems all converged in Washburn's version of physical anthropology. The anthropologist should see not primitive forms and features, but adaptational complexes. The adaptational complex is a concept embedded in a specific practice of vision; the body is rebuilt and redrawn for measurement or dissection by behavioral hypotheses. Therefore, the scientist should look for models of aspects of ways of life. Kitten skulls could be examined experimentally to understand primate facial growth not because cats were primitive to hominids, but because the felines could work to model a problem. Cat facial growth simulated aspects of another kind of growth. The cat face was an abstraction for analysis, not a simpler form in a sequence. Modeling, not enchaining, was the key scientific operation in experimental physical anthropology.

The year after the Cold Spring Harbor symposium, Washburn (1951b) published "The New Physical Anthropology," codifying the polemic and research program joining physical anthropology and the modern evolutionary synthesis, which should be read with the paper on "The Strategy of Physical Anthropology" presented in 1952 at a Wenner-Gren Foundation symposium on Anthropology Today.[29][Figure 8.2] From their titles on, these papers adopted the rhetoric of a new age, a revolutionary break with an unfortunate past, when opinion could not be sifted from science, and racism and many other ills could result. The sense of revolution was heightened by the claim that the changes were not yet evident in the literature, but were at "the cutting edge" of general and international conversations in discipline-founding settings. Washburn's scientific way of life was a programmatic one; he was the spokesperson and organizational gatekeeper for resources to effect research programs. And for Washburn, physical anthropology's task was to catch up with what evolutionary zoology had accomplished fifteen years earlier.

What did the collective and individual hominid body look like disassembled and reconstructed by these analytical tools? The individual hominid body was "factored" into adaptational complexes molded by selection and evolving unevenly. For hominids, there were three regions of interest, each related to a behavioral complex: the arms and thorax (brachiation), the pelvis and legs (bipedalism), and head (enlarged brain from tool-using way of life). Each complex could be further analyzed into functional subunits, like the upper and lower ilium studied as a function of sex differences, where birth canal width and striding gait opposed each other. Conditions present to make a trait emerge had to be imagined and modeled if possible (Washburn 1951b). The functionalism of Washburn's anthropology through the late 1970s cannot be overemphasized. Parts had to be related to each other to reveal functional meanings; the body was a system. Clinical reasoning (diagnosis) and developmental methods provided the major practices for the research program.

How was this program institutionalized in the years around 1950? Funded by the Wenner-Gren Foundation, Washburn, then the secretary of the American

Figure 8.2   Sherwood Washburn explicating the "new physical anthropology" to colleagues at the Wenner-Gren Symposium on "Anthropology Today," in 1952. From left, the visible faces are Floyd Lounsbury, Irving Rouse, Carlos Monge, José Cruxent, William Caudill, Sherwood Washburn. Property of the Wenner-Gren Foundation for Anthropological Research, Inc., New York, NY. Published with permission.

Association of Physical Anthropologists organized the Summer Seminars in Physical Anthropology from 1945–52. He organized publication of the Summer Seminars' results in the *Yearbook of Physical Anthropology* (Baker and Eveleth 1982) and edited the Viking Fund publications for Wenner-Gren from 1955–60. Alliances with Simpson and Dobzhansky in evolutionary theory and with Schultz in primate comparative anatomy were also facilitated by the international, networking conference system initiated and funded by Wenner-Gren. In 1949, with his ten-day "Heritage of Conquest" conference, Sol Tax had generated the model for these large Wenner-Gren conferences spanning many days, for which people distributed papers in advance.[30] Washburn organized two major conferences—"The Social Life of Early Man" in 1959 and "Classification and Human Evolution" in 1962—and attended others. Washburn and Tax planned the Wenner-Gren "Anthropology Today" conference in 1952.

Before National Science Foundation money was available for physical anthropology and prehistory, and even well after the mid-1950s, the Wenner-Gren Foundation was a force in institutionalizing "the new physical anthropology." Washburn traveled to East and South Africa in 1948 on a Viking Fund grant from Wenner-Gren. This trip put him in a position to advise on further projects. From 1951 to

1961, with 61 % of its research budget, the Foundation sponsored forty-seven research projects and conferences in paleontology/Early Man studies. Starting in 1952, the Wenner-Gren Foundation funded the Early Man in Africa program, an effort to identify and coordinate the burgeoning paleontology and archaeology of hominid and hominoid fossils in Africa. Washburn traveled through Africa again in 1953 evaluating foundation expenditures for the program and attending a Wenner-Gren London conference on Early Man in Africa.[31] Reporting on personnel and centers of activity in Africa, Washburn and Clark Howell (who received his doctorate under Washburn in 1953) had considerable influence at Wenner-Gren. Beginning in 1954 Washburn directed a series of articles in the *American Anthropologist* on results of the Early Man in Africa project; Howell was the first author.

The first phase of Early Man in Africa aimed to identify younger workers, prepare status reports on data, identify promising lines of research, and establish rapport with existing investigators by providing grants-in-aid. Simultaneous with the first stage, phase two introduced European and American workers to the scene, including eventually many United States graduate students in the Washburn and Howell networks. Phase three was to inaugurate coordinated international team research. The idea for multidisciplinary team projects sustained over several years grew partly from the Leakeys' practice at Olduvai Gorge in East Africa, first explored by Louis Leakey in 1931 and the site of the famous discovery by Mary Leakey in 1959 of the australopithecine fossil, *Zinjanthropus*. Beginning in 1966, the first full multidisciplinary project was put together by Clark Howell, then at the University of Chicago and later at the University of California at Berkeley, at Omo in Ethiopia. From both federal and private sources, U.S.-funded paleoanthropology from 1959–81 cost about $4 million, yielding around 600 fossils (Boaz 1982: 253). "Paleoanthropology" is Clark Howell's word for the multidisciplinary discourse grounded in these international undertakings with deep roots among the scientists and institutions in former African white settler colonies and western scientific centers. Most of the fossils enlisted in contemporary debates about human origins have been found since World War II, especially since 1960. Most of what is known about the lives of monkeys and apes has been found out in the same period. The discourse of physical anthropology makes claims about the species as a whole in a world riven with deadly divisions and promising differences. The claims have been about universals; the speakers have been quite particular, precisely positioned, deeply engaged in a world system in which the bones of our ancestors and the living bodies of our zoological relatives are actors in the reconstruction of "man." The material matrix was established for the masculine scientific birth of Man the Hunter. [Figure 8.3]

The Hunting Way of Life

Up to this point in Washburn's writing and organizing, the specific hominizing adaptive complex was quite generally linked to notions of bipedalism and tools. But in the mid- and late 1950s, that adaptational complex began to have much more precise content; it became the hunting way of life. The idea of hunting driving the evolution of "man" was, of course, old. But by the 1960s, that way of life became the dominant theme of a complex research and graduate training program, in which field studies of wild primates and of living human hunter-gatherers were

THE FAR SIDE   By GARY LARSON

Early vegetarians returning from the kill.

Figure 8.3 "Early vegetarians returning from the kill." *Far Side* cartoon by Gary Larson. Published by permission of Chronicle Features, San Francisco, California.

joined to the ongoing work on African hominid and hominoid fossils. Beginning in the middle 1950s, psychology and primate behavioral field studies provided the two new disciplinary and research components necessary to Washburn's hunting hypothesis, complementing comparative functional anatomy in the tool kit for scientifically constructing the human way of life. Through psychology and primate field studies, in joining the "new physical anthropology" to the evolutionary synthesis, Washburn translated the population of the modern synthesis into the structural-functional social group grounding his analysis of the hunting way of life. The translation proceeded by way of the inherently ambiguous and therefore fruitful mediating concept of behavior, through which locomotion, tool use, and speech all found their root in a subsistence adaptation to a specific environment. The extraordinary imaginary and mythic body of the original male hunter on the African plain of human genesis acquired respectable, visible clothing in the quite ordinary behavioral sciences of the academic way of life.

In 1955 and 1956, two meetings were sponsored jointly by the American Psychological Association and the Society for the Study of Evolution, with the support of the Rockefeller Foundation and the National Science Foundation, to bring comparative psychology into the modern synthesis. Anne Roe from psychology and Simpson from evolutionary theory were leading figures behind the influential meetings; many participants have been met before in this chapter, including Washburn, Bernhard Rensch, Ernst Mayr, and Julian Huxley. New players from European ethology and United States psychology were or became major figures in the history of primate studies, including C.R. Carpenter from Pennsylvania State University, Robert Hinde from Cambridge, Harry Harlow from the University of Wisconsin, and Frank Beach, who was to join Washburn in establishing an experimental animal behavior station at Berkeley when Washburn moved there from Chicago in 1958. It was a meeting of notables to authorize not a single theory, but a field of interlinked interpretations dominating post-war evolutionary behavioral science within an "interactionist paradigm" for at least the coming decade.

In this context, Simpson's definition of *population* and Sherwood Washburn and his Chicago student Virginia Avis's translation of the paleontologist-zoologist's population into the behavioral psychologist-anthropologist's *social group*, via the hunting hypothesis, was portentous for the subsequent history of primate field studies as part of a therapeutic physical anthropology enmeshed in the discursive politics authorizing universal man. By Simpson's definition, a population was a unit of common ancestry; approximately contiguous in space; similar or coordinated in ecological role, somatic characters, and behavior; and reproductively continuous over many generations (1958: 531–32). Simpson stressed that evolution occurred in populations, and that adaptation was a remote effect, meaningful only in so far as it contributed to reproductive success. While this formulation could have led to a de-emphasis on adaptation, focus on the concept of behavior in an environment strongly determined by psychologists highlighted the adaptationist possibilities.

But there was considerable slippage in what behavior and adaptation meant to the different interpretive communities. Simpson argued that the principal problem of evolutionary biology was the origin of adaptation (1958: 521). From here it was a short step to redefining the key problem of evolutionary behavioral science (including Washburn's influential physical anthropology) as the origin of adaptive behavior in a functioning social group, with adaptation seen more in terms of the integration of groups than in terms of differential reproductive success. With "group integration" as the core scientific object of knowledge grounding the all-important "sharing way of life," "behavior" would link the adaptationist physical anthropology and therapeutic medical psychiatry into a common research program and public discourse on modern crises. However, if an adaptationist approach were taken without substituting the functionalist integrated group for the Darwinian population, then "differential reproductive success"—fatally for a scientific humanism committed to the "sharing way of life" in the heterosexual reproductive family and in the integrated family of man—might lead instead to an emphasis on competition, individualism, antagonistic difference, and game theoretic views of life as a problem in strategic decision making. Sociobiologists have been even more resolutely adaptationist than Washburn's circle, but in sociobiological accounts each adaptation has tended to be ascribed to a *particular* genetic basis within a logic of competitive individualism. As a result, the story of the genetically based global adaptation—the "sharing way of life," rooted in the humanizing, originary drama of male bonding, nuclear family, and male provisioning by the bipedal, tool-using, hunting ancient hominids—is no longer so easily available.

Because of the relation between biology and culture embedded in the post-war doctrines of human nature to which both Washburn and Simpson subscribed in the name of the modern evolutionary synthesis, there was an important restriction on where the concept of biological adaptation could be applied in the analysis of human social groups. The kind of adaptation which the biological scientist was authorized to speak about had to be located at the origin; it had to be a founding adaptation, on the basis of which doctrines of human equipotentiality and liberal democracy could rest secure. In the context in which Simpson and Washburn were writing, their authority on adaptation could not transcend the construction of the space of origins. Mediated by behavioral science, functionalist *social* science took over beyond the border between nature and culture, biology and society. Functionalist *behavioral* science was precisely at the boundary, mediating biology and society in an interactionist paradigm. The founding adaptation had the task of enabling the human way of life, for which culture was defined as "man's" nature. As Julian Huxley phrased it, culture was the human adaptation. Limits on that adaptation, on its range of meanings, on its forms of organization, i.e., the grounding syntax of human nature, had to be stabilized before the transfer across the border, where the heteroglossia of historical languages took over. On the ground of that founding adaptational complex, cultural relativism was safe; it could not produce a doctrine of natural, biological superiority of some human groups over others, nor could it result in legitimating practices in human society contrary to the "human way of life." The free world's "open society," as well as "the family," would be scientifically grounded in the original, functionally organized personal and collective body of man. Difference was tamed, and plasticity was nicely constrained in the matrix of its generative adaptive process.

In this context, the juxtaposition of concerns in Simpson's paper on "Behavior and Evolution" makes sense. In the midst of surveying approaches to defining populations, summarizing genetic mechanisms, taxonomizing behaviors, and envisioning adaptations in the zoological world, Simpson was enmeshed in a political discourse on human nature. The code is not difficult to decipher. Simpson wrote that "man" was an anxious being because he was a conflictual product of the old and the new, of "deep-seated biological characteristics and recent strong cultural controls" (1958: 519). Evolutionary science could not be transferred into the space of culture except analogically; but by its explanation of the origin of the human adaptation, evolutionary biology could speak to the contemporary "anxieties." The founding adaptation was about process and behavior; it laid the basis for scientific humanism. Here is where Washburn's translation of the population into the social group at the founding moment of the human way of life should be examined, because here is where the collaboration of a physical anthropologist and a psychiatrist in institutionalizing the next decade's primate studies had its justification.

Washburn and Avis opened their paper on the "Evolution of Human Behavior" by reminding the reader that human biology was the prerequisite of development of an elaborate way of life, for which the differentiations separating humans and apes were bipedalism, tool use, and speech (1958: 421). Since features critical to behavior do not fossilize, and since behavior is precisely what must be explained, imaginative reconstruction was explicitly legitimated as part of the scientific craft.

Arguing that much in human life is shared with other primates, Washburn and Avis were particularly committed to a close human-ape relationship, written into the shared brachiating body in papers from the 1950s and later written also into the evolutionary clocks set by molecular taxonomy.[32] Before the australopithecine fossils, there was no way to move from the brachiating arboreal vegetarian to a tool-using hunter, i.e., "man" (1958: 426). *Australopithecus* provided a bipedal organism, who probably used, if not made, tools, setting up the self-making dialectic of culture and body that defined the disciplinary space of the humanist discourse called physical anthropology.

Two conceptually linked changes in the australopithecine hominid body, one for the male body and one for the female, bore witness to the power of the self-making dialectic. The small canine teeth of the australopithecine fossils were taken as witness to the transfer of male group defense and internal dominance arrangements from the biological to the cultural tool. Weapons, the iconic tools, were no longer built into the body, freeing man both for the possibility of disarmament and for the possibility of a self-made apocalypse. Freedom and choice in relation to the female was marked by a different anatomy and physiology. To be human, the female body had to have "lost" estrus in order to ground a species-defining characteristic of universal man—"the family." In the 1950s, Washburn sounded the salient tone for universal man of this specifically human female sexual physiology; i.e, the tone of nearly constant female "receptivity" enabling human forms of "limitation of male sexual activity" (Washburn and Avis 1958: 423), or phrased later in a less masculinist way, "the loss of estrus creates the possibility of choice in sexual behavior" (WG: Sherwood Washburn to Paul Fejos, 14 April 1961).

Washburn and his students always argued that sex was not the basis of primate, including human, sociality (Lee and Lancaster 1965). But sexual difference did ground a fundamental functional adaptive complex, reproductive social relations, which at the human level had to be structured for creatures who could also act in liberal political theory, i.e., make contracts on the basis of a specific kind of equality. Sexuality and technology, each marked in the appropriately gendered bodies, from which physical marks of animality had been erased, were both part of a logic of culture, which rested on the transfer of social causation from the determined body to the disembodied realm of free choice. Physical anthropology was witness to this transfer. Ironically, for a science born in the eighteenth- to mid-twentieth-century practices of marking the body of man to produce the bodies stained by sex and race, post-World War II evolutionary physical anthropology took up the task of erasing these marks to create a more versatile, biologically authorized universal man for the open society. As a result of a species-defining loss, sex and weapon moved from the anatomically marked bodies of female and male animals to universal, unmarked human bodies. The social abstraction, gender, supplanted biological sex for both male and female, as culture, the realm of deliberate action and choice, began its reorganization of the body. As a consequence, meaningful racial difference disappeared in the same narrative logic that saw human female sexual physiology and male "technological physiology" removed from the biological body.

The australopithecines were intermediates. Washburn and Avis argued that their tool use was probably to extend plant foods. Furthermore, they argued that to attain the human degree of evolution, some device to enable a mother to carry her ever more dependent child was likely to have been essential Thus, well before later feminist debates, two of the key elements for arguing the primacy of ideologically and functionally female-coded "technological" innovations in the fundamental humanizing adaptations were already part of the reconstruction. However, in the Washburn narrative, vegetarian and child-focused tool using by the intermediate forms simply set the stage for the genus-defining action to follow. It was in imagining the parameters of life in this final humanizing adaptational complex, hunting, that the most decisive shift was made from the neo-Darwinian concept of population to the functionalist notion of the social group, organized by differentiated psychological needs and social roles.

The hunting of large animals characterized all of the genus *Homo*. The fundamental social consequence of hunting was a new kind of social cooperation—1) among males, and 2) from males to the group in food sharing. Economic interdependence followed from hunting. The territorial consequence was simultaneously profoundly psychological: the world became the territory, as the bounded horizons of vegetarian primate cousins broke open for those engaged in the self-making hunt, where "carnivorous curiosity and aggression have been added to the inquisitiveness and dominance striving of the ape" (Washburn and Avis 1958: 434). The masculinism of these formulations has been widely remarked and criticized. Their fundamental liberal scientific humanism, engaged in the constructions of man in Cold War struggles for doctrines of human nature grounding racial equality and liberal democracy, has been less noticed. The contradictory creature produced through the hominizing behavior of hunting was a natural global citizen, as well as a natural neo-imperialist; a natural political man, as well as a natural sadist; a natural providential father and reliable colleague, as well as a natural male supremacist. His plasticity defined him; his most fundamental pleasure threatened him with extinction. He was father to himself (Delaney 1986). At the origin, man was made as his own creator and destroyer, fully free, constrained only by a nature that made him equal to his brothers and responsible for his fate. Hunting is the act of human procreation, founding at once the nuclear family and the family of man—and laying the foundation for the uneasy technological-family discourses of a nuclear era.

Washburn's later formulations of Man the Hunter did not substantially alter these themes. They were most hardened and widely disseminated in the volume from the Wenner-Gren Man the Hunter Symposium in 1966, organized by former Washburn students Irven DeVore and Richard Lee. The symposium repeatedly undermined its title by recognizing the difficulty of defining "hunting" and by paying attention to "gathering" (Lee and DeVore 1968: 4–7). In discussion, both Washburn and DeVore emphasized the primate matrifocal "family's" continuity over generations, which had been highlighted in recent primate field studies. But if a matricentric group helped, below the human level, to organize interpersonal and intergenerational social life, it was simply to make all ready for the fertilizing principle of the hunting adaptation to produce "the family" and male-bonded, expansive public life.

To this end William Laughlin, a fellow former Hooton student, emphasized that hunting was not a mere subsistence category, but a way of life that served as the "integrating schedule" of the nervous system. Full of cybernetic language about programing, in which the fully abstracted tool, "information," ideologically displaced material tools, as these had displaced the body's equipment. Laughlin's article (1968) could stand as a caricature of idealist, masculinist physical anthropology, in which the intellect, ideas, information, and the brain are the rarified products of the self-reproductive predatory activity which is synonymous with being human. Abstractions are the real booty in the hunting way of life. Gathering is about local foods; hunting is about universal principles. Washburn's co-authored contribution eschewed the rhetoric of cybernetic functionalism, but agreed that the ground of human unity, of the possibility of human universals, is the hunting way of life, which was contrasted at some length to agriculture. Hunting was the "total adaptation" whose major consequence was a world view, indeed precisely the human psychological nature for a global view; hunting demanded "all the human changes," from caring for the sick, to making man the enemy of all other animals in the creation of the concept of the wild, to the basing of art on the artifacts of war, to the love of killing, to male-male cooperation (Washburn and Lancaster 1968: 296–302). Human flexibility—the all-important trait of "plasticity" that assured equipotentiality without threatening hierarchy—followed from a dietary innovation that can only be compared to potent sacrifice, where the animal is consumed to make the man. The population, in which adaptation had meaning only in relation to the discipline of differential reproduction, was far from this anti-racist, imperialist, patriarchal, utopian adaptive complex. Differential rates of reproduction, only remotely tied to adaptation in Washburn's integrating functionalist sense, could not ground the scientific production of universal human nature. All of history, including all of racial differentiation and all of cultural difference, even the great mutation of settled agriculture, paled into insignificance in the face of the truly human way of life, pregnant with ultimate threat and ultimate promise. All men, all those who reproduce themselves with their tools, were equal in this primal and universal matrix of masculinist reproduction of the species. Hunting was not about getting enough vitamin B12.

This extraordinary picture of what it meant to be human had to become part of an anti-racist program in teaching, reaching from schools through university, at the height of the period in which the nations of the Third World gained independence. In "The Curriculum in Physical Anthropology," Washburn summarized fifteen years' of experience for a volume on *The Teaching of Anthropology*, a strategic part of institutionalization of teaching in a period of rapid disciplinary expansion. Washburn noted that "at least 75 percent of the social studies of the last sixty years were a complete waste of time" (1963b: 43) because students were trained in error. Racism was not mentioned, only *error*. So the solution proposed was rhetorically named *science*, not politics. Washburn argued that the synthetic theory "puts evolutionary problems in the form in which culture is important" (1963b: 42). Stressing a similar theme in "The Study of Race," from his 1962 presidential address to the American Anthropological Association, Washburn argued that no race has evolved to fit modern conditions; we are all threatened, with no biological guarantees of racial fitness. Here, Washburn faced politics directly, arguing that IQ tests reflect discarded genetic thinking, differential death rates are due to discrimination, and no society realizes the genetic potential of its members. Potential, plasticity, human universals, and shared threat were the threads of this web of post-war biological humanism, the legacy for Man the Hunter's descendants.

The Study of Primate Behavior

All of the components of the Washburn explanatory program were in place before studies of groups of living primates emerged as a major capstone to his academic edifice. These studies grew from unplanned observations of baboons during the 1955 Pan-African Congress, followed by three months of baboon observations in 1956. Washburn actually did little observing of living primates in the field; but his students at Chicago and Berkeley defined a major body of primatological practice. Prior to that time, only two Ph.D.s (one of them Washburn's) had been earned in anthropology departments for some branch of primatology since the publication of Robert and Ada Yerkes's monumental tome, *The Great Apes*, in 1929. Between 1960 and 1979, there were to be 161 more (although only a portion of them were behavioral field studies)—with others still in departments of psychology and biology or zoology. Of the first nineteen of those in anthropology, fifteen were supervised by Washburn; of forty-seven behavioral primatological doctorates in anthropology prior to 1979, twenty were granted at Berkeley, which also awarded fourteen of the forty-two anatomical primate Ph.D.s.[33] My list of thirty-two primate behavioral or anatomical doctorates granted by Berkeley from Washburn's arrival until his retirement in 1980 includes sixteen women. Although his former student, Phyllis Jay (later Phyllis Dolhinow), was the principal primate behavior thesis supervisor after she joined the faculty in 1966, most of these students also worked with Washburn. The nearest competitors were Chicago, where the Washburn influence remained after he left, and Harvard, under his former student, Irven DeVore.

Following his observations of baboons in Southern Rhodesia, Washburn spent a year (1956–57) at the Center for Advanced Study in the Behavioral Sciences organizing his data and working out the implications of the addition of primate field studies to his framework. Placing living primates in an already developed logic of experimental model systems for comparative analysis, Washburn used them to model particular functional complexes and to highlight differences as well as similarities in the ways these functional complexes were integrated into simian or hominid ways of life. "Closed" baboon society modeled a contrast to the "open," potentially global, hominid subsistence adaptation, where family and home grounded the wanderings of the world hunter (thereby coding the border between animal and man in terms of the poles of Cold War ideological discourse).

Back at Chicago Washburn reorganized courses around functional categories ("walking, eating, mating, thinking"), rather than the previous chronological approach to human evolution. The discourse on origins acquired the syntax of gerunds—turning teleological substantives (thought, bipedalism, etc.) into behavioral processes.In his proposal for the study of the "Evolution of Human Behavior" to the Ford Foundation, which funded his program from 1958–61, Washburn argued that this change emphasized "evolution primarily as a method of understanding human behavior, (rather than the study of evolution being to determine man's place in nature)."[34] Observations of behavior, comparative functional anatomy, and experimental manipulation were the legs of a pedagogical and research tripod, all expressed in the grant rhetoric of revolutionary change in the study of human evolution.

Living forms had to be studied before fossils; study of diet should precede examinations of teeth. In this context, Washburn's program privileged an exploration of the relevance of a "hunting past" to the psychology of "modern man." Experiments involving manipulation of rat diets, compared to studies showing alteration in human physical parameters across generations as a function of migration, would inculcate in graduate students the doctrine of mammalian, including human, plasticity. The rats were neither globally nor hierarchically related to humans; but they were an abstract, partial, and *anti*-reductionist model, which allowed the critical translation from skull measurements to allegorical discourse. "The experiments would show that the plasticity of man has far reaching consequences, whether we are concerned with evolution, race, migration, or the administration of schools and colleges."[35] The pedagogical experiment led readily into the technical translations of the same moral discourse on the human way of life into particular new research projects undertaken by Washburn's graduate students. Here studies of joints-posture-locomotion, brain-skull, and diet-teeth-face were embedded in an encompassing humanist, behavioral discourse. Field work on wild primates was a strand in this web.

In his first statement to the Ford Foundation, Washburn stressed the practical applications of his research program; but he did not yet envision their social and psychiatric relevance. After his move to Berkeley in the summer, his progress report insisted "that the investigations of the behavior of baboons in Africa will have by far the greatest practical applications. Our earlier work on the behavior of baboons has proved of value to psychologists and psychiatrists."[36] Dr. David A. Hamburg, Chief of Adult Psychiatry at the National Institutes of Health, was listed for the first time as a consultant for Washburn's program. That connection would grow, punctuated by Hamburg and Washburn's joint planning of the Primate Project at the Center for Advanced Study in the Behavioral Sciences in 1962–63. Central in the planning of that year, which foregrounded and synthesized early primate field studies, was Irven DeVore, whose 1959 studies of baboons in Kenya were funded by the Ford Foundation grant. After his first term at the behavioral sciences center, Washburn had approached Tax to find a social anthropologist to put on the baboons, who could model some of the consequences of the hypothesized crucial move to the open savannah in the hominid drama. Tax suggested DeVore, with whom he had worked in "action anthropology" on the Fox Project; and DeVore, without a single course in physical anthropology, found himself participating in Washburn's Chicago primate seminar and then spending 1200 hours mostly in Nairobi National Park in 1959 watching baboons for his 1962 doctoral thesis. "My marching orders were very straight-forward. 'DeVore, you've absorbed Murdock, Radcliffe-Brown, and Malinowski. Go out and tell us what it's like with the baboons.' "[37]

Although DeVore had no systematic training in field biology, he had been influenced by W.C Allee—a theorist of cooperation as a basic biological principle, Quaker sympathizer, and member of the foundational Chicago school of animal ecology. DeVore related that his fundamentalist Texas religious background made the religious aspects of Allee's approach appealing (interview). Allee helped contextualize DeVore's approach to male dominance hierarchies. Literally, male dominance hierarchies *animated* baboon society, gave it life. Competition was a proximate means to the larger end: coherent social structure and group survival. The dominance hierarchy of adult males was the independent variable around which the dependent variables, like female rank, were claimed to be organized. In the Washburn tradition's terms, the male dominance hierarchy was the crucial adaptational complex enabling the primate social group. But how was the social group to be described? Here, it was not the Chicago school's organismic community ecology, but the psychologically preoccupied social science practiced at the Yale Institute of Human Relations in the late 1930s and 1940s and a kind of general "Radcliffe-Brown and

Durkheim for the masses" that focused what could be seen in the field in Kenya in 1959.[38] The problem was to establish functional social integration; its specific relation to nuanced theoretical strands in social anthropology was distinctly subsidiary. Although mentioned in Washburn students' references and histories of primatology, social theory was scarcely visible in DeVore's published field study or dissertation except in the bibliography. The names of the social scientists tended more to lend authority than to supply analytic equipment to the rhetoric of the early primate field studies. Durkheim and Radcliffe-Brown were named to authorize a group-structural functionalism that had its own powerful lineage in biological discourse; the Malinowski-Murdock strand in Washburn and DeVore's interpretations emphasized individual needs and functional behavior.[39]

One of the credentials DeVore brought to Washburn was that he had "absorbed" George Peter Murdock. With roots in Malinowski's versions of functionalism, Murdock also shared the basic approaches of the Rockefeller Foundation-funded Institute of Human Relations, which had been established to bring together sociology, psychology, and anthropology in a social therapeutics of everyday life in conflicted capitalist culture.[40] Murdock's basic point was that there exists a "universal cultural pattern," whose explanation cannot be in history, but must be "sought in the fundamental biological and psychological nature of man and in the universal conditions of human existence" (Murdock 1968: 232)—ideas that were widespread in behavioral and social sciences of the 1950s and 1960s. He claimed an "essentially psychological character of the processes and products of cultural change" (1968: 239). For Washburn's program, "the common denominators may be regarded as the result of the gathering and hunting way of life having dominated 99 percent of human history" (Washburn and Jay 1968: x). This is the familiar frame of Washburn's scientific humanism, an account of human nature that evaded history, relegating it to the recent laminar residues of the geological scale. "Man" was self-made, and yet "human nature" was safely unchangeable by history. The moment of origins, the boundary between nature and culture, and the operation of constraints were thus the chief objects of this scientific narrative.

In order to understand the "universal cultural pattern," the scientist would study principles of learning, looking for limitations of the range of potential responses set by "human nature." Not surprisingly, in continuity with nineteenth-century evolutionary anthropologists, Murdock saw the "nuclear family" of father, mother, and child as a constant of the basic cultural pattern. "In contrast to many lower animals, the father is always a member of the human family. . . . Man has never discovered an adequate substitute for the family, and all Utopian attempts at its abolition have spectacularly failed" (Murdock 1968: 244–5). The institution of marriage and division of labor by sex were ensured by the powerful principle of limited possibilities. Human universals rested on the principle of paternity as understood in European-derived societies, so nicely analyzed by Carol Delaney (1986) to mean what Aristotle meant—masculine reproductive potency; masculine formal, final, and efficient causality. As a subject of behavioral science, Man is self-made, father of the species and the guarantor of human nature. The family of man depended on the human family, an indispensable ambiguity facilitating the junction of discourses of technology, heterosexuality, and reproduction. Not about maternal function except derivatively and dependently, "the family" is about the Father, an independent variable, a cause, not a function. Baboons, who were not human, but models for comparison and contrast, did not have "the family": they had the animating principle of male dominance surrounding, literally, the mother-infant bond. But that alone could not create the "open" human society, the 99 % of human history enabled by exogamy, economic exchange, and male-bonded, predatory world travel. Man the hunter, the framing presence within which DeVore's baboon story took shape, ensured the "universal culture pattern," including the limits to the possibilities for a non-paternal field of human unity.

DeVore did not entirely neglect female and infant behavior. His first reports from the field were full of descriptions of the infant as a principal attractive center for the troop; and he and Phyllis Jay published early and simultaneously on mother-infant behavior, a major integrating axis of the structured social group.[41] But DeVore's writing was not fundamentally about development. By contrast, Jay has organized much of her interpretive strategy around the processes of development, still within the frame of the structured social group. Because she worked with langurs, whose life in the trees was given as the reason for the lesser importance of male dominance and its supposed protective function, Jay's account was freed from the narrative constraints of the hunting hypothesis, and her developmentalism was elaborated in the context of functionalist role theory that emphasized sex differences. Looking to biological theorists with a strong developmentalist and holist cast, like T.C. Schneirla (1950), she was interested in adaptation in the social psychological sense, in role learning, and in rates of social change. Normal versus stressed behavior were the poles of analysis: for example, rapid rates of new male takeovers of established troops in circumstances judged to be crowded could lead to social stress and pathological behavior, like males killing infants. Jay ordered her early papers around social bonds that organized and maintained group stability, beginning with the infant and working up to adult role behavior. Dominance hierarchies figured in her analysis, but were subordinated to a developmental and role theory perspective on group integration.[42]

The connection between the consequences of the hunting way of life and the contrastive primate studies was quite direct in time and personnel. Asking Washburn to take over the planning of the Wenner-Gren conference on The Social Life of Early Man in 1958, Paul Fejos wrote that he particularly wanted Washburn's knowledge and interest on "sub-human behavior," especially baboons. In his responses, Washburn was excited about what he saw as new ideas for the conference, prominently including the ability to show "sex differences in attitudes" to be a direct consequence of the hunting way of life, shown by studies on "ease of learning." Material on hormones, personality, authority, and territory all fit in that frame. The conference considered at length the connections of the baboon way of life, anticipated from DeVore's study, with possible behavior of the australopithecines; hunting, bipedalism, and tool use were the turning points in the comparisons. But it was not until the second Wenner-Gren conference Washburn organized that he was satisfied with the demonstration of the power of the hunting hypothesis. De-Vore, who played a large role in planning that conference, chairing the session on Primate Social Behavior, later wrote Fejos that the successful final conference was the first of its kind since the 1950 Cold Spring Harbor Symposium that launched the new physical anthropology.[43]

Washburn's influence at the Wenner Gren Foundation, where 74% of the budget between 1965–80 went for primatology and early human research (Baker and

Eveleth 1982: 42), was particularly consequential for the propagation of his orientation. Following the conference of 1962, the foundation funded eight primate conferences between 1965 and 1982. The first was organized by Phyllis Jay in 1965; followed by John Ellefson in 1968; Russell Tuttle in 1970; Jane Lancaster in 1975, with the New York Academy of Sciences, co-organized by Horst Steklis and Steven Harnad; Mary Ellen Morbeck in 1976; and Shirley Strum in 1978—all of the leading figures were Washburn or Dolhinow students. The ape conference held in 1974 was planned by Jane Goodall and David Hamburg, a Washburn associate throughout this period (Hamburg and McCown 1979). [Figure 8.4] Not until the 1977 meeting planned by Irwin Bernstein was a Wenner-Gren primate conference initiated completely outside the Washburn network, followed by another (on infanticide) in 1982, initiated by Washburn-program opponents, Sarah Blaffer Hrdy and Glen Hausfater. But Washburn had built well; the vision linking the hunting way of life and observational study of primate behavior was institutionalized in students, jobs, publications, and foundation support.

Although after 1970, the number of primate students not associated with Washburn increased, his students dominated the field in the United States in the 1960s and have remained a force well beyond. They got important jobs during the period of the subject's institutional consolidation (up to 1975), e.g., at the Davis, Berkeley, San Diego, and Santa Cruz campuses of the University of California, and at Harvard,

Figure 8.4 The Wenner-Gren Symposium on "The Behavior of the Great Apes," Burg Wartenstein, Austria, July 20–28, 1974. Seated (from the left): Jo Van Orshoven, Itani Junichiro, Lita Osmundsen. Front row: Konishi Masakazu, Roger Fouts, David Horr, Richard Davenport, David Hamburg (Co-organizer), Jane Goodall (Co-organizer), Robert Hinde, Elizabeth McCown, David Bygott, Biruté Galdikas-Brindamour, Patrick McGinnis, Nishida Toshisada. Back row: Peter Rodman, William Mason, Irven DeVore, Emil Menzel, William McGrew, Dian Fossey, John MacKinnon. Property of the Wenner-Gren Foundation for Anthropological Research, Inc., New York, NY. Published with permission.

Chicago, Stanford, Pennsylvania, Oregon, and Texas. They in turn supervised the dissertations of a large percentage of the next wave of field workers. While the dense institutional network associated with Washburn does not indicate a tight intellectual community, it does indicate patterns of access to careers, recognition, and publication, as well as considerable shared explanatory background and assumptions about field practice, which were to be the basis of both criticism and revolt.

Nuclear Culture, Human Obsolescence, and the Psychiatry of Stress

The discourses of Cold War, nuclear technology, global urbanization, ecological crisis, sexual politics, and racism were written into the bodies of early man and living primates. These discourses intersected most densely in the psychiatry of stressed universal man—threatened now with intolerable rates of change and with evolutionary and ideological obsolescence. These were the discourses that structured Stanford's undergraduate Human Biology major, a popular 1970s program preparing students to define and then address controversial politics as bio-social issues. David Hamburg was a founder and teacher in the program, and other members of his Department of Psychiatry played an important part.

Earning his M.D. in 1947, David Hamburg was early drawn to the study of human responses to stress. As a captain in the United States Army, Hamburg delivered a paper at the 1953 Symposium on Stress held by the National Research Council Division of Medical Sciences at the Army Medical Service Graduate School at Walter Reed Hospital, where he began his research career.[44] As a leader in opening up research on human coping behavior under stress, Hamburg became interested in the evolutionary origins of the human stress response; and meeting Washburn at the Center for Advanced Study in the Behavioral Sciences in 1957–58, he developed his psychiatric interpretations within a shared evolutionary doctrine of adaptation and behavior.[45] From 1950 to the mid–1960s, Hamburg conducted laboratory and clinical investigations into the relations of hormonal homeostasis to stress and emotional stability in people. Periods of transition, like adolescence and menopause, had particular significance in a research framework privileging responses to rapid rates of change, i.e., stress. Hamburg used his considerable institutional power and administrative talent to encourage the development of psychiatry as an experimental behavioral science nationally (Hamburg 1970). After serving as chief of Adult Psychiatry at the National Institute of Mental Health from 1958–61, Hamburg chaired the Psychiatry Department of Stanford Medical School until leaving to become president of the National Academy of Sciences Institute of Medicine in 1975. At Stanford, he founded a new department of psychiatry focused on the biology of mental illness and participated in the Laboratory of Stress and Conflict. With his approach to the evolutionary origins of human stress responses deeply tied to Washburn's design for man, David Hamburg's itinerary is a map of the constructions of post-World War II biological humanism.

The Human Biology Program was one of many United States universities' responses to the student politics of the 1960s, especially on environmental and Vietnam War issues. Both of these areas of student protest cut deeply into ideologies of progress through science and technology and into the marriage between the human family and post-nuclear technological discourses. One programmatic response was to emphasize "social responsibility of scientists" and the importance of socially committed scientific education in programs leaving intact ideologies of scientific objectivity while emphasizing service and social relevance. Stanford's program aimed "to prepare policy makers and citizens who have an understanding of biological principles" by focusing on "the complex relationship of man with nature, exemplified by the dilemmas of medical-social policy, population problems, pollution of the environment and conservation of resources needed by the species."[46] Economics, psychology, primate behavior, human genetics, the environment of man, human sexuality, energy utilization, bio-social aspects of birth control, resources and the physical environment, and political processes and human biology were typical courses. Hamburg taught a course on "behavior as adaptation" that compared "adaptive patterns" of primates, human hunter-gatherers, agriculturalists, and industrial societies. Current crises appeared in this frame as failures of adaptation, where the evolutionary and psychiatric meanings of adaptation graded into each other in the fruitful ambiguity essential to the narrative field in which universal man traveled. This important ambiguity was enhanced by another critically ambiguous concept bridging the biological, technological, and social: stress.

"Stress" emerged from the endocrinology of the 1930s and the medical and psychiatric practice of World War II as a dominant integrating concept for post-war social and personal life. "Stressed systems" were those in which adaptive mechanisms, especially communication processes, had gone awry. Fruitfully ambiguous, allowing therapeutic interpretation to slide easily among physical (endocrinological, neural), psychological, and social domains of experience, stress was part of a technological discourse in which the organism became a particular kind of communications system, strongly analogous to the cybernetic machines that emerged from the war to reorganize ideological discourse and significant sectors of state, industrial, and military practice. "The management of stress" has been a major idiom of techno-humanist discourse in medicine, popular human potential movements, personnel policies, and everyday constructions of experience for broad groups of the population. In that idiom, life appeared as a problem of time and information management; "adaptation" appeared as a process of ordering of rates of flow, of ergonomic engineering. Utilization of information at boundaries and transitions was a critical capacity of communication systems potentially subject to "stress" resulting in communication breakdown.

In an evolutionary context, stress idiom was part of an anxious discourse about nuclear war, environmental destruction, unprecedented population growth, sexual and racial conflict. Was "man," evolved in the face-to-face hunter-gatherer societies of the open savannah and the rich hunting societies of the ice ages, "obsolete" in the face of his own creations? Had social evolution proceeded too fast for the organically and psychologically conservative human species? Could science offer any counsel to aid a stressed species faced with the final failure of adaptation, i.e., extinction from the consequences of his own tool/weapon? Would failures of communication and inability to utilize information effectively overwhelm a species defined by its plasticity? What were the limits to that ability to learn? How could the "sharing way of life" be reaffirmed in the global conditions of Cold War, decolonization, and urban crisis? Was universal man, barely born in post-war evolutionary biology, burdened from the start with an intolerable stress load? Stress idioms were part of a universalizing discourse about the species, in which psychiatry, a therapeutics of stressed systems, bridged the biological and behavioral sciences to improve communication and assess adaptive flaws. In this frame, a Human Biology Program could become a popular liberal response to intense political struggles on campuses. This was the context in which Hamburg and Goodall collaborated in joining Gombe East and Gombe West at Stanford in the early 1970s.

An iconic product of Hamburg's Stanford years emerged from the 1969 Conference on Coping and Adaptation, in which Washburn participated.[47] The conference privileged transitional experiences—personal and developmental, social, and evolutionary. The organizers had proposed their meeting in March, and by May 1968 the meanings of coping and adaptation had taken on an unforeseen urgency on campus in the wake of the spring 1968 urban and campus demonstrations—particularly for university science departments seen to be deeply implicated in late-capitalist wars and social and environmental exploitation. The "developmental crises" of menopause and adolescence allegorically illustrated coping strategies at times of rapid change; the moody teenager and middle-aged woman figured a global transition for universal man in technological society. The human genetic and cultural heritage was rich with ways of "cop[ing] with stress in a broad range of social and psychological habitats." Could there be nonviolent ways of meeting global stress and international conflict? Hamburg and his colleagues in psychiatry aimed to make "these questions not only relevant but researchable" (Coelho and Adams 1974: xxiv).

For David Hamburg, an important dimension of that research program was evolutionary, and here primate studies figured prominently. "Thus an understanding of the behavior of the chimpanzee may contribute to a fuller understanding of the origins of human behavior and adaptive response in the face of new and potentially threatening situations" (Hamburg et al 1974: 407). Hamburg and his colleagues believed that justification was clear to study a species which is intelligent and social, with enduring attachments and "family ties" and a long life span, extensive communication repertoire, ability to make and use tools, hunt cooperatively, and to share. Chimpanzees too faced adolescence, characterized by "striking changes in behavioral patterns with clear sex differences" (1974: 407). In addition, chimpanzees presented a "primate equivalent of what is clearly a major current social problem for man, namely destructive aggression" (1974: 408). Teenagers, menopausal women, and aggressive chimpanzees all functioned, through their concrete social status as research subjects, as metaphors in the anxious discourse of technological humanism, which turned on the unspoken question: would adult white ruling class men, who were not research subjects, finally destroy the planet in nuclear war or environmental degradation? Could understanding the hormonal, neural, behavioral, and social webs of life transitions in youngsters, older women, and apes shed oblique light on the future of universal man? But the chimpanzee work could only trace a baseline for human response to stress. For humans, the rate of cultural change had far outstripped any genetic change, precisely the kind of rate difference that had broken the tie between race and culture for the framers of the UNESCO race statements in 1950 and 1951. The rate difference that authorized universal man also introduced intolerable stress. All humans were ill-designed for the stresses of late twentieth-century technological cultures; "there may be some respects in which modern man is obsolete."[48] "Man meets the problems of the atomic age with the biology of hunter-gatherers" (Washburn et al 1974: 7).

If evolutionary physical anthropology played midwife at the international birth of this universal man, so poorly equipped for nuclear culture, stress psychiatry had the task of therapeutic management and redesign. Consistent with a strong liberal ideology within scientific humanism, in Hamburg's lecture on "Man and Nature" at his debut as president of the Carnegie Corporation in 1983, universal man became scientific man. In hope of a way out of irresolvable conflict, the future of the creature with the biology of the hunter-gatherer was to form the family of international science. "The scientific community is the closest approximation our species has so far constructed of a single, interdependent, mutually respectful worldwide family. . . . So too the spirit of science must be brought to bear on this crucial problem of nuclear conflict." The sharing way of life would be recuperated in a science that would "transcend its traditional boundaries and achieve a level of mutual understanding, innovation, and cooperation among its disciplines rarely achieved in the past" (in Lederberg 1983: 431). The human family and the family of man would be achieved in the scientific family, beyond the conflicts of gender, race, and nation that haunted those other imaginations of humankind. This is a discourse that Andrew Carnegie, the philanthropist and author of the *Gospel of Wealth*, would have endowed. Perhaps it is best forgotten that the Manhattan Project stands as one of the greatest interdisciplinary cooperative scientific projects of the time that called forth the birth of biological universal man. Birthing the family of man in the nuclear age, the Manhattan Project brought forth the bomb, christened Big Boy, which earned its father, J.R. Oppenheimer, a Father of the Year award. Reproductive and technological discourses converged with a vengeance.[49]

Sociobiology, Feminist Anthropology, and the Fate of the Washburn Plan

In the last quarter of the twentieth century, the "Washburn plan's" universal scientific man, birthed in a deeply flawed but important struggle against scientific racism, was not the only version in western bio-anthropological discourse and primate studies of human lives, projects, and travel plans. The alternatives were also sources of ambiguous promise and threat embedded in other versions of being human.

In retrospect, some former Washburn students described their sense of having been part of a plan, in which their project was to explore a "functional adaptive complex," a part of the body or mind of Man the Hunter or his primate cousins. The list of doctoral thesis topics validates this impression, but most impressive is the long series of sessions full of Berkeley students' papers at the meetings of the American Anthropological Association or the American Association for the Advancement of Science. For example, an all-day forum at the AAA in 1963 had twelve Washburn students: Adrienne Zihlman spoke on range and behavior, Judy Shirek on diet and behavior, Phyllis Jay on dominance, Suzanne Chevalier on mother-infant behavior, Suzanne Ripley on maternal behavior in langurs, Jane Lancaster and Richard Lee on the annual reproductive cycle in primates, Donald Lindberg on social play in nonhuman primates, Roger Simmons on size of primate groups, Russell Tuttle on primate hands, Donald Sade on grooming patterns in free-ranging rhesus, and John Ellefson on primate psychological constraints. The character of a "plan" was even more evident in the 1966 AAAS Design for Man symposium, which explicitly considered "the human plan" from the functional vantage points of "The Bipedal Plan" (Adrienne Zihlman), "The Skillful Hand" (Richard van Horn), "Biology of Language" (Jane Lancaster), "The Evolution of Emotion" (David Hamburg), and in the privileged unmarked category, "Summary" (Sherwood Washburn, speaking on the hunting way of life and the 99% thesis).

However, from the beginning many social anthropologists were far from convinced about the fruitfulness of the Washburn-linked strategies for constructing the human way of life and its comparative primate ways of life. Hints of trouble from the anthropological clan outside physical anthropology were plain in criticism Washburn students encountered in their other courses and in a not-so-underground anonymous early 1960s parody (by R.L. Murphy), entitled "Man Makes Himself?" The conflation of the family of man and the human family also evoked the objections of those social anthropologists who saw in the neo-evolutionary, post-World War II discourse the tired preoccupations of nineteenth-century writing on "the family." Zoologists, for their part, were far from always happy with the physical anthropological program for primates typified by the Washburn tradition's work.[50]

The Harvard Bushman (San) Project (1963–74), co-directed by DeVore and Richard Lee, illustrates one fault line in the shifting scientific/ideological ground that destabilized the "Washburn plan" and UNESCO's universal man.[51] Lee's master's thesis at Toronto was on "Primate Behavior and the Origin of Incest," and DeVore, Washburn, and Jay encouraged him to go to Berkeley, originally for monkey research. With Washburn's approval, he then leapt at the chance to study hunter-gatherers of the Kalahari desert. He wrote his doctoral dissertation on the subsistence ecology of the !Kung, before DeVore got him his first job at Harvard. The San !Kung were preeminently UNESCO's universal man, hunter-gatherers on the open savannah, living *en famille* the human way of life. Early in the !Kung project, Lee was keen on the evolution of human behavior theme, with the !Kung modeling the crucial 99% of the human way of life, not as primitives, but as fully, indeed normatively, human. But his observations of South African labor recruitment transformed his image of the Garden of Eden into an historical space (Lee 1985). Hunter-gatherers heading for the mines changed the weight of that laminar 1%, making the difference between looking for a "universal culture pattern" and learning to describe barbed wire fences, in order to ground a different scientific politics, one based on "solidarity" rather than "universality."

During the same period, DeVore's position also evolved, as he became a convert to sociobiology—which became a major alternative account for authorizing human fundamentals, as competitive individualism and technicist narratives of life grew in ideological and philosophical importance in the later 1970s. E.O. Wilson's, *Sociobiology, the New Synthesis* (1975), highlighted in its subtitle the fraternal heresy in his ambition to go beyond the "modern synthesis" of men like Dobzhansky and Simpson, whose versions of biological humanism appeared to be a barrier to the inclusion of the evolution of human social behavior in a Darwinian explanation.[52] Cybernetic functionalism luxuriated very broadly, contributing to a kind of technicist revision of the humanisms forged in the early years after the Second World War. Wilson's new synthesis could not have been constructed without its cybernetic doctrine of the organism-machine and its related commitment to biology as a technological communications engineering science (Haraway 1981–82). Cooperation was still the central problem for sociobiologists, but the explanatory strategy for accounting for it veered sharply away from the social functionalism of men like Washburn.

There have been several versions of sociobiology; and many, including DeVore's, are in partial opposition to Wilson's. But they all emphasize a genetic calculus and strategic model of rationality deeply indebted to high-technology war culture. Yet simultaneously, sociobiological models have provided abundant ideological-scientific resources for breaking up previously "natural" units, from the organism, to the heterosexual couple, to the adaptationist-functionalist social group, to the mother-infant dyad. This ambiguous fracturing and recomposing of units, and problematizing of natural statuses in a bio-technical discourse potentially hyper-conscious of its own strategies of construction, has been employed by sociobiological/socioecological feminists like Sarah Blaffer Hrdy and Barbara Smuts to retheorize what its means to be human by retheorizing what it means to be female.[53] In that context, the title of a Smuts manuscript, written for a volume appropriately to be called *The Aggressive Female*, conveys the flavor of early 1980s translations into a bio-social scientific rhetoric of the preceding decade's feminisms: "Sisterhood is Powerful." In Smuts' rhetoric, the hard, game-theoretical languages of optimization strategies and genetic investment tend to give way to detailed arguments about friendship among non-sexually allied male and female baboons and about logics of female sources of power in sociobiological worlds.

The 1970s were also the time of another major challenge to the Washburn UNESCO conception of Universal Man: Woman the Gatherer. Developed from within the Washburn network by two Berkeley Ph.D.s, Adrienne Zihlman and Nancy Tanner, gathering woman emerged from a contentious marriage of Euro-American feminism and biological humanism in the traditions that nurtured Hamburg and Washburn. But like most apparently simply liberal amendments to the false universalism of unmarked categories, this female striding into history with a baby sling and a digging stick and founding language in the company of other females had radical implications that fundamentally undermined her father-tongue's myth system. Her tools and her company challenged the stories of hominization in a theater where the "first beautiful objects" were the staples of masculinist fantasy—"efficient high speed weapons," emblems of the wedding of functionality and abstraction in cultures whose aesthetics perhaps owe more to war than to any other social practice (Washburn and Lancaster 1968: 298).

Woman the gatherer appeared as a serious organizing hypothesis within physical anthropology at a moment analogous to early man in Africa's umbilical tie to the United Nations' founding documents on race and human rights. Rooted in the struggles for decolonization and women's liberation, the United Nations Decade for Women began in Mexico City in 1975 and concluded in Nairobi in 1985. This was also the decade in which the scientific credentials earned by white western feminists, with the help of funds made available by the space race and the Cold War, allowed them to challenge limited but important aspects of the falsely universal discourses of masculinist life and social sciences. In the international "discovery" of the multiple forms of women's agency and voice, symbolized and enacted in the official and non-governmental meetings of the U.N. Decade for Women and in the array of women's movements around the world that made the Decade necessary, other imaginations and politics of human ways of life were available to scientific research programs.

Woman the gatherer was enlivened within a broad western feminist discourse about the sex/gender system that had been systematically invisible within the confines of masculinist biological humanism and its associated sciences.[54] The analytical concept of the sex/gender system for reordering sciences and related discourses on "human" unity and differences was clarified, adopted, and finally challenged by heterogeneous women's movements in the 1970s and 1980s. The pattern may be seen as analogous to the social struggles against fascism in the 1940s and the long processes of decolonization sharpening in the 1950s and 1960s, which highlighted, used, and finally undermined the discursive orderings of the nature/culture system. Nature was the resource to the productions of culture; it was a universalizing and productionist version of the mind/body distinction (Strathern 1980). Sex was the resource to the productions of gender; the bridge between body and mind was similarly behavioral science. The emergence of the sex/gender system as an explicitly theorized object of knowledge helped undermine the mid-century biological humanism of the nature/culture distinction built into the 1950s UNESCO statements and physical anthropology. The biological humanism which was midwife to the U.N. "family of man" contained resources appropriable, if hardly perfect, in the anti-colonial struggles of the 1950s. But this humanism was much less adaptable in the scientific and political contestations against male domination.

By 1975, the elements of the Washburn version of universal man were dispersed, contested sharply and often discarded or incorporated into other narrative fields with very different outcomes. Physical anthropology's 1950s relocation of discourse about primitivity onto monkeys and apes made little sense to primate students concerned with behavioral ecology tied much more to zoology than to anthropology. The social group as the key adaptational complex, subject to stress and failures of communication, was displaced as the central scientific object of knowledge by sociobiology's ascendant versions of genetic inclusive fitness maximization strategies. In this account of natural philosophy, universal man looked more like yuppies of both sexes than the inheritor of the sharing way of life. Feminist challenges to the hunting hypothesis undercut the fundamental organizing axis of male dynamism in hominization. And the biological characterization of universal man, grounding a shared human nature promising a permanent break between race and culture at the cost of heeding the constraints on the obsolescent hunter-gatherer in nuclear society, looked like a denial of cultural reinvention and post-colonial difference from the point of view of dynamic tendencies in ethnography (Clifford and Marcus 1986). The modern synthesis of evolutionary theory had to contend both with the genetic hyper-adaptationism of many strands of sociobiology, the "new synthesis," and with promising hypotheses of non-adaptationist mechanisms of evolutionary change. The narrative field of the Bridgewater Treatises of the twentieth century, which sustained through the 1960s the scientific humanism of the 1950 and 1951 UNESCO documents, was fundamentally strained by the hyper-technological discourses of ascendant molecular biologies and their bio-technical versions of what it means to be human. And in the context of the U.N. Decade for Women, Man the Hunter, once authorized by the human rights and anti-racist documents of the first decade of the United Nations, seemed a poor bearer of the new liberatory discourses and politics on sexual and racial difference.

Africa remains at the heart of international science and politics; but Early Man in Africa, bearer of western imaginations of origins and ends, seems in the 1980s to be more about the simulations of media images, Lucy in the Sky with Diamonds, than about the first steps of the species at the dawn of a self-making history. The animals and the fossils are locked into an international cultural politics of representation and access to the means to tell a common human story on a continent that has given us "racism's last word," apartheid (Derrida 1985). The South African child-fossil, the Taung skull, stands at the end of this chapter as a part looking out onto a whole dissected by apartheid's National Security Management System, sustained by the communications machines of a high-tech world system. Remodeling a human way of life in the 1990s will perhaps mean more dismantling than managing and rebuilding the sciences of stressed communication systems.

9

METAPHORS INTO HARDWARE: HARRY HARLOW AND THE TECHNOLOGY OF LOVE

> The cloth surrogate mother was literally born, or perhaps we should say baptized, in 1957 in the belly of a Boeing stratocruiser high over Detroit during a Northwest Airlines champagne flight. Whether or not it was an immaculate conception, it was a virginal birth. The senior author turned to look out the window and saw the cloth surrogate mother sitting beside him with all her bold and barren charms. The author quickly outlined the researches and drafted the text and verses which would form the basis of his American Psychological Association Presidential address a year later. The research implications and possibilities seemed to be immediately obvious, and they were subsequently brought to full fruition by three wise men—one of whom was a woman. (Harlow et al 1971: 539)
>
> Sadism demands a story. (Mulvey 1975: 14)
>
> You see, I don't want to go down in history as the father of the cloth mother. (Harlow, quoted in Suomi and LeRoy 1982: 326)

Father Knows Best

The recipient of the highest awards and public acclaim that his science could offer, Harry F. Harlow (1905–1981) was a master narrator. He could design and build experimental apparatus and model the bodies and minds of monkeys to tell the major stories of his culture and his historical moment. One story is about the prolific nature of science, about the reproduction of authorship in the virile children of the mind, about the rabbit-like fecundity and virus-like repetitiveness of the curious modern object called scientific research. And like viruses and sadists, Harlow's repetitiveness was innovative and visionary. The images, words, and people issuing from the primate laboratories at the University of Wisconsin in Madison were key actors in mid-twentieth-century U.S. sociotechnical orders.

In an exemplary career in comparative psychology, Harlow authored over 300 books and articles, founded two major research laboratories, delivered innumerable popular lectures for audiences from listeners of the Voice of America Forum to sexologists and college students taking introductory psychology, and produced 36 Ph.D.s. His laboratory was a model for successful primate husbandry, as he (and his students, wives, post-docs, research associates, secretaries, technicians, animal caretakers, primate import entrepreneurs, architects, janitors, and donors) created and managed one of the first large captive indoor breeding colonies of monkeys. The colony was the site of production of disease-free rhesus infants as components of a major comparative psychology testing industry measuring simian mind and

231

emotion in the service of liberal, humanizing reforms in social services, education, psychiatry, and family life. Establishing procedures between 1955 and 1960, Harlow and his colleagues produced monkeys for research at a scale that dwarfed Robert Yerkes's 1930s chimpanzee breeding station. Harlow's laboratory breeding populations were the settler colonies to the field primatologist's social practice of natural resource extraction. Harlow told the story of the superiority of the artificial and fully controlled, of the laboratory that supplanted and surpassed nature in productive power.[1]

Another major Harlow story is the closely related one of the second birth, that self-birthing in western humanist myths, in which the hero completes his task, thereby fathering himself and figuring the appalling immortality that was denied in the first mortal, maternal birth. In this chapter's opening text, Harlow narrates this story in the burlesque style that characterized his popular writing and erupted uncontrollably in his more disciplined professional texts. It is a parodic story about mirrors and self-reproduction in the alcohol-saturated belly of a late capitalist commercial stratocruiser flying over the capital of the American automotive industry in 1957. It is also a story that illuminates the conjoined scaffolding of sadism and masculinism sustaining many narratives of scientific authorship and the adventure of doing science.[2]

The senior author of the popular *American Scientist* article, Harry Harlow told of looking out the window by his seat and seeing his reflection, his double sitting beside him, "with all her bold and barren charms." His reflection was his product, the artifact that would make him immortal for the audience of his scientific peers and his public. He was fixed in and by his own gaze. His double is a seductive, sterile woman, who, incorporated into his "text and verses" for a presidential address, becomes a prolific, if minimal, mother. The surrogate, substitute self was a scientific inscription device, birthing the textual children that promise continuance and fame to their authors. Described more fully below, the surrogate mothers were a series of variations of cloth-covered or wire mesh, milk-dispensing or dry, rocking or stationary, endowed with head and face or only a trunk, "mothers" to which otherwise isolated baby monkeys had access.

The surrogate mother at her richest had only one site of lactation, another in the long series of functional artifactual jokes issuing from Harlow's lab. "In devising this surrogate mother, we were dependent neither upon the capriciousness of evolutionary processes nor upon mutations produced by chance radioactive fallout. Instead, we designed the mother surrogate in terms of modern human-engineering principles. We produced a perfectly proportioned, streamlined body stripped of unnecessary bulges and appendices. Redundancy in the surrogate mother's system was avoided by reducing the number of breasts from two to one" (Harlow and Mears 1979: 106). Harlow's virginal self reflection in the airplane was an efficient, androgynous android—bold and barren, a modern mother. In union with and double of the scientist, she is reproductive, but never redundant. The series of artifactual mothers not only reflected Harlow's vision. The mothers, and their laboratory hardware successors used to study psychopathology, also birthed a series of graduate students. Their writing that gave them authority as psychologists emerged from their practice within the social system of experimenter, infant rhesus, and technical constructs.[3]

The quest narrative containing Harlow's story of his vision of the self/surrogate,

"From Thought to Therapy," credits the timely aid of birthing assistants, i.e., other wise men in the school of midwifery accredited by Socrates—one of whom was a woman, in this case, his co-author wife, Margaret K. Harlow. The status of the woman in the laboratory, the iconic female man, has seldom been as clearly put. A doctoral research scientist, Margaret Harlow became a midwife in the reflective birth of her husband as father to himself-as-mother, a scientific hero.[4] This is masculinist narrative in its most skeletal form. No wonder the style of the Harlow labs reads as burlesque parody. The simultaneously literal and jocular quality of Harlow prose is part of its fascination, as the scientist translates metaphor into hardware.

Indeed, that translation is the primary quest of the Harlow laboratory, as both "nature" and "society" are rewritten into productive technical-scientific artifacts in a limitless itinerary called research. Good artifacts produce "families of problems" (Harlow and Mears 1979: 125) and concatenations of research projects, as language and technical construct endlessly generate each other in a process of translation and inscription. Harlow's papers are often narrated as travel stories, with serendipitous discoveries and unexpected turns in the road, leading in Harlow's case from psychological learning experiments to study of affectional systems in maturation to simulations of human psychopathology and therapy. The road is magical, a kind of labyrinth, with many detours promising the golden reward of endlessly fruitful scientific reproduction, the specific kind of immortality mythologized in laboratory travel literature. "We have never completely forsaken any major research goal once we pursued it, and we are still searching for the end of the rainbow—even though we have already found our fair share of research gold" (Harlow et al 1971: 538). Continued funding for further research, the mythic generative power of gold, is only one of the meanings for the common scientific narrative strategy Harlow appeals to in this quotation. To practice science offers the ideal travel experience.

Reinforcing the heroic masculinist narrative of self-birthing is the forceps of sadism. It is important to stress that the sadism does not lie, at least not originally, in the fact of causing repeated pain to animals in the course of experiments. Rather, the sadism is the organizer of the narrative plot and part of the material apparatus for the cultural production of meanings; sadism is about meanings produced by particular structures of vision, not about pain. In fact, sadism is about pleasure in vision; it is an erotic visual discipline for self-objectification. It is the forceps of the children of the mind and the eye. The visual technology typically works by reducing the other to a flawless, perfectly controlled mirror of the self, in a dialectic of master-slave set out clearly by Hegel (1979) in *Phenomenology of Spirit* and studied extensively in political theory and psychoanalysis.[5] Sadism produces the self as fetish, an endlessly repetitive project. Sadism is a shadow twin to modern humanism, a fact well understood by de Sade and Foucault. Harlow's lab was about the fulfillment of primate potential, not about the agony of research animals. Their pain was *at most* an unfortunate byproduct, to be eliminated if possible. Indeed, Harlow's experimental subjects were probably better off than the overwhelming number of laboratory primates in the same period, more a dismal commentary on lab culture than a tribute to the Wisconsin practices.[6] Sadism is about the structure of scientific vision, in which the body becomes a rhetoric, a persuasive language linked to social practice. The final cause, or telos, of that practice is the production of the unmarked abstract universal, man.

To borrow from modern film theory, sadism in Harlow's laboratory is an effect of the "scientific appparatus." In "The Violence of Rhetoric," Teresa de Lauretis links Yurij Lotman's theory of plot typology to Laura Mulvey's analysis of the structure of narrative pleasure in dominant Hollywood film (de Lauretis 1987: 43–45). Lotman argued that in the mythical text there are only two characters, the hero and the limit of his action or the space through which he moves. The hero is the creator of differences and as such is structurally male; the female is both the space for and the resistance to marking, "a topos, a resistance, matrix and matter" (1987: 44). To be the originator of differences is to be the author—the defining position for the unmarked gender, the masculine. In Hollywood cinema, a visual discipline, the female/woman is fixed in the position of icon, spectacle, the one looked at, in which the subject sees the objectification of *his* action and subjectivity. She is his work. The female becomes his product, his reflection, perfectly mirroring his fantasy of himself, as all traces of her resistance are ground away. Her matter is formed by his activity, the plot of Aristotelian philosophy so basic to biology. The narrative of this objectification is the plot of sadism; it is its story. Not an exhaustive account of desire in the laboratory, the story is nonetheless part of the lab's apparatus of bodily production.

How to look is built into the spectacle, as aggression and anxiety are transmuted into the gold of the perfect image, the simulation or copy that exceeds the original, whose independent existence may in any case, in this narrative, be doubted. Visual inspection, always the privileged form of knowing for western scientists, shows only the reflection, copy, substitute, fetish, in an endless chain of image-signifiers. In stark contrast to the codes of the *National Geographic*, neither nature nor author is represented in Harlow's iconography; only the inanimate surrogates and monkeys in their fully cultural form as experimental inhabitants of a laboratory colony are pictured in the scientific texts. The monkeys are even stripped of their "rhesus-specificity"; they are models and substitutes for human infants. The mythic dialectic between author and limit/space/nature has been objectified in the image, whose story or history is repressed in the timeless quality of the scientific artifact.

One may look at this image safely, without emotion, or rather, with the disciplined emotion of disinterest, proof of one's accomplishment (Keller 1985). As Harlow repeatedly said about the value of the surrogate mothers, "We have absolute experimental control over them."[7] Laura Mulvey discussed this kind of control, in the context of filmic imaging, with the aid of psychoanalytic theory. In that perspective, the body of the (real) mother signifies the threat of powerlessness (castration), and so the image of the mother can rouse anxiety. The dis-ease can be deferred by the production of an ideal image, a fetish, a substitute, a spectacle.[8] There is no end to the process of deferring anxiety; it is endlessly reproductive of knowledge, self, and pleasure. No wonder Harlow's work was presented to television audiences in 1960 by the popular scientific CBS Conquest Series, under the perfect title, *Mother Love*, in which the narrator solemnly intoned, "The man in charge is Harry Harlow." His task was to show what an infant's love for its mother really is.[9] The answer was, in unintended irony, *touch*, the bridging of distance. Touch was not visual; it was the sensuous clinging of the nearly totally isolated monkey to the remnant of social life present in the cage, the cloth-covered surrogate. Voiced by the deep-toned narrator, the concluding moral of the TV program was, of course, that "until man knows himself, he can know nothing." The mother surrogate and infant isolation experiments in the Harlow labs were part of this stunning project of humanistic self-knowledge.[10] It was in this context that one must see the TV scientists talking calmly to the camera about love, while the visual field behind and around them is full of self-clutching, autistic infant monkeys, experimentally produced to show the "touching" adequacy of a mother surrogate in a liberal, rational society. Sadism demands a story.

Harry Harlow did, in fact, become known as the father of the cloth mother, as "her" images populated numerous textbooks, scientific articles, and films, becoming part of the iconography of psychology through the 1960s and 1970s.[11] It was a complex scientific maternity, one addressed to the "social problem" of human motherhood in the context of western economic and sexual rearrangments of the post-World War II period.[12] Harlow produced a particularly American scientific mediation of the specific historical construction of white, middle-class gender, embodied in the semiotic and technical apparati of sexuality and reproduction. Stimulated by his meeting with the psychiatrist John Bowlby at the Stanford Center for Advanced Study in the Behavioral Sciences in 1958 (an alternate origin story to the reflective self-birth in a jet over Detroit), Harlow began devising experiments to test Bowlby's theories about infant-mother separation in the genesis of infant emotional disorders.[13] The initial impetus for Bowlby's work was the condition of war orphans in institutions in Europe after World War II. By the 1950s, mother-infant separation studies had a quite different context. Harlow's research on the mother surrogates was originally received in the liberalizing context freeing white, middle-class ("human") mothers from the tyranny of doctrines of the infant's need for non-working, full time, "natural" mothers.[14] The freedom was perhaps more about exchanges on the labor market than about women's reproductive freedom. However, Harlow used the popular mother surrogate work to argue for a more humane understanding of primate (including human) mothering, in which social influences were shown to be more important than behaviorist stimulus-response or instinct/drive reduction models of development.

A critic of drive reduction theories throughout the 1950s, Harlow showed in learning experiments that curiosity was a motivator, and a more powerful one than food. The "drives" of hunger, fear, or rage inhibited more complex intellectual and emotional development, for which social interactions and curiosity were both goal and reward.[15] Harlow was consistently critical of behaviorist and psychoanalytic versions of reductionism. His own "reductionism," his stripped-down, efficient, artifactual rhetoric, was about translation and social action, not about "mechanistic" models. The mother surrogate was a literal social construct, modeling social interaction, which destabilized the privilege of the natural object, including mothers theorized by either behaviorists or drive reduction psychoanalysts.

As a "humanizer of mothering," Harlow told of the development of "affectional systems," beginning with an adequate maternal and infant love that could be achieved by working mothers (and fathers sharing child care), moving through peer friendships, and ending reassuringly (to whom?) in effective heterosexuality in nuclear families, where the paternal affectional system flourished. This liberal message reached white, middle-class, working mothers in the same decade that produced Betty Friedan's *The Feminine Mystique*. Harlow's lab was designed to be "humanist": "If one searches for data that are socially significant as well as statistically significant, a laboratory should be more than a concrete dungeon. . . . The social environment of the laboratory should stimulate all of its inhabitants, including the experimenter, to achieve their full intellectual potentials" (Harlow and Mears 1979: 162). Again, sadism is inherently about the development of potential; it is integral to the vision of man. "The challenging problem is to produce laboratory environments in which the feral animal transcends its feral capabilities. What man did for man he should be able to do for monkeys" (1979: 5).

So, with the mother surrogates, Harlow and his colleagues demonstrated that contact comfort, not primary drive reduction of hunger, was critical to an infant primate's earliest healthy emotional development. The "real mother" remained superior to the "man-made" one, ensuring in the rhesus case, for example, that the infant become a placer rather than a spreader of feces (Harlow et al 1971: 540). But the superiority had a functional basis, and so one open to future social engineering. However his message was received by wage-earning, white women, Harlow's writing is replete with ambivalence about the working wife. It is an ambivalence built, by reversal, into the laboratory logic of replacement of the mythic maternal function by its equally mythic paternal function. Harlow argued that women threatened to replace men in the public space of the economy; so his laboratory cloth surrogate showed that fathers could replace mothers in the home. He burlesqued as always: "But I have also given full thought to possible practical applications. The socioeconomic demands of the present and the threatened socioeconomic demands of the future have led the American woman to displace, or threaten to displace, the American man in science and industry. If this process continues, the problem of child-rearing practices faces us with startling clarity. It is cheering, in view of this trend, to realize that the American male is physically endowed with all the really essential equipment to compete with the American female on equal terms in one essential activity: the rearing of infants" (Harlow and Mears 1979: 125). If women threaten to replace men, then mothers can be scientifically shown to be redundant. Substitution is dangerous if you lack the technology of reproduction: the laboratory.[16]

The surrogate man-made mothers translated controversial social issues into laboratory artifacts; in this sense, the surrogate mother was a member of a generative series of Harlow technology reconstructing metaphor into hardware. Verbal and artifactual rhetorics were both enlisted in these key translations and reflections effected by the scientific research laboratory. Women's and infants' lives became the focus of intense scientific constructions, simulations, and substitutions in the University of Wisconsin Psychology Primate Laboratory. Women and human infants were translated first into the denaturalized bodies of rhesus monkeys and then into the jocular hardware of laboratory constructs in a chain of signifying practices crucial to post-war reproductive politics.

The Surrogate Mother and her Mutants

Harry Harlow's laboratory had its origin in the history of mental testing in comparative psychology. He had been trained by masters of the field at Stanford in the late 1920s: Calvin Stone, Lewis Terman, and Walter Miles, all contemporaries and peers of Robert Yerkes. Harlow married one of the subjects of Terman's "gifted children" study. Comparative psychologists have been extraordinarily creative in devising testing situations and technology; the testing industry is central to the production of social order in liberal societies, where the prescriptions of scientific management must be reconciled with ideologies of democracy. Harlow's lab's creativity in generating testing technology did not begin with the surrogate mother envisioned in an alcoholic smog over Detroit. Finishing his Ph.D. in 1930, Harlow took up an academic position as assistant professor at the University of Wisconsin. He intended to study the workhorse of experimental psychology, the rat; but the Psychology Department's animal laboratory had been torn down the previous summer. Harlow ended up studying monkeys at the Madison Villas Park Zoo. Over many years and many transformations of laboratory space and scale, the Harlow primate study facilities became the university Primate Laboratory and the federally funded Regional Primate Research Center, which he directed from 1964–71.

Harlow's first major contribution to laboratory artifacts was the development of the Wisconsin General Test Apparatus (WGTA), which allowed efficient presentation and variation of discrimination problems to experimental subjects. The backbone of monkey intelligence tests, "(t)he WGTA brought to the study of primate learning capabilities what Henry Ford's assembly line brought to manufacturing: a means by which a large number of discrete learning tests could be rapidly presented in highly standardized fashion to subject after subject" (Suomi and LeRoy 1982: 320). The learning tests were integrated with cortical localization and lesion studies and with ontogenetic studies of motivation and learning. It was these series of tests over more than 25 years that established that primates learn to learn (i.e., form "learning sets" or strategies for solving problems similar to those they already know how to solve), that curiosity motivates learning better than food (the work that attacked drive reduction theory), and that large numbers of monkey infants can be hand-reared in partial isolation to provide disease-free subjects for psychological measurement of the ontogeny of learning. The WGTA positioned experimenter and monkey face-to-face, but vision was not symmetrical. The experimenter viewed the monkey through a one-way screen. The monkey sat in a cage in front of a stimulus tray on which various problems were set and rearranged objects on the tray to give answers.

This simple architectural arrangement was remarkably flexible in the kinds of problems that could be raised. The testing situation allowed Harlow and his colleagues to bring into and resolve in their laboratory a previously heterogeneous lot of questions and claims about learning that had been vexatious to many animal and human psychologists. The WGTA embodied the central logic of the laboratory: a means of multiplying, displacing, and condensing phenomena so as to interest diverse constituencies (Latour 1983). In enlisting these constituencies, the WGTA moved from the Harlow lab and into other worlds, e.g., in the testing of learning deficits among retarded human children. The WGTA was prolific and reliable, generating an endless and varied progeny of scientists and science.[17] The discovery of learning sets justified Harlow's election to the National Academy of Sciences in 1951. It was out of this history that the mother surrogate work was possible.

In an expansion of the Primate Laboratory in the mid-1950s, the Harlow team needed many more young monkeys for the learning and brain localization research; and so they decided to breed their own animals, both to cut down on disease in the colony and to get a "research effective" if not "cost effective" experimental subject (Suomi and LeRoy 1982: 327).[18] This decision was inadvertently one source of the study of the effects of social isolation, because infants were separated as soon as possible after birth, reducing the mental trauma (whose?) and danger of injury to both animals and people of repeated separations resisted by real mothers and babies and giving experimenters easy access at their convenience. Beginning in the mid-1950s with about 40 individuals, the infants were kept singly in cages, able to see other animals and lab staff, but in touch only with their own bodies and a cloth diaper in the enclosure. These animals showed gross behavioral trouble, staring into space, clutching themselves, and performing unending stereotypic automatic movements.

As adults, neither males nor females knew how to have reproductive sex; and the females who did become pregnant reentered the story of the lab as themselves part of the experimental apparatus to produce psychopathology in their young. They were "rejecting mothers." Since they rarely became pregnant by "natural" means, a disconcerting development in a would-be breeding colony, the experimenters intervened. A device was used to immobilize a female while artificially inseminating her. In the normal ribald humor of the laboratory, granted over $1 million by the National Institutes of Mental Health to study the nature of primate love, this piece of lab equipment was, in print, called the "rape rack"; "we resorted to an apparatus affectionately termed the rape rack, which we leave to the reader's imagination" (Harlow et al 1971: 545). After this "procedure," the females then became a natural-technical object of knowledge called the "motherless monkey mother," the most effective in a sorority of "evil mothers" engineered by the Harlow lab to study psychopathologies for the NIMH. The isolated and technologically raped rhesus mothers were members of the family of maternal surrogates; they were constructs for the literal translation of metaphors into hardware—or in 1980s techno-speak, wetware. "Not even in our most devious dreams could we have designed a surrogate as evil as these real monkey mothers" (Harlow and Mears 1979: 220). Misogyny is deeply implicated in the dream structure of laboratory culture; misogyny is built into the objects of everyday life in laboratory practice, including the bodies of the animals, the jokes in the publications, and the shape of the equipment. Misogyny is the result of discipline in Foucault's sense; it is a productive—and well-funded—discursive practice.[19]

The other "evil mothers" at the Wisconsin Primate Laboratory were not animate; rather, they were variant replicants of the original cloth mother. The hypothesis inspiring the design of these mothers was that maternal rejection caused psychopathology. The literal translations of this social metaphor all involved surrogates "designed to repel clinging infants . . . One surrogate blasted its babies with compressed air, another tried to shake the infant off its chest, a third possessed an embedded catapault which periodically sent the infant flying, while the fourth carried concealed brass spikes beneath her ventral surface which would emerge upon schedule or demand" (Harlow et al 1971: 543). The last mother was elsewhere called the "iron maiden" (Harlow and Mears 1979: 220), perhaps unknowingly recalling the famous torture device of that name, said to be used by the Countess Elizabeth Bathory in sixteenth-century Hungary to kill young girls in her erotic practices, in stories fundamental to homophobic vampire literature (Baty 1985). "Fortified by the knowledge of the powers of our mechanical mother (the original cloth surrogate), we tried to create pathological disorders in the infant by changing our cozy mother into a cruel caricature of herself" (Harlow and Mears 1979: 220). None of the evil surrogates provided useful models of human psychopathology, at least not of the kind intended in the laboratory, though they might, obviously, be said to provide good models of psychopathology among experimenters. But before turning to experimental designs that did work according to plan and their meta-phoric logic, let us move from the evil mothers to the normal—to the original cloth and wire surrogates used to study not psychopathology, but the maternal and infant affectional systems. [Figure 9.1]

The basic design was simple. The mother was made from a block of wood, covered with sponge rubber, and wrapped with terry cloth. A light bulb behind her provided heat, and a round doll head with bulbous eyes, smiling mouth, and big ears topped the figure, giving the impression of a clown face. "The result was a mother, soft, warm, and tender, a mother with infinite patience, a mother available 24 hours a day . . . Furthermore, we designed a mother-machine with maximal maintenance efficiency, since failure of any system or function could be resolved by the simple substitution of black boxes and new component parts. It is our opinion that we engineered a very superior monkey mother, although this position is not held

Figure 9.1 Infant rhesus monkey with cloth and wire "surrogate mothers." Harlow Primate Laboratory, University of Wisconsin. Published with permission of Helen A. LeRoy.

universally by the monkey fathers" (Harlow and Mears 1979: 106). The inadequate mother constructed for the same series of experiments showing the role of contact comfort was a wire mesh creature, also with a light bulb and an optional "breast," but with a square head and eyes, and no mouth or ears, giving the impression to adult experimenters and readers of bleak joylessness. But "(t)o a baby, all maternal faces are beautiful. A mother's face that will stop a clock will not stop an infant" (1979: 132).

These creatures allowed endless repetition with variation, all to test normal mothering and infant response, i.e., variables in "maternal efficiency" conceived in an engineering frame. So there duly appeared among the maternal androids a cloth surrogate with ice water pumped through "her" body (named predictably the "ice cold mother"). This creature's opposite was the "warm woman" or "hot mama." Neonatal monkeys preferred the warm surface; and in the case of the icy cloth surface, they recoiled to a far corner of the cage and never approached it ("her") again. "There is only one social affliction worse than an ice-cold wife, and that is an ice-cold mother" (Harlow et al 1971: 540). Surrogates were constructed with gentle rocking motion, without heads ("simplified surrogates"—"designed to get maximal personality out of minimal mother"), with reversed heads, and so on (Harlow and Mears 1979: 135). Simplified surrogates were made with various coverings, from cotton socks to linoleum and sandpaper. Infants were put with these "mothers" on various separation schedules and in diverse situations, such as the open field test box full of toys and/or fear stimuli. Monkeys' intellectual and emotional development was documented in detail in relation to each replicant and each context. Any student could find a variable to test, a social metaphor to build. The reproductivity of the lab was amply proved. The National Science Medal awarded to Harlow in 1967 showed at least that.[20]

Harlow and his colleagues isolated five affectional systems for study in various kinds of lab apparatus: infant love and maternal love, peer love, heterosexual love, and paternal love. The systems grew out of each other in various ways, and the monkey work was always closely associated with the mental health institutions. The study of paternal love is of particular interest for this chapter because its elucidation required a particularly intriguing artifact: the nuclear family apparatus. [Figure 9.2] If ever there was a device designed to let animals exceed their feral achievements, this was it. "The nuclear family apparatus . . . is a redesigned, redefined, replanned, and magnified playpen apparatus where four pairs of male and female macaques live with their offspring in a condition of blissful monogamy. In the nuclear family apparatus each and every male has physical access to his own female and communicative access to all others" (Harlow et al 1971: 541). The monogamous father became an iconic Harlow natural-technical object of knowledge in a period of great concern for "the family" that characterized suburban America in the 1960s. The nuclear family apparatus was part of the incitement to discourse about sex and gender in the privileged biopolitical arena where power is embodied in modern societies. The apparatus made literal the theoretical concept of obligatory heterosexuality that would emerge from the mid-1970s in Euro-American feminist theory.[21] The laboratory rhesus monkeys, as always, complied in the production of discourse in the rhetorics of their own pliable bodies.

Each infant in the nuclear family apparatus, a planned social environment worthy of Disney World, had access to the whole neighborhood, including his or her own

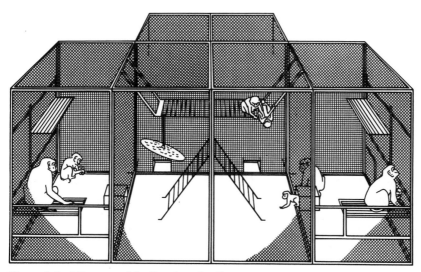

Figure 9.2   Diagram of the "nuclear family apparatus." Harlow Primate Laboratory, University of Wisconsin. Published with permission of Helen A. LeRoy.

father. "Their parents, however, always remained at home together" (Suomi and LeRoy 1982: 323). The apparatus was the final summation of the previously disassembled affectional systems, in which the stripped-down, isolated monkey was gradually reconstructed to wholeness. The nuclear family apparatus embodied the telos in the study of love. Margaret and Harry Harlow designed an apparatus to optimize an infant's exposure to all the postulated affectional systems, in an architectural creation of the putatively "normal"—but never quite sufficiently in evidence to set mental health experts and other policy makers at ease—"human" nuclear family. The monkeys responded to this social opportunity beautifully. The fathers were nicely social with the babies and showed that they had a function in family life: threatening external enemies (experimenters mostly, Harlow recognized, in his always honest jokes). The fathers had enough other functions to give solace to the advocates of family togetherness. They prevented mothers from abusing babies or abandoning infants (a fairly difficult task in the spaces provided). Many nuclear fathers played with infants more than many mothers, even without housework as a maternal excuse. Pre-adolescent monkeys of both sexes could be studied with their younger siblings to see the development of sex differences in nurturing behavior and of precursors of paternal behavior. The nuclear family apparatus makes plain the simple truth that the primate body is a discursive construct and therefore a literal reality, not the other way around.

Perhaps best of all for professional middle class audiences, the young monkeys raised in this normative utopia responded by becoming smarter than other monkeys. The nuclear monkey children outdid all their laboratory and wild peers on intelligence tests. The superiority only emerged in the most complicated intellectual work the monkeys were asked to do. Harlow speculated that the enriched environment, the nuclear family, produced confident primate children, ready to excel, in metaphoric contrast to lower class human children whose family deprivations might result in impaired personalities and so low achievement. Daniel Patrick Moynihan (1965) could have done his famous report on the "pathological," mother-centered black family at the Wisconsin Primate Laboratory (compare Stack 1974). To Harlow's credit, he recognized that his hypothesis for explaining the performance of his smart monkeys "is as difficult to test as is the hypothesis that middle-class children excel intellectually over lower-class children because of their environmental advantages" (Harlow et al 1971: 543).

From the utopia of the nuclear family apparatus, let us conclude by looking at one of Harlow's last legacies to laboratory hardware: called by its designer "the well of despair" or "the vertical chamber apparatus," developed not only to achieve total social isolation, including visual contact, but explicitly to reproduce the state of utter hopelessness described as characterizing human depression (1971: 546). Designed in the late 1960s, the pit was an effort to augment the depression created by earlier forms of social isolation. As always, the justification was the search for adequate models of human suffering and applicable therapies, including social therapies, drugs, and electroshock. If the monkeys could be utterly reduced to despair, could they be reconstituted by the addition of identifiable agents, including other monkeys? Can humans be reconstituted with similar technologies? Although Harlow recognized that the lab's normal monkeys were in fact partial social isolates, he and others looked from the mid-1950s for better experimental designs for infant isolation, including both schedules and apparatuses for confinement. A graduate student, Guy Rowland, standardized an isolation protocol for the Wisconsin laboratory (Rowland 1964); his device allowed monkeys to be raised from birth without seeing any other animal or part of an animal except the experimenters' arms and hands, and those only in the first 15 days of life. The experimenter could watch the monkey through a one-way vision screen, and learning tests could be administered by remote control. Monkeys could be released from the apparatus at varying ages and tested for recovery once they were allowed specific kinds of social contact. Full 12-month isolates could serve as a control group, since they never showed any vestige of social ability or signaling for the many years they were maintained in the lab (Harlow et al 1971: 545).

The vertical chamber apparatus' innovation was its production of depression without any social separation schedules. The device was a tall stainless steel chamber with smooth sides sloping to a wire mesh platform above a rounded steel bottom. Although they could move through the three dimensions of the chamber, typically after a few days the monkeys remained huddled in a tight ball at the bottom. "Chamberings" for a few weeks efficiently produced acute experimental depressions, which could then be treated. W.T. McKinney, a psychiatrist, joined Harlow's team to help develop this exciting work. For the scope of his work on love, depression, and its treatment, Harlow received the most prestigious award in the field of psychiatry, the Kittay Scientific Award.[22]

The literal quality of the imaginative social unconscious nurtured in the Harlow laboratory was never more in evidence than with the well of despair. For the historian of science trying to avoid simplisitic interpretive strategies, the metaphoric translations in the Harlow lab are defeating in their isomorphisms. The punishment schedule for the analyst looking for ways to read the Harlow lab differently seems analogous to the "learned helplessness apparatus" introduced at Wisconsin about 1977, with which monkeys were electrically shocked whether getting correct answers or not.Reading the Harlow literature and tracing its modes of production is profoundly painful; joking seems the only available way to stay in the story, as it compulsively repeats itself over more than 40 years of exuberant experimental design. Joking is a reading strategy, as it was Harlow's writing strategy. Harlow's verbal jokes insist on the reader's attention to his lab's practice of seemingly literal translation and interconversion of artifact, word, social realities, and ideologies. But even the cooler technical language in the lab's publications plays persuasively with the construct rhetorics to tell the twinned stories of masculinism and sadism in the structure of bodies, technologies, and other laboratory signs embedded in the contest for primate meanings. The monkeys and artifacts were the plot space, the matrix, for the heroic story of Harlow's quest and self-reflection. Self-clutching in despair or self-confident in nuclear utopia, the monkeys gave back the language with which they were built, and it was a language of scientific love. But it was not literal, after all; the monkeys' and artifacts' seeming immediacy and self-evidence is the last joke in the story. Their reflective communication was coded with all the power of the discipline of psychology in the biopolitical era. In the measure of love is the literalization of sexual politics.

10

THE BIO-POLITICS OF A
MULTICULTURAL FIELD

The Mirror and the Mask: The Drama of Japanese
Primates

Japanese field study of an indigenous monkey inaugurated post-World War II
naturalistic studies of nonhuman primates. The origin of the post-war primate
story is within non-western narrative fields. In the beginning, Japanese primatology was both autonomous and autochthonous—but not innocent, not without history. Human and animal, the actors and authors appeared on an island stage
that was not set by the story of Paradise Lost. The Japanese monkeys became part
of a complex cultural story of a domestic science and a native scientific identity for
an industrial power in the "E/east."

Japanese primate studies originated in 1948 among a group of animal ecologists,
including Imanishi Kinji, who had earned a Doctor of Science from Kyoto University
in 1940. In the first generation of Japanese to pursue studies of animals in their
natural environments, Imanishi led expeditions to several areas outside Japan. The
Japanese primatologists were well aware of the western work, and they cited Yerkes,
Carpenter, and others with appreciation and critical evaluation. But the Japanese
forged an independent primatology, whose characteristics were part of cultural
narratives just as they have been in the west. As tropical primates have been mirrors
for western humans, domestic Japanese monkeys have been mirrors for their skilled
indigenous observers.

Before turning to primates, Imanishi studied Japanese wild horses. In 1950

Imanishi and Miyade Denzaburo formed the Primates Research Group. Provisioning began in 1952, about the time of the establishment of the Kyoto University Anthropology Research Group. In Tokyo a medically oriented Experimental Animal Research Committee started up, and in 1956 the Kyoto and Tokyo groups established the Japan Monkey Center, followed in 1967 by the national Primate Research Institute of Kyoto University. By 1961, more than 20 Japanese macaque groups had been provisioned and brought into systematic observation.[1] In 1961 the Japanese began langur monkey studies, with the cooperation of Indian scientists in India, and launched the Kyoto University African Primatological Expedition. In 1965 they initiated a long-term chimpanzee project in Tanzania's Mahale Mountains.[2] In 1972, Japanese workers started groundwork for their study of pygmy chimpanzees in Zaire.[3]

Westerners were unaware of the Japanese work until after 1956. About 1957, Yale medical primatologist Gertrude van Wagenen found a book with a picture of monkeys in a bookstore in Japan. She wrote Stuart Altmann at the National Institutes of Health and then sent him the book, which was *Japanese Monkeys in Takasaki-yama* by Itani Junichiro (1954).[4] The Jesuit John Frisch, Sherwood Washburn's graduate student at the Univeristy of Chicago, translated the book and subsequently wrote a description of Japanese primate studies for western readers. Altmann had several Japanese articles translated and published.[5] Japanese workers were invited to the Primate Year at the Center for Advanced Study in the Behavioral Sciences in 1962–63. The National Institutes of Mental Health funded the first English translation of Japanese primate studies, and after that the Rockefeller Foundation sponsored translation from the Japanese journal *Primates*, published in English after the first two issues, reflecting international language politics. From about 1959–60, Japanese and western workers have been in regular contact, despite on-going difficulties of linguistic, cultural, and scientific communication.

National Primates

It is ideologically and technically relevant that the Japanese studies were initiated on a species, *Macaca fuscata*, native to Japan. Although Japan had been a colonial power since the late nineteenth century, Japanese founding frameworks for watching monkeys and apes did not depend on the structure of colonial discourse—that complex search for the primitive, authentic, and lost self, sought in the baroque dialectic between the wildly free and subordinated other. Rather, Japanese monkeys have been a part of the construction of a specifically Japanese scientific cultural identity. Constructing that identity has been a major theme in recent Japanese social, cultural, and intellectual history.[6]

Japanese monkeys might be viewed as actors in a Kabuki drama or a Noh performance. Their stylized social gestures and intricate rule-ordered lives are like dramatic masks that necessarily both conceal and reveal complex cultural meanings about what it means to be simultaneously social, indigenous, and individual for Japanese observers. Seeking the truth of nature underneath the thin, often obscuring layer of culture, the westerner tends to see in our primate kin a deeper shared animal nature. In contrast, perhaps, Japanese primate observers have seen simian masks expressing the essential double-sidedness of the relations of individual and society and of knower and known.[7] It is not a "truer" nature behind the mask that is sought within a Japanese cultural frame; nature is not the bare face behind the mask of culture. Instead, the figure of the Japanese dramatic mask alludes to a powerful abstract stylization of the specific social intricacies and profoundly individual qualities that pattern primate life. The dramatic mask in Japan is a figure of the co-determining relations of inside and outside and of the subtle reversals of position that change inside into outside. Masks cannot be stripped away to reveal the truth; rather the mask is a figure of the two-sidedness of the structure of life, person, and society. I am suggesting that Japanese primatologists constructed simians as masks in these culturally specific senses. Japanese monkeys were crafted as objects of knowledge showing the structure of an interactionist, relational, contextual self in a highly differentiated social world like that inhabited by their skilled observers (Smith 1983: 68–105). "Nature" was made into an object of study in Japanese primatology, but it was nature as a social object, as itself composed of conventional social processes and specifically positioned actors, that intrigued early Japanese monkey watchers.

Let us follow some of the issues raised by this highly modern, richly traditional, indigenous science. Through a gesture appropriate to the western structures of appropriation of the "other" that this book has not escaped, Japanese primatology can be mined for resources to illuminate questions about gender, feminism, and orientalism. The early ethnographies of Japanese monkeys can be useful distorting mirrors for Euro-American feminists tangled in a culturally given story about nature and organic unity.

The search for the untouched heart of nature seems not to be a dream shared by the Japanese with western observers. In a dissertation and other writing on different forms of anthropomorphism in Japanese and western primatology, Pamela Asquith investigates the historical and philosophical roots of Japan's extensive studies of the social lives of nonhuman primates.[8] She argues that the particular related boundaries between human and animal and between mind and body, so crucial to western Greek and Judeo-Christian mythology and to derivative ideologies of scientific objectivity, are not part of Buddhist or Confucian Japanese cultural heritage. While concern with status, personality, social change and stability, and leadership pervades Japanese primate studies, the split between observer and observed, so crucial to the western quest for a healing touch across the breach, is missing.

When the human-animal boundary is not culturally crucial, two things change which matter immensely to the themes of this book: First, "nature" cannot be constructed as a health spa for the ills of industrial society; i.e., the tortured negotiation of touch with the representatives of a region from which "man" is banned does not dominate popular consciousness and covertly inform positivist science. A corollary of this first point is that "woman" will not be symbolically and socially required to cross taboo lines in scientific primate studies, in order to allow the field to include certain kinds of practices and theories coded as empathic or intuitive in the west, and so suspect for scientific men. The crucial question of identification with the "object" of study will not have the same gender load. It is hard to know what effect large numbers of Japanese women might have had on Japanese primate studies, which have been more male-dominated than in the United States.[9] The *particular* scientific gender-coding characterizing western analysis of the relations of women and science are not meaningful in the Japanese context. The second point is that the reliability of scientific knowledge does not depend on enforcing the boundary against the forbidden desire of touch with nature. The dialectic of touch and transcendence in "Teddy Bear Patriarchy" is not a Japanese story. The eros, the politics, and the epistemology of primate science are culturally and historically specific.

Instead, as Asquith (1981) argues in her interpretations of the publications of Japanese primatologists Imanishi Kinji and Kawai Masao, strong Japanese cultural sources posit a "unity" of human beings and animals.[10] But it is hardly a unity that would be comfortable to those who seek in mystifications of the "Oriental" a solution to the forms of alienation built into western scientific and social practice; it is not a unity innocent of cruelty and power. For example, Kawai explains why children's throwing stones at and bullying animals is not considered unacceptable in Japanese culture. Relations between species are like relations between older and younger siblings or parents and children. What westerners might consider violence might be seen as acceptable "family" behavior. Kawai argues that the western (apparently) moral stance rested on a particular notion of superiority to the animals that enjoins human stewardship. The story of Adam's commission in the Garden as planetary park ranger, with the special power to name his charges, is not indigenous to Japan.

In Buddhist traditions, the series of possible transformations among animals, gods, and humans sets up a horizontal system of relationship. Again, lest a western reader seek in the absence of the kind of hierarchy familiar in the great chain of being a solution to western problems, Japanese notions of continuum are perfectly consistent with what westerners would see as quite cruel behavior to animals and to disabled people, based on the available notion of reincarnation in an animal, female, or deformed body as punishment for past imperfections (Asquith 1981: 353). Buddhist approaches can easily insist on the polluting aspects of animals, women, and the body. The ease of interchange among beings can be a source of considerable danger and anxiety. Similarly, within Confucian cultural resources in their Japanese form, the idea of a continuum between animals and humans is not inconsistent with a hierarchy within the unity. What is excluded is the idea of "special creation" or Christian stewardship that has been critical in the western history of natural history, evolutionary biology, and conservation (Asquith 1981: 352). Buddhist ideals of compassion can ground relationships to animals, women, or other suffering beings in ways similar to masculinist human stewardship in Christian cultures, and certainly Japanese relations with human and nonhuman entities do not exclude, but insist on dominance and subordination within a social and ontological unity.

Stressing the troubling or contradictory dimensions for westerners of Japanese cultural tendencies in human-animal relations is not to claim that the Japanese are cruel to animals or to categories of humans, but is rather a caution against the cannibalistic western logic that readily constructs other cultural possibilities as re-sources for western needs and action. It is important not to make Japanese primatol-ogy "other" in that sense, so that important differences can be appreciated, rather than mystified. In addition to the relevance for primate studies of Japanese reinven-tions of Buddhism and Confucianism, those early importations from Japan's first "West" (i.e., China and Korea), there is also another relevant "traditional" strand in Japanese cultural reinventions: Shinto and its emphasis on matrilineality and the importance and power of mothers in social life. But it is not simply the obviously "religious" or "cultural" stories that frame the drama of Japanese monkeys' lives.

The narratives of modern sciences themselves are equally "traditional" and equally "reinvented" from the point of view of the post-World War II constructions of primate studies.

Asquith emphasizes that Japanese primatology, as it was developed by Imanishi and his co-workers, was neither traditional nor especially congenial to other mid-century Japanese scientists. To make nature an *object* of study is precisely not a Japanese move, and all of Japanese modern science has been in tension with groups crudely lumped as "traditionalists." Neither was Imanishi's science, with its emphasis on identification with the animals and on a disciplined subjectivity, congenial to "westernizers" or "modernists," who regarded Japanese cultural sources as unproductive for the development of modern science. Imanishi's Japanese scientific contemporaries initially rejected his primatology as excessively subjective or anthropomorphic (Asquith 1981: 350–51).

Asquith argues that Imanishi's approach was an original, "individualistic" synthesis of western and Japanese strands. Such "individualistic" groups arose in several areas of natural science with the lifting of severe regulations on research (accepting western technology but rejecting western philosophies of nature and science) at the end of Meiji Japan in the early twentieth century. And after World War II, the Japanese have forged a nationally specific organization of research and cognitive style in science. Imanishi's and Kawai's version of the synthesis has been part of the framework for a productive, collective Japanese branch of life science in primatology involving several institutions, international study sites, state support, and the training of graduate students in numbers rivaling the largest western institutions.

What have been the special characteristics of Japanese studies of primates? First, several commentators have remarked on the early introduction of "provisionization" (Imanishi 1963: 70) of the free-ranging animals with food, thereby systematically increasing the observability of the highly mobile animals and altering the relation of oberver and observed that compromises the "wild" or "natural" status of the animals from western points of view. The "wild" status of the primates for the Japanese referred more to their running away from observers than to an essential character that would offer epistemological and symbolic guarantees. In the Japanese view, provisionization expressed an exchange or relationship that *already* existed (Asquith 1981: Chpt. 6).

A contemporary western tourist in Japan is impressed by the *visibly* constructed nature which is loved and cultivated. The cultivation or domestication is not only found in the Japanese gardens, which have become such a popular commodity in western architecture and landscaping. An ancient gnarled tree will be duly marked with a plaque as a national treasure. A famous natural scene, the changing seasons, trees in blossom: all these are deep in literary and other aesthetic practices. Nature is not just beyond the frontier, just beneath the crust of culture, the animal as opposed to the human. Nature is not presented as valuable because it is wild in Teddy Roosevelt's sense.[11] Nature is an aesthetic value and understood to require careful tending, arrangement, and rearrangement. Nature in Japan is a *work* of art. A provisionized monkey troop was seen to be in the process of domestication (Imanishi 1963: 70); that process in no way violated its natural status, which rested not on non-interference but on the particular quality of relationship.

Western obsrvers have experimentally fed animals in the field, prominently in Carpenter's and all following work on Cayo Santiago and in Goodall's chimpanzee studies, but overwhelmingly, westerners try to maintain—or apologize for violating—a "neutral" relation to the animals, which is believed to minimize the "interference" of the observer with the natural character of the object of knowledge that will be so important to the epistemological status of the resulting report. Deliberate experiment is one thing; "uncontrolled" interaction with the animals quite another. The repeated allegories of violation of this rule in western primate studies are the stuff of informal culture, popular presentation, admonitory stories of polluted science, or personal idiosyncracy that should never be allowed to interfere with real science. Studies of provisionized animals do not have quite the same status for western observers as observations of a previously undescribed species in an area far from human activity. To study monkeys on Cayo Santiago is surely valuable, and much cheaper in a period of declining funding for primate studies, but nothing like the pleasures of the Kibale forest in Uganda or Amboseli National Park in Kenya. The fully controlled laboratory and the fully "natural" field situation ground the most reliable science for westerners. The intermediate zones have been areas of great nervousness. Pollution of a study is easy if the boundary between human and animal is incautiously crossed (Douglas 1966). Lapses into "anthropomorphisms" are rigorously defended against in the field, and the apparatus of the animal laboratory can be read as an elaborate defense against boundary trangressions.

Western scientific women in particular must negotiate their relation to these taboos. On the one hand, many have deliberately taken advantage of the greater latitude for women in western culture to acknowledge emotional exchange with the animals and to affirm the importance of identification or empathy in a way that they believe improves the research. On the other hand, these same women, as well as many who allow no greater identification than their male peers, repeatedly report having to guard against incautious admission or cultivation of their feelings, in order to be respected scientifically or to avoid being labeled "naturally" intuitive.

The second marked characteristic of Japanese primatology has been Japanese workers' renown for their individual identifications of large numbers of animals. It has not been atypical for all the Japanese associated with a macaque study site to recognize by face at a glance more than 200 individuals, most of whom are named (and numbered). Western observers also pride themselves on recognizing animals, and many western scientists name their study animals rather than number them, but the comprehensive, detailed catalogue of the lives of *every* animal in a group is a Japanese trademark. It is instructive to compare the North American Stuart Altmann's tattooing of each rhesus monkey on Cayo Santiago in 1956 and his subsequent system of data collection with the field system of the Japanese. Both recorded detailed information on individual animals, and both pursued long-term research at stable sites. But Altmann's report was severely "objective," while Japanese writing of that period was highly "ethnographic," a kind of interpretation of primate cultures.[12] Altmann's report was at least as laden with cultural values as the Japanese accounts, and in particular I read both literatures of that period to fit comfortably with culturally specific, scientifically coded masculinist concepts and practices.[13]

The skill of recognizing the unmarked animals is associated with a strong sense of the individual animals' "personalities." Practiced western observers are also renowned for recognition feats, for example, Iain Douglas-Hamilton with elephants and Linda Fedigan with Japanese macaques transplanted to Texas. But the story surrounding Douglas-Hamilton emphasizes the photographs of each elephant and patient memorizing of each nick in the ear or another discrete, tattoo-like natural marker, not the complex "personality" of a physiognomy (I. and O. Douglas-Hamilton 1975). The "technology" of identification is narratively emphasized despite the book's overwhelming story of the hero scientist's and his photographer wife's personal touch with a nature dangerous to ordinary people. The issue of recognition, like each of the "special" characters of Japanese primatology, also is differentiated by gender for western observers—in a way that puts western women and Japanese (mostly male) observers in a common symbolic location.

For example, Linda Marie Fedigan notes her "misgivings about female empathy" partly as a response to an annoying experience that warned her about the difficulty of women's being credited with their hard-won scientific accomplishments. After "having put many hours of effort into learning to identify the individual female monkeys of a large group, my ability was dismissed as being inherent in my sex, by a respected and senior male colleague" (Fedigan 1984: 308). Conversely, Irven DeVore reported disappointment that *his* naming and recognition of many animals other than the dominant males went unremarked in the partly gender-determined narrative of his responsibility for masculinist theories of male dominance supposedly based in his failure to see most individuals.[14] The point is not that western women, western men, and Japanese scientists have different, mysterious ways of learning the identities of particular animals; all the ways involve quite ordinary hard work and concentration. But the feats or failures of identification carry different cultural meanings according to the powerful markers of national and gender identities. From a structurally western point of view, to be Japanese is more than a "nationality"; it is to be a representative of the "Orient." White women and the Japanese live in the "East," where Man is not alienated. The *narrative* is more revealing than the method. Similarly, the narrative that leads historians of primatology, including this one, to construct allegories of identification or alienation in discussing the neutral subject of "scientific method" is as important as any technical protocol.

The dialectical relationship of the specific personal character of each animal with the whole social system of relationships has been the basis of a special concept (*specia*), introduced by Imanishi into the primate literature, but never taken very seriously by puzzled western primate scientists. Asquith related the Japanese arguments that their technique of personal recognition was bound up with a different approach to evolutionary theory and the concept of adaptation. The theory of natural selection was not understood as the center of Darwinism in western countries until well into the twentieth century, and different national understandings of Darwin and evolution have received considerable attention from historians of biology.

Evolutionary theory in Japan was worked into culturally specific patterns too. The founding Japanese primatologists had no difficulty with the fact of evolution, but their questions and resulting explanatory systems were directed to sociology and social anthropology, and not to questions of fitness and strategies of adaptation. Most Japanese primate writing stressed what western observers were more likely to see as "proximate" explanation, i.e., the social interactions themselves, their patterns of change and specificity, while neglecting "theory" or the "ultimate" explanation of adaptive strategies and reproductive fitness maximization.

Imanishi explained the concept of *specia* to "denote the aggregation of all individuals belonging to the same species which occupies a definite area on the globe." In addition, he proposed the term *oikia* to refer to the organization of higher verte-brates into particular aggregations denoted usually by terms like family or troop, i.e., by terms stressing social organization.[15] Both terms were constructed within a "culture and personality" framework for studying primates. Japanese workers in this period used definitions of culture suggested by the Boasian culture and personality theorists, and they explicitly compared their observations to issues raised by Marga-ret Mead.[16] Asquith points out how difficult, or more, superfluous, the concepts of *specia* and *oikia* have seemed to western primatologists. But Imanishi's suggestions are a window onto Japanese constructions of nature. The social conventions of the monkeys were precisely the natural-technical object of knowledge constructed by the Japanese observers. The gap between natural and conventional is not the crucial issue; but the subtle social surface, the mask, is the essence of the *specia*. The *specia*, specific to an area of the globe, and the *oikia*, specific to a particular troop or monkey kin-organized group, are concepts grounding Japanese concern with the rootedness, i.e., with the native status, of a social entity. Nature as "wild" does not concern the Japanese in the way it does Euro-Americans, but nature as "native" is a matter of great concern. Focus on *specificity* in all its layers of meaning—individual personality, collective culture, habitat, troop identity, kin lineage—is an operation for building primate mirrors for a people preoccupied not with the colonial "other," but with the problem of establishing its uniqueness in the context of a history of extraordinary cultural importations and reinventions, from Buddhism to Dar-winism.[17]

The third trait of Japanese primate studies from the beginning was collective, long-term study of several groups of the same species. They did not consider it remarkable for ten observers to study in detail the demographic and social histories of a group of monkeys for ten years and beyond. This orientation persisted in their expansion of primate observations to species outside Japan. While long-term, collective sites have come to characterize western primate studies as well, initially a person who wanted to make a mark on the field sought an "untouched" species in an "undisturbed" environment. Western stories about the origins of the long-term sites tend to de-emphasize the presence of many people and to stress the role of an individual founder. The popular stories of Jane Goodall and Dian Fossey are perhaps the most extreme version of this common narrative. The tensions among generations of western workers between those who founded sites, and in some sense see those sites as their personal achievement in making a section of "nature" into an object of knowledge, as well as a personal spiritual resource, and later workers, described by founders as seeking only data and not really caring about the animals, surfaced in my interviews. The quality of western scientific primatology as travel and quest literature, wherein the individual hero brings back a prize valuable for the whole community, is a covert but important dimension that seems foreign to the social and symbolic organization of Japanese primate studies. Similarly, the story of the isolated western observer who stays too long and loses his western rationality—or his life—in a bodily touch with nature which went too far does not seem present in Japan's narratives of natural history.[18]

The final aspect of Japanese method is a philosophic synthesis of the meaning of the practices of provisionization, individual identification, and long-term collective work. Asquith presented in English for the first time Kawai's *Nihonzaru no seitai* (*Life of Japanese Monkeys*, 1969), in which he proposes the concept of *kyokan* ("feel-one") to designate the particular method and attitude resulting from feelings of mutual relations, personal attachment, and shared life with the animals *as the foundation of reliable scientific knowledge*. *Kyokan* means "becoming fused with the monkeys' lives where, through an intuitive channel, feelings are mutually exchanged" (Asquith 1981: 343). The "sympathetic method," not to be confused with a western romanticized organicism that excludes power and violence, is crucial to the question of "objectivity." "It is our view, however, that by positively entering the group, by making contact at some level, objectivity can be established. It is on this basis that the experimental method can be introduced into natural behaviour study and which makes scientific analysis possible ... It is probably permissible to describe the method of the Primates Research Group as 'the new subjectivity' " (Kawai, in Asquith 1981: 346).

In her interviews with over 40 Japanese primatologists, Asquith found that no one but Kawai used the term *kyokan*. The other workers attributed the word to Kawai's eccentricity, and Kawai noted his uniqueness on the point. But perhaps only the word, and not the underlying positions on the subtlety of the connections of subjectivity and objectivity, is at issue. Asquith reported that Itani, Professor of the Laboratory of Human Evolution, preferred to call their method "anthropomorphic," stressing their assumption that since monkeys have "minds" of some sort, some kind of empathetic method would be reasonable and likely required to understand simian societies.[19]

Females: A Site of Discourse on Social Order

Both the *kyokan* methodological attitude and the possible cultural tones from Shinto sources sounded in several of the early discoveries of the Japanese primatologists raise interesting questions for a history concerned with the relations of feminism, women, females, and life sciences. Western primatology's "early" notorious focus on males to the exclusion of females and consequent masculinist interpretations of primate life have been noted in many historical accounts of the field. Many of these historical claims are part of a reassurance that the problem is well in hand now, with both women and men properly chastened by the lesson of the bad old days. Heroes who led the way out of biased science differ in the various accounts. Feminist-inclined women tend to emphasize the entry of their conspecifics (congenerics?) into the field in the context of a political women's movement,[20] and others look to ordinary progress in the self-cleansing scientific history of ideas.[21]

The early influential presence of western women in primatology makes the case complex. It is impossible to separate cleanly their stories from those of their male peers in the same explanatory traditions and moments of primatology. For example, Sherwood Washburn's first two students of primate social life, Irven DeVore and Phyllis Jay, both theorized about male dominance hierarchies as organizers of social cooperation, although differences that could be considered gender related were also prominent. The two principals do not have the same retrospective opinion about their own and each other's work on these issues, and my reading of their early papers adds a third and dissimilar interpretation from either of theirs. Also, current sociobiological male and female theorists emphasize female biology and behavior for excellent reasons, which are *now* seen to be built into neo-Darwinian selection theory. Why now? Is it women who matter? or feminism? or neither? or both? and matter to what?

In the 1950s the Japanese were reporting that the basic structure of Japanese macaque society was "matrilineal"; i.e., organized around descent groups of hierarchically related groups of females and their offspring (Kawai 1958). The rank of a male depended more on his mother's rank than on his individual exploits, and the rank of a daughter was stably predicted from her birth order and matri-lineage.[22] Upsets could occur, but it did not make sense to see troop structure as the function of dominant males, who nonetheless filled a leadership role-function. Imanishi emphasized that for a male to be accepted into the core group of an *oikia*, so as to be part of the cluster of dominant males and females with their young, he had first to be accepted by the dominant females (Imanishi 1963: 79). He speculated that a psychological process of identification with high-status males was available for the male young of a high-status female. Those males would have the proper confident attitude that would lead to acceptance and high rank later. Males without the crucial maternal history would be permanently peripheralized and possibly leave the troop. Kawamura Syunzo (1958) also reported on a female-led troop, the since famous Minoo-B group, referred to as having a "matriarchal social order." Itani reported on frequent "paternal care" of young by males, especially those of unstable rank who appeared to perform this role as "a sign of interest in the central part of the troop" and to establish their position (Itani 1963: 94). Caring for youngsters and being accepted by important females seemed a better route to success for ambitious young males than fighting their high-status seniors.

Itani (1958) and Kawamura (1963) described the propagation of new food-preparation habits through the social structure of seaside troops fed sweet potatoes and grain, which some animals proceeded to wash and float, respectively. In a famous case, the change in troop habits was observed to be the fruit of the lavatory insight of a juvenile female, who washed her sweet potato free of grit and so spared her teeth. [Figures 10.1, 10.2] The other youngsters and females got the idea from her in a pattern that flowed through the female-lineage hierarchies. Sub-adult males got the important dental-care idea last because they were most peripheralized from the core group of females and young. However, if a top central male did get a new idea, his practice would spread even more quickly through the society. But social

Figure 10.1 Japanese macaques wash sweet potatoes in the ocean. © Mori Umeyo. Courtesy of Nishida Toshisada.

Figure 10.2 Monkeys feeding outside the park office at Awajishima. © Pamela Asquith.

and technical innovation emerged from the practices of youngsters and their moth-ers.[23] The reports were not about the question of sex in cultural innovation, but about the processes of "tradition" or "protocultural behavior" in a nonhuman species. Read retrospectively in relation to western debates about gender and sci-ence, the description of females stands out in comparison to the early male-authored western accounts.

That all of these events in the history of a monkey species entered popular and technical literatures in several languages itself speaks to complex interest in the question of female "power." The matrilineal troop structure data were particularly damaging for the male dominance hierarchy explanations of macaques, and these data were cited by some western (male and female) primatologists as early as they were published in English (Losos 1985). Both Japanese and western men were interested in reporting what females and youngsters did, and these observers were capable of relating what they saw to explanations of how primate societies worked. Japanese men, not members of a culture famous for its congeniality to women, at least in the last millenium or so, made these points in their original reports in the 1950s. What else must be said about these narratives?

First, they *are* narratives—culturally important stories with plots, heros, obstacles, and achievements.[24] The presence of females, even powerful females, hardly makes a narrative women-centered, much less "feminist." And to be *Woman*-centered and *women*-centered are quite different things. In Japanese (and western) primatology the bulk of discussion about female-centered organization was directed toward understanding *male* life patterns and to explaining social conservation.

However, there is a strong Japanese cultural preoccupation with mothering and with mother-son relations that has different tones from western versions. Japanese historians stress the importance of Shinto tradition, with its female shamans and major female deities, especially the sun goddess, Amaterasu. The absence of a monotheistic, patriarchal father god changes men's imagination about women and females. Contemporary Japanese culture is replete with fiction, films, varieties of pornography, and social commentary on the ideology of mother-determined Japanese men's lives (Buruma 1985: 1–63). In complex interplay with Japanese histories of gender domination and conflict, these stories affect how both women and men conceive the social and natural worlds. Neither the stories nor the social and natural worlds split in the same way as in the west.

In addition, Japanese women are traditionally regarded as the source of the most crucial innovation for a modern people concerned about its native roots: Japanese women, from the period of Lady Murasaki's *Tale of Genji* in Heian Japan, originated Japanese written literature, while the men were writing in Chinese. Japanese women, as mothers at the node of kin groups, but also as literary figures, are at the sources of Japanese historical accounts of what it means to be Japanese, to be native, to have one's own language.[25] For the Japanese, the issue, echoing in primate studies as in other cultural practices, seems to be less the *origin* of language, the boundary between animal and human, than the *uniqueness* of one's own language, the bound-ary between native and stranger, the subtle play between conservation and change. Females and women seem socially and imaginatively to be located at dense intersec-tions of meaning for these issues in Japan. I am not arguing that there was a direct determination between these broad cultural patterns and what Kawai or Kawamura saw Japanese monkeys doing in the 1950s. But the narrative field—the story-laden quality of observation, description, and explanation general to the life sciences, and especially primatology—for any particular Japanese account was structured by different axes of meaning and possibility. The period since Japan's defeat in World War II has been a critical one for Japanese cultural reinvention of identities in extraordinarily complex patterns of "scientific" and "traditional," native and foreign, revolutionary and conservative. In that context, the primatology of a native species may be read for its stories about gender, innovation, and conservation.

For westerners, it is not remarkable that social stability was reported to rest on female-offspring networks. Nineteenth-century doctrines of the organization of the female around the uterus required that and more. Scientific attention to females has not been lacking. Indeed, the reproductive and nurturing female body has been constructed as a core scientific object of knowledge co-extensive with the history of biology as a discourse on systems of production and reproduction.[26] Women and females appear as scientists and as objects of scientific attention in the wake of twentieth-century global women's political movements; but the kinds of stories, the kinds narrative fields of meanings, in which those beings appear were profoundly restructured. Feminism, as well as primatology, is a story-telling practice.

The percentage of women among Japanese observers is among the lowest in the world. The 1980 membership lists of the International Primatological Society indicate that about 9% of the Japanese members were women. Japanese women primatologists have not had the visibility of their western counterparts. But from very early there were women; for example, Mori Umeyo's mother was also a primate scientist, resulting in probably the first mother-daughter lineage of primate watchers.[27] Mori Umeyo was a participant in Shirley Strum's 1976 Wenner-Gren baboon conference. Senior Japanese male scientists have commented in publications on the relevance of women observers to the content and accuracy of research.

For example, Pamela Asquith quotes Kawai Masao on his frustration at not being able to remember well enough the identities of females in troops he was observing. Kawai admitted,

> We had always found it more difficult to distinguish among females as we could not see any particular differences among them. However a female researcher who joined our study could recognize individual females easily and understood their behaviour, personality, and emotional life better than that of male monkeys. . . . The reason why the study of female monkeys has lagged behind is that the researchers have all been men. I had never before thought that female monkeys and women could immediately understand each other; but there has been some barrier between female monkeys and male observers. This revelation made me feel I had touched upon the essence of the feel-one method; at the same time I realized the importance of the phenomenon.[28]

Would the unique "female researcher" have seen her results as the fruit of natural affinity, or would she with Fedigan have remembered hours of work, perhaps invisible to the men because most of her life was invisible? Was the barrier between Kawai and the female monkeys a lack of natural affinity or the social history of patriarchy?

Kawai went further, and Asquith, a western woman writing a history of primatology, noted his usage in a masterpiece of understatement in a modest footnote.

Like Yerkes before him, in a different culture which still somehow fostered male researchers' feelings of annoyance at female animals seen to take advantage of their sexual politics to curry undeserved favor, Kawai described an incident in which females chased off a high-ranking male, relying on the implicit support of a still higher ranking male. Kawai perceived the females as "swaggering" and "taking advantage of" their relation to the ascendant male.[29] In *Adam's Rib*, written between 1941 and 1946, and so the first feminist commentary on the history of primatology, Ruth Herschberger (1970 [1948]: 5–15) had the chimpanzee Josie comment on Yerkes's similar perceptions in the 1930s. Herschberger, prescient about the dilemma of male observers isolated from women in the field, titled her first chapter ironically, "How To Tell a Woman from a Man." Between seeing no differences among females and perceiving their actions in terms of his own solidarity with the offended male, Kawai's description of matrilineal rank structure, supposedly crucial to deposing male dominance hierarchy theory as the explanation of the organization of macaque primate society, makes this reader remember that matrilineal organization is eminently compatible with male power and masculinist standpoints in social theory. *The River Ki*, a contemporary popular Japanese fiction written by a major Japanese woman author, makes the same point (Ariyoshi 1980).

My concluding moral is simple: Holism, appreciation of intuitive method, presence of "matriarchal" myth systems and histories of women's cultural innovation, cultivation of emotional and cognitive connection between humans and animals, absence of dualist splits in objects of knowledge, qualitative method subtly integrated with rigorous and long-term quantification, extensive attention to the female social organization as the infrastructure grounding more visible male activities, and lack of culturally reinforced fear of loss of personal boundaries in loving scientific attention to the world are all perfectly compatible with masculinism in epistemology and male dominance in politics. The lessons of Japanese primatology for current analyses in western feminist philosophical and social studies of science are clear on these points. Western science suggests the same message, but perhaps it is more obvious in the context of sharp cultural difference, the sharpest, in fact—that between the mythic world historical structures called "West" and "East."

Both nineteenth- and twentieth-century western feminist theorists have argued that the self-contained, autonomous, western masculinist self, and the knowledge of the world he produces, is somehow truly opposed by women's putatively less rigid selves, issuing messages in another voice. The knowledge of the world imagined to come from this point of view, called women's, is variously described as more holist, less hostile to the body and to nature, promising a healing touch for alienation in industrial societies, and able to circumvent condemnation to the endless chain of substitutions for elusive true knowledge of nature, sometimes called a fully human knowledge. Psychoanalytic theory, history of science, and cognitive psychology are some of the disciplinary tools feminists have used to make these arguments.[30]

In complex sympathy with these claims, the feminist philosopher of science Sandra Harding (1986) noted that contemporary criticism of "western" science by Afro-American theorist Vernon Dixon aligns him neatly with organicist ideologies of contemporary white women. Elizabeth Fee (1986) systematically showed how oppositional movements in science in the last twenty years have echoed each others' rhetoric and ideologies about science, while each movement claimed to ground its insights in an historically unique, sufficient, and necessary experience of domination, mediated by "dualist" science. In these arguments, there is little or no analysis of the historical and textual forms of power and violence built into "holist," "non-western" frameworks. It has become clear to critics of anthropology how the observed and described peoples are turned into resources for the solution of other people's dilemmas (Fabian 1983). Those studied are regularly found to have just those properties that the writer's culture lacks and needs or fears and rejects. It is this structure that defines the logical move that constructs what will count as "primitive," "natural," "other."

Feminist theory has repeatedly replicated this "naturalizing" structure of discourse in its own oppositional constructions. It is at least odd that the *kyokan* method described in Japanese primatology (Kawai 1969), a description of the moment of transcendence of gender in a western white woman's relation to cells she was observing (Keller 1983), a male activist's description of Native Americans' relation to nature before white violation (Means 1980), and an Afro-American man's discussion of organicism (Dixon 1976)—not to mention the sordid history of organicism and rejection of "dualism" in explicitly racist, fascist twentieth-century movements—make similar ideological and analytical moves (Fee 1986). That each of these constructions could be seen as eccentric or exceptional, e.g., not representative of most Japanese workers' views of their method or of women geneticists' approaches to sub-cellular structure, does not weaken their power as privileged allegory about oppositional practice.

What is the generative structure of oppositional discourse that insists on privileging "unity" at the expense of painful self-critical analyses of power and violence in one's own politics? There is at least a century-long tradition of Japanese writing about the difference of its philosophy of science compared to the procedures of western science, which have played such a crucial role in modern Japanese history. This tradition is an important kind of oppositional literature produced within a culture that both maintained its national independence and adapted the modern sciences to its traditions and institutions, so as to challenge western scientific hegemony very successfully in critical fields, including particle physics and micro-electronics. But Japan's complex oppositional-assimilationist approach to the sciences has *not* found a natural unity innocent of power and domination.

Japanese monkeys and those who watch them are inserted in that oppositional-assimilationist discourse on science, in a story-telling practice about themselves as industrial peoples who construct nature as an object of knowledge. The eccentric discourse on *kyokan* (and the wider related discourse on "authentic" Japanese approaches to the problem of western "objectivity") is one product of that practice; it is a discourse on superiority and inferiority couched in a code about unity. Perhaps as a response to marginalization in its myriad forms, the "orientalization" of the self seems built into global systems of oppositional discourse about nationality, race, sex, and class. I read Kawai's text on *kyokan* as an oppositional discourse to the dominant western ideologies of scientific method. Kawai's text runs parallel for part of its course to the privileging of organicism in western feminism. Both are oppositional formulations in the face of a dominant scientific ideology and practice, perceived as western in the one case and as masculinist in the other. And both posit a "new subjectivity" in response to the cultural domination of "objectivity." Both are discourses about a "native" or original unity. Both also have much to say about the charged symbolic and social status of mothers. Both are laced with structurally suppressed internal complexities about power. Perhaps the chief lesson of Japanese primates for feminist theories of science is that living in the "East"—no matter whether that place is found inside a cell, in the right half of the brain, in the Sacred Hills of Dakota, in mothering before or beyond patriarchy, or on Koshima Island in a matrilineal *Macaca fuscata* colony—is no solution for living in the "West."

Supplies of Sacred Monkeys: Primatology in India

The history of post-war India's independent and collaborative primatology is intimately linked to monkeys in the Cold War; to the demand for monkeys in foreign biomedical research; to Indian monkeys' dynamic historical status as "sacred" rather than "natural"; to the history of Indian scientific doctoral education; and to different forms of colonialism, neo-colonialism, and the timing of independence in India compared to African states. Let us begin by placing monkeys in nuclear culture.

In 1987 in the United States a fiction film called *Project X* offered a utopian tale of nonhuman primates in U.S. military research in the Cold War. The film effected a series of displacements and condensations of popular images of other primates and people that highlight the material and semiotic stakes in the construction of primate myths. In the mid-1950s, India briefly embargoed exports of monkeys because they were being used in nuclear war research. Embargoes and refusals of research permission run like a red thread through the history of Indian primatology, marking the politics of decolonization and national independence. Shifting the story of using simians in nuclear radiation experiments from India to Africa, *Project X* was about a young chimpanzee kidnapped out of Africa in a scene reminiscent of the slave trade, once again displacing racial anxieties in U.S. white culture onto the anthropoid apes.

Once in the U.S., the young ape found himself in an American Sign Language experiment in a psychology laboratory, under the benign surrogate maternal tutelage of a white young woman Ph.D. student (Helen Hunt). Good science was coded unmistakably female. The orphan ape and scientist bonded in parent-child relation, and the child ape acquired human language from the mother. But the chimp was stripped from his mother a second time by the law of the father, this time in the form of termination of the mother's research grant. The child was secretly sold to an Air Force research unit, where science was coded unambiguously male and evil. A renegade airman (Matthew Broderick), assigned to the top secret Project X as punishment for various high jinks, slowly discovered the real methods of the experiments. Chimps were trained in jet plane simulators to fly and then graduated to another test area, where they were subjected to a series of irradiations. Their decreasing ability to continue flying was recorded in order to predict the capabilities of human pilots in nuclear war. The airman rebelled, found the mother-scientist, and the two of them tried to rescue the chimp. They formed a nuclear family in a great hurry.

But the child would achieve independence even more quickly. The language-adept chimp taught AMESLAN to the other chimpanzees; and they began their own rebellion, spurred by witnessing the death agony of an irradiated comrade. With language, knowledge of mortality, and a skill learned in the service, the chimps broke out of the facility and stole an airplane, which the hero chimp-son piloted to freedom. He crash-landed in the Florida Everglades, a post-nuclear Eden and a mythic encampment of escaped slaves, symbols of liberation from bondage. The chimps were not found by the project officials, who believed the animals to be dead. But the now heterosexually bonded airman and scientist knew their no-longer-animal son and his community to be safely back in nature. (Perhaps the chimps will be able to make a living from the regional drug trade.)

Finding *Project X* more plausible than I ever had *Totem and Taboo*, I enjoyed this disgraceful film, crying in all the right spots and suppressing all critical impulses in a full indulgence of edenic nostalgia and identification with the noble animals and the happy couple refounding a true family—until the final irony pressed home the terms of the construction of this nuclear, colonial, and compulsively heterosexual kinship myth. The chimpanzee actors, forever in my mind's eye happy in the everglades, actually lived in an institution called Sabo's Chimps in Amenia, New York. From the pre-licensing check in 1982 through a report in July 1986, David Sabo's establishment was repeatedly cited for non-compliance with the Animal Welfare Act.[31] The chimps were reported to live in dark, filthy, contaminated conditions, from which they were removed to be filmed in scenes of primal liberation for audiences that paid taxes for decades for the military use of several thousand nonhuman primates, surrogates for man in fact and in symbol. Mostly Indian monkeys, these "other" primates have prefigured the apocalypse—literally. But in Hollywood film they have been raptured out of the end of history (because they learned to fly on a simulator and to talk with a surrogate mother) to an Eden near Disney World. Meanwhile, Sabo refused permission in 1985 for a USDA compliance officer to photograph the captive chimpanzees.

Let us reverse a key dehistoricizing displacement of *Project X* and move back to India from Africa and to rhesus monkeys from chimpanzees, in order to trace one part of a national and collaborative Indian primatology in resistance to Cold War and colonial constructions of nature and science. The narrative begins in the United States Armed Forces Radiobiology Research Institute in Bethesda, Maryland. In the new-speak languages—multinational parodies of cultural reinvention—that have proliferated in official discourse since the atomic blasts on Japanese cities in August, 1945, that institution is charged in the name of defense with preparing the United States to fight and survive a nuclear war. Funded by the Defense Nuclear Agency, the AFRRI has killed over 2000 rhesus monkeys since 1966 by lethal exposures to radiation, in order to study in a "human surrogate" the time-course and symptoms of decline in the ability to perform tasks, like flying a jet airplane, after severe radiation exposure. The monkeys were trained by electric shock to perform tasks like running continually on a treadmill. Their progressive inability to do so, instead collapsing and convulsing on the experimental chamber where they were shocked repeatedly, was monitored in a "death watch." These experiments were part of U.S. neutron bomb development. The Director of the Defense Nuclear Agency explained to a California senator in 1977 that the "experiments" were "essential to the medical support of the Department of Defense" and that "to the best of our knowledge, the animals experience no pain in the radiation experiments, though some of them die."[32] This is the rhetoric of extraterrestrialism, where science and science fiction are indistinguishable.

Dr. Donald Barnes, who lost his job in 1979 for questioning the necessity of similar routines at the Air Forces's School of Aerospace Medicine in San Antonio, Texas, where about 4000 monkeys have been killed since 1957, when the first nonhuman primates were placed in tubes at various distances from atomic test sites, described his work: "In years past, I was ordered to keep a death watch on these irradiated monkeys, which meant, simply, to see what happened until they died of radiation injury. Do you have any idea how miserable it is to die from radiation injury? I do, I've seen so many monkeys go through it" (McGreal 1981). Monkeys captured from countries such as India, Bangladesh, Malaysia, Indonesia, Kenya, and Bolivia have been used in experiments to test effects of nuclear blast and fall out. They have also been used at Fort Dietrick, Maryland, to test biological warfare agents. Third World simians prefigure the cyborg world of nuclear war.

Many things could be said about this nuclear primate science, where monkeys act as "man's" surrogates in simulations of the apocalypse—after it has already been visited on two "virgin" (previously unbombed) populous cities and on thousands of enlisted men in test blasts over the deserts of the U.S. Southwest. But there is one fact around which my story turns: In using monkeys from India in military research, the United States directly violated a formal agreement, which it had entered into in 1955 with the independent government of India, that specifically prohibited the use of rhesus monkeys in atomic blast experiments and space research. In 1978, following wide publicity by activists from the International Primate Protection League (IPPL), including Dr. S.M. Mohnot, a zoologist at the University of Jodhpur, about the military research, the Indian government again embargoed all further export of monkeys. Soon after, prodded by Dr. Zakir Husain, member of IPPL and active in his nation's Wildlife and Zoological Societies, Bangladesh followed suit, banning export of its monkeys for any purpose. Despite regular pressure by official U.S. agencies, including the Department of State, the embargoes remained in force.[33] Extractive colonialism, while not over in the history of primate studies, has been constrained in the post-World War II order. Emblematic of this struggle, "a United States-sponsored resolution before the World Health Assembly calling on monkey habitat countries to export animals to western research facilities, was withdrawn when African nations threatened to denounce military experimentation on monkeys on the World Health Assembly floor."[34]

In the interaction of people's lives in close proximity to monkeys and of the politics and science of the aborted monkey harvest lie seeds of India's national primatology. Simians interact with people continually and in culturally specific ways in twentieth-century India. Species indigenous to India include rhesus, bonnet, lion-tailed and pigtail macaques, langurs, and Hoolock gibbons. In Hindu scriptures, the *Ramayana* records the monkey god, Hanuman, rescuing Sita, wife of Rama, the reincarnation of Lord Vishnu. Hanuman is primarily represented as a langur, but has come to stand for all monkeys, who share then a sacred status (Southwick and Siddiqi 1985). But to be "sacred" is not the same as being "natural" in the western sense; in particular, to be sacred is not the same as being "wild." Indian monkeys interact, and have interacted for centuries, with agricultural people, more and more intensively as the human population has grown; and they also interact with urban dwellers in markets and with people in temples and at road sides. The interactions are a complex mix of tolerance, competition, exploitation, and mutualism among all the species and cultures concerned. Hardly always harmonious, the historical integration of monkeys with people in India, in practical affairs of everyday life, makes cultural nonsense of the notion of primates as revealing the secret and primitive nature of "man." Indian monkeys are sacred animals protected by Hindu belief and practice, but they are not at the western boundary between nature and culture.

Widely distributed over Asia, the rhesus macaque has proven remarkably adaptable to changing forms of interaction with humans. Its population center is in North India. As pressures for "cropping" rhesus for western biomedicine intensified in the 1950s, and Indian voices suggested conflicting positions in relation to this trade, the absence of basic knowledge about the demography, habitat distribution, foraging habits, and social organization of the rhesus monkey became important both to western and Indian scientists. Population estimates of the rhesus ranged as high as twenty million, and most commentators believed the extraction of 100,000 animals a year could be sustained. Colleagues in the study of the behavior and demography of Indian rhesus since 1959, Charles Southwick, first on a Fulbright Postdoctoral Fellowship at Aligarh University and then affiliated with the Johns Hopkins School of Hygiene and Public Health, and M. Farooq Siddiqi, a geographer from Aligarh University, recently summarized the questions with which they began what became twenty-five years of work together.[35] Could that large harvest be maintained? What did Indian citizens think of the export of the monkeys? What regulated rhesus reproduction and population levels? How many rhesus were there, and what were the population trends? The collaboration between U.S. and Indian scientists to answer these questions is a window onto the multicultural construction of a primate field.

Southwick represented the convergence of interests in population dynamics and regulation from his employment in a school of public health—an institution concerned with human population control, pest management, disease vectors, and maximizing production of and access to important medical research species—and from his graduate training in behavioral ecology with John Emlen in Wisconsin. Emlen himself had gotten interested in population ecology while he was rat control director in Baltimore during World War II doing alternate service as a conscientious objector.[36] While he was at the Johns Hopkins School of Hygiene and Public Health in those years, he read Raymond Pearl, Alfred Lotka, and L.J. Reed, all founders of ecological analysis of carrying capacity. Later, Emlen had doubts about David Lack's (1954, reprinted 1966) influential theories of population regulation in natural bird populations primarily by resource limitation—theories that have influenced many primate socioecologists. Instead, Evelyn Hutchinson's student Robert MacArthur, also a consummate bird expert, informed Emlen's approaches at Wisconsin. Southwick wrote his doctoral thesis in 1953 on growth and control of confined mouse populations supplied with unlimited food. While in Emlen's lab, Southwick also came under the influence of N. E. Collias, a former doctoral student of W.C. Allee at the University of Chicago (Allee et al 1949). Collias was in Emlen's lab as a post-doc in the late 1940s. In 1951, with Collias, Southwick did his first primate field work—a study of population density and social organization in howler monkeys on Barro Colorado Island in the Panama Canal Zone.[37] Southwick came to his collaboration in India from a milieu emphasizing understanding population regulation from the naturalistic study of wild populations.

The extensive list of Indian doctoral colleagues with whom Southwick worked emphasizes an important historical difference between India and Africa that impacts on issues like ongoing research permissions for foreign scientists and the degree of control exercised by Indian national scientists over the kind and conditions of research done in their country. Indian scientists had both different educational experiences under British colonial rule than did Africans in the white settler colonies and different cultural relations to the landscape and the animals on it. Their social background did not lead them to flee the "bush" for careers in law or medicine in the city. To study primates in India was not to "return to the bush." The class, race, caste, and colonial social relations of production of Ph.D.s have not been the same in India and Africa; while by 1985 there was still not a single paper in the primate literature authored by doctoral black African Ph.D.s, Indian Ph.D.s had authored papers since the 1950s.[38] In Africa, the primate literature was produced by white colonists and western foreign scientists under no pressure until well after independence to develop scientific, collegial relations with black Africans.[39] African primates, including the people imagined as wildlife, modeled the "origin of man" for European-derived culture. India has been used to model not the "origin of man," but the "origin of civilization." Both are forms of "othering" for western symbolic operations, but their differences matter. India symbolically is one of the west's birthplaces; Sanskrit is a western mother-tongue.

The scientists of Aligarh University in northern India were concerned with the intersection of demands on the land and on the cultures of the people represented by the harvest of 100,000 rhesus per year, by intense post-colonial agricultural and economic development, and by human population increase. The focus of the research undertaken by Siddiqi and Southwick was from the start about the *interaction* of rhesus and people, not about the primate or the "primitive" on the *boundary* between nature and culture: "We were particularly interested in population ecology and social behavior, and the effects of changing ecological, economic, and social conditions on these populations" (Southwick et al 1980: 152). The Aligarh scientists' mode of analysis, in collaboration with Southwick, was to construct the object of study as a system of production, studied with the techniques of demography and resource management, comparing rates of gain and loss of various segments of populations, annual turnover, and the like. Social behavior, constructed as a dimension in the system of production/reproduction, was scrutinized for its relevance to natural systems of population regulation—or to their absence or breakdown under excess pressure, e.g., from too much trapping or harassment from villagers.

The researchers found deep ambivalence among local populations toward the monkeys. Villagers were usually unwilling to kill the animals, which retained their sacred status, but they were often pleased to see them trapped and removed. But the change of land-holding practices from small farmers to agribusiness alters these relationships. More importantly, this economic and political change has meant the destruction of immense amounts of rhesus habitat in forests. Managers of large agribusiness interests do not often concern themselves with the traditional rights of either local people or local monkeys to food. Reforestation usually emphasizes monoculture stands of commercial trees, not the complex production of the original mixed stands. Not surprisingly, the general conclusion from the demographic studies of primate populations in India is that they have suffered alarming decline since the first reliable counts in 1959.[40] Indian scientists were worried about decline before the export embargo of 1978, adding to their motivation to stop the trade from the misuse of the animals in war research in the U.S. In 1977, India mounted The National Primate Survey of India, led by K.K. Tiwari, R.P. Mukherjee, and G.U. Kurup, and involving dozens of scientists and fieldworkers. Taking five years to complete, it documented the depleted state of Indian primate fauna (Southwick and Siddiqi 1985).

If India may represent in my story a post-colonial nation with a sophisticated national primatology and the political and technical ability to restrain western biomedical and military hegemony over its own inhabitants, human and animal, it must also stand for the consequences of new systems of domination, tied intimately to the post-war multinational order. It is ironic that probably more of India's primates have died from the consequences of agriculture in its recent multinational capitalist form as agribusiness, than from their irradiation in military laboratories. For monkeys, as for people, the bomb and the green revolution, like disease, are survival issues that simultaneously constrain and enable the forms which knowledge may take. From the first decade after World War II to the present, Indian primatology may be unpacked to show how modern war, advanced technologies, and biomedical institutions—social relations frozen into the bomb, the seed, and the vaccine—are maps of post-World War II material and semiotic fields of force shaping multinational primate studies.

The Heart of Africa: Nations, Dreams, and Apes

In the 1980s the mountain gorillas of central Africa were inhabitants of three independent nations: Zaire, Uganda, and Rwanda. The modern fate of these animals is closely entwined with the practices and discourses of colonization, decolonization, and neo-imperialism. Representations of mountain gorillas have been icons for tales of world systems, from Paul du Chaillu's 1861 rewritten-to-order, bone-chilling accounts of his masculine encounter with the fearsome beasts to the news reporting of Dian Fossey's grisly machete murder in December 1985 at the Karisoke Research Center, which she founded in Rwanda's Parc National des Volcans in the late 1960s.[41] The stories of encounter with Africa's mountain gorillas and with the heterogeneous people who claim them as theirs are versions of what can count as nature for nineteenth and twentieth-century Europeans, Euro-Americans, and—finally—Africans. These are tales full of graveyards, trackers, scientists, park guards, poachers, guerrillas, television cameras, animal and human media stars, and tourists. Here, tangled social relations are mediated by the representations and actual "pro-filmic" lives of members of threatened species—human and nonhuman—in a post-colonial anti-Eden. The gorillas' "family lives" have been resources for discourse on the "family of man" from Westermarck's marital reveries and Freud's primal hoard to a modern U.S. graduate student's data on paternal care of the young, cast in the frame of retheorizing nature as a strategic field of (genetic and academic) investment possibilities in late-capitalist scientific cultures.

In 1957, when graduate student and accomplished naturalist, George Schaller, walked into the office of the man who would sponsor his doctoral thesis on the mountain gorilla, John Emlen, the eminent American ornithologist and behavioral ecologist, the gorillas of the Parc Albert in the heart of Africa were still the wards of the Belgian government. Scion of a wealthy and influential Boston family and Harvard ape anatomist and naturalist, Harold Coolidge, who had seen the mountain gorilla on Harvard's 1934 George Vanderbilt African Expedition, paved the way with the officials of the Belgian Congo for the independent Schaller.[42] Coolidge helped arrange funding, and Schaller and Emlen undertook the gorilla study at his suggestion. Money came from the National Science Foundation and the New York Zoological Society, headed by Fairfield Osborn, son of the president of the American Museum of Natural History in the time of Carl Akeley's gorilla expeditions in 1921 and 1926. In February, 1959, Emlen and Schaller, with their wives, flew from New York to Africa.

While, determinedly unarmed, Schaller chronicled the lives of near-human relatives in the first African national park, the colonial order of primatology was coming to an end. In September, 1960, as he descended the mountain on which Akeley was buried, the exhilarating, dangerous, and complex events of Congolese independence from Belgium had made continuing the research impossible. Belgium's rule had been fatally shaken by World War II and the events following, which had exposed Congolese workers and soldiers to worldwide social and intellectual currents leading to decolonization. Agitation against exploitation of labor, racial discrimination, unequal education, and above all, the demand for "Uhuru," freedom, in the Congo would restructure the lives of visiting scientists, African citizens, and gorillas.

In the disorder and fear of the final events leading to independence, Belgian park officials fled and did not return. On August 3, a 23 year-old agronomy student, Anicet Mburanumve, was appointed the new head of the park. On August 13, with armed park guards, Mburanumve shot and rounded up cattle of encroaching Watusi herders, and a few days later he sent porters to get Schaller safely off the mountain.[43] The two months in which the gorilla reserve on Mt. Mikeno and Mt. Karisimbi in the Virunga Volcanoes had been unattended had seen a large incursion of illegal cattle-grazing, a portent of the complex and often tragic politics that have engulfed the lands where nonhuman primates live with people throughout Africa. From this point on, the decisions about what would count as nature, and for whom, could not be made in Brussels. The immensely contradictory legacy of western constructions of nature as the primitive and original heart of the world, where people and animals were "wild," included both the parks and the deep disaffection from them of many African peoples. Healing that disaffection has become critical to the survival of many nonhuman primates. The politics of such a healing are the politics of the survival of internally divided people, many of whom need land and have good reason to regard the parks and game reserves as colonial violations. The construction of nature as an imaginative resource to heal the ills of western industrial peoples was harsh medicine for many Africans. It is still not clear whether the effects of that medicine will finally lead to the survival or extinction of the gorillas. That will finally depend on transforming the social relations that made the gorilla's initial study as models for the "family of man" possible. Ironically, a crucial part of those transformations enables the conversion of representations of mountain gorillas from the secretive and hidden worthy brother of Man the Hunter and his double, the heroic scientist-explorer, into a spectacle for sale in the world's largest single industry—tourism.

People like the young Anicet Mburanumve regarded the parks as "national heritages belonging to all of those who form a nation—to the whole population, not to a single clan or tribe pretending to have rights of first occupancy or of land-ownership."[44] But the fate of the gorillas of the Virungas is even more complex, depending on the actions of independent nations, contending groups within each nation, and the conflicting pressures of global corporations and scientific institutions. The politics of the decolonized parks throughout the Third World have been part of the internal and international politics of nation-building and of conservation and exploitation of resources. Perhaps, in 1961 Tanganyika's first president, Julius Nyerere, got closer to naming—for both independent Africans and westerners— the value of the nature that was constructed in places like the Serengeti and the Virunga Volcanoes: "I am personally not very interested in animals. I do not want to spend my holidays watching crocodiles. Nevertheless, I am entirely in favor of their survival. I believe that after diamonds and sisal, wild animals will provide Tanganyika with its greatest source of income. Thousands of Americans and Europeans have the strange urge to see these animals" (quoted in Nash 1982: 42). Historically a western construction, "nature" has become complexly global in its significance. In the late twentieth century, nature must be lucrative to endure. The survival of primates, people and animals, depends on a dialectic of love and money, both of which have been built into global scientific and popular primatology.

The National Geographic TV special, *Gorilla* (1981), gives a glimpse of how field primatology is done in the wake of these issues. The last sections of the film bring the viewer to attention with shots of a serious army of park police armed with modern automatic rifles and drilling in military formations. These people patrol the Rwandan section of the Virunga park system, the Parc National des Volcans, where Dian Fossey conducted her work with the mountain gorilla for 13 years, beginning in 1967. In 1980 her controversial role in enforcing park regulations pressured her to leave Africa and conduct lecture tours in the United States. Fossey began in the Zairoese section of the gorilla habitat, where Akeley and Schaller had preceded her. She was forced to flee that area in July 1967, by Zairoese soldiers involved in a rebellion in Kivu Province. In Rwanda, establishing the Karisoke Research Center, Fossey used controversial direct action tactics to deter local poachers who harrassed, trapped, and killed gorillas. She bitterly opposed the schemes to protect the gorillas through making them a foreign tourist resource. Her intense efforts to protect and to be "alone" with personally named, wild gorillas in what was for her, and for the west before her, a dream land (Fossey 1983: 1) gave way to painful political difficulties, as well as to a different organization of scientific research.

That organization was both more collective, involving institutionally organized, long-term research where no one person would ever have the role Fossey once did, and more individualistic, involving students coming to get their thesis data and perhaps having little further emotional or structural relation to the project. The emotional organization of primate work changed with the political, financial, and organizational shifts. The people who founded these sites have often found the shifts wounding. After Fossey's departure, another philosophy of conservation, called by Fossey "theoretical conservation," took precedence over anarchist direct action, or as Fossey called it, "active conservation" (Fossey 1983: 57–59). "Theoretical conservation" meant the systematic promotion of tourism, education, and state police action to make the parks safe and profitable. Profitable to whom remains contested.

Gorilla shows a scientific couple, Amy Vedder and William Weber, both former doctoral students in primate behavioral ecology with John Emlen at the University of Wisconsin and both representative of the next generation of primatologists after Fossey and Schaller. Vedder and Weber were among the dozen or so Ph.D. students who did their field research at the Karisoke Research Center from the early 1970s, helping collect long-term records on the animals and beginning to publish their dissertations and papers that belong to a "problem-oriented" rather than first survey literature. In the film Amy Vedder habituates a gorilla to the presence of white tourists paying for a day excursion to see a "wild" gorilla. Vedder whispers to them about gorilla behavioral ecology as they watch intently and snap their cameras. The framing context is the previous panoramas and close-ups of "encroaching" black farmers at the edge of the park. The narration is about the pressure of human population and the disappearing gorillas, reduced to fewer than 300 in this region since Akeley's days. The day tourists and the TV viewers are treated to sumptuous shots of gorillas glimpsed through the rich foliage. We learn that tourism increased park revenues by more than 100% in one year. We do not learn what farmers adjacent to the park got. We can hope that the profits from plane reservations and hotel bills went less to multinational corporations and more to local Rwandese business. From other sources, we learn that by 1985, gorilla-related tourism was the fourth largest source of foreign exchange for Rwanda. More than 6000 people per year paid for the experience of "direct vision" of even post-colonial nature. We also learn that the gorilla population, still fragile, appears to be increasing slowly and that Rwandese in Kigali speak with pride of *their* gorillas.

Bill Weber is introduced as the education coordinator of the Mountain Gorilla Project, founded by several international conservation organizations. Hoping to build a conservation ethic, he travels around Rwanda teaching school children about gorillas with slide shows and films. None of the children has ever seen a gorilla. We hear snatches of Weber's French, the language of instruction, and like the park itself, the "vestige of colonial time" (sound track). Along with Kinyarwandu, French is an official national language. Weber constructs a conservation ethic by enlisting the *identification* of the young students with the large mammals. He elaborates on the "family" of the gorilla, the similarity of their lives and ours, across cultures and species, and he tries to relate their families to the social settings of the children. Weber speaks directly to the National Geographic camera, pleading for the animals, not for his need to be physically with them "in nature"; he may never return to Rwanda. But he pleads for the importance of the "need to know they exist" (soundtrack). The last shot shows a little white child's hand reaching through the mesh netting of a cage to touch a tiny black gorilla baby inside.[45]

This stunning 1981 closing shot recodes the colonial discourse of race onto a post-colonial restrained African animal and an exploring western human. The film's explicit message was precisely to deconstruct that code, but the continuing latent logic of racism throws the viewer backwards and forwards into the spectacle of the graveyard of Africa's Eden. The gorilla sanctuaries of Rwanda and Zaire are not only the contested terrain of the living; they are also memorials to those who have died in the service of the west's colonization and recolonization of the image of nature. Carl Akeley's violated grave in Zaire must be remembered in the same all-too-real dream field with Fossey's and her gorilla siblings' graves in Rwanda. In 1977, poachers in the Rwanda park killed and decapitated a male gorilla, named Digit by Fossey. Digit became an international symbol of Fossey's relation to nature, and the Digit Fund supported her extraordinary anti-poaching measures. Digit himself was buried in a graveyard of gorilla casualties in the struggle (Fossey 1986: 10–15). Fossey was found murdered in December, 1985. She had made passionate enemies everywhere—among western scientists, Rwandese officials, and local people. Her passion was for the individual animals, the species, and a construction of nature. An American graduate student, Wayne McGuire, studying paternal care among the gorillas, was tried for the murder *in absentia* by a Rwandan court, convicted, and sentenced to death by firing squad on evidence that made his culture's readers wide-eyed with disbelief. To them the evidence pointed to irate poachers. The student's academic career was in shambles, his data terminated, as the nature that generated them was barred to him. Fossey herself was buried in the gorilla graveyard, next to the bodies of mutilated animals representing the family of man.

From the first *National Geographic* article by Fossey on the mountain gorillas, her life was made into a media event; and her death confirmed the commodity value of her play in the theater of nature in the African Eden and of the twentieth-century discourses on extinction and exterminism.[46] By autumn of 1987, Warner Books had a book out on her life by a popular nature writer (Mowat 1987). In summer 1987 the *International Primate Protection League Newsletter* reported that Warner Bros. and Universal Studios had joined together to make the feature film of Fossey's life, played by Sigourney Weaver, famous for her role in the science horror films, *Alien* (1979) and *Aliens* (1986).[47] A hitch loomed in filming the gorillas: Mountain gorillas are not trained for Hollywood, so the studios tried to get permission to use captive young chimpanzees, dressed as gorillas, to play the two orphans Dian Fossey tried to protect from export to a European zoo. Piling irony on bitter irony, the IPPL reported that one of the chimps appeared to have been illegally sold to Hollywood Animal Rentals by a U.S. zoo, whose only excuse for holding the animals is their responsibility for breeding an endangered species. Adult gorillas in the film were to be played by the same humans who so successfully simulated the apes fostering Tarzan's cinematic rebirth in *Greystoke*.

The film was released in 1988. Its best actor was surely the mime who played Digit. The film was full of quotations from the previous National Geographic Society footage. National Geographic, not the contemporary national park in Africa, provided the original "nature" for this film. Perhaps the most interesting aspect of the commercial cinema was the traffic between shots in which gorilla actors were either real animals in Rwanda, captive animals on the movie set, human mimes in gorilla suits, or puppets. The mimes knew how to act because of Fossey's research and writing, but most of all because of Robert Campbell's previous National Geographic film narratives. The real gorillas did not know how to act in front of the camera in the 1980s. In a postmodern frame, there is bitter irony in the recreation of both Fossey and the mountain gorillas as copies of murdered originals and copies of previous films and photographs. Fossey died trying to hold onto her culture's dream of an original and timeless nature.

In the ongoing lives off-camera of people and animals, L. Rwelekana, a long-time and skilled field assistant at Karisoke Research Center, died in a Rwandese prison, held for the murder of Fossey but never tried and, according to western gorilla researchers from the site, certainly innocent. Unlike Fossey's end and McGuire's murder charge and trial *in absentia*, Rwelekana's death did not become the subject of a *Vanity Fair* or *Discover* special; and he was completely absent from the commercial film. His death was less easily assimilated into the colonial nostalgia and Banana Republic fashion clothing lines of 1980s neo-imperialist culture. Meanwhile, recorded in the sepia tones of the ubiquitous colonial-nostalgic aesthetic appropriate to the age of heroes like Colonel Oliver North and the Nicaraguan Freedom Fighters, Carl Akeley's life and African adventures were featured in the February 1987, 100th anniversary issue of *Sports Afield* (found for me in a U.S. doctor's office by a friend waiting to be examined for a possible AIDS-related infection, a disease killing many people in Rwanda). That night's dreams emerged from the soft tones of *Sports Afield*'s memorial to the great taxidermist and photographer, "The Man Who Put Africa on Display." Nations, dreams, apes, and viruses ran together in a condensed dream logic of post-colonial paranoia, where Africa remains the heart of the west's unconscious, denied history still in its practices of representations, from fashion clothing to conservation tourism to the images still brought *Out of Africa*.[48]

Access to Camping in the Malagasy Republic

In 1960 the French colonial protectorate of Madagascar became the independent Malagasy Republic. This 1000 mile-long island in the Indian Ocean off the east coast of Africa is home to the prosimian lemurs, ancient symbols of lust in western culture, unique species of animals most Malagasy people have never seen, and since the 1950s, subjects of modern primatology. A student of lemur social behavior, who did her first field work on Madagascar from 1962–63, Alison Jolly may help us thread our way further into the tissues of race, nationality, gender, and decolonization that structure the narrative field of primatology.

Jolly, then Alison Bishop, finished her Ph.D. in Zoology at Yale University in 1962, writing a thesis on the use of the hand in lower primates. Having distinctly not enjoyed the work she was supposed to be doing with sponges in the laboratory, she relates that she fell in love with John Buettner-Janusch's captive lemur colony. Buettner-Janusch kept about fifty prosimians for his work linking biochemistry and anthropology, but Jolly's interests were the evolution and nature of intelligence. In this she was a spiritual offspring of the eminent Yale ecologist, G. Evelyn Hutchinson, who encouraged her interests in Jean Piaget, Alfred North Whitehead, and Konrad Lorenz. Hutchinson had exercised his usual brand of teaching during the summer Jolly did research for him on a book they planned to write jointly: His basic advice was to go read interesting things.[49] Hutchinson's world enlarged and confirmed the young Alison Bishop's growing preoccupation with the ontogenetic and phylogenetic development of intelligence that has pervaded her work. She moved her attention to Buettner-Janusch's lemur cages. Richard Andrew served as the chair of Jolly's thesis committee. A doctoral student in the early 1950s of ethologically trained Robert Hinde at Cambridge University's Madingley Ornithological Field Station, Andrew used the Yale lemurs for behavioral studies.

This web of connections from anthropology, ethology, and ecology led Jolly to London in April 1962, to the special symposium on primates of the London Zoological Society, where she heard Jane Goodall give her first scientific paper.[50] Jolly was deeply impressed by Goodall and excited by the evidence of interest in what felt like a new field. Its newness was underlined for her when Solly Zuckerman defensively rejected the data presented by K.R.L. Hall and Jane Goodall that undermined his earlier conclusions. Chair of the session, Zuckerman announced that all his work had been confirmed (patently false) and that he had to remark "on a few presumed facts which I have heard today [which] have indeed been little better than anecdote picked up second hand" (Zuckerman 1963: 119–22). Perhaps he had not recently reread his work from the 1930s.[51]

With National Science Foundation money, Jolly went to Madagascar for about one year of field study of the unique prosimian lemurs native to that island (Jolly 1966). She produced a book-length preliminary survey, giving for each of several species a general description, taste of the ecology, description of individual and group behavior, and account of relations with other species. She recalled, "I didn't know anything; I looked at everything . . . and had lots of money for all sorts of things I didn't know how to use—cameras, landrover, etc. . . . There was money available—and a general feeling Americans should go abroad" (Jolly, interview 1982). Passing through Paris, she sought out and asked advice from Jean-Jacques Petter and Arlette Petter-Rousseaux, who had done their doctoral studies together on their honeymoon, in the late 1950s, under François Bourlière of the Physiology Department of the Faculté de Medicine. In a twist on the common trope of the married field team, Jean-Jacques did the diurnal lemurs, and Arlette studied their nocturnal cousins. The French couple found themselves in Madagascar, still France's protectorate when they began. World War II and its aftermath set in motion forces that ended direct colonial domination throughout what would come to be called the Third World and that also brought young Europeans and Euro-Americans in large numbers to this decolonizing area. The intersection determined much of primate studies.

Jolly emphasized her naiveté about the political situation of the post-colonial Malagasy Republic. In retrospect, from experiences like collaborative work with Malagasy biological colleagues on a subsequent trip and her international experience, due partly to her husband's work as an economist in Uganda and Zambia, Jolly analyzed how her work had been affected by that post-colonial reality. In 1962, she had found the total French substrate in a nominally independent country a little surprising, but very convenient. Like every primatologist I have interviewed who studied in the decolonizing world in the decade of the 1960s, Jolly took for granted going to a French planter for support, speaking French with everyone (her college majors had been zoology and French literature, a subject her father also taught), and seeking French officials for all sorts of matters. In 1964 she began attempts to provide some material on local animals for Malagasy schools, when she learned that Malagasy children who went to school learned about European rabbits in their books, had likely never heard of lemurs, and certainly had not seen their pictures. The only available lemur photographs in the Malagasy Republic were on the covers of match books for tourists, priced too high for most Malagasy to purchase.[52] Expertise about lemurs was a Malagasy export, a part of the extractive processes of colonialism, the prize taken home by the western traveler in exotic lands.

In the critical period of the victory of Malagasy nationalism between 1972–75, the rule by French interests was broken. "The government changed from one that still had French civil servants in every office to one that is visibly Malagasy" (Jolly 1980: 7). French scientists were told to leave. They did—with a vengeance, taking with them everything except some cars and the buildings they were in. Jolly reports one French physical oceanographer's embarrassment at his government's policy; he felt he had been made into a carpet merchant. The Malagasy responded by denying all research visas between 1975 and 1983. "We go in as tourists or teachers in the university; most of us just don't go." Jolly herself made a series of short visits.[53] In sympathy with Malagasy policy, Jolly summarized their position:

> Why, they asked, do missions from other countries not stop, visit, and lecture at the university of Tananarive as a matter of course? You would not expect a party of Russian scientists to turn up in Texas, train their binoculars on the whooping cranes, and never so much as nod at American colleagues or officials. You do not find be-telephotoed Texans in Szechuan oggling pandas without permission of the Chinese. Why should foreign scientists come to Madagascar and often act as though the Malagasy do not exist?

Jolly quoted Dr. Etienne Rakotomaria, Director of Scientific and Technical Research in the Malagasy Republic, speaking at a meeting of biologists in 1975: "Scientists will only be allowed to work here if they arrange reciprocal benefits for Malagasy colleagues. The people in this room know that Malagasy nature is a world heritage. We are not sure that others realize it is our heritage" (Jolly 1980: 7). Those words marked the end of the material conditions of western "pure science" in Madagascar; that was a science practiced in the frame of a specific historical construction of nature. But Rakotomaria's words marked a possible beginning for a Malagasy-controlled reconstruction of what nature and conservation must mean for national, ecological, and cultural survival. What might a post-colonial reinvention of nature be like?

For Alison Jolly, the decolonized conditions for studying the lemurs she loved since graduate school intersected in curiously fruitful ways with another major system of social relations structuring primate field studies: gender and marriage. But not gender and marriage in the abstract; primatologists have not come proportionately from all social classes among Euro-Americans. Access to some wealth, both from birth and marriage, may have been particularly crucial for women who have done primate studies. Just as the French expulsion from Madagascar is not typical of all field research stories in the decolonizing world, neither is Jolly's relation to "personal life" identical to other researchers'. But the slogan coined in the Euro-American women's liberation movement in these same years, "the personal is the political," suggests a perspective for seeing both how the specifics of Madagascar illuminate a systemic history of race and nationality and how the particulars of Jolly's juggling of career and marriage indicate the social determinations of gender and class.

Jolly gave birth to her first of four children while she was writing *Lemur Behavior* in Kampala. She was 27 years old and already had her Ph.D. and a year of post-doctoral field work, culminating in an important book. She reported that her husband was supportive of her work, emotionally and materially, and his job took them to places where she wanted to be—Uganda, Zambia, Sussex, New York. Alison Jolly was able to travel to many kinds of habitats (Barro Colorado, India, East Africa, Madagascar) and to "do snatches of research" and some part-time teaching (Zambia, Sussex, Cambridge). Those "snatches" were precisely what were imposed on all visiting scientists who would study Malagasy flora and fauna after 1975. Primarily, Jolly wrote books synthesizing others' research or books contributing to culturally and politically sensitive conservation efforts in the Malagasy Republic.[54]

Jolly narrated experiencing a great deal of pleasure in this pattern. Partly in view of her mother's modeling a self-employed career at home as a professional mural painter, Jolly reported she never doubted that she would have a career and that her hybrid arrangement *was* a career. That combination of securities, which could include self-esteem from children and from practicing a science requiring international travel, was race-, gender- and class-specific, as well as specific to heterosexual women who could draw on a white male middle-class income because of the *institution* of marriage. Following such a pattern, men would be much less likely to see what they were doing as a satisfying *professional* career, and less culturally and economically secure women would likely not have the freedom to take the significance and reality of their scientific work for granted. The long history of women in less industrialized biologies, prominently in natural history, displays both openings for women in science through hybrid amateur-professional "styles" and also linked barriers to high status as a tenured academic and principle investigator on grants (Rossiter 1982). Women of color do not have the same history of access to the production of science through "amateur" or "self-employed" practices in natural history. They were more likely to be its objects than its practitioners.[55] Jolly's account of how her kind of life in science was materially possible included a rare explicit recognition of the convergences of large historical determinations, the positioning of social subjects, and intimate personal choices.

There has been an intense quality of "craft production" in the social conditions of Jolly's work that depended on these intersections of marriage, class, race, educational background of herself and her parents, the state of disciplinary development of primatology when she left graduate school, the colonial relations of science in Madagascar when she began her field work, the compatibility of decolonized patterns with her later working situation, and her achievements before marriage and children that established her credibility with publishers and research and conservation institutions with which she affiliated. Jolly's narrative stressed that women entering biology under different and more common circumstances were less likely to be successful craft producers of science and more likely to disappear, as many of those who started school with her did (interview).

Primatology has been more based on "craft production" than many twentieth-century biologies, continuing to depend on quite idiosyncratic individuals and field sites and less on the industrialized, routinized conditions that have characterized molecular biology, for example. There has been a strong trend away from craft organization of primatology, e.g., with growing needs for major computing and technologically sophisticated laboratory facilities, complete with their elaborate hierarchical division of labor (armies of post-docs, technicians, cleaners), but not to the same degree as for the laboratory-based biologies. Field primatology has been less likely to be successfully "industrialized" and more likely to see its funding cut in favor of those biologies more relevant to the multinational commodification of nature. That social fact has made more cracks for middle- and upper-class Euro-American women without white male-typical careers to fit into. The same social fact has kept conservation work—of a kind sensitive to the politics of decolonization— a poor stepsister of "modern" biology, in the priorities of the paternal state. Conservation-oriented primatology remains still more organized along "craft" lines than other areas of the discourse. These lines converge in social practices by and cultural coding of women in the field.

These themes came together in Alison Jolly's writing of *A World Like Our Own: Man and Nature in Madagascar*, based on a World Wildlife Fund-sponsored five month trip, accompanied about half the time by a western wildlife photographer, Russ Kinne. "Because we travelled with Malagasy colleagues, we were received as friends in villages, with traditional warmth and hospitality" (Jolly 1980: xv). The book is deliberately polyphonic, structured with audible opposing points of view and privileging the voices of villagers who illegally eat lemurs and fell forests, as well as urban Malagasy scientists' concerns and those of herself as a Euro-American conservationist. There is no fiction of neutrality, and, despite the title's invocation of the hoary spirits of Man and Nature, also no fiction of "man" in the abstract responsible for plundering the earth. Nor is there the fiction of "nature" with meanings unmediated by people's relations with each other. The dialectic of universalism and particularism in the book invites readers to understand the constructed and struggled for nature of knowledge—and of nature itself. "This is the novelists' way of reaching a conclusion, not the scientists', but the argument seems far too important to leave to faceless statistics" (xv). Jolly echoes the themes of contemporary concerns about dialogic ethnography and heteroglossic texts; this is an approach to nature possible only after the shock of decolonization (Clifford and Marcus 1986).

The ties of gender, science, and decolonization are evident in Jolly's book. In her travels through the western deciduous dry woodlands, she was accompanied by a botanist from the University of Tulear, Rachel Rabesandratana, who showed Alison her flowers as Jolly showed her lemurs. [Figure 10.3] "After all, Alison, if you can park your children for this part of the trip, so can I. My husband will just pay a little less attention to his Marine Research Station for the next few days and a little more to Haja, Hary, Naina, Noro, and Hobilalao" (204). With this introductory allegory of women's shared conditions, Jolly takes the reader camping with them in a woodland near Bekily, but only after Rachel, "the city woman in her formal white lamba [bought] surplus rice from the village grandmother. I wondered who really knew the right path: Rachel and I with our university educations, our travel to many continents, our juggling of children and intellectual career, or this old woman whose grandchildren tilled the fields of her grandfather, so that her generous hands might fill our cups" (214). The lightly nostalgic allusion to grandmothers and their wisdom

Figure 10.3   Jolly watching Rabesandratana gathering medicinal plants. © Russ Kinne. Comstock.

has a curious function in this text: it disrupted the unity of women just established in the story of the parked children. Alison and Rachel do not inhabit the same world as the woman selling the rice, nameless as she must be in her textual function of marking the allochronic time of tradition.

But the unity of Alison and Rachel is problematic too; international science and temporarily abandoned children cannot hold them together. Jolly pitched her orange nylon tent, casually asking Rabesandratana if she had ever camped in her travels through Europe. "Of course not. . . . We visited cathedrals, not forests. Here it would be considered impolite if not insane to refuse village hospitality to sleep out among the ghosts and the wild boars, dragging one's own house like a snail" (217). Jolly teaches the reader: "Rachel had again neatly punctured the bubble of our provinciality. 'Everyone's' memory of camping out is only a western memory. If our own enthusiasm for the wild began with the nineteenth-century Romantics, access to the wild began to spread with Lord Baden-Powell and Ernest Thompson Seton and Henry Ford. The wild as a separate place, nature as a separate entity 'to go back to,' is the cumbersome Western concept" (217). But the reader is not allowed to rest here either, in a pleasurable indulgence in sophisticated rejection of the bad and parochial "west": Rachel ended her lecture on cathedrals and camping with the emphatic point that she and her husband lacked good camping equipment in her student days in Europe. "But I assure you, I enjoy this life" (217). These are the nuances out of which the story of primates in Madagascar is told in this very "womanly" text, which refuses closure and insists on situated experience. Rachel's and Alison's voices on nature were interrupted by another voice: the cliques and squawks of the forked lemur, an elusive, nocturnal, nectar-browsing denison unique to the western Malagasy woodlands. If the lemurs were to stay audible to scientists, they would have to become part of the complex *historical* conversation among the differently situated members of the primate order—perhaps by noisily interrupting the allegories of union of man and nature that ground narrative traditions in natural history.

In "Passionate Detachment," discussing women's filmmaking practices, film theorist Annette Kuhn (1982) contrasts the voices of feminine and feminist texts. In Kuhn's terms, Jolly's text, insisting on multiplicity and disrupted closure, is "feminine," as francophone theorists have elaborated the idea of "l'écriture féminin." The convergence of Jolly's practice with that of self-referential and deconstructive ethnography highlights the global social basis of textual politics—the struggle between the neo-imperialist system of multinationalism, where the west and its international allies are still hegemonic, and the contesting claims of multiculturalism, rooted in counter-hegemonic cultural reinvention (Clifford 1988). In Jolly, there is a further concern: "multispecies-ism." In a kind of Whiteheadian "concrescence" worthy of her reading of that philosopher in graduate school with Evelyn Hutchinson, Jolly's textual politics converged with what she named as the most important priority for those, like herself, trying to preserve a world full of animals amidst a multicultural, power-charged "conversation" composed of specifically differing voices: the priority was disarmament.[56]

But in spite of the ultimate "ecological" issue of nuclear war, tied to the imagery of primates blasted into space, the daily work of a conservationist remains the complex global and local politics of participating in reconstructing the meanings and social relations of "nature" within a decolonizing world. This is a world in which intensified immiseration, unbearable debt burdens, neo-imperialist and state socialist development debacles, and continuing nationalist, internationalist, and humanist hopes (and some achievements) to end the unacceptable inequalities in life and death for people all bear directly on the life chances of a lemur, baobab tree, or forested watershed essential to the next decade's rice crop or clean water project for cities and villages. Inheritors of their own history and natural history, there is no way for westerners to participate innocently in "reinventing" nature in a world untouched by western hegemony. Western forms of love and knowledge of nature have been profoundly colonial; knowledge of how that has been so cannot be allowed to degenerate into an excuse for losing an historical capacity to know, love, and act in relation to the strange and dynamic category still somehow able to be called "nature." If it once was, nature is no longer simply a western epistemological and social imposition. Like other languages of the colonizer that have been reinvented for other conversations, the languages of nature have become polyglot and international. Rather than loss of innocence being an excuse for *not* participating in these life and death conversations, this loss is a necessary but not sufficient condition for taking part usefully.

Beginning in 1983, and decisively in 1985, the politics of naturalistic research and conservation changed again in the Malagasy Republic. In the face of the Ethiopian famine and the growing social and political threats to the fundamental ecological structure that must sustain life throughout sub-saharan Africa, the Malagasy government began to emphasize conservation issues directly. That decision included granting research visas once again to foreign scientists, but in a carefully controlled manner that maximized the relation of foreigners' research to national scientific needs and national independence. Exposing the World Bank's anti-environmental development loan policies, as part of its overall strategies that could only deepen the fundamental conditions of inequality, provided a somewhat more favorable climate for setting terms of development aid sensitive to the consequences for watersheds, soil conservation, overall national natural resources, and the relation of these to social justice.

In late 1985 the Madagascar government hosted an International Conference on Conservation for Development. The Malagasy Minister of Livestock, Fisheries, and Forests, Joseph Randrianasolo, placed the conference in relation to the last one held there in 1970, which had emphasized the uniqueness, beauty, and scientific interest of Malagasy flora and fauna. In 1985 the stakes could not be accommodated in the dehistoricizing categories of aesthetics and pure science. "Now we want to manage our resources so that Madagascar can be self-sufficient in food and fuelwood."[57] The ties of forest, water, and rice were in the foreground. It is these ties setting up the terms of local and global "conversation" that might make it possible for lemurs, foreign and national lemur biologists, agriculture, and Malagasy rice growers and eaters all to thrive. Without that conversation, nature will not be decolonized and reinvented; it will be destroyed. Continuing more than twenty-five years of non-innocent relation to Madagascar, Jolly has taken a modest, concrete part in these developments.[58] She continues to write about and take part in the politics, ecology, and behavior of human and nonhuman primates in a web of specific historical relations that include agencies of the United Nations; international conservation organizations; the World Bank; Malagasy scientists, officials, and villagers; western scientists and popular science readers; and the flora and fauna of

Madagascar. When she fell in love with a colony of captive lemurs at a major scientific university in a New England town about 1960, Jolly could have had no notion that all the terms of that relation—love, colony, science, primate—were punning signs pregnant with what would become her life's personal and professional obligation: participating in negotiating the terms on which love of nature could be part of the solution to, rather than part of the imposition of, colonial domination and environmental destruction. [Figure 10.4]

Figure 10.4   A ringtail lemur (*Lemur catta*) mother and young. Copyright Phillip Coffey. Courtesy of Alison Jolly.

Part Three

The Politics of Being Female:
Primatology is a Genre of Feminist Theory

.

11

WOMEN'S PLACE IS
IN THE JUNGLE

[If] the rhetoric of conviction were to be replaced by the rhetoric of mockery, if the topics of the patient construction of the images of redemption were to be replaced by the topics of the impatient dismantling and upsetting of every holy and venerable image—oh, on that day even you, William, and all your knowledge, would be swept away! (Eco 1983: 476)

Rouse ye, my people, shake off torpor, impeach the dread boss monkey and reconstruct the Happy Family. (Mark Twain)

The first lines quoted here, from Umberto Eco's *Name of the Rose*, were spat out by the aged and blind Medieval librarian, Jorge, who was desperately trying to prevent the discovery of the lost book of Aristotle's *Rhetoric*, on laughter. Laughter, thought Jorge, was inimical to the salvation of the soul; it was the devil's work, leading the Christian away from serious contemplation of the truth. William thought rather that laughter was an indispensible tool in the pursuit of understanding. Mark Twain and P.T. Barnum would have agreed. The second quotation is from "Barnum's First Speech in Congress."[1] Twain composed his version of what Barnum surely would have said, had he been elected in his bid for Congress. The tradition of hoax in American history is indelibly associated with P.T. Barnum. Hoax was a popular nineteenth-century form of entertainment that tested the intelligence of the audience; it was less a form of deception than a form of interrogation and an invitation to find the flaw in an apparent natural truth. Hoax assumed greater confidence in the active intelligence of the audience than did the more reverent television nature special. The relation of hoax and popular natural history is unnervingly close. A reminder of this relationship is particularly salutary in approaching the political and biological science of being female. Hoax and natural history both have deep roots in democratic and populist histories, but the practice of hoax more seriously resists the closures of those hegemonic discourses on nature in which each being finds its ordained place. Ordination has been generally bad for the health of females. Feminists—women and men—and women—feminist and not—trace a fine line as scientists drawing and redrawing the objects of biological and medical knowledge marked female. Females as natural-technical objects of biological discourse are not unlike the conundrum in P. T. Barnum's early museum that tested the viewer's credulity and acuity—a mermaid composed of the head and torso of a mummified monkey stiched to the tail of a large fish. Our problem will be to find the evidence of stitchery without ripping out the patterns in the lives of females—fish, monkeys, or scientists. Complemented by a ready suspicion for the flaw in apparent natural truths, laughter is an indispensable tool in deconstructions of the bio-politics of being female. Suspicion and irony are basic to feminist reinscriptions of nature's text.

Redrawing Sex: Ruby Tuesday Testifies on the BBC

A personal interview and television documentary with an authentic witness of the ape-to-human transition could only be imagined in America, with its extraordinary cultural appreciation of personal testimony about human potential. From the Great Awakenings of the eighteenth and nineteenth centuries through the proliferation of human potential therapies of the twentieth century, Americans have known how to make experience talk. "Ruby" is a freeze-thawed woman of the Pliocene, fortuitously recovered for science by a team of paleoanthropologists, with the famil-iar names of John D. Hansom, Roderick Luckey, and Aaron Killjoy. A creation of Adrienne Zihlman and Jerold Lowenstein (1983), Ruby makes a single appearance, an interview conducted in the British Museum, in which she makes her account of her life available to an international public. (Zihlman and Lowenstein, her husband, are Euro-American evolutionary scientists, one a Ph.D specialist in primate compar-ative functional anatomy and the other an M.D. with interests in biochemical aspects of fossil interpretation.) Ruby had learned enough modern language to make the interview possible, although ambiguity about many key events at this fateful time of transitions results from the difficulties of translation in this unique ethnographic encounter. But Ruby seemed reassuringly adaptable in the world in which she woke up, even to seeing career possibilities in her status (as long as she stayed unmarried) and to getting around London on a moped.

What provoked this article, the interview with "Ruby"? That question cannot be answered directly—a rhetorical problem structuring all of Part III because "the politics of being female" are at the origin of western order, including scientific accounts of what it means to be human, to be female, and to be an organism. But women's authorship of those politics is not, literally, "original." Females and women have been sites for the construction of others' discourses (Mani 1987). In the west—including western science in all its foundational mythic moments of origin with "the Greeks" or at the great instauration of "the scientific revolution" or at the moment of Darwin's transformative account of "the origin of species"—to be female has been to be a pretext, not an author and a subject of history. To be female has been to be Woman, the plot space for male potency (de Lauretis 1987). Marked as female, the social group of women comes into a scientific world already crowded with the presence of the word; women's stories are perceived to begin *in medias res*, a rather difficult position for a putative origin account. *Actual* men's accounts, of course, also always begin in the middle, within a web of sustaining and limiting discourses. It is only Man's account which produces the illusion of the start of the conversation, and he is not real, despite his world historical power. But actual men's accounts have been taken as the premise for further discourse more easily than actual women's writing. This is not a small matter. How may the people known simultaneously as women and as scientists—an oxymoronic social subject only beginning to break down—intervene in the construction of the potent natural-technical objects of knowledge called females?

I like to use the image of "Milton's daughters," borrowing from the feminist literary critics, Susan Gubar and Sandra Gilbert (1979), because I am compelled by their account of the female children of John Milton, reading the Bible to their blind father, the great Protestant poet of *Paradise Lost* and civil patriarch of the English Revolution, participating in the seventeenth-century debates that founded the terms of male citizenship in much of the modern western world, while crafting the corresponding epic of female incapacity in his portrait of Eve. Who knows what stories those daughters were imagining? Gubar and Gilbert, painfully aware of the power of Milton's writing in crafting the English language and bounding anglophone imaginations of origins, set out to account for English-speaking women's literary productions in terms of alternate strategies available to Milton's daughters. They could attempt to ignore the father's story, starting completely anew; or they could pretend to misunderstand, to remain faithful, but still to say something more freeing for themselves in the structure of narrative. Either way, they risked looking foolish, as if they did not know the rudiments about human potential and had forgotten that their role was to read the story, not write it.

Women practicing the highly narrative life sciences and human sciences of biology and anthropology have been in a similar social and linguistic position.[2] Some of the women practicing primatology who have contributed to this section and to other parts of the book have objected to the image of "daughters," seeing their contribution unmarked by gender and not derivative from the authorizing masculinist accounts. I do not think scientific women have been that lucky. The law of the father might be a myth, but its very real potency is hard to deconstruct. Perhaps it is more important to remember that the father was blind than that he bestowed the first names, or even the capacity to name, in the primate order. And perhaps this generation of daughters and sons can shatter the logics and hoaxes of kinship so that Ruby can move safely around the streets of London. But Ruby cannot come first; Zihlman's and Lowenstein's parodic construction was a response to a son who was very faithful to the fathers. The tale begins with interpretations of the recent reappearance in the paleoanthropological field in Hadar, Ethiopia, of a diminutive, ancient (say 3 million-year-old) hominid grandmother—of erect and bipedal habit, but small mind—named by her adamic founders after Lucy in the Sky with Diamonds (Johanson and Edey 1981). Lucy could be Lucien, but let's give her her sex, since it is crucial to the story at hand. The paucity of African names in paleoanthropological and primate literature speaks volumes about the limitations of Adam's claim to species fatherhood.

Lucy's nearly complete skeleton was dug out of the earth by the skilled hands of a brotherhood that recognized in her and associated skeletons a resource for re-establishing potent masculinist versions of "The Origin of Man" (Lovejoy 1981). Lucy was quickly made into a hominid mother and faithful wife, a more efficient reproducing machine than her apish sisters and a reliable, if poorly upholstered, sex doll. These are the qualities essential to the male-dominant, "monogamous," heterosexual family, named "the family" with mind-numbing regularity. Lucy's bones were incorporated into a scientific fetish-fantasy, dubbed irreverently the "love and joy" hypothesis in Sarah Hrdy's response (Hrdy and Bennett 1981: 7). What makes Lovejoy's interpretations of Lucy "masculinist," as opposed to simply distasteful and controversial for his scientific opponents? Is the "masculinism" simply shoddy science in this case? Or is the problem deeper? The answer lies in Lovejoy's unwitting discipleship to the father of biology, Aristotle. Lovejoy's "Origin of Man" is enmeshed in the narrative of active, potent, dynamic, self-realizing manhood achieving humanity through reproductive politics: paternity is the key to humanity.[3] And paternity is a world historical achievement. Maternity is inherently conservative and requires husbanding to become truly fruitful, to move from animal to human. Standard in western masculinist accounts, *disconnection* from the category "nature" is essential to man's natural place: human self realization (transcendence, culture) requires it. Here is the node where nature/culture and sex/gender intersect. In the transition to a savannah-mosaic environment, the narrative of matrifocal, female-centered worlds of apes had to give way to the more dynamic "human" family. "In the proposed hominid reproductive strategy, the process of pair bonding would not only lead to direct involvment of males in the survivorship of offspring[;] in primates as intelligent as extant hominoids, it would establish paternity, and thus lead to a gradual replacement of the matrifocal group by a 'bifocal' one—the primitive nuclear family" (Lovejoy 1981: 347–8).

Nothing a female could do could lead the species across the hominoid-hominid boundary; she was already doing the best nature allowed. "She would have to devote more energy to parenting. But natural selection has already perfected her maternal skills over the millions of years her ancestors have occupied West Africa. There is, however, an untapped pool of reproductive energy in most primate species—the male" (Lovejoy 1984: 26). After all, Tarzan was adequately mothered by an ape. In Lovejoy's account, through provisioning his now pair-bonded and sedentary mate at a home base with the fruits of plant and small animal gathering, a male could lead the species across the boundary to the origin of man in the assurance of fatherhood. Lovejoy gave up hunting to mark manhood, but he could not dispense with paternity. The species had reason to stand upright at last, even if not too efficiently at the start. Man was on the long lonesome road.

And women's place in this revolution is where it was imagined cross-racially in a fair section of U.S. 1960s politics—prone. As Lovejoy put it, women did not "lose" estrus; they constantly display its signs. For the new strategy to succeed, "the female must remain constantly attractive to the male. . . . While the mystery of bipedality has not been completely solved, the motive is becoming apparent" (1984: 28). Males arose to stand upright so that this better success at provisioning would be rewarded by making them into Man, a father, desiring subject, and author. Small wonder that Lovejoy cited his brother-colleague for evidence that "[human] females are continually sexually receptive" (Lovejoy 1981: 346). The footnote read "D.C. Johanson, personal communication" (1981: 350).

Why did serious scientists need to respond to this story? Zihlman was involved with her own research and publication, attempting to establish the authority of stories quite different from Lovejoy's, some of them involving Lucy. It took time to write about Lovejoy, just as it took space in this book, and Lovejoy has not taken the time to write in detail about the interpretations of Zihlman. His decision not to cite Zihlman's substantial and directly pertinent technical analyses, in a paper replete with references, including the one cited above crediting his colleague's sexual boasting, effectively obscured from the readers of the 1981 *Science* featured story her significant work on bipedalism, sexual dimorphism, and reconstructions of hominid social and reproductive behavior at the crucial boundary. The *Science* article is the point: Lovejoy's story and his involvement with immensely important fossils cannot be ignored. Milton's daughters do not have that luxury. But they do have a weapon more potent than the undecidably lost or omnipresent signs of estrus. They write.

Zihlman responded with Jerold Lowenstein in the parodic, but seriously intended interview with the restored *Australopithecus* female fossil: "A Few Words with Ruby." Ruby got her name from the Ruby Tuesday of the Rolling Stones. She discussed the social-reproductive lives of her group, as well as her relationship with her discoverers' scientific friend, Dr. Aaron Killjoy. "Ruby sighed, 'One thing hasn't changed in three million years. Males still think sex explains everything' " (1983: 83). Ruby was slated for a busy schedule under the patronage of science, including a BBC documentary called *Ruby, Woman of the Pliocene*. But she took time to describe her life in terms reminiscent of a contemporary species, *Pan paniscus*, the pygmy chimpanzee, Zihlman's favored model species for studying origins. The essentials of Ruby's account include active, mobile hominid females, even when carrying babies; food sharing patterns emerging from matrifocal social organization, with selection for more sociable males in that context; and open and flexible social groups. Food played a larger role than sex.

But aside from the specifics, there is a formal difference in the Zihlman story, both in the interview with Ruby and elsewhere.[4] There is no origin of the family. There is no chasm, no expulsion from the Garden, no dramatic boundary crossing. The Miocene/Pliocene boundary is depicted as less hostile, more as an opening of possibility for which *paniscus*-like hominoids were ready, socially and physically. There is no narrative of a time of innocence in a forest, followed by a time of trial on the dry plain, calling out the heroics of reproductive politics. The basic narratives of causality depend less on the antagonistic dialectic of nature/culture that generates the dramatic stories of the west and its others. In the western sense, there is simply less drama. Zihlman's stories regularly do not generate "others" as raw material for crucial transitions to higher stages. This is not a result of "moral superiority" or "genius"; it is an historical possibility made available by political-scientific struggle to generate coherent accounts of connection. One object of knowledge that falls away in these accounts is "the family." In a sense there is nothing to explain, no primal scene, whose tragic consequences escalate into history, no civilization and its discontents, no cascading repressions. No wonder the reproductive politics look different.

Coloring within the Lines: A Pedagogy of Primate Relationships

These basic narrative strategies constrain Zihlman's accounts of both physical and social parameters of human evolution. They are iconically represented in her *Human Evolution Coloring Book* (1982a) illustration of Lucy and her relatives, the pygmy chimp and a human. "Man" here is a female, literally standing for the general condition, a visual jolt in the illustration, even allowing Lucy's probable sex. The outline of a tall human figure contains a twinned ape, one half of whom is a pygmy chimp female; the other half, joined at the midline, is a reconstruction of Lucy. The three figures share several body boundaries, while differentiated in degrees of bipedal specialization and other particulars that mark the boundary between hominoid and hominid. There is a play of similarity and difference among the two genera of hominids, *Homo* and *Australopithecus*, and the chimpanzee species, *Pan paniscus*. They model each other in an invitation to the student to color their common space. Boundaries in Zihlman's accounts suggest zones of transition rather than the inversions of dualist stories. [Figure 11.1]

The visual icon of Lucy, the *paniscus* female, and the modern human female torso is a play of spaces and boundaries, some of them shared, some of them discontinuous. The coloring book is a social object for teaching, just as the museum diorama and National Geographic special were. In the act of coloring inside the lines set by the authority of the author, the medical or physical anthropology student gets a sense of the bio-politics of being primate. Zihlman's book scrupulously illustrates *human*, i.e., general, points with *specific*, i.e., marked bodies belonging to particular sexes, species, cultures, and races. Her attention to representing something of the diversity of primates, human and nonhuman, creates an odd book that looks cluttered with the particular. People looking through it for the first time have sometimes complained that it is a feminist polemic filled with only females. Exactly one half the representations of the animal and human bodies where sex/gender can be distinguished are female. Therefore, it looks like a new oppressive totality, not science, but ideology. Learning a different way to see means starting again at the beginning, with a child's object, a coloring book. Color Lucy Ruby.

In myriad mundane ways, primatology is a practice for the negotiation of the possibility of community, of a public world, of rational action. It is the negotiation of the time of origins, the origin of the family, the boundary between self and other, hominid and hominoid, human and animal. Primatology is about the principle of action, mutability, change, energy, about the possibility and constraints of politics. The reading of Lucy's bones is about all those things. In other times and places, people might have cast Lucy's bones in the rituals of necromancy for purposes western observers called "magical." But western people cast her bones into "scientific" patterns for insight into a human future made problematic by the very material working-out of the western stories of apocalypse and transcendence.

The United Nations Decade for Women: An International Field of Differences

Universal man walked out of UNESCO House in Paris in the early 1950s, left his footprints in the volcanic ash in his travels, and turned up fossilized in East Africa only moments later, while his living descendents on the border between Botswana and South Africa modeled the sharing way of life and hope for a future in nuclear times. Others of his descendents left their footprints in moon dust. But the terms of existence of the western version of universal man came unraveled in global processes of social transformation, including decolonization, women's movements for emancipation, and the stakes of Cold War projected into the Third World. Primatology was established as an international field in this global setting. With the

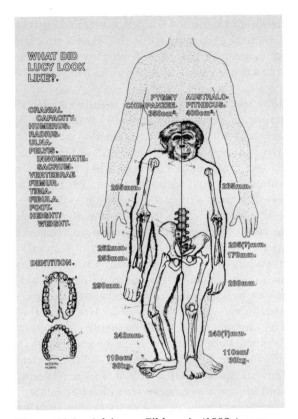

Figure 11.1   Adrienne Zihlman's (1982a) conception of the relations among a pygmy chimpanzee,
the australopithecine fossil named "Lucy," and a
modern human. This coloring exercise in an educational publication teaches that the living species most
like the human hypothetical ancestor is the pygmy
chimpanzee. Published by permission of Adrienne
L. Zihlman. In a very different construction of origins, the "discoverer" of the fossil Lucy, Donald
Johanson of the Institute of Human Origins, has
joined with the designer of E.T., Jonathon Horton,
and museum exhibit designer, Kevin O'Farrell, to
create the prototypes for a line of rubber dolls—
Lucy, her "husband" Lorcan, and their children Lon-
nog, Lifi, and Liban. This "first family" will also star
in a book of high adventure on the North African
Savannah, to be published by Villard Books. (*The
Scientist,* 23 January 1989, p. 3)

other biologies and anthropologies, primate studies have been charged with the
construction and reconstruction of the great marked bodies of western scientific
narratives. It is not simply coincidental that the reworkings of what counts as female
in primate studies since the early 1970s have been accomplished in concert with
worldwide reworkings of what the differences and similarities within and among

women might be—and might mean for any practice enforcing what counts as human and what counts as natural. A large part of the shared problematic has been the effort to reconstruct descriptive practices. Seeing women and females differently does not come easily to those raised with the visualizing technologies of universal man.

The United Nations had to respond to the manifestations of the revolution in gender that is occurring all over the planet in very inhomogeneous, contradictory, and internally contentious ways. The list minimally includes a global explosion of women's collective and personal agitation and self-expression; revolutions in demography; multiform crises in family and other forms of gender-structured collective life; deep questions about sexuality and structures of desire; vast scandals in the sexual division of labor and in access to the goods of life, like food, credit, machines, and effective literacies; the intersections of race, gender, and class in establishing the life chances of women and children; major new actors in the production of knowledge and culture; consequences of women's simultaneous centrality and official invisibility in subsistence work, including agriculture, and high technology work, including the communications industries; and the complex relations of neo-imperialism, nationalism, radical and conservative revolutionary movements, and feminism. In many very different situations in the last twenty years, autonomous and self-conscious women's movements emerged to contest the terms of women's lives. Women both affirmed their heterogeneous selves as a world-changing reality and deconstructed Woman as a culturally parochial myth implicated in world systems of domination. Symbolic of the widespread emergence of new social subjects and collective organization, which had enabled the identification of issues and needs previously invisible and unspeakable, the United Nations established 1975–85 as the Decade for Women.[5] The concluding meeting in 1985 was in Nairobi, scene of the 1955 Pan-African Congress in physical anthropology that helped institutionalize the man-the-hunter approach to scientific craftings of what it means to be human. The 1955 Congress was dominated by Europeans and by descendents of white settlers in the colonies from Africa to America. The meetings in Nairobi in 1985, even the official U.N. meeting, not to mention the exuberant forums of the Non-Governmental Organizations, were quite another matter. The contradictions, tentative connections, analyses, and hopes among women globally and locally were enacted in a world of palpable difference. No conception of universal man could emerge. The U.N. Decade for Women perhaps should be read as a "post-modern" phenomenon, offering a vision of possible connection and hope for global futures only on condition of accepting the permanent refusals of closure of identities, adequacy of descriptions, and master narratives about what it means to be female, woman, and human.

The decade from 1975–85 was also a period of uneven discovery of that lesson in some of the strands of primate studies. I would like to interpret primatology's scientific politics of being female from within the field of possibility that the women in Nairobi in 1985 tentatively constructed by taking responsibility for power-charged difference and contradiction as the grounds for feminism. That is, I approach the upcoming analyses from a science-fictional standpoint, or a utopian moment of hope. The hope is that in the de/reconstructions of woman and female going on internationally in science and politics, there is emerging a field for envisioning fruitfully contradictory and multiple possibilities for new links between knowledge and power, for new apparatuses of bodily production for craftily reinventing what it means to be—always situated, always specified—human.

In that context, the use of primate studies by women in a multinationally authored text from the U.N. Decade for Women is emblematic (New Internationalist Cooperative 1985). *Women, a World Report* embodies in a particularly interesting way the problematic and constructed connections, *versus* natural identities or unities, among women as an emerging contradictory collective social subject: Buchi Emecheta, a Nigerian, living and writing in London, wrote the section on women and education in the United States; Angela Davis, a U.S. black and feminist theorist and activist, wrote on women and sex in Egypt; Nawal el Saadawi, an Egyptian feminist theorist and activist, wrote on women and politics in the United Kingdom; Elena Poniatowski, a woman of French-Polish and Mexican parentage who writes on Latin American politics and popular culture, did the essay on sex in Australia. The book was the result of an international exchange of women writers, with women from poor countries invited to write about women in the rich world and vice versa. Debbie Taylor's general introduction tries to tame the chaos of reports and data from the U.N. Decade for Women on the topics of family, agriculture, industrialization, health, sex, education, politics. The first citation in the section on sex was to Meredith Small's edited volume on *Female Primates* (1984), to support the *World Report*'s reconstruction of the relation between female/woman in the context of the emergence of global feminism. The Small book was used to stress the active agency of females, their self-organizing activity and social centrality. In this origin story, females are not resources for males, but organizers of action, sexual and otherwise.

Part III is about feminist contests for the meanings of primatology, embedded in the enabling contradictions which structure feminist discourse in European and Euro-American, high-technology, capitalist culture in the late twentieth century. Pervaded by and reproducing the very logics of domination and appropriation it struggles against, this feminism nonetheless resists, destabilizes, contradicts, and restructures its generative discourses, including biology and anthropology. "Discourse can be both an instrument and an effect of power, but also a hindrance, a stumbling-block, a point of resistance and a starting point for an opposing strategy" (Foucault 1978: 101). In an ironic twist on the logics of the life and human sciences, the marked body becomes the self-activating body; female/colonized/laboring/animal bodies become citizen and the scandal requires theoretical and practical transformation.

My argument has several strands. First, European and Euro-American feminism and primatology are both western and sexualized discourses inheriting the structuring logics of hierarchical appropriation proper to "human" (western "man's") self-formation. Second, feminist theory and primatology are synergistically deeply implicated in the production of biology and anthropology, those discourses in which both female and woman have been pivotal natural-technical objects of knowledge. Third, reconstructing primatology's technical and popular stories is a serious form of feminist practice, and stories are reconstructed in the elaboration of multiple kinds of stakes and practices in social life. And finally, primatology changes the possible meanings of many feminisms. All women's practice is not feminist, and men's practice can be feminist. It is a logical mistake, a category error rooted in fundamental repressions, to translate an exploration of gender and science into the sociology of women and science (Keller 1985). Feminism and primatology are each science and politics, producers of fact and fiction, technical and social disciplines. They are both social practices for writing stories about who "we" are and for policing boundaries and structuring fields for achieving that identity. The chapters of Part III are about the intersection of feminism and the sciences of monkeys and apes since about 1970.

My contention is that the intersection—coupled with other aspects of the "decolonization of nature" that have restructured the discourses of biology and anthropology, as well as other practices of international politics—destabilizes the narrative fields that gave rise to both primatology and feminism, thereby generating the possibility of new stories not strangled by the same logics of appropriation and domination, but also not innocent of the workings of power and desire, including new exclusions. But the intervention must work from within, constrained and enabled by the fields of power and knowledge that make discourse eminently material.

Sex and the West

Time has been "other" in western primatology; the past, the animal, the female, nature: these are the contested zones in the allochronic discourse of primatology. But by the middle of the 1970s, that sense of time and place, which had been dependent on western hegemony for its maintenance, showed signs of cracking open to allow a different scientific narrative structure. Part III will explore some of those cracks, and the fields of meanings and practices they generate, in the context of the dense intersection of western feminism, multicultural and global feminism, decolonization, and multinational capitalism. The focus of the examination of cracked and transformed narrative fields will be Euro-American women's constructions of what may count as *female* from the early 1970s to the mid-1980s—a potent version of what may now count as *nature* for late industrial people.

Insight into the symbolic structure and social power of the western branch of primatology is to be gained by focusing an analytical lens on a particular pair of intersecting axes, concerned with sex and the west, two story operators constitutive of the origin stories told by the Judeo-Christian segment of the "peoples of the Book." These patriarchal, monotheistic children of the father-God learned to read the Book of Nature written in the ciphers of number in the founding times of the Scientific Revolution. These are the people who transformed salvation history into natural history, and then constructed biology and anthropology on its stage. The life sciences have from their birth been inherently dramatic—story-laden, as well as theory-laden. Facts are always theory-laden; theories are value-laden; therefore, facts are value-laden (Young 1977). Scientific processes and products are value-laden in a particular way bound up with persistent stories that are themselves social-material forces, as well as responsive to other social forces and relations. The human sciences have been the sciences of man—the narratives of Christian in *Pilgrim's Progress* transmuted by capitalism's dynamic of accumulation. And western primatology has remained a discourse at the boundary between life and human sciences, a mediating discourse for establishing the "human" place in "nature." First the sons told the stories, and then the daughters. Both have been consumed with the topics—and tropes—of the origin of the family, the state, and the individual.

Gender: The Sexual Politics of a Word[6]

Let us step back and re-look at meanings of sex and gender in order to be able to evaluate what practices in primate studies might suggest for feminist theory. Although they are closely related, nature/culture and sex/gender are not identical orders of difference in inherited western frames or in recent feminist innovations. Gender is the politics of sex, the ordering of sexual difference, just as culture designates the political realization of natural materials. But nature/culture and sex/gender are separate axes in a field. Their relationship to each other can and does take many forms, generate many versions of stories, ground and legitimate many political positions about the original questions.

The field organized by and around these mobile, dynamic, productive axes is a discursive field; i.e., it is about language, especially writing and other forms of signification, such as filmmaking and museum display. It is a field of meanings. Such a field is not in opposition to "social" and "bodily" determinants; it is one kind of material determinant, influencing the course of people's lives and influenced by them. Stories are material practices; boundary conditions are not just structuralist fantasies, but potent aspects of daily life. Discourses are not only social products, they have fundamental social effects. They are modes of power. The life and human sciences are powerful actors in an age of bio-politics, in which the management of the efficiencies of bodies is a major constructive practice. Scientific discourses both bound and generate conditions of daily life for millions. To contest for origin stories is a form of social action.

However, the argument of the upcoming sections of *Primate Visions* is not only that primatology is structured as western and sexualized discourse, with all the consequent contradictions, limits, and possibilities for intervention and reconstruction. I am also concerned to examine feminism in relation to these same complex historical, cultural logics and practices. Feminism as western political theory can be said to begin at the same historical moment and for the same historical reasons as the discourses of biology and anthropology, with roots in the eighteenth century and flowering in the nineteenth century. In this period the organism—animal, personal, and social—became the privileged natural-technical object of knowledge. Organisms were structured by the principles of the division of labor. The special efficiencies derived from the separations and functional management of the new scientific entities called race, sex, and class had particularly strong effects.

The female animal emerged as a condensed focus of medical and other practice in the late eighteenth and nineteenth centuries, as woman emerged as the nub of social theory.[7] The marked bodies of race, class, and sex have been at the center, not the margins, of knowledge in modern conditions. These bodies are made to speak because a great deal depends on their active management. The biological body is historically specific; the biological organism is a particular cultural form of appropriation-conversation, not the unmediated natural truth of the body. Functionalism emerged as the ruling logic of the discourses of bio-politics. Primatology is inconceivable without the logic of functionalism and its complex theories of adaptation and specialization of the marked bodies. The mind/body logic and its social dominations are old in western culture, but its specific bio-political forms are found in modern discourses in biologies and anthropologies. The organism is the historically specific form of the body as scientific object of knowledge from the late eighteenth century until the mid-twentieth century. That is a key reason why females and women, far from being ignored in biology and anthropology, became the locus of highly productive discourses and other social practices. One finds not the absence of female/woman in the age of bio-politics, but their fruitful ubiquity, under the logics and social conditions of masculinist appropriation, as a question of the enhancement of social and organic efficiencies. How have those conditions changed in the last twenty years in the context of a worldwide revolution in the positions of women and other "others" recasting the terms of marking bodies?

Gender is a concept developed to contest the naturalization of sexual difference in multiple arenas of struggle. Feminist theory and practice around gender seek to explain and change historical systems of sexual difference, whereby "men" and "women" are socially constituted and positioned in relations of hierarchy and antagonism. The complex analytical and political tension between the paired binary concepts of sex and gender ties feminist theories of gender closely to the constructions and reconstructions of the natural sciences, especially the life sciences. These discourses are central to western social technologies for mapping the distinction between nature and society, history, or culture. These fields of knowledge and power map the scope for dreams and projects of social action. The biology of sex helps construct a shared sense of possibility and limitation. Part of the reconstruction of gender is the remapping of biological sex. Biology is an historical discourse, not the body itself.

Constructing Female

Because it is a celebration of primate females and the women who made them visible, i.e., a construction of a "we," *Female Primates: Studies by Women Primatologists* (Small 1984) deserves a full analysis, but I will look at only two pieces for their strategy in introducing subsequent papers and so framing the enterprise. These essays are cautionary examples of tempting but problematic narrative moves in recasting western stories about females in primatology. [Figure 11.2] Each piece raises the question of the difference it makes that women do primatology focused on female animals, but each also adopts a philosophy of science and an ideology of progressive improvement of knowledge that block further investigation of an epistemic field structured by sex and gender. From the point of view of the framing pieces, "male bias" exists but can be corrected fairly simply. There is no need for dangerously political social relations within primatology and no need for the matter to challenge the practitioners' "native" account of how knowledge is made, at least not in public. Bias cancels bias; cumulative knowledge emerges. The root reasons given, however, hint at a stronger position: only bias ("empathy") permits certain "real" phenomena to be knowable, or only explanation from the point of view of one group, not the point of view of an illusory whole which actually masks an interested part, gets at the "real" world. In this case, bias or point of view turn out to be the social and epistemic operator, sex-gender. The major scientific-political question is how such a potent point of view is constructed. In the construction of the female animal, the primatologist is also reconstructed, given a new genealogy. But the rebirth is within the boundaries of the "west," within its ubiquitous web of nature-culture. Primatology here remains simian orientalism.

Jane Lancaster, a Ph.D. student of Sherwood Washburn at UC Berkeley in 1967

Figure 11.2   Portraits in the Primate Order. Drawing by Linda Straw Coelho. Published with permission. This image appeared on T-shirts and on the cover of the official program of the IXth Congress of the International Primatological Society, Atlanta, Georgia, 1982. Part of the effort to provide nonconventional visual images of primates, Coelho's drawings illustrated the cover and chapters of Fedigan's *Primate Paradigms* (1982).

and a senior student of primate behavior from anthropological points of view, introduced the volume as a whole. The introduction is remarkable for its adherence to sociobiological and socioecological perspectives; it is a sign of the triumphant status of those explanatory frameworks in evolutionary biology, including primatology in the mid-1980s. Within that frame, Lancaster looks at primate field studies to understand four areas of sexual dimorphism: "sex differences in dominance, mating behavior and sexual assertiveness, attachment to home range and the natal group, and the ecological and social correlates of sex difference and body size" (Lancaster 1984: 7–8). In each case, the point is that "females too do x." It turns out that 1) females are competitive and take dominance seriously; 2) females too wander and are not embodiments of social attachment and conservatism; 3) females too are sexually assertive; and 4) females have energy demands in their lives as great as those of males. Focus is on females and not on "the species as evolving as an amorphous whole. We explore the social world of females rather than that of the social group.... We learn to understand the reproductive strategies of females and to balance these strategies against those pursued by males of their social systems. ... At last we are coming to a point of balance where the behaviors and adaptations of the sexes are equally weighted" (1984: 8). Finding females means disrupting a previous whole, now called "amorphous," rather than the achieved potential of the species. Feminism absolutely requires breaking up some versions of a "we" and constructing others.

Lancaster's is a very interesting construction of a "we," where the boundary between female animal and woman primatologist is blurred, ambiguous. The deliberately ambiguous title of the whole volume is echoed again and again: "we" are all female primates here, outside of history in the original Garden. That Garden naturally turns out to be in the liberal "west." Competition, mobility, sexuality, and energy: these are the marks of individuality, of value, of first or primate citizenship. "Balance" is equality in these matters, hard-won from specific attention to the point of view not of the "amorphous whole" but of "the social world of females." Lancaster's is an origin story about property in the individual body; it is a classic entry in the large text of liberal political theory, rewritten in the language of reproductive strategy. Sex and mind again are mutually determining. In the reconstruction of the female primate as an active generator of primate society through active sexuality, physical mobility, energetic demands on self and environment, and social competition, the woman "primatologist," i.e., female (human) nature, is reconstructed to have the capacity to be a citizen, a member of a public "we," one who constructs public knowledge, a scientist. Science is very sexy, a question of eros and power. Appropriately, this "we" is born in an origin story, a time machine for beginning history, therefore outside history.

Thelma Rowell, a senior zoologist at the University of California at Berkeley who has played a major role in disrupting stories about primate social behavior, especially stories about dominance (Rowell 1974), was invited to introduce the first subcollection of papers called Mothers, Infants, and Adolescents. Rowell's message, as always, was about complexity. She is not hesitant to point out the legacy of male bias in primatology, e.g., in the classification of females as juvenile or adult exclusively as a function of their capacity to breed, while males were categorized by a whole series of stages grounded in social as well as minimal reproductive functions. "For that matter, there is little recognition of continued social development in human females, which for most purposes are also classified as either juvenile or old enough to breed. In contrast, continued social development following puberty in males was recognized in the earliest studies of primate social behavior, just as the stages of seniority are often formally recognized among men. This dual standard has, I think, delayed our understanding of primate social organization" (Rowell 1984: 14). She points out the merit of the following papers in seeing the primate world from the "female monkey's point of view" and thereby "challenging accepted explanations." She goes further, writing, "I have a feeling it is easier for females to empathize with females, and that empathy is a covertly accepted aspect of primate studies—because it produces results" (1984: 16).

But she backs off from exploring unsettling implications of these positions about the structuring of the observer determining the possibility of seeing. Instead, because males identify with males and females with females and primatology attracts both human genders, the result is additive, canceling out "bias" and leading toward

cumulative progress: "The resulting stereoscopic picture of social behavior of primates is more sophisticated than that current for other groups [of mammals]" (1984:
16). But the stories are not stereoscopic, where the images from separated eyes are
interpreted by a higher nervous center; they are disruptive and restructuring of
fields of knowledge and practice. The reader is not an optic tectum, but a party to
the fray, so hope for higher integration from that source is futile.

Further, "empathy" produces results in human anthropology as well, forming
part of a very mixed legacy that includes universalizing, identification, and denial
of difference, as the "other" is appropriated to the explanatory strategy of the
writer. Empathy is part of the western scientific tool kit, kept in constant productive
tension with its twin, objectivity. Empathy is coded dark, covert or implicit, and
objectivity light, acknowledged or explicit. But each constructs the other in the
history of modern "western" science, just as nature-culture and woman-man are
mutually constructed in a logic of appropriation and progress. When Lancaster
wants to see "balance" and Rowell writes about a "stereoscopic picture," they simultaneously raise and dismiss the messy matter of scientific constructions of sex and
gender as objects of knowledge and as conditions of knowing. Official (or native)
philosophies of science among researchers obscure the complexity of their practice
and the politics of "our" knowledge.

Counting

About how many women practice primatology for a living? That question is
difficult to answer for many reasons. My focus is primarily on field primatology,
i.e., studies of wild or semi-free ranging but provisioned animals in an environment
that can be epistemically constructed to be "natural," a possible scene of evolutionary
origins. But primatology is both a laboratory and field science that crosses dozens of
disciplinary boundaries in zoology, ecology, anthropology, psychology, parasitology,
biomedical research, psychiatry, conservation, demography, and so on. There are
three major professional associations to which primatologists from the United States
are likely to belong, but many of the individuals who have made major contributions
and who allowed me to interview them and have access to their unpublished papers
do not appear on the membership lists ever or for several years at a time.

The membership lists from the American Association of Physical Anthropology
(AAPA), the American Society of Primatologists (ASP), and the International Primatological Society (IPS) around 1980 suggest the level of participation of women in
field primatology. These global disciplinary counts would give a minimum picture,
because there is good reason to believe women are more heavily represented in
field primatology than in exclusively laboratory-based practices, and they have
been more authoritative in field primatology whatever their numbers. There is no
absolute division between field and lab, but there is a sometimes tense difference of
emphasis, despite the official doctrine that naturalistic studies require complementary laboratory studies with their greater power of experimental manipulation. I
am here ignoring the large issue of skewed emphasis from concentrating on North
Americans. With a few important exceptions, the authoritative spokeswomen and
the largest numbers of women in primatology have been United States nationals,
trained in U.S. institutions, and/or employed in U.S. institutions. The overwhelming

majority, relatively more than for other biological sciences, have been white, although that might be changing a bit.

In 1977–78, the IPS (founded 1966) roster listed 751 members, of whom 382 listed U.S. addresses, 92 U.K. addresses, 115 Japanese, 14 African (10 from South Africa), and 151 other locations. In the IPS, overall women were 20% of the membership: 22% of the U.S. total, 22% of the British, 9% of the Japanese, and 24% of the "other." Many individuals could not be identified by gender from initials; they were left out of these calculations, probably resulting in understating the representation of women. By subdiscipline, women accounted for 22% of the anthropologists, 12% of the medical researchers, 27% of the psychologists, 19% of those involved primarily with zoos or wildlife conservation, 19% of the zoologists or ecologists, 25% in other categories. Percentages of total membership of both men and women that could be ascribed to these subdisciplines are 17% in anthropology, 20% in medicine, 16% in psychology, 3% in zoos and wildlife, 9% in zoology or ecology, and 25% in other areas. In the IPS, but not in life sciences as a whole in the United States, women are strikingly disproportionately present in zoos and wildlife areas, as well as in zoology and ecology. If most people working in zoo and wildlife conservation were counted as environmental scientists in National Science Foundation statistics, the figure in the IPS is striking. The NSF's 1986 *Women and Minorities in Science and Engineering* shows that only 10% of all environmental scientists are women, while 2% of women scientists and engineers are environmental scientists.

The 1979 roster of the AAPA (founded 1918) lists 1200 persons, about 26% of whom appear to be women. The 1980 roster of the ASP (founded 1966) lists 445 individuals, of whom only 23 are other than U.S. citizens; these are largely Canadian. About 30% of the ASP in 1980 were women, including 45% of those who gave an anthropology-related address. Academic jobs for primatologists, whatever their discipline of training, are often in anthropology departments. Most of these people have done field studies of primates, although not necessarily outside a laboratory colony. In the ASP, about 24% of the psychologists were women, 36% of those in zoos or wildlife conservation, 20% of the total in zoology/ecology, and 47% of those whose interests intersect with psychiatry (compared to 11% in the IPS). Women primatologists appeared to be trained in and/or have jobs in anthropology in higher proportions than their overall representation in the ASP, but not in a vastly higher proportion than their presence in anthropology as a whole in the U.S. While women in the ASP were in zoology, ecology, zoos, and wildlife conservation at levels near or below their total proportion in the society, again they were in these areas much more frequently than expected from national statistics covering all similar sciences. This statistical picture is consistent with the prominence of women authoring the literature of field primatology, compared to related fields of zoology, mammalogy, and conservation. U.S. women primatologists appear to be more likely to join the American society than the international association, compared to U.S. men. In 1984, the percentage of members of the ASP who were women had increased to about 36% (Hrdy 1986: 136).

It is difficult to estimate how many primatologists from the U.S. about 1980 might have done field studies as a major part of their contribution to the discipline, and how many of those might be women. But, using the specialty distributions from both the IPS and ASP in 1980, it seems safe to judge that not much more than a third of people who could call themselves primatologists have done field studies. If about a third to a half of the members of the ASP did field studies, that would be somewhere near 150–225 people, including about 50–75 women. These figures are probably high. The point is that around 1980, there were many fewer than 100 women in the U.S. who could be called field primatologists. Discounting the pre-World War II studies by a handful of men, field work substantially began about 1960. The original men and women were mostly alive in 1980, and many remain productive and active members of the primate professional associations. There is not much reason to suppose that the cumulative number of U.S. women who have ever published field studies in primatology reaches 100. Adding women from other nations would not be likely to increase that figure by as much as 50%. These crude estimates are meant only to show that there cannot be more than a few dozen women, and maybe three times that many men, writing the publications in all of modern field primatology. There are many more people involved in *producing* modern field primatology than in writing it; but the field is tiny compared to the biomedical, biotechnological, and molecular biologies, not to mention psychology or anthropology as a whole.

For comparison, the January 1986 National Science Foundation publication *Women and Minorities in Science and Engineering* notes that by 1984 in the U.S. women represented about 25% of all employed social, life, physical, and mathematical scientists; only 3% of employed engineers; and 13% of employed scientists and engineers overall (Contrast that with women's figure of 49% of all professional and related workers, disregarding stratification in what counts as professional, let alone "related.") In the U.S. in 1986, there were about 4.6 million employed scientists and engineers (NSF *Data Book*, 1987). About 11% of women scientists and engineers have a Ph.D., compared to 19% of the men. Partly because their rates of employment in science and engineering have been increasing faster than men's, 60% of women in 1984 reported less than 10 years of experience, compared to 27% of men. About 85% of growth in employment of women doctoral scientists from 1973–1978 was in life sciences, social sciences, and psychology; but from the mid-1970s the rate of increase of women in engineering and in computer sciences was much higher than for social sciences. Life sciences (16%), social sciences (18%), and psychology (17%) account for 51% of women scientists and engineers, compared to 19% of men. Women comprise about 40% of all psychologists, 30% of social scientists, 25% of life scientists, and 11% of environmental scientists. This is the pool from which primatologists come, and they come in numbers roughly characteristic of other life and social sciences. Compared to these unspectacular showings, women primate scientists stand out slightly overall, with intriguing differences in specialty choices. Judged by publications, women's impact in primate studies has been greater than their numbers, compared to most other areas of anthropology and possibly all other areas of biology.

A statistical consideration of the intersection of gender and race in science and engineering complements earlier arguments in *Primate Visions* on the exclusion of women of color from field biologies generally and primate studies specifically. To simplify, let me illustrate the point by looking only at life sciences and at figures for U.S. black women. Overall, black women make up 5% of all U.S. women in science and engineering; that adds up to about 25,000 black women compared to 512,000 women of all races. Among doctoral women scientists and engineers, only 3% were black, totalling about 1,400 women. That is, black women are even less likely to survive than white women through the doctorate in the heavily race and gender-skewed upper levels of the science system. When they do survive and get jobs in academic institutions, they achieve tenure at higher rates than white women. Women make up 12% of all white scientists and engineers, 25% of all black scientists and engineers, 7% of Native Americans, and 14% of Asian-Americans. The figure for Hispanic women was reported differently in a separate discussion; Hispanic women make up 3% of women scientists and engineers; Hispanic men make up 2.1% of all men scientists and engineers. Black women scientists are about twice as likely compared to white women to choose a mathematical specialty (7%, 4%); and white women are almost twice as likely compared to Afro-American women to choose life sciences (16%, 9%). In the entire U.S., there are about 2,500 black women life scientists at all degree levels. Assuming that *doctoral* black women are in the life sciences at the same percentage as black women with other credentials (9%), there are fewer than 125 Ph.D.-holding black women in life sciences in the U.S. (9% of 1400), compared to about 15,000 white women. That comes down to a greater than 100:1 chance that a particular woman doctoral life scientist will be white. It is not surprising that doctoral women studying monkeys and apes for a living are white. These facts impoverish all of American science, but in a field as culture and story-laden as primatology, the implications are closer to the heart of the science. That fact must be kept in mind throughout "The Politics of Being Female." The figures would be slightly different for psychology or anthropology, the other relevant degree fields for recruiting primatologists. Even with these odds, there are U.S. black women publishing in field primatology, e.g., Clara Jones, who earned her Ph.D. at Cornell in 1978.[8]

Gender also has intersected with class in determining access to crafting narratives in primate studies. Among the women I interviewed, themes of class privilege partially compensating for gender handicaps to give access to the practice of science surfaced repeatedly. Race was usually a silent privilege for both men and women among the people who cooperated with my study. Women informants described the effects of class and its links with their gender in practical terms, such as being able to travel abroad studying as a kind of tourist-scientist in an archaeological dig in order to help resolve a period of academic crisis that included overt sexist discrimination in a previous university situation; being able to fund field work from family sources when grants did not come through; being able to draw on the academic self-esteem and knowledge of options made possible by the model of an academic professional parent, including a mother; and being able to sustain an independent career, perhaps as a research associate without much or any salary, in the face of sexist treatment by individuals and institutions, while publishing exten-sively to become attractive at a higher level of the academic job market later. Money from parents and/or from upper middle-class husbands helped some women sustain heterodox careers, including having children earlier than would otherwise be possi-ble for a professional woman—or having children at all. To have children is an act that can scuttle a woman scientist's career if she is held to a male-norm career clock and/or if she must rely on one income, even a middle class, academic income. On the other hand, several of my women informants, but none of the men, described marriage and children as a factor in delaying finishing degrees and especially in delaying establishing post-doctoral research programs. Others described practical barriers and personal reservations about seeking a doctorate at desirable institutions where a professionally senior husband was a faculty member in the only relevant departments. For the men, the pattern was more likely to include a wife, not necessarily a scientist, helping in the early field research, although a couple of the women also had that experience. Also, for men in my sample, gender-linked access to the means to be a scientist included veterans benefits.

The prominence of married couples publishing together in the primate literature is striking. Likely missing many, I have counted over 65 such couples (not all in naturalistic field studies) in a publishing field with a few hundred authors in the post-World War II period. Obviously, marriage does not exhaust the kinds of intimacy that are important in the ways primate networks function professionally; but it is a particularly public form, with inherited loaded meanings, especially for the women, both those with access to the mixed, but substantial, blessings of heterosexual privilege and those without such access. Marriage has systematically different effects for men and women, whether they are married or single. Even with new post-World War II—and especially post-Sputnik and post-women's movement—options permitting more degrees of freedom in a gender-structured society, including giving access to the practice of science, women continue to need to arrange their lives in relation to the consequences of the *institution* of heterosexual marriage, whether they are married or not. The need is not always accompanied by the possibility. The women spoke about the issues of both class and gender, and about their links, more spontaneously, possibly in response to my gender and gender-skewed way of interviewing, but also possibly because the men were less explicitly aware of how their work was structured by their gender, in concrete detail rather than in sociological abstraction. Gender, race, sexuality, and class are not less structuring for men, or white people, or heterosexuals, or middle class people. Rather, the structure is less visible, and it takes a different kind of work to learn to see than for those at a less privileged node in a complex, hierarchically organized field.

I do not have evidence to claim that men and women primatologists have come from different class backgrounds, but I am suggesting that class and race might affect them differently as a function of gender. My interviews left me with questions, not answers: Did anglophone white women need to be in a relatively higher class position than white men to find themselves in primate studies between 1960 and 1985? Post-war primate studies grew up with a vast expansion of science funding nationally and internationally. Did that fact deeply alter the class composition of natural history fields compared to the first half of the twentieth century? Both men and women in primate studies have needed philanthropic and public science funding, and both have needed and sought independent career jobs. Both benefited from the explosion of material support for jobs and research in primate studies in the 1960s and early 1970s. Both suffered from the constrictions in jobs and funding, or at least lower rates of increase, since the mid-1970s. My point is not that white men and women inhabit totally different material worlds, but that they experience "common" worlds differently on the basis of gender.

Publishing

The explosive growth of post-World War II primatology has overlapped the "second wave" of the Euro-American women's movements. Young women and men entering primatology in those years could not be unaware that their field was contested from the "outside," in gender politics and much else. It was also contested from the "inside." One result was the explosion of writing by women on primate society and behavior, both for popular and professional readerships. The following catalogue, consisting only of books, not the major form of publishing especially in natural sciences, is not exhaustive; but it gives the flavor of abundance and a chronology showing the steady rise in women's production of primatology. Looking only at women-authored material in this way obscures the work by several men that also contributed to deconstructing what counts as female and woman.

Nadie Kohts, *Untersuchungen über die Erkenntnissfähigkeiten des Schimpansen aus dem zoopsychologischen Laboratorium des Museum Darwinianum in Moskau* (1923), opens my list in order immediately to transgress the categories of American, post-World War II, and field primatology, and also just to honor an important predecessor in the appreciation of primate mind.[9] Robert Yerkes and Ada Yerkes's *The Great Apes* (1929) marks the frequent role of the officially non-scientist wife, who contributed substantially to the production of the primate text. Belle Benchley's *My Friends the Apes* (1942) is probably also little known except to the aficionada/os of apes, but she marks several categories important to gender in primatology: zoo work, lay status, success in ape breeding.[10] These three pre-World War II names underline my inability to find a single book, popular or professional, about primates by a doctoral woman scientist in the *world* before the 1960s. There are several by men.

Then comes the best known name of all, beginning a chronological list of women professional biologists and anthropologists writing for many audiences: Jane van Lawick Goodall, *My Friends the Wild Chimpanzees* (1967) followed by *In the Shadow of Man* (1971) and *The Chimpanzees of Gombe* (1986); Thelma Rowell, *Social Behaviour of Monkeys* (1972); Alison Jolly, *Lemur Behavior* (1966), *The Evolution of Primate Behavior* (1972, 2nd ed. 1985), and *A World Like Our Own* (1980); Jane Lancaster, *Primate Behavior and the Emergence of Human Culture* (1975); Sarah Blaffer Hrdy, *Langurs of Abu* (1977) and *The Woman That Never Evolved* (1981); Alison Richard, *Behavioral Variation* (1978) and *Primates in Nature* (1985); Jeanne Altmann, *Baboon Mothers and Infants* (1980); Katie Milton, *The Foraging Strategies of Howler Monkeys* (1980); Nancy Tanner, *On Becoming Human* (1981); Linda Marie Fedigan, *Primate Paradigms* (1982); Adrienne Zihlman, *The Human Evolution Coloring Book* (1982); Dian Fossey, *Gorillas in the Mist* (1983); Barbara Smuts, *Sex and Friendship in Baboons* (1985); Shirley Strum, *Almost Human* (1987). This list includes only books singly authored by doctoral women in the field.[11]

A larger picture emerges if we consider the profusion of books focused on debates about sex and gender which take serious account of the work by women primatologists and reconstructed men primatologists. Every one of these books is part of a large international social struggle, especially from the 1960s on, about the political-symbolic-social structure, history (natural and otherwise), and future of Woman/women. The political struggles are not context to the written texts. The women's movements, for example, are not the "outside" to some other "inside." The written texts are part of the political struggle, but a struggle conducted with very specific "scientific" means, including possible stories in the narrative field of primatology. By definition, the origin point has to be outside the history I will tell, therefore consider first the unique, renegade pre-1960s classic, a book that is to female primates and feminist primatology as Simone de Beauvoir's *Second Sex* is to feminist theory of the second wave: Ruth Herschberger, *Adam's Rib* (1948); it was reissued in paper in 1970, hardly an accidental date. Herschberger's dedication of the book to G. Evelyn Hutchinson recalls the crucial importance of pro-feminist men in the pre-history of feminist struggles for science.

A title from the 1960s gives the starting point for thinking about females with regard to (zoological) class, but note how the field expands through the 1970s, when maternal behavior is no longer the totally constraining definition of what it means to be female: Harriet Rheingold, ed., *Maternal Behavior in Mammals* (1963); Elaine Morgan, *The Descent of Woman* (1972); Carol Travis, ed., *The Female Experience* (1973); Rayna Rapp Reiter, ed., *Toward an Anthropology of Women* (1975), with the classic paper by Sally Linton, "Woman the gatherer: Male bias in anthropology" and an influential paper by Lila Leibowitz, "Perspectives on the evolution of sex differences"; Evelyn Reed, *Woman's Evolution* (1975); M. Kay Martin and Barbara Voorhies, *Female of the Species* (1975), dedicated to Margaret Mead; Ruby Rohrlich Leavitt, *Peacable Primates and Gentle People* (1975); Cynthia Moss, *Portraits in the Wild* (1975); Bettyann Kevles, *Watching the Wild Apes* (1976) and *Females of the Species* (1986); H. Katchadourian, ed., *Human Sexuality: A Comparative and Developmental Perspective* (1978); Lila Leibowitz, *Females, Males, Families: A Biosocial Approach* (1978); Lionel Tiger and Heather Fowler, eds. *Female Hierarchies* (1978); W. Miller and Lucille Newman, eds., *The First Child and Family Formation* (1978); Elizabeth Fisher, *Woman's Creation: Sexual Evolution and the Shaping of Society* (1979); Frances Dahlberg, ed., *Woman the Gatherer* (1981); Helen Fisher, *The Sex Contract: The Evolution of Human Behavior* (1982); Ruth Bleier, *Science and Gender: A Critique of Biology and Its Theories on Women* (1984); Ardea Skybreak, *Of Primeval Steps and Future Leaps* (1984).[12] This list contains travel literature, textbooks, conference publications, children's literature, academic feminist theory, popular polemics, and science reporting. All are part of the apparatus of production of the primate body. It would be a mistake to leave out science fiction, which is both influenced by and an influence on the struggles over sex and gender in primatology, e.g., Jean Auel's *Clan of the Cave Bear* (1980), Marge Piercy's *Woman on the Edge of Time* (1976), and the bio-fiction of C.J. Cherryh and James Tiptree, Jr. (Carolyn Jane Cherry and Alice Sheldon).

This catalogue is heterogeneous from several points of view—political allegiance, intended audience, credentials of the authors and editors, publishing format, genre, etc. Interestingly, it is nationally and racially homogeneous; this point matters in view of the universalizing tendency of the literature, which repeatedly seeks to be about the nature of "Woman." No one could claim from any of the lists in this paper that white U.S. women occupy a unified ideological space or are in any simple sense "in opposition" to masculinist positions, much less to men. But it should also be impossible to miss the collective impact of these public stories: new lines of force are present in the primate field. It has become impossible to hear the same silences in any text. The narrative field has been restructured by a polyphony rising from alalia to heteroglossia. In the practice of telling important origin stories among peoples of the Book, women now also speak in tongues, imagining female within a reinvented language (Elgin 1984).

Volumes edited or co-edited by professional primatological women produce another long list that begins with the publication of the papers from Phyllis Jay's 1965 Wenner-Gren conference (Jay 1968). For this study, the list ends with a spate of books published in the mid-1980s. These books mark the prestige of "sociobiological theory" in primate studies and the complex place of sociobiology in the crafting of self-consciously pro-female and often feminist accounts of primate, and indeed vertebrate, evolution, behavior, and ecology.[13] Meredith Small's collection, *Female Primates: Studies by Women Primatologists* (1984) is explicitly a kind of celebration of female primates, human and animal, in collaboration to write primatology. It was published as Volume 4 of Monographs in Primatology under the scrutiny of a nine member editorial board, only one of whom (Jeanne Altmann) is a woman. The editor was a graduate student, and she was explicitly encouraged by her advisor, Peter Rodman. *Female Primates* includes twenty-one women and one man (as a co-author) among its authors. The range of concerns includes post-menopausal animals, female adolescence, female sexual exuberance, feeding strategies, mating systems explained from the point of view of female biology as the independent variable, and much else. Any notion that the book might be pollutingly popular should be nipped by its style.

Sites

The portrait of publishing and rough numerical representation needs to be complemented by a brief review of the major institutions that have produced women scientists in the field. Women's professional practice in field primatology has meant access, submission, and contribution to the institutional means of producing knowledge. Despite the National Geographic's imagery of "Jane" alone in the jungle with the apes, a Ph.D. is bestowed for *social* work, often experienced as lonely and sometimes named as alienated, in a different sort of jungle where monkeys and apes are transcribed into texts, or more recently coded onto tape. Women did not earn Ph.D.s for research on primate behavior randomly from all possible doctorate-granting institutions where people did such work. For example, Harry Harlow's laboratories at the University of Wisconsin were particularly impoverished in human female doctoral fauna. That fact contributes to the pattern of more women in the field than in lab-based psychological primatology. Two universities were initially crucial, the Anthropology Department of the University of California at Berkeley and the Sub-Department of Animal Behaviour of Cambridge University at Madingley. By the 1970s, Stanford University, with its ties to Gombe and its captive chimpanzee colony, and Harvard University's program in physical anthropology became important.

From 1958 to 1980 at least 18 women earned Ph.D.s for work on primate evolution and behavior at the University of California Berkeley in a program deeply influenced by Sherwood Washburn's plans for reconstructing physical anthropology. Many of those women were the students of Washburn's former student, Phyllis (Jay) Dolhinow, who joined UCB's faculty in 1966. The UCB program has been famous for its unusually large number of women students in the early years of post-war primatology. The role of Washburn in the accomplishments of his students is controversial and many other figures were crucial to their intellectual formation, e.g., Peter Marler, Frank Beach, and Thelma Rowell. But the program founded and sustained by Washburn's power in physical anthropology was the route to credentials for the U.S. women until the late 1970s, as well as most of the men through the 1960s. The women often formed cohorts with each other that have influenced their professional lives over several years. Key names include Adrienne

Zihlman, Jane Lancaster, Mary Ellen Morbeck, Suzanne Chevalier-Skolnikoff, Naomi Bishop, Suzanne Ripley, and Shirley Strum. Many of the UCB women have been leaders in reconstructions of sex and gender in primate story telling. They provided peer cohorts for each other during graduate school and have formed critical support networks in later professional life. Their relationships with their male student peers are an important part of the story. There are several "generations" of UCB primate women, not to mention individual heterogeneity, and generalizations are tricky. Their strengths and limitations are controversial and are germane to the debates about explanatory powers of sociobiology and socioecology compared to evolutionary structural functionalism.

Like Washburn, Robert Hinde of Cambridge has sponsored the doctoral work of a significant number of the important women primatologists, including Jane Goodall and Dian Fossey. At least as important has been the work of Thelma Rowell, an early Hinde student, who, after a period at Makerere University in Uganda, moved to the Zoology Department at UCB, where her presence made a major difference to the primate students in the Anthropology Department as well, perhaps especially the women. My interview informants argued that Goodall and Rowell were critical to Hinde's theoretical and methodological development, leading him to see beyond Lorenz and Tinbergen to the complexity and individuality of primate behavior. Including Ph.D. students and post-doctoral associates, since 1959 about 15–20 women primatologists have been associated with Hinde's laboratory at Madingley. Crucial names also include Dorothy Cheney, Kelly Stewart, Carol Berman, and Phyllis Lee. The approach of his lab may be followed in a recent volume of essays (Hinde 1983). Many of these students were Americans who earned their Ph.D.s in the U.S. and did post-doctoral work associated with Madingley or vice versa. Networks of institutions and researchers are probably more useful way for tracing primate lineages than dissertation advisors. Crucial in these networks are long-term field sites, like Gombe, Amboseli, Gilgil, Cayo Santiago, and others.

Stanford University was for a time at a nodal point of institutions and field sites. Stanford women, former undergraduates as well as graduate students, have important ties with other central institutions grounding primate research. In addition to Gombe, they have worked at Harvard, Cambridge, U.C. Davis, Kekopey Ranch at Gilgil, Amboseli and the University of Chicago, the Rockefeller University's research station at Millbrook, and other places. Their ties with each other and male peers were crucial to setting up a second primate year at the Center for Advanced Study in the Behavioral Sciences in 1983–84 to produce a volume reflecting the recently ascendant explanatory frameworks. Reconstructed female animals, as well as women primatologists, occupy very active positions in that text (Smuts et al 1987).

Irven DeVore was a dominant figure in Harvard's program in physical anthropology after he finished his Ph.D. in 1962, from many accounts as Washburn's favored son. DeVore's undergraduate course in primate behavior at Harvard has been immensely popular, and since DeVore's famous "conversion" to sociobiology (to Washburn's great dismay) in the 1970s, that course and his graduate primate seminar have been important institutional mechanisms for reproducing the explanatory strategy in younger workers. Robert Trivers's tutelage of DeVore was pivotal. It appears that in the first sociobiological years the seminars were classically "male-dominated," by faculty and students. But then the name of Sarah Blaffer Hrdy began to appear in print and in my informants' accounts. An unrepentant sociobiologist, she centered females in her accounts in ways that destabilized generalizations about what "sociobiology" must say about female animals or human women. She is also an unrepentant feminist, greatly admired by the reviewer of her *Woman That Never Evolved* in *Off Our Backs*, the major radical national feminist newsprint publication in the U.S., and greatly criticized by many feminist opponents of liberal political theory, including its sociobiological variants. Women students like Patricia Whitten and Nancy Nicolson who came to Harvard for graduate work after Hrdy consistently named her presence as a crucial supportive factor in their own confidence and intellectual power. They formed cohorts with each other and regarded Hrdy as a kind of elder sister. These networks ground much of the currently interesting reconstructions of primate females and primate society as a disrupted "whole." Hrdy became a professor at U.C. Davis in the 1980s, another institution that has become a nodal point for reconstruction of female primates and education of women primatologists, such as Joan Silk, Margaret Clarke, Ardith Eudey, Linda Scott, and Meredith Small. [Figure 11.3]

One last locus is the savannah baboon research project in Amboseli National Park in Kenya and the Department of Biology (Allee Laboratory of Animal Behavior) at the University of Chicago, where Jeanne Altmann and Stuart Altmann have worked since 1970. Jeanne Altmann has been important in primate field studies since she began working with Stuart Altmann in the early 1960s, but she earned her Ph.D. only in 1979, with a dissertation ("Ecology of Motherhood and Early Infancy")

Figure 11.3   Langur female relatives and their young on Mount Abu, India. Published with permission of Sarah Blaffer Hrdy/Anthrophoto.

submitted to the University of Chicago Committee on Human Development. Initially, Jeanne Altmann, without a Ph.D., was rarely invited to conferences unless her husband was also invited. Progressively, she became a power in her own right in the field. Jeanne Altmann was cited by my younger informants as a significant node in developing "invisible colleges" among women.

It is time to turn to the case studies for exploring the complex intersections and divergences of feminism and primate studies in the practice of North American white women. The intersection works not by replacing feminist stories for masculinist ones, or scientific stories for ideological ones, truths for representations, but by restructuring the whole field of possible stories. To destabilize a story field, one must do many things, such as write computer programs, argue for different data collection protocols, take photographs, consult on national science policy bodies, write high school texts, publish in the right journals, etc. Even to imagine destabilizing stories, one must be formed at a social moment when change is possible, when people are producing different meanings in many other areas of life. Destabilization is a collective undertaking. Within the altered field structure, new dominations are possible, but so might be something else. Although my focus will be on the practice of western white women in primatology and changing constructions of what it means to be female in biology and physical anthropology, these matters concern both men and women, males and females, as actors, authors, and subjects of the stories. That women usually took the lead in the reconstructions was not a *natural* result of their sex; it was an *historical* product of their positioning in particular cognitive and political structures of science, race, and gender.

12

JEANNE ALTMANN:
TIME-ENERGY BUDGETS OF
DUAL CAREER MOTHERING

The true situation may be the opposite of the apparent one.
(J. Altmann 1980: 169)

J eanne Altmann has organized her practice of watching baboons around the consequences of this central premise. Originally in mathematics and a veteran of primate field studies since 1963 at the Masai-Amboseli Game Reserve, later Kenya's Amboseli National Park, Jeanne Altmann listed first in the category of Major Research Interests, on her *curriculum vitae* in the 1980s, "non-experimental research design." In her work, that deceptively simple term moved to the center of authoritative practice and away from being perceivable as a slip of the tongue, nearly a contradiction in terms for field workers without access to the interventionist techniques of the true experimentalist, who alone could really achieve the sacred state of doing "research design," rather than mere observation and description. By the time she published her first paper solely under her own name in 1974, called modestly "Observational Study of Behavior: Sampling Methods," Altmann was widely recognized among field workers for setting standards that progressively had to be met if one's descriptions and explanations of regularities emerging from field studies of monkeys and apes were to be taken seriously. Altmann was not the first field worker to worry whether the way she and others got their data in the first place could allow them to address the questions they thought they were asking.[1] But she decisively analyzed and codified the issues at a turning point in what could count as an explanation in field primatology. This matter has been of fundamental significance in destabilizing what could count as a female, and especially in problematizing that endlessly remystified state, primate motherhood. Working out from that pivotal 1974 paper back to about 1960 and forward to the early 1980s, we can

304

unpack how Jeanne Altmann's career and writing may serve as an allegory for feminist queries about gender and science, i.e., about relationships between positioning as a gendered social subject and the production of knowledge and philosophy of science. The word *female* hardly appears in the 1974 text; nowhere is there a hint of explicit political analysis of the sort that might appear in a feminist journal; and the language of argument proceeds in the masculine generic (the observer, he). How may one reposition such a text as a pivotal document in the primatological politics of being female?

First, what did the "Sampling Methods" paper say and what was its history before and after publication? [Figure 12.1] The published paper became one of the most cited in the modern animal behavior literature, and even so it remains undercited, as a function of its terms (especially focal-animal sampling) and arguments becoming part of the common sense of subsequent field workers. Drawing from the laboratory-based work in experimental animal psychology and from methodological discussions in quantitatively oriented human social and behavioral sciences, Altmann surveyed seven sampling methods, which had been or could be practiced by field workers, for their major strengths and weaknesses in relation to the sorts of questions an observer would like to ask. Summarizing how many kinds of questions might be applied to data sets obtained in a particular way, she recommended when to use a given method and outlined how to go about it. For example, she showed how sampling *ad libitum* was primarily of heuristic value and able to give records of rare, significant events. But such sampling could not be used to address questions about

Figure 12.1   Philip Muruthi, a member of the baboon project field staff responsible for collecting data, teaches Moi University students about baboon conservation and research in Amboseli. Courtesy of Jeanne Altmann.

rates or durations of events or states; nor could it allow one to compare subgroups around variables of interest, such as percent of time males and females of particular age-sex categories might spend eating, fighting, grooming, or lolling about. This seemingly innocent point had one important consequence for rereading the previous primate literature: an observer could say very little about what males specifically do from *ad libitum* observations of males, no matter how extensive such observations were.

Less obvious and more unsettling was her elucidation of the potential of a widely used sampling method called "one-zero," in which the occurrence or non-occurrence of an event was recorded. This method had been widely used by sophisticated, key people in human and animal studies, including primatology. Altmann's persuasive discussion was followed by her devastating conclusion about recommended uses of the method, which stands as a model of succinct scientific prose: "none" (1974: 261). Some senior people, for example, Hans Kummer and Robert Hinde, reacted better than others to the consequences of that conclusion for the status of their previous work and for their previous advice to their students.

On the other hand, "focal-animal sampling" could result in data appropriate for examining sequential restraints on particular kinds of events or conditions, percent of time spent doing something or being in a given state, as well as rates, durations, and nearest neighbor relations. In short, if a worker structured sampling in this manner, the chances were considerably higher that comparative questions could be framed and even answered in such a way as to stabilize what could count as scientific knowledge about contentious issues. "Focal animal sampling" was Altmann's term, proposed to codify and foreground a practice in which all occurrences of a specified action or interaction would be recorded for a pre-set sample period, the length of which was always recorded, as well as the time the sample individual or group was actually in view during the sample period. It was crucial to attempt to follow the focal unit (group or individual) to whatever extent possible during each pre-set period. This little point could cost years of one's life if the study species lived in trees and moved around at night, as well as if the study species presented much easier situations for observation by an earthbound biped scientist. The schedule of whom to sample could be determined by a range of procedures, from random number tables to sampling all members of a class of interest, e.g., neonates, recent immigrants, or adolescent females. Length of the sampling period would be designed to relate meaningfully to the kinds of states or events carved out of the flux of monkey existence for analytic scrutiny. Obviously, comparative statements, such as what one group of animals did more than another, required comparative data. Making this matter explicit foregrounded previous *implicit* comparative statements that had taken the form of group *a* does *y*, without remarking on group *b* in relation to *y*.

The sampling procedure was not presented as a panacea; it was not proposed to replace the process of asking questions in the first place nor as a substitute for construction and interpretation generally. The always receding mirage of human-culture-free categories of purely biological interest could not appear magically from a table of random numbers, a clock, and properly selected sample units. Biology remained a human culture-specific discourse, and not the body of nature itself. But neither was Altmann's paper a narrowly technical intervention that bypassed the permanently contentious problem of crafting ways to see the primate world. The paper destabilized a great deal of previous primate field research, including that by senior and powerful people. The embarrassing truth was that many of the regularly cited field studies especially before the mid-1970s both gathered and analyzed data in a way that did not justify the conclusions reached. Appreciating that fact, Altmann made a savvy intervention that enabled a number of skeptical moves, including her own and others' reconsiderations of sex/gender issues in primatology. Strategic deployment of the arguments in the 1974 paper could turn the dramatic foreground into unfortunate sampling error. We shall see how Jeanne Altmann's feminism in science often turned on such reduction of high drama to a more mundane reality, the "real truth" behind the dramatic "appearances." Her feminism was operation- alized as a keen sense of critical method, with consequences for allowable narratives about women scientists and about animals. Since, in both technical and popular literatures, drama in narratives of primate lives turned on highly visible sex and violence, a new script depended on different visual filters, like a set schedule for where to look when.

Many smart people have made skeptical analyses that prove unassimilable by the community whose practices need reforming. Why did Jeanne Altmann's interven- tion succeed in changing practice widely? After all, at the time she published that paper, she would not even have a doctorate for another five years. She was married to primatologist Stuart Altmann, and she had co-authored their 1970 monograph based on about 1500 hours of observation at Amboseli during 1963–64. She had served as a kind of statistical clerk-consultant during the period of data analysis for her husband's 1950s doctoral studies on Cayo Santiago, and they had worked and published as full collaborators ever since. Therefore, her lack of a doctorate did not translate into lack of standing among people who knew how work was done at Amboseli. Other important early people in primate studies had unorthodox starts, especially the women (Goodall and Fossey most prominently). Still, until after 1974, Jeanne Altmann could not be invited independently to conferences; and her social positioning as a wife/mother/woman could be selectively used to obscure her matu- rity as a scientist. She had no institutionally validated training in evolutionary or behavioral biology, and an undergraduate degree in mathematics (1962) and a Masters of Arts in Teaching (mathematics, 1970) did not convey high scientific status either. Her official position was Research Associate in the Biology Department of the University of Chicago (where Stuart was Professor from 1970), a category with mixed benefits for the many women ambiguously privileged to hold it.[2] While one would hope they were not irrelevant, the merits of an argument alone seem insufficient guarantors of authority in the face of serious obstacles.

What did Altmann do to be heard? Let me use that question to elucidate her approach to the issue of being simultaneously a scientist and a woman with explicit feminist concerns about women in science, about approaches to improving scientific accounts, and about the analyses other feminists sometimes offer of the relation of gender and knowledge practices.

The 1974 paper succeeded partly because of the way it came to publication. Altmann circulated drafts of "Sampling Methods" for a long time before publishing it. Asked to prepare a short piece on different ways of analyzing data, she began writing the paper in 1969 for a conference Stuart Altmann organized at the Ameri- can Association for the Advancement of Science. She quickly decided the problem was not really how people analyzed data; mathematical sophistication in statistics was not the key problem, even though statistics for analyzing non-experimental data were comparatively poorly developed and many primate observers were woefully deficient in even simple statistical expertise. The problem was how people *collected* data. That is, the problem was *conceptualization*. Considering at the time whether to go back to graduate school, Altmann articulated her goals and intellectual motivation in primate field studies in the context of producing her methods paper. The first version was read at the AAAS in 1970; the paper then circulated for nearly three years in the explicit context of seeking to make it usable, rather than using it dramatically to trash someone else's work. It continued to be revised after it was submitted to *Behaviour* around the end of 1972. She waited significant periods of time for people's responses to the drafts; circulated like this, the paper was widely influential before its official publication. Before and after its publication, much of the response came from graduate students designing their work. Trained in the period of rapid growth of primate studies in the early and mid-1970s, this generation of students would codify what would count as "problem-oriented" field research over and against many of the earlier studies that could be labeled "initial survey" or merely "natural history." Stories from these students not infrequently highlighted their perception that their approaches were in opposition to the practices of their official advisers. Jeanne Altmann would get letters from students in the field with little training asking for methodological help; she gave it. Progressively, she became a node in a network or "invisible college" of field workers, including a growing network of young women. Jeanne Altmann became simultaneously a senior mentor and a peer contributor to reformulations of what could count as female for scientists and for their research subjects, animal and human.[3]

Altmann herself named an important mentor: Beatrice Whiting at Harvard's Laboratory of Human Development, where Jeanne Altmann had worked as a statistical clerk, coding and analyzing data in 1959–60 for the Six-Culture Study of Child Behavior. Subject to a normal gender-reproducing academic practice, Beatrice Whiting had been a core person in this famous study in quantitative social-behavioral comparative research at Harvard from the mid-1950s, as well as in much other important work, but she was granted tenure at Harvard only shortly before her retirement many years later. The people associated with this project developed the data and techniques of recording, coding, and comparing behavior observed in different "natural" settings.[4] They worked out procedures for getting inter-observer reliability, handling large numbers of variables, and carving out categories meaningful to the kind of analysis privileged in the study (deeply rooted in this case in Malinowskian functionalism and in learning theories of Hull and Dollard). The Six-Culture Study people learned how to get the strong and consistent relations between variables that they needed to say anything statistically significant about their small sample sizes. They learned how to get and process data that could answer specific questions about behavior as a measure of personality and culture. The study used five-minute focal sampling, focused on categories crafted to be transcultural in previous social-behavioral discourse (e.g., self-reliance, achievement, aggression, nurturance, submission), mapped daily rounds of children, etc. They devised twelve categories to compare across many variables in six cultures. The kind of knowledge they constructed was *built* to produce replicability across "cultures."

That is, the Six-Culture Study required what Marilyn Strathern (1987) called a "modernist" aesthetic concept of culture, in which a "culture" is carved out of the unruly world as an object of knowledge like a modernist work of art—a unit perceived to have its own internal, architectural principles of coherence. Jeanne Altmann operated out of this aesthetic and epistemology when she drew from the Six-Culture field studies to improve simian field work as a modernist scientific production.[5] Strathern's analysis exploring the fault line between modernism and postmodernism in ethnography and anthropology—in which the high stakes are the authorization or prohibition to craft *comparative* knowledge across "cultures," from some epistemologically grounded vantage point *either* inside, outside, or in dialogical relation with any unit of analysis—made the crucial observation that it is not the written ethnography that is parallel to the work of art as object-of-knowledge, but the *culture*. The Romantic and modernist natural-technical objects of knowledge, in science and in other cultural practice, stand on one side of this divide. The postmodernist formation stands on the other side, with its "anti-aesthetic" of permanently split, problematized, always receding and deferred "objects" of knowledge and practice, including signs, organisms, selves, and cultures. Whether scientific analysis could ever be postmodernist becomes a compelling question within this frame. What would stable, replicable, cumulative knowledge about non-units look like? Is the discourse of modern immunology and genetics a hint? (Haraway forthcoming a). The issue is not method—technical versus interpretive, quantitative versus qualitative, reductive versus holist, etc.—but the structure (or anti-structure) of the "object" allowed to materialize in discourse. That is, the issue is how it is possible to imagine the world, including oneself as observer-commentator-writer, to be articulated.

This mutation in western aesthetics and epistemology relates paradoxically and fruitfully to the kind of relation between feminism and science that Jeanne Altmann has articulated in her publications, journal editing, scientific meetings, public lectures, and interviews. The paradox is particularly interesting as it emerges in the kind of beings females turn out to be as objects of knowledge and women turn out to be as scientists. It turns around the tension between "identification" versus "problematization" or deconstruction as a strategy of naming (stabilizing in knowledge practices) female/woman.[6] Altmann formed herself scientifically out of a practice that rested on modernist, coherent objects of knowledge; simultaneously, her practice of relating gender and science turned on disrupting modernist conceptions of sex, gender, and the relation between them. In order to be a feminist and a biologist, Jeanne Altmann foregrounded as method and as object of knowledge *non-identification* and *non-coherence* in the charged field where sex /gender intersects science.

Altmann emphasized in interviews her view that a feminist is discredited in her persona as scientist by directly putting "politics," e.g., one's gendered social identity, into "science." The feminist is not only discredited in the eyes of other scientists, but even worse, in her or his own self-assessment as author of knowledge recognizable as scientific. The first disjunction (non-identity), then, is between who one constructs oneself to be and what one does to intervene in knowledge production, including science. "Being" does not ground knowledge, at least not until "being" has been made into a strategic, *built* site generating interrogation, not identification. This is a deceptively simple matter. For example, because she is simultaneously annoyed as a woman and a scientist by the primate literature's overtly sexist accounts of categories called leadership and control, the one thing she specifically should not do is substitute the mirror-image reverse account or method, except perhaps as caricature to show the sad status as science of the original masculinist version. The move of simple reversal is epistemically weak—even if it is occasionally rhetorically useful to make previous knowledge move into the satisfying category of cautionary example of bad previous bias, indeed, unfortunate pseudo-science. The powerful move is to displace and destabilize what may count in the relevant discursive community as knowledge.

Substitutes do not destabilize; they replicate. Substitutes are tied to identifications. Male-centered and masculinist accounts cannot or may not be replaced by female-centered and feminist accounts provided by women writing from the point of view of themselves being females or women. The confusing language of sex and gender here is a symptom of the muddled quality of identification as an epistemological strategy. But category formation at all the powerful nodes in the apparatus of knowledge production can, perhaps, be thoroughly destabilized. If those unsettled conditions can be produced in a discursive field, like primatology, then fundamental change—transformation in the generative field—is just possible. Destabilizing the positions in a discursive field and disrupting categories for identification might be a more powerful feminist strategy than "speaking as a woman."

There are several concrete examples of category destabilization in Altmann's "micro" practice. How to shift knowledge-power relations: foreground micro practice, not high drama; effect destabilization, not substitution. As American editor of *Animal Behaviour* from 1978–83, she entered into lengthy negotiations with authors of manuscripts who used categories like "rape" to describe their animal subjects' behavior. She did not invoke her political beliefs as the reason for using her editorial power to negotiate manuscript revisions. Like her approach in the "Sampling Methods" paper, she "simply" learned how to show an author that his or her argument was better served by a different practice of category formation. In the process, the manuscript's argument might become something quite different. At the very least previously stable matters become contestable. The feminist scientist does not know the truth in advance by virtue of standing on the high moral and political ground; rather, she or he contests for how to construct what can authoritatively count as the case about the world within the lumpy cultures shared by the scientists' diverse constituencies. Altmann's procedure was to extend the category of bad science to cover use of previously correct categories (or, as in the case of "rape" in the sociobiology debates, incipiently correct categories). The author could, of course, respond that the scientific argument required the word *rape*, but it might be difficult to be convincing; the cost could be made too high. Convincing authors, rather than enforcing political correctness, had the added benefit of making them into reproducers of restructured categories in their subsequent editing. "In other words, I made science serve me, just as other people with other points of view do. . . . It's not like we didn't have the animals there to look at and get data on."[7]

The animals are material-semiotic actors in the apparatus of bodily production.[8] They are not "pre-discursive bodies" just waiting to validate or invalidate some discursive practice, nor are they blank screens waiting for people's cultural projections. The animals are active participants in the constitution of what may count as scientific knowledge. From the point of view of the biologist's purposes, the animals resist, enable, disrupt, engage, constrain, and display. They act and signify, and like all action and signification, their yield no unique, univocal, unconstructed "facts" waiting to be collected. The animals in behavioral biology are not transparent; they are dense. Like words, machines, equations, institutions, generic writing conventions, people, and landscapes, the animals have specific kinds of solidity in the apparatus of bodily production.

Altmann held a workshop at the Animal Behavioral Society meetings at which the participants listed all the terms in their science which seemed to raise problems. "One of the things that comes out right away is that the terms are either about violence or sex, almost all of them. . . . And we tend to emphasize that, male or female difference aside; that's part of the primate tradition also" (interview).Faced with such drama, such a deep narrative tradition, effective micro-practice would favor deliberate reduction of the emotional temperature built into words and deliberate decisions about which controversies to foreground scientifically.

For example, to center the debate on the biological meanings of infanticide among primates too easily plays into the culturally overdetermined lust for sexualized violence. This could be dangerous territory on which to build feminist approaches to science, as a practitioner of science or as an historian and culture critic. It is too easy to replicate, not destabilize; it depends on questions of power in the control of public meanings, whatever the intentions of participants in discourse. The point is not that infanticide is biologically meaningless and should not be debated; it may well be a quite appropriate topic of study. The point is about *drama* and choices about what to popularize, what to focus on in scholarly discussions of controversy, i.e., how to foreground and how to get the power to foreground.[9]

In 1981 Altmann and Martha McClintock, who had been on the editorial board of *Animal Behaviour* during Altmann's tenure as editor, presented a working session on women in science, in which each woman tried to unpack how she framed questions in her science as a function of "all those people I was and am" (interview). "All those people" did not additively produce a masterful single self, the woman-feminist-mother-scientist whose privileged point of view could illuminate nature. Altmann was more interested in the partial conversations and disjunctions. In some sense, "those people" were strangers to each other, and no single authentic self could organize them into final coherence.[10] One goal was to explore the limits of shared experience with groups of colleagues. Altmann developed a public non-unitary self concept to talk about the non-congruence (not necessarily opposition) among "selves," such as oneself as mother, feminist, or scientist. What made up the strategic site called "woman" was a process akin to juggling—keeping several realities in precariously patterned motion and building strength to see the world that way. How Altmann used "all those people" to construct a scientific point of view takes us the next step into the paradox of her status as woman and scientist and into her non-isomorphic feminist and scientific strategies of uncoupling and problematizing identification between females and women. Here, "being" a woman becomes a constructed site for interrogating meanings, a kind of paradigm for a possible strategic site to produce better science, not a resting place in a unitary female body grounding "woman's" experience as nature grounds culture. It was a feminist move—in both Altmann's and my senses—to forbid that resting place. In my terms, that move permits feminist science. Only then could female bodies, in Altmann's case, baboon mothers at Amboseli, become the subjects of scientific investigation capable of restructuring basic meanings of females in biological discourse.

Altmann did not intend especially to study mothers and infants at Amboseli. She had several compelling reasons not to want to do that. Because of strong American ideologies encouraging identification and testimony about experience, ideologies that were put into action in some of the more flagrant masculinist accounts in primate studies and in some kinds of widely publicized feminist practice, for a woman to study females was overdetermined, too much the "natural" thing to do. Several of my women informants commented on their hesitation to take females seriously as subjects for their research; it would cost them scientific authority and it would not feel like a chosen topic. This hesitance cannot all be ascribed to the devaluing of anything female in science, internalized with a vengeance by women. There seems also to be a feminist "instinct" at work in this specific cultural context to avoid reproducing gender in an over-determined identification with females. Altmann also remarked that she had had two human infants, for whose care she was chiefly responsible until after feminist women's movements and her return to graduate school led her and her husband to re-evaluate distribution of domestic work well after the children were out of early childhood. Altmann had little desire to spend long hours watching what mothers and babies ate. What made her finally do more than a little of that kind of work?

"And then increasingly, it was screaming at me, 'These are the most interesting individuals; this has the greatest evolutionary impact; this is where the ecological pressures are' " (interview). Retrospectively, she restructured the background into the foreground metaphorically by an appeal to a scream from the outside, a kind of call to reason from nature. The process that culminated in publication of her doctoral thesis as *Baboon Mothers and Infants* in 1980 restructured the foreground and background by a rich set of metaphor-generating, research-design, data-gathering, mathematical-modeling, collegial-networking, and writing practices.

The previous foreground concerned evolutionary theorists' widespread belief in large differences among primate males in reproductive success. Within Darwinian evolutionary biology there is a ruling explanation in "the last instance"; it is differential reproductive success. Natural selection is a process whose beginning and end is differences in gene frequencies among populations and different genetic makeup of individual organisms within populations. If, in a sexually reproducing species, females have about equal chances of leaving viable offspring, there will not be much difference among females in their contribution to the gene pool. If, on the contrary, the males have vastly different reproductive success rates among themselves, the "evolutionary action," generation of difference,[11] must come primarily from them. Part of destabilizing the notions of *stable* male dominance hierarchies in organisms like baboons was shaking up arguments about dominant males' greater success in breeding in terms of evolutionary significance.[12]

Jeanne Altmann was among those who began to argue that primatologists had vastly overrated differential success rates in males and vastly underrated them in females. At least part of the problem was the preference for high drama, embodied in such things as faulty data collection practices paying too much attention to murder, sex, and mayhem. Nonetheless, data had accumulated that males change status relatively rapidly, and most males move through many relations to opportunities to breed in their lifetimes, so that a thin time slice into baboon life would not show the large degree of evening out of chances over the years. Neither would a thin time slice show the kinds of differences and their pervasive consequences among females. By lowering the temperature and thickening the time slice, females were recast as generators of diversity in evolutionary terms. The basic plot and cultural form were given by the constraints of Darwinian theory, but the scientist-dramatist constructed who counts as an actor, a privileged status in western mythic, philosophical, political, and scientific narrative.

Networking with colleagues was one of the practices for shifting attention of researchers and positions of subjects of research in the explanations. Jeanne Altmann was part of a crucial resource in primatology—a long term study site.[13] From hers and Stuart Altmann's initial work in 1963–64 on, several months in 1969, and then from 1971 on, the Amboseli baboon project people had accumulated systematic data and developed mathematical models to deal with demography, social behavior, and ecology. Special studies had been done on dominance relations and reproduction, parasitology, feeding behavior and time budgets, home range utilization, dominance and nutrition, and so on. Various mathematical techniques were explored to develop a theory of optimal diets, mating simulation models, etc. Detailed genealogical charts exist for a large number of animals habituated to being observed. Every person who worked at Amboseli contributed to the common monitoring projects, as well as gathered data for her or his own project. The common data were developed into computerized data banks at the University of Chicago. The baboon mothers and infants study could rely on all of this to contextualize its own data. There were also networks beyond Amboseli, for example, the kinds of work many other women and also some men were doing in the 1970s to retheorize the activities of women and females in both biology and social science. Jeanne Altmann's citation network made this kind of enabling "intertextuality" visible.[14] Altmann corresponded with Patricia Draper, whose work on ecology and demography crucial to women's lives was part of the Harvard !Kung study, and borrowed her manuscript. Altmann's thesis was written for the Committe on Human Development at Chicago, partly to avoid enrolling in the Biology Department where her husband was a professor, but also partly because the Committee offered relevant networks and resources. These networks were material-semiotic mechanisms for making comparisons, for enhancing some similarities and muting some differences across a discursive terrain. Drama can be more varied when the tradition is rich.

A fundamental metaphor grounded Altmann's organization of *Baboon Mothers and Infants*: dual career mothering. This metaphor contained the tension between identification and problematization that marked the central paradox of her resolution of gender and science. "Juggling" was the key word. Female primates' lives seemed to Altmann defined by juggling, by doing several things at once. For example, her study contained no separate category for child care because she could see no such separation into clean categories in her subjects' lives; there was instead managing competing demands of such things as nursing, eating, carrying, disciplining, socializing with other adults, and paying attention to opportunities and dangers. Altmann turned her deconstruction of identification among women and between women and females in science, which from another point of view could be seen as a complex *kind* of identification, into a way of naming the multiplicities she invoked to account for baboon mothers and so retell basic evolutionary dramas.

Class and race are silent parts of the metaphor of dual career mothering. Most mothers in U.S. society have multiple tasks, identities, lives. But a career is a class specific privilege; and naming multiplicity in this way inadvertently foregrounds a professional middle class and throws into the background, e.g., generations of black women in the U.S. who had to raise children while bringing in money from sub-working-class jobs. Foregrounding women's, especially mothers', multiplicities can also decenter everyone else's. It is a problematic strategic decision, not a naturally appropriate feminist act, to construct mothers as paradigm cases of a certain kind of juggling, related to certain arguments about selves in western culture. Where what mothers do has been counted as background noise, the matter to the form of social life, the resource for the act of generators of diversity, such foregrounding is a feminist move. The move seems justified even across species, making woman / female a jointly unmarked, paradigm case of "juggling," and de-emphasizing differences among women and among species. But the move loses legitimacy when the paradigm is fixed as Nature, rather than as a tool involved in a paradoxical relation between identification and deconstruction.

"Budget" was the other key metaphor in Altmann's reconstruction of females. "You could call it budgets or something else . . . The issue is hierarchy of demands and the immediate consequences of these demands. What is and is not flexible in one's life is to me terribly fascinating, terribly important biologically, and [also important] for my experience as a human being, a woman, a mother" (interview). Altman did not invoke multiplicity for its own exuberant sake, but coupled it to a constraining metaphor of budgets. Budgets were a means of investigating degrees of freedom. Altmann's frame did not privilege unlimited possibilities and transcendence of limits in narratives about paradigm primate selves. Hers was not a narrative of spacemen freed from gravity, generators of diversity escaping from necessity. This insistence on limits is another aspect congruent with contemporary feminism and other critiques of certain western humanisms linked to science and technology. But her budgets had another source as well, one much more explicitly in the analysis: a 150+ year history of time-energy studies in social sciences. Altmann crafted an ergonomics of baboon mothering, a kind of analysis only comprehensible within the history of discourses on labor in industrial social systems. Evolutionary biology contains a myriad of contents and forms indebted to the enabling social history of industrial capitalism, e.g., investment strategies and efficiency management sciences. The point in locating a metaphor is not to discard it as polluted, but to show how it is enabled and in turn enables specific kinds of accounts. A metaphor will destabilize some similarities and rely, often silently, on others. This point is fundamental to understanding discursive reconstruction of female/woman.

Both the form and the content of *Baboon Mothers and Infants* enact the fruitful tension between identification and deconstruction/problematization which structure Altmann's feminism, scientific epistemology, field practice, writing, and the relations among them all. The first sentence of her report on her 15 month field study in 1975–76 set the emotional dramatic temperature low and the expectation of focus on dailiness and material complexity high: "One of the first things that strikes anyone who follows a group of baboons across the African savannahs is how much time and energy they spend just making a living" (1980: 1). Drawing from the seven previous years of background data on Amboseli baboon groups, the study focused on Alto's group, numbering 47 individuals in 1975.

The language—prose, graph, metaphor, narrative, mathematical equation—is fascinating. Alto, the oldest and highest ranking female, died in 1976, leaving a 17 month old daughter, Alice, who slept in contact with her mother's associate, Peter (probably not Alice's father), and her sister, Dotty, for the next few months after Alto's death. Peter was a fully adult male, who was "particularly gentle with infants. . . . [and] was closely associated with Juma when Judy was born" (1980: 204, 213). This kind of prose about Alto, Alice, Judy, Juma, and Peter cohabits a text with language like that in the following discussion of sociobiological debates about parent-offspring conflict rooted in different genetic "interests": "In terms of population genetics models, I am suggesting that the benefit of maternal care to the infant is not a linear function of cost to the mother but rather approximates a linear function of the log of cost and reaches an asymptote or upper bound." The next sentence is: "These immediately detrimental effects of any attempt by an infant to demand more of its mother, combined with an infant's increasing abilities to care for itself, abilities whose development is crucial to the infant's eventual independence, suggest that parent-offspring genetic conflict of interest may arise infrequently as a relevant variable in many real-life situations" (1980: 182). Even if conflict of (genetic) interest existed in a given situation, Altmann's analysis of costs leads her to suggest that alternatives to conflict, such as cooperation and compromise, are likely to predominate at the *behavioral* level (1980: 185). The text is full of detailed description and figures summarizing data on demography, ecology and maternal time budgets, social relationship analysis (e.g., a matrix of the percentage of time an individual was within two meters of each mother during her infant's first three months of life; graphs of rates of interactive approaches to and by mothers during specified times in reproductive history; rates of spatial displacements; calculations of glance rates; etc.), infant's use of space, maternal style, weaning and independence, dominance position and economies of attention, and evolutionary models of parental investment. Throughout, data-rich sophisticated analytical moves intersect abruptly with compellingly simple prose about active baboon subjects living intensely interesting, usually low drama lives in Alto's group.

What are these kinds of heterogeneous, closely intertwined prose doing? What is the relation between content and form, and in what sense *is* the form the content? (White 1987). The analytical moves and the intimate prose confirm each other and destabilize each other at the same time. In one sense, there is nothing unusual about Altmann's text; scientific writing is inherently heterogeneous, making use of several kinds of languages in abrupt interaction and requiring a range of literacies from readers. But Altmann's text crafts a particularly tight relation between content and form, in which preferences for multiplicity and complexity, strategic moves to foreground the previously obscure and underrated, and substitution of compelling simplicity for high drama in the writing itself and in its message all converge to produce a powerful effect. The effect allows Altmann to move across species and cultures to reconstruct meanings of female and woman without seeming to flatten contradictions and particularity; it allows her to foreground motherhood without remystifying it; it allows her to insist that the reader acknowledge the status of baboons as subjects with lives of their own without surrendering objective scientific knowledge practices. And finally, it allows her to enact her feminism as a scientist. Altmann's feminism is an enactment of "generic" heterogeneity. In her writing, mobilizing the apparatus of bodily production, the tension among the heterogeneous genres and signifiers that make up this text is intrinsic to rewriting sex and gender in the intertextual discourses of biology and politics.

13

LINDA MARIE FEDIGAN:
MODELS FOR INTERVENTION

[I]f people become what they think they are, *what* they think
they are is exceedingly important. (Fedigan 1982: 7)

The problem comes in knowing which unanswerable questions to ask. (Pilbeam 1980: 268; quoted in Fedigan 1986:
58)

I t would be difficult to find a primatologist more conscious than Linda Marie
Fedigan of the delicate determinations of what may count as primate science
and as nature by the niceties of words and publishing practices. Acutely sensitive to models, metaphors, mirrors, and the histories of controversies buried in
narrative details, Fedigan negotiates a complex feminist and scientific stance in her
own publishing practices. Fedigan, an anthropologist at the University of Alberta
in Canada, turns her sense of language into an argument for the active, historical
status of the objects/subjects of primate scientific knowledge—both the animals and
the people. This perception of the animal-human relation is close to the core of
critical interventions in the political science of being female within the knowledge-
producing traditions examined in this book. The issue goes beyond discussions of
"empathy" or "identification" with the animals by the researchers, whether or not
women engage in such practices more than men, whether men and women "empathize" more with same-sex animals, and so on. Reviewing *Female Primates*, Fedigan
expressed her "misgivings about the wider implications of female empathy": "I well
remember my dismay when, having put many hours of effort into learning to
identify the individual female monkeys of a large group, my ability was dismissed
as being inherent in my sex, by a respected and senior male colleague" (Fedigan
1984: 308). But in her and Larry Fedigan's draft chapter and resource guide
for an American Anthropological Association Project on Gender and Curriculum,
Fedigan allowed more room for cross-species, same-sex empathy in the context of
the western women's movement and its associated privileging of non-reductionist

316

modes of knowing that insist on complexity and non-binary, interactionist relationships of subjects and objects of knowledge. Empathy is both part of a code in a
narrative and a culturally specific way to *construct* what counts as experience. The
relevance of "empathy" as a problematic, gender-charged scientific practice was less
Fedigan's point than the salience of the debate on empathy to an insistence on
seeing primates—people and animals—as complex co-actors in the crafting of
knowledge at particular historical moments.[1]

Perceived to have the status of actor, animals are not epistemologically or ontologically passive "resources" for the production of science; "sex" is not resource for the
production of gender; "nature" is not resource for the production of culture.
Neither does the animals' status of actor allow the scientist to make the complementary move to appropriation in traditional stories about authoritative western knowledge of nature: direct knowledge through union. Without invoking unmediated
"union" with the organism as the alternative, Fedigan makes a critique of the
"productionist" bias in evolutionary and anthropological explanation and its role in
the constant privileging of scarcity, resource competition, simplistic models of power
as social dominance, and impoverished ideas of what animals and scientists do in
the potent space called "the field."

Partly as a consequence of this critique, Fedigan privileges a disjunction in key
western bioanthropological origin stories between sexual difference and human
nature. Whether its practitioners like it or not, primate science generates these origin
stories for wide audiences. In Fedigan's accounts, the drama of sexual difference is
not to be found played out at the hominoid-hominid boundary, where in other
scripts the biology of heightened sexual difference becomes the originary raw
material for the linguistic and technological productions of gender, culture, and
species man. Faithful to this disjunction, Fedigan maintains a critical tension among
the internally heterogeneous identities and positions of female, woman, and scientist
as makers of models, objects of knowledge, and generators of metaphor. Like
Altmann, she walks a tightrope between female-centered and deconstructive accounts. Suspicious of and critically sympathetic to empathy as a method of feminist
or scientific knowledge, paradoxically committed both to foregrounding female
animals and to refusing sexual difference as a basis for science, and locating herself
in a citation web of scientific, feminist, and other critical discourse, Fedigan models
modes of intervention to recast inherited stories. Her goal is to encourage data-
rich, language-sensitive, and theoretically powerful accounts of primate lives, i.e.,
"good" science. She puts her discussions in places that both facilitate and set implicit
rules for their adoption by several of the audiences that mine primatology for
authorization of their origin stories about human and animal lives, including other
feminists, primatologists, and teachers of general anthropology courses. Let us see
how she attempts to do "good science" through writing and publishing practices in
the context of a career as a physical anthropologist whose authority is rooted in
field studies of free-ranging monkeys.

Born in 1949, Fedigan earned her Ph.D. under former Sherwood Washburn
student, Claud Bramblett, at the University of Texas at Austin in 1974 for a study
of social roles in a transplanted troop of Japanese monkeys living on the Arashiyama
West Primate Research Station at LaMoca, Texas (Fedigan 1976). Throughout her
writing the tradition of structural-functional analysis and role theory is central and
frames her use of other explanatory strategies, e.g., from socioecology, evolutionary

population biology, or sociobiology. A gift from the Japanese, the Arashiyama West troop immigrated to Texas in 1972, complete with 18 years of geneological records and other invaluable data collected by Japanese primatologists. The monkeys had recently separated from the main troop and formed a complete "natural" unit. About 1968 Kawamura raised the possibility of the transplant to a visitor in Japan, John Emlen, an important figure in facilitating international primate field studies beginning with his initial involvement with George Schaller in the late 1950s. The troop could be valuable for training American students, as well as for Japanese students; and the late 1960s was an active period of funding support for establishing primate field studies. It proved difficult to find a home for the Japanese monkeys, however, and three years elapsed before the monkeys' release in 1972 on the Dryden ranch near Laredo, Texas.[2] After Edward Dryden's death two years later, a search began for another home for the transplanted colonists, resulting in their move 60 miles north to Dilley in 1980. Already by the early 1970s, in the context of declining rates of federal civilian science budget increases, the United States National Science Foundation had become wary of funding primate field research and expensive naturalistic colonies. Laboratory science has a structural advantage in times of resource competition over work with a naturalistic bias.

At LaMoca in 1979 Fedigan wrote the manuscript published in 1982 as *Primate Paradigms: Sex Roles and Social Bonds*, her first large intervention in the sexual and linguistic politics of primatology. The book came out with the Montreal press, Eden, in its women's studies list two years after it was first advertised in the University of Chicago Press feminist studies journal *Signs*. Fedigan had wanted to call her book *Primate Mirrors: Reflections on Sex Differences in Behaviour*; but losing that title, she compromised to sidestep an embarrassingly flashy title.[3] Avoiding the connotations of "roots," with that term's tones of a continuous basic biology underlying culture, "mirrors" would have emphasized the process of historically located human researchers actively polishing reflective surfaces that returned the images of their own societies and bodies in their pictures of the animal other. "Paradigms" had a more scientific sound, while still foregrounding a sense of alternatives and choices, especially after Thomas Kuhn's (1962) famous book on paradigm contests in scientific revolutions.

Fedigan's text avoided relativism—any account is merely the culturally biased product of scientists and most stories are equally good—while insisting on the inherently story- and history-laden mediations of primate science. Repeating a theme often invoked by texts that seek to displace the authority of previous accounts, her fundamental bias was that the more adequate explanation is not necessarily the simplest one.[4] The constitutive interpretive processes of science were not allowed to fall into the background. Thus, the author constructed her authority on the permanently contestable grounds of overtly interpretive and critical discourse, where language and history were not permitted to lose their intrusive, material omnipresence in knowledge projects.

Natural scientific rhetoric is normally much more ambivalent about such writing strategies. In the semiotician Umberto Eco's sense, scientific texts are rarely "open texts," inviting construction of the meanings dialectically in interaction of reader and text. Scientific conventions are full of ways to close off interpretation while still producing the effect for readers of having read a self-critical text. In her paper on the meaning of the category "domestic labor" in western anthropological analysis,

Marilyn Strathern (1986a) discussed the kind of reflexivity that has emerged in anthropological texts. Relative to other western discourses, the problem of securing vigilance over ethnocentrism in social and cultural anthropology has resulted in a literature replete with ways to problematize its own categories and to foreground the complex strategies of identification and distancing that anthropologists invoke in both field practice and writing. Physical anthropology, on the other hand, inherited many more of its writing practices from the natural sciences, especially biology and medicine. In the conventions of natural science, critical reflexivity is not theorized to be about ethnocentrism in the relation of subject to subject. Ethnocentrism might be defined as the stance facilitated when specific human beings are in unequal interaction with and represent other people. But in the natural sciences, critical reflexivity is theorized to be about relations of subject to object, i.e., about possible bias of individual observers or of any possible human observers in the face of nature *it*self. In biology, the prophylaxis for such dangers is conventionally neither a dialogic or open text nor a text's foregrounding its own categorical cultural specificity, but the production of the objectivity effect through emphasizing strategies of impersonal distancing. In the conventions of natural science from which modern physical anthropology inherits its principal rhetorics, universality achieved through impersonal objectivity is the cure for bias, not specificity achieved through heightening history-bound contestations for meanings. Fedigan's bioanthropological prose is anomalous in its cautious deliberate use of some of the conventions of cultural anthropology and fiction, practices that produce much more semiotically open texts.

The audiences invited into Fedigan's book highlight its status as a contestable terrain, which she mapped with uneven constraints for contending interpretations while inviting readers' simultaneous reworking of meanings. That is, the text maintained its authority by building in interpretive constraints in the context of deconstructive discussions, e.g., of the history of sexual selection theory.[5] Besides students and non-specialists of all types, Fedigan particularly hoped for an audience of "specialists in Primatology and Women's Studies" (Fedigan 1982: i). Not only were these fields capitalized, their practitioners were *both* marked by the noun *specialists*. The Women's Studies audience was not the amateur consumer of science conveyed by the professional producer group of Primatologists. Fedigan constructed both equally as audiences and interlocutors in her first book. In effect, Fedigan constructed herself as author accountable to both constituencies in the interests of good science. She wrote from and to both groups of authoritative readers.

The shifting locations of female animals in the published text and in its preceding 1979 manuscript can serve as an icon of the productive tensions that structure Fedigan's writing. In the earlier version the tensions were less evident. The book was to be about "science in general, primates in particular, and female nature especially. . . . This book is an effort to present my view of female primates" (manuscript, p. 1). The manuscript emphasized its location in debates about women's changing places in Euro-American societies and the need for a sense of alternatives in the face of reigning masculinist doctrines of human and female nature. A forcefully anti-reductionist philosophy of science was presented explicitly in the context of privileging complexity in order to tease out more female user-friendly programs for human and animal nature. The manuscript was thoroughly comparative, devoting large sections to male behavior, but its strategy of unsettling the authority of male-centered accounts was mainly one of foregrounding female animals. By the

1982 book, Fedigan's writing was much more destabilizing, complex, and deconstructive. Keeping a strong focus on the legacy of sexist scientific language and the inescapability of social practices in the constitution of what can be perceived as scientific, the book worked less by reversals and showed the effects of deepening debates within both primatological and feminist discourse on sexual difference. While maintaining a distinction between sex and gender—a distinction orthodox in both feminist and social/cultural anthropological discourse, but often elided and misunderstood among biological primatologists—Fedigan's emphasis was more and more to deconstruct both the terms. Male and female, masculine and feminine were mutually constituting dichotomies that obscured both the animals and the scientists' actual practices in area after area of investigation.[6] The unyielding center of her analysis of each topic was the discriminating, active roles of all the animals in a social group. Fedigan presented the data to reveal the scandal of the category of sexual differences constantly silencing similarities. Repeatedly, females previously consigned to a category of resource or matrix emerged from an analysis as active generators of lives and meanings. Similarily, scientists were actively reinterpreting the explanatory possibilities made available out of insistently named, specific histories. Within that frame more complex pictures of power and knowledge (among both the animals and the scientists) could emerge.[7]

The photographic essay in *Primate Paradigms*, as well as in other texts that emphasize the animals' individuality, complexity, and agency—in short, their status as western actors—is a concentrated locus of meanings.[8] Old animals can be seen fully engaged in daily life; monkeys are shown enjoying swimming, interacting in all sorts of contexts, communicating with a wide range of facial expressions, and so on. The photos invite a reader's identification with attractive, interesting, and only slightly exotic other selves. Often the photographs are explicitly portraits, individualized presentations of a coherent subject composed by western standards. The visual representations in these books are much more than pictures of warm, fuzzy animals; they are an essential part of the books' arguments about boundaries of self and community and relations of observers to animal subjects. These photographs recall Akeley's dioramas; they are arguments about ways of life, not decoration for some real text located elsewhere. [Figure 13.1]

The particular kind of monkey self that Fedigan authorized emerged most clearly in her critique of those sociobiological models of explanation that depend reductively upon self-maximizing strategic reasoning, constant adaptationism, and strict scarcity assumptions (Fedigan 1982: 288–306). She limited her discussion to kin selection, reciprocal altruism, and parental investment in relation to sociobiological accounts of sex differences. She began by grounding particular evaluations within a discussion of opposed views of the biological world structured by constant competition, pushing organisms toward optimization for extreme efficiency, compared to a world with a great deal of slack, full of diversity structured by many processes loosely related or unrelated to strict Darwinian selection.[9]

Fedigan stressed the extreme difficulty of measuring the key "costs and benefits" in evolutionary arguments and the naive measures too often taken to resolve the fundamental disputes. She showed how sociobiologists relegate an alternate explanation to the category of "proximate mechanism" when the crucial testing of alternate hypotheses has not been done. Fedigan did not dismiss sociobiological arguments or refuse to use them carefully in much of her work, but she foregrounded

Figure 13.1   Linda Marie Fedigan and her Japanese macaque subjects in Texas. Courtesy of Linda Fedigan. This image was the author photograph for *Primate Paradigms*.

the yawning gaps between theorizing and empirical demonstration and the overdetermined attractiveness of possessive individualist, scarcity, and adaptationist assumptions, all of which blunt the critical spirit within just that community that claims to be particularly good at theory. In bringing this discussion to bear on the argument over whether or not sociobiology is sexist, Fedigan refused to take the easy path of limiting the argument to the human terrain. Preoccupation with the naturalistic fallacy and human sociobiology deflect attention from the object of Fedigan's emphasis. "The issue here is whether sociobiological theory *does* accurately and usefully reflect and predict animal behavior, or does it represent attempts to develop biological justifications of the belief, rather than the fact, that females are in some way inferior? Thus sexism may lie not only in statements by experts as to what *ought* to be, but, more importantly, in the manner in which they decide and determine what *is*" (1982: 301). Fedigan locates the most important part of the debate in the *biological* discourse and refuses the fact/value dichotomy. For scientific and feminist reasons, which are neither independent nor identical, sex, i.e., what it means to be biologically female, is as critical as what it means to be located in gendered cultural space as a man or a woman.

Approaching the issue through unpacking assumptions in Robert Trivers's (1972) parental investment models and George Williams' theories of original sexual asymmetry, Fedigan emphasized specific traps in investment logic and pointed out that different ways of assessing costs do not lead to the assumption that males invest less than females in reproduction. The notion that males invest less is the assumption that makes females limiting resources for male strategists and leads to replicating traditional western forms of sexism in the biological discourse. Not finding the arguments to be convincing about adaptive advantages of sexual reproduction at either the gene or individual level of selection, Fedigan foregrounded the possibility that sexual reproduction—recognized widely within evolutionary theory as anomalous—may not be particularly *adaptive*, simply *possible* within the degrees of freedom allowed in a fairly permissive biological world. Displacing adaptationism from accounts of the origin of sexual difference would fundamentally change the bio-politics of being female in western stories. It is parallel to displacing the origin of the heterosexual nuclear family from the origin of what it means to be human. Fedigan made both these moves, and for each she drew from both feminist and scientific debates and writings.[10]

Fedigan judged that there was structural sexism in Williams's (1975) and Maynard-Smith's (1978) particular accounts of the origins of sexual reproduction and of anisogamy (unequally sized reproductive cells). "It seems to me that the above model for the origin and maintenance of differentially sized gametes justifies the charge that this area of modern evolutionary thought *begins* with the assumption that females lose in competition with males, and then reasons backwards to a plausible biological origin for female inferiority (the primeval conflict and loss by larger gametes [the limiting resources])" (Fedigan 1982: 305). Fedigan concluded that sociobiological evolutionary theory was not inherently sexist, in the sense that females need not be assumed intrinsically "losers in competition between the sexes" (1982: 306). That remains a cultural assumption built into much biological discourse, but not essential to Darwinian evolutionary argument. Both the "loser" and the "competition" terms can be destabilized. Whether viewed as an ideology or as a scientific preference, adaptationism is not necessarily sexist, whatever its other problems; but in combination with assumptions of original female inferiority, it is a powerful tool to reproduce females as resources for male action—symbolically, scientifically, and socially. That structural material-semiotic fragment is replicated at all the key origin points in western sexist discourse, from accounts of the beginning of sexual reproduction, to reconstructions of the origin of "the family," to theories of the structure of desire and action in narrative itself. That is why intervening in the fundamental rules for telling any possible story matters. What makes some stories possible and others almost impossible to tell? "If people become what they think they are, *what* they think they are is exceedingly important" (1982: 7).

Fedigan has published several papers on the Texan Japanese monkeys, but two other essays seem to me to be at the heart of her importance to the bio-politics of being female. Both are general reviews published in annual volumes likely to be consulted by teachers looking for a good overview of major topics for writing a lecture, by colleagues wanting a good summary of literatures and arguments for situating their own research and writing, by graduate students studying for orals, or by seminar students looking for paper topics. These papers were placed where their arguments would be likely to spread beyond narrow circles, to be reproduced in adjoining discourses, to help mold the common sense of young scholars. Fedigan's negotiation of a dialectically structured feminist and scientific stance in her publishing practices is particularly clear in key moves in her reviews of contentious issues: first, studies of dominance and second, the role of models in reconstructing stories of passage across the hominoid-hominid border.

"Dominance and Reproductive Success" appeared in the *Yearbook of Physical Anthropology* in 1983. At its heart was a discussion of the model of "priority of access to estrus females," on which has hung the biological meaning of male dominance interactions and hierarchies. The fundamental question is whether there exists for a given species a relation, and if so how strong, between competition for social dominance and differential reproductive success. The relation between agonistic dominance and reproductive success has been more studied in primates than in any other taxon. The gendered culture-bound preoccupation of it all could strike an off-planet anthropologist as flatly monomaniacal.[11] As usual, Fedigan began with an extensive historical commentary. From it emerged an odd scandal: the easiest aspect of differential reproductive success to measure has rarely been measured. Most of the attention has been directed to male reproductive success, where paternity is usually very hard to gauge, whereas it is fairly straightforward to get at reproductive success for females. Measuring lifetime differentials still requires years of detailed records, and going beyond one generation throws the question of female differential reproductive success into the same quandary as single-generation-focused male measures. Females have both male and female children. Fedigan's tables show that since Carpenter's 1942 paper, there have been more than 20 studies of reproductive success in male primates (15 since 1974), and since the first effort to measure females in 1974, there have been 8 queries about them.

Common in sexual selection theory from Darwin on, the problematic *assumption* of greater variance in male reproductive success is part of the problem. But equally troubling has been deciding what really constitutes biologically meaningful measures of dominance, i.e., whether dominance hierarchies (rather than interactions) are artifacts introduced by the observer or insisted upon by the animals, easy to see, very hard to see, impossible to miss but not very important to much else, or subtle but critical to everything. But let us forge ahead. In the midst of an embarrassing fullness of publications on the priority of access to estrus females model, which has the annoying limitation of looking only at *mating* to assess male reproductive success, Fedigan highlights the nub of the matter. Both the view that female primates are objects, i.e., "passive resources like peanuts or water," and the view that the female's best reproductive strategy is to choose male "winners" in the dominance stakes were developed in the absence of systematic research into actual processes of female choice in primates—despite the fact that female choice was central to Darwin's original model (Fedigan 1983: 112–13). Fedigan concludes not that dominance is non-existent or trivial—that remains a culturally complex, empirical matter—but that "our thinking about [dominance] has not done justice to the complexities of social power and the manner in which it operates in primate societies" (1983: 113). She outlines various ways of improving the model, such as taking account of age, residency, and demographic structure in thinking about alternative reproductive strategies. She emphasizes the reasons to study female reproductive strategies, which requires recognition that females do compete with each other and assessment of reproductive costs as well as benefits of high rank. Demographic and environmental variables are critical, and female and male reproductive strategies should be modeled as life history phenomena. Dominance issues, then, become so contextualized that they lose the quality of neurotic fixations at the heart of evolutionary arguments. Any "simple" relations of dominance to reproductive success or attention to only one sex disappear into cautionary reminders of poor science. In this review, Fedigan articulates the emerging standards of a new generation of women and men primatologists, for whom questions about the biology of being female restructure basic evolutionary theory and basic field practice.

This articulation has been enabled by complexly interrelated feminist and biological discourses, with their available moves for unpacking story structures and objects of knowledge and drawing new ones. Attention to differential female reproductive patterns is mandated by both western feminist and sociobiological, socioecological logics. Put into productive interaction, including tension, with each other, these two sources of reformulation of evolutionary stories can have unexpected effects for science and for feminism. Fedigan's performance of the normal scientific function of writing a critical review, that is, an action unambiguously from the "inside" of scientific practice, effects the slight genetic mutations and subtle warps in the rules of form in both scientific and feminist texts that give us hints of what one might mean by a feminist critical reformation of what may count as knowledge—including what may count as natural scientific accounts of objects of knowledge and of the plots in which they make sense, encompassing such things as females and evolutionary narratives.

Feminist projects in science are distinctly not feminine science. Rather, what might be problematically named feminist science in local practices destabilizes meanings of both sex and gender and restructures the "science/gender system" as it maps the field of meanings ordered by the axes of nature/science and sex/gender (Keller 1987). It is specifically the permanent tension between construction and deconstruction, identification moves and destabilization moves, that I see, not as uniquely feminist, but as inherent to feminism—and to science. Both feminist and scientific discourses are critical projects built in order to destabilize and reimagine their methods and objects of knowledge in complex power fields. Addressed to each other, western feminist and scientific discourses warp each other's story fields and redraw possible positions for claiming to know something about the world, including gendered social space and sexed bodies.

So far, in relation to the natural-technical objects of knowledge in primate studies, Fedigan's attention has been on the sex side of the sex/gender distinction. What are the biology and politics of being female or male? How might redescribing what counts as biology change what is peceived as "not-biology"? What are the politics of difference in drawing marks of distinction on the body of nature? But in the complex fabric of primatology, sex and gender interdigitate through the setting of boundaries between the charged western categories of nature and culture, itself a process deeply influenced by the histories of colonialism, decolonization, racism, anti-racism, and feminism. Primatology works as a technology of gender, as well as a technology of sex (de Lauretis 1987). Meanings of gender are produced by discourse about boundaries, transitions, and origins. Positioning within gendered social space for historical, scientific women and men is effected by the system of scientific production, an apparatus of bodily production, including, prominently, publishing practices. Gender is part of the apparatus of scientific production, and in turn is replicated or destabilized through it. This is evident in a particular local scientific practice—building models of human evolution.

Fedigan published "The Changing Role of Women in Models of Human Evolution" in the *Annual Review of Anthropology* in 1986. Her concern was with women as both objects and producers of evolutionary accounts of the potent transition from animal to human in western scientific story fields. That transition also marks the boundary between sex and gender, from female or male to woman or man, as modern western people have defined the distinction. The development of gender in each individual, embodied social subject re-enacts that originary transition. Gender is narrated as one of the products of human evolution, so how gender and sex are marked at the boundary between hominoid and hominid sets constraints on basic cultural stories about what it means to be a man or woman. Is there sharp sexual difference? Is it antagonistic? Complementary? Adaptive? Insignificant? Flexible? Fixed? Linked or unlinked to reproduction and production? Can the heterosexual reproductive imperative be relaxed in knowledge-power fields enough to permit escaping the binary restriction on sex and gender? What kinds of evolutionary narratives could have more than two sexes or genders? Acknowledging that many primatologists want nothing to do with the use of their science to build models for human evolution within anthropology, much less the use of their science to trace out some of the explicitly political and science fictional questions in the above list, Fedigan insisted that the appropriations will go on anyway. She wishes to have some power to set the rules, to manage the constraints of what may be called a *scientific* narrative and to suggest imaginative possibilities for narratives in other genres. To have such power, one cannot refrain from narrativizing. There is a cultural privilege for discovering the "original blueprint" in the archive of nature (Fedigan 1986: 26). So she sketches an historical account of both how (originary and so putatively universal) women's roles in various models have changed and how middle class Euro-american women and men, living specifically located social lives, have written different stories about the history of nature, and especially of the traffic between nature and culture.

In the context of a detailed genealogy of models about human evolution in the nineteenth and twentieth centuries, written in a narrative form to emphasize possible scientific progress but also inherited biases and mistakes, Fedigan constructs a citation pattern and privileges forms of argument that stand in sharp contrast to another paper on primate models for human evolution composed about the same time, "The Reconstruction of Hominid Behavioral Evolution through Strategic Modeling" (Tooby and DeVore 1987). Three contrasts stand out: 1) their respective constructions of a web of relevant literature and debates and the basic form of their own narratives as they construct an image of what science is, 2) their approach to the questions of sexual difference at the potent origins and boundaries, and 3) their treatment of a specific model, Adrienne Zihlman's.

Fedigan's citations draw constantly from literatures and debates marked as both scientific and feminist. She highlights the immersion of evolutionary science in western cultural history, and she uses recent literary scholarship to analyze the structure of narrative in evolutionary modeling. None of this destroys the authority of her scientific discourse; it constructs a different model of authority with a strong premium on explicit discussion of deeply constitutive cultural mediation, including mediations of her own writing. Several kinds of reasoning are given historical location, to show how they might make sense to constituencies inside certain discursive traditions. She does not reduce models of legitimate reason and explanation to agonistic strategic reasoning. Her histories do not lump together all feminist interventions into evolutionary arguments, but differentiate among contending feminist accounts and lineages of arguments. Fedigan remembers that moves inside biology had origins in particular papers published in other contexts, e.g., Sally

Linton's 1971 essay on "Male Bias in Anthropology."[12] Fedigan lived through the history, and her paper foregrounds a changing feminist discourse in interaction with changing bio-anthropological modeling. Although her paper on dominance and reproductive success foregrounded life history perspectives, demographic variables, and environmental constraints in assessing fitness costs and benefits, in this paper on modeling she does not cite widely from the socioecology literature.

Tooby and DeVore could not have written a more contrasting paper, forgetting for a moment the issue of content—i.e., which models the authors' believe in or think worth the cost of tax dollars and people's careers to organize the resources to test—and looking only at form, reference, and privileged models of how to argue. Initially, like Fedigan, Tooby and DeVore acknowledge the difficulty of modeling human evolution, and both believe it nonetheless remains an important *scientific* endeavor. But rather than locating the matter in cultural and intellectual history of nineteenth through twentieth-century evolutionary discourse and in problems of narrative and representation, Tooby and DeVore appeal to the potent mythic history of great physical science and its grand models of heaven and earth, from Kepler, to Newton, and Einstein, in whose company they place Darwin (1987: 184). Advocating what they call "conceptual models" that privilege "strong deductive principles" claimed to be available now to biology in recent theories of selection (inclusive fitness, game theory, evolutionary stable strategies, etc.), Tooby and DeVore affiliate their own form of argument to Maxwell, Newton, and Einstein (1987: 189). All of the papers they cite as providing "theory" or "strong conceptual models" are by men, and none is contextualized in cultural history (1987: 190). Their genealogy is a patriline of great founding fathers, not a map of messy literatures intersecting in local practices through which sciences and, not so incidentally, genders are built. Their paper is located at the end of a conference volume; it alone is headed by the label "Theoretical Issues," a status thereby denied all the other papers, which become examples of this or that model, but not theory itself. The book places Nancy Tanner's essay, "Gathering by Females," at the furthest remove from Theory.

The conference celebrated and re-enacted the founding symposium organized by Sherwood Washburn on "The Social Life of Early Man" a quarter century before, a meeting in which DeVore played a large role and which was basic to Washburn's promotion of the man-the-hunter hypothesis. That is the memory written into the Tooby and DeVore paper, not the early mimeographs published as Linton's 1971 objections to the fruits of the Washburn project, including DeVore's work on baboons in particular. Tooby and DeVore, like Fedigan, are conscious that feminism was part of the challenge to the man-the-hunter perspective, but the feminism they invoke has no history after its initial formulations and no wider citations outside the models they will criticize. First they praise feminism for its "vigorous advocacy of the importance of women in all areas, and its emphasis on the ubiquity of male biases and pretensions" (1987: 213). But then, when they discuss reasons for the fall from grace of the man-the-hunter models, this "vigorous advocacy" (which never reached the level of being one of the constituents of real science, only of cultural influence on science, even if it was salutary) linguistically has become "feminist revisionism" (1987: 222). The linguistic stage had been set for counter-revisionism, this time with claims of full consideration of the always equal importance of females in evolutionary processes.

The hunting hypothesis this time will not be rooted in Washburn's structural functionalism, but in Maynard Smith's evolutionary stable strategies, i.e., in the privileging of a particular model of evolutionary action in sociobiology called strategic reasoning (1987: 222–26). All the important actors, from gene to organism are recast in evolutionary narratives as strategists in a vast game whose stakes are reproductive fitness, i.e., staying in the game as long as possible. In a sophisticated and cogently argued synthetic paper, which could not properly be dismissed as simplistic sociobiology, Tooby and DeVore contrast what they call referential models with conceptual models, and within the latter category foreground "strategic modeling." Exclusive reliance on this kind of modeling encourages just the move Fedigan criticized as part of an adaptationist bias: constant invocation of the distinction between proximate means and ultimate evolutionary ends (Tooby and DeVore 1987: 235) to facilitate assumptions about untested relations of biological processes to fitness. But Tooby and DeVore are very inclusive in the kinds of processes and variables that strategic modelers would have to consider in a "species' distinctive system of adaptation"; and their list is very similar to Fedigan's in her dominance and reproductive success paper, emphasizing comprehensive life historical, demographic, environmental, and organismic variables (1987: 201).

What separates Fedigan from Tooby and DeVore in their approach to modeling is not their listing of what should be included by the late 1980s in evolutionary arguments, but their linguistic and social pictures of the form "theory" takes, and not so incidentally, whose theories and models are the most data-rich and cognitively powerful. Tooby and DeVore's exclusive reliance on strategic reasoning and strategic modeling (1987: 190–200), with its associated genealogy of how science advances, subverts their own desire to be able to come to terms with the richness of variables they invoke on page 201. The world of theory and the world itself become exercises in military-like strategies with tight constraints. The scarcity and competition assumptions tightly rule every biological process. No wonder the language itself militarizes the field of biological interaction and the cognitive processes of human reasoning, including what will count as scientific theory.

> A significant fraction of any species' environment consists of constantly co-evolving organisms; fitness gains by one species frequently occur at the expense of other species that are preyed upon, defended against, or parasitized . . . The defenses of plant and animal prey species can be circumvented by "surprise" attacks . . . Goal-oriented actions by humans, shaped to suit the particular situation, constitute surprise attacks. We accomplish this by conceptually abstracting from a situation a model of what manipulations are necessary to achieve proximate goals that correlate with fitness. (1987: 209)

Fedigan's and Tooby and DeVore's models of science and webs of reference deeply influence how they deal with the issue of sameness and difference, including sexual difference at the potent originary boundaries between animal and human, nature and culture. We have already looked at the theories of the origin of sexual reproduction itself and at how theorizing that anomalous innovation as adaptive or not may become part of wider meanings about sexual difference and biological female-male equality or inequality. The modeling papers focus on the hominoid-hominid transition: Is this transformation theorized as a period of sharp breaks and intensified adaptive differences between males and females? Or do the theoretical narratives downplay adaptive, originary sexual difference? The narrative stakes are again about degrees of freedom in the natural world serving as allegories about human freedom in the cultural world.

Fedigan's treatment of two authors, Lila Leibowitz (1983) and Richard Potts (1984), demonstrates her stakes in theorizing reduced sexual difference and deferring its salience to processes and moments outside the theater of origins of what it means to be human. Criticizing Owen Lovejoy's scenario of human origins, Fedigan discussed "another recent, but much less widely noted, model of human origins" that began with a similar question to Lovejoy's, but suggested a fundamentally different answer (Fedigan 1986: 38). Agreeing with Lovejoy that early humans hovered near extinction, Leibowitz argued that the sexual division of labor was nonetheless likely a very late human invention, so that for most of human evolution males and females engaged in largely similar productive activities facilitated by the key human invention, food-getting tools. In the narrative code of evolution accounts, late means less fundamental to human nature. Anything recent must surely be less "biological" and so easier to change. That structure is a narrative code, not a fact about change.

The productionist emphasis of Leibowitz's account is partly rooted in her marxism. Rather than speeding up baby production through male provisioning of sedentary females, Lovejoy's projection, Leibowitz imagined humans hedging against marginal replacement rates through accumulating surplus food, resulting in the beginnings of exchange. Fedigan does not argue that Leibowitz's account is necessarily correct, but that it shows at least as good a grasp of the thin evidence as Lovejoy's. Most forcefully, Fedigan argues that Leibowitz's heterodox account "is one of the very few attempts to strip away the remaining assumption common to all models, that sex differences must have been significant in the earliest stages of human evolution. . . . Yet it can be very enlightening to think through what we have assumed to be the less probable solution. Could characteristic human societies have originated without a sexual division of labor beyond that directly related to insemination, gestation, and lactation? Could some behavioral invention, characteristic of neither males nor females and requiring equivalent participation, have been the catalytic event that set humans moving along their own distinctive evolutionary path?" (Fedigan 1986: 38).

These concerns inform Fedigan's discussion of Potts's re-examination of "sharing" or "home base" models based on animal bone accumulations associated with human stone tool remains. He argued that these stone-bone assemblages do not necessarily imply a home base, where hominids lived for periods of time. Rather, the assemblages, with marks indicating that both large carnivores and hominids had been active at these locations, indicate precisely that it is unlikely that hominids lived there. Potts saw the sites as tool caches and processing sites. Fedigan concludes, "One implication of this new understanding of bone-and-artifact associations for early hominid sex roles is clear: if there is not evidence for home bases where the sick and the dependent waited for the well and the productive, then perhaps we can finally free our minds of the image of dawn-age women and children waiting at campsites for the return of their provisioners" (1986: 57).

Fedigan is looking for a narrative of human evolution that is not about the origin of home and family, a most unnatural preference. "Women and homes have been inextricably linked in our cultural imagery, and thus the shaking loose from the home base focus for early hominid social life may allow our imaginations to turn to alternative scenarios" (1986: 57). Women, the home, the family, and reproduction inhabit a common conceptual space, differentiated from the oppositely gendered regions of the public and production. Here, Fedigan joins other feminist anthropological criticism that has highlighted the western cultural specificity of notions of "home," the "domestic," and women's analytic assignment to locations in these curious theoretical spaces, which are imagined to have materialized in the species-making zone between hominoids and hominids (Rosaldo 1980; Strathern 1986a: 21).

Tooby and DeVore also discuss Potts's ideas, as well as the broader question of what kinds of similarities and differences between animal and human they imagine must be explained in the process of hominization. Their narrative preferences are predictably quite different from Fedigan's, and neither would allow the easy resolution that the field is so unconstrained by data that any story is finally as good as any other. Fedigan is more explicit about just how story-laden theory in this area must be, but she does not give up the field to taste for that reason. In a sophisticated way, Tooby and DeVore insist on a drama about differences—between animal and human and between male and female (Tooby and DeVore 1987: 207–9, 231). A unique human mating system remains basic to stories which make sense to them; the origin of the family, with the word recoded into the technical vocabulary of sociobiology, is what needs to be explained. Gender, heterosexuality, and deep splits between animal and human are at the core of hominization, whether in DeVore's former structural-functionalist and quasi-group adaptationist accounts or in his new sociobiological, strategic modeling arguments. Tooby and DeVore found Potts's discussion as persuasive as Fedigan did, but for them it raised quite different possibilities—that home bases were still salient but were separate from the processing sites, that the most important aspect of Potts' view was the implication of intense hominid direct or indirect competition with the large carnivores, and that what needs explaining if hominids left tool caches all over the place is how they protected their labor investment and scarce resources from other hominid groups. That is, Tooby and DeVore saw in Potts's arguments a way to rephrase the problem of "intergroup relations," a phrase I hear as a euphemism for the origin of human war (1987: 219–21).

Tooby and DeVore define war as "a high degree of male coalitional intergroup aggression" and locate it at the potent zone of "unique hominid divergences" (1987: 207–8). War and sex are closely linked in the narrative. Many features of "the unique human mating system" require explanation for Tooby and DeVore. "However, perhaps the single major question facing hominid social theory is the reconciliation of group life with high MPI [male paternal investment] and sexual exclusivity [human quasi-monogamy]. What role intergroup conflict (as in lions and chimpanzees) may have played in this process is another crucial question" (1987: 231–32). How *have* humans been able to live with each other? The question continues to prevail over all others in a narrative frame ordered by strategic reasoning, adaptationism, possessive individualism, and profound antagonism at all the key moments of transition, from the development of the first gametes to the origin of high MPI and sexual exclusivity among tool-making hominids. Here lies the scandal of difference of all kinds, including its incarnations in differential reproductive success, in western stories of self and community, one and other, same and different.

Not surprisingly, Fedigan and Tooby and DeVore have utterly different evaluations of the models for human evolution suggested by the physical anthropologist, Adrienne Zihlman. The advocates of strategic modeling write, "We know that certain hypotheses, such as Lovejoy's (1981) and Zihlman and Tanner's (1978) cannot be true, because they violate validated principles of evolutionary biology" (Tooby and DeVore 1987: 236). To violate the already validated surely *sounds* like bad science. They cite Zihlman again in a brief allusion to her suggestions about the pygmy chimpanzee modeling certain features important to human evolution, but nothing after that (Zihlman et al 1978). Referring to *six* Zihlman contributions after 1978, Fedigan concludes, "Zihlman's most recent publications on the gathering model attempt to account for more of the data from all sources than any other model I have seen, and yet her interpretation of early hominid life has received no more attention from the paleoanthropologists than other less 'data-based' models."[13] What is happening here? To begin to sort out this question, it is time to turn to Adrienne Zihlman's career and writing—not finally to judge if she or DeVore and Tooby are right in an imaginary disinterested extraterrestrial court, but to understand again how the politics of being female are part of the stakes in the scientific contest for what may count as human. Fedigan has modeled publishing strategies as interventions into the apparatus of bodily production, and her textbook and reviews have sharpened our sense of the particular grinding of primate mirrors with tools of narrative and metaphor. Zihlman will take us the next step into the bio-politics of difference.

14

ADRIENNE ZIHLMAN:
THE PALEOANTHROPOLOGY OF
SEX AND GENDER

I suspect our role was also a crisis millions of years ago. (Zihlman 1985b: 2)

As in most things in life, the debate centers on two themes: food and sex; or to give it a proper academic tone: diet and reproduction. (1985b: 3)

Hunting is often presented without precursors, as if it too came out of a bad headache, like Minerva springing from the head of Zeus. (1985b: 11)

These, then, are the True Male and the True Female, the average, the typical, the normal, and to judge by a look around us, possibly the extinct. For these representatives of the basic differences between the sexes appear to have been put together by calipers and glue rather than by the shakier hands of Mother Nature. (Herschberger 1970 [1948]: 2)

Science, Politics, and Stories

One story is not as good as another. This book is about what enables and what constrains a particular kind of story-telling practice—scientific narrative in a field of extreme boundary disputes, among many differently situated writers and readers that range from popular filmmakers to mathematical population biologists, comparative anatomists, and molecular biologists. The boundary disputes written into the bodies of primates—fossil and living, human and nonhuman—involve the major themes of modern history, from decolonization to nuclear war to feminism. In the context of unprecedented expansion and proliferation of sciences and science-based technologies, the disputes fundamentally involve constraints and enabling histories specific to the kind of cultural material practice that primate science is: bones, animal activities, discourses in adjoining sciences like cultural anthropology and medicine, and accumulation of many years of inhomogeneous data on dozens of species from laboratory and field. Attention to narrative is not instead of attention to science, but is emphasized in order to understand a particular kind of scientific practice that remains intrinsically story-laden—as a condition of doing good science.

A primate female, a white woman, a late twentieth-century scientist, and a U.S. middle-class feminist, Adrienne Zihlman's craft is to build reconstructions of hominoid and hominid evolution. Her profession is to sketch the originary zones for contested ways of life. There is no way her scientific practice could escape the particular pleasures and dangers embedded in the story-laden sciences. In the 1970s, she was a principal generator of a being called "woman the gatherer," often regarded as an oppositional mirror twin of "man the hunter," a feminist version of a prior masculinist account of human nature. Seen from this perspective, woman the gatherer could only be the flotsam surfacing from the submerged continent of subjugated knowledges. Her appearance could only be explained by *cultural*, not *scientific*, genealogies. Thus, it is claimed that she was caused to appear in the paleontological record by feminism, not science; such appearances might be salutary, but must be blessedly ephemeral, not finally authorized by the evidence. She was called forth by one story to counter another. Compare them to see equivalent bias, include the best of both in the interest of balance, and then get back to the real work of science, in this case, paleoanthropology.

Zihlman has resisted this representation of the genesis and ontology of woman the gatherer, both in the geologic past and the political present. She has refused the claim that woman the gatherer and man the hunter are dichotomous opposites, and she has insisted that her position is grounded in better science than competing models and narratives of primate evolution. Woman the gatherer is not an unchanging ideological apparition reified in the bones of fossils and the bodies and lives of modern chimpanzees and human subsistence foragers, but a flexible being occupying one node in a strong scientific synthesis of the evidence constructed by the material practice of physical anthropology. Let us construct a narrative of entwined fates for woman the gatherer and woman the scientist in Adrienne Zihlman's professional life in order to deepen our exploration of primatology as simultaneously a modern life science and a genre of feminist theory.[1]

Entering the Anthropology Department at the University of California in 1962, Zihlman began her graduate study of physical anthropology at the height of Sherwood Washburn's programme for joining functional comparative anatomy and field studies of living monkeys and apes to explore models for the human way of life. She earned her Ph.D. in 1967 for a thesis on bipedalism that was embedded in the man-the-hunter hypothesis. Comparing living normal and pathological human gaits with living chimpanzee locomotion and reconstructing australopithecine locomotion from fossil material, her basic argument was that the earliest hominids were effective, but not efficient bipeds. *Australopithecus* had a different locomotor pattern from *Homo*. Drawing on the principle that changes in behavior precede changes in morphology, her discussion emphasized selection for the capacity to walk long distances on the African savannah as part of the hunting way of life.

One dozen Washburn students completed their Ph.D.s in 1967, including Claud Bramblett, Linda Fedigan's dissertation superviser at Texas; Donald Lindburg, whose early emphasis on female choice in rhesus mating patterns quickly became important to Zihlman and others; and Jane Lancaster, who also later located herself at the intersection of feminism and primate biosocial science. Washburn students who overlapped significantly with Zihlman included Phyllis Jay, finishing up her Chicago thesis on langurs; Theodore Grand, a close friend whose approach to functional anatomy was important to Zihlman; Ralph Holloway, interested in the evolution of the brain and social behavior; Richard Lee, whose studies of the subsistence ecology of the Kalahari San people were crucial to challenges of the species-defining role of hunting in human evolution; and Suzanne Ripley, from whom Zihlman took her first course in primate behavior in 1963 and who wrote a long, single-spaced, typed letter in response to Zihlman's paper, on chimpanzees and the origin of tool use, to advise the younger woman to establish her own intellectual viewpoints independent of Washburn's. There were also Suzanne Chevalier-Skolnikoff, Kathleen Gibson, Mary Marzke, Alan Mann, Donald Sade, Paul Simonds, Russell Tuttle, Judy Shirek, John Ellefson, and Andrew Wilson. Vincent Sarich, whose work on molecular taxonomy would greatly influence Washburn's and Zihlman's claims about the importance of the human-chimpanzee relationship, was a beginning graduate student in the 1962 course on Fossil Man for which Zihlman was a reader. Washburn was an influential teacher for Zihlman. His insistence on behavioral anatomy has remained at the center of her approach to modeling in hominid evolution.

In both its collegial and antagonistic aspects, Washburn's wider network was crucial to Zihlman's initial research and professional foundation. Approved by Lita Osmundsen, an important supporter for the young Zihlman, a Wenner-Gren grant funded her first research trip to study the hominid post-cranial fossil collections of South Africa and visit sites in East Africa. Washburn had made connections for her with C.K. Brain at the Transvaal Museum, and in Africa she formed collegial relations with Phillip Tobias and other actors in the reconstruction of human origins.[2] She participated in a seminar with John Napier and his students in London before leaving for Africa in 1966; Napier provided a letter of introduction to Louis Leakey, whose relations with Washburn, and so with those identified with him, were strained.

Mary Leakey was practically the only visible woman active in paleoanthropology. Zihlman admired her work, but they were not close colleagues. Unlike the women who did field studies of living primates, who had a large peer group of other women, some of whom were modestly powerful in the growing discourse, Zihlman entered what is still a profoundly male-dominated branch of physical anthropology. She took her first job at the young University of California campus at Santa Cruz in 1967, where she was encouraged to apply by Richard Randolph, a Santa Cruz faculty member who had recently completed a Berkeley cultural anthropology Ph.D. With the rudiments of a professional network but still naive about publishing strategies and the networks of competition and authority in her field, Zihlman left Berkeley intellectually grounded in functional comparative anatomy, the modern evolutionary synthesis, and the results of field studies of social groups of living primates. She also left just before the life-changing intellectual and social challenges of the Women's Liberation Movement.

Zihlman had been conscious of the importance of women as models, teachers, and peers before. Like thousands of other young anthropologists, she narrated a personal bibliogenesis in her freshman reading of Margaret Mead's *Coming of Age in Samoa* and Ruth Benedict's *Patterns of Culture*. Dorothy Kascube taught Zihlman undergraduate linguistics at the University of Colorado; the ethnographer, Ruth Underhill, lectured in a 1961 Summer Institute on American Indians; and Laura Nader taught Zihlman social organization at Berkeley. While male teachers, like Jack Kelso, were also remembered as fundamental, the charisma and encouragement of these women were important. Zihlman emphasized their modeling a possible but difficult woman's life as a scientist, such as Nader nursing her first child while teaching in a major university and Kascube juggling two small children with research and teaching. In the 1960s, these models indicated possibility, rather than signifying the unequal terms of success in male-career pattern sciences for women socially responsible for child care. Lita Osmundsen had also given birth to twins shortly before she and Zihlman met. Osmundsen's combination of a major position at the Wenner-Gren Foundation with a "normal" heterosexual, white, middle-class woman's "personal" life mattered to Zihlman.[3] Osmundsen remained a supporter and friend. As she was for many of the students, Alice Davis, the administrative assistant at UCB, was part of Zihlman's support network in graduate school. Zihlman took more or less for granted that about half of Washburn's students were women; and despite sexist remarks from some men about "Washburn's harem," she believed men and women were treated similarly in this program.

At Santa Cruz, Zihlman began to reconstruct her views of gender and science in specifically feminist terms. That meant not so much finding sexist inequalities in one's personal history,[4] but learning a new way for a physical anthropologist to view the human way of life and its species history. That view should affect how both men and women in biosocial anthropology practice their science. In 1970, Zihlman heard Sally Linton's paper, "Woman the Gatherer: Male Bias in Anthropology" at the American Anthropology Association meetings in San Diego. Many of the elements Zihlman would develop in the next few years were sketched in that "oogonial" paper (Zihlman 1985b: 11). Empowered by the women's movement, Linton asked *scientific* questions and criticized *bias* in her science that had resulted from producing knowledge almost exclusively from men's points of view. She gave a *cultural* genealogy to what had seemed scientific, and suggested a dynamic scientific centrality for what had seemed unimportant, unchanging, or just absent. Linton's particular object of criticism was the Washburn and Lancaster (1968) formulation of the man the hunter hypothesis and its linked view of the human way of life, but the implications of her argument were much broader than one evolutionary narrative. Linton focused on the skills and activities of females in the hominoid-hominid transition. She made the crucial logical point that has been lost in subsequent discussions of "woman the gatherer": "hunting cannot explain its own origin" (Slocum 1975: 43). Linton argued for attending to the extension of the already strong mother-infant bond in primate societies to females' increasing the scope of their foraging for infants who remained dependent for longer periods. Linton insisted on the crucial importance of the Darwinian notion of female sexual choice years before that point was stressed by sociobiologists. She argued that men's sexual control of women must be seen as a modern *institution*, not a natural fact rooted in our animal past. Variable length consort-like associations among early hominids seemed much more likely to Linton than permanent pair bonds; such a mating pattern was more or less continuous with those of close human relatives and not unlike those of many living humans. Matrifocal groups composed of both sexes and "cultural inventions" for food gathering and sharing and child care seemed the most likely patterns before large animal hunting could have become possible later.

Linton understood the ideological and mythological core of the debate about the "first tools"; so she shifted the linguistic terrain from those originary tool-weapons that are so potent in masculinist, western technologically preoccupied cultures. Shifting from "tools" and "weapons" to "cultural inventions" led to emphasis on "two of the earliest and most important cultural inventions[:] containers to hold the products of gathering and some sort of sling or net to carry babies" (1975: 46). Not ignoring male competition for females, Linton did not grant it species-making power. "It could easily have been handled in the usual way for primates—according to male status relationships already worked out—and need not be pictured as particularly violent or extreme" (1975: 47).

Placing more weight on the process of asking questions than on any particular answers, which must remain open to construction of evidence and other normal scientific constraints on validity, Linton stressed that her whole argument came from a simple maneuver—*asking* what females were doing. She did not think she asked because she was a female/woman; American anthropologist women did not know how to ask such things any more than men did, until they consciously learned to see differently in the course of a major historical social movement. In the late 1980s, asking what females do might seem inescapable for an evolutionary scientist. If it is inescapable now, part of common sense, at least a major part of the reason resides not in the female sex of people like Linton, but in their coming to a specific consciousness about the constitutive relations of science and gender, as well as science and other positional markers.

"But political consciousness, whether among women, blacks, American Indians, or any other group, leads to re-examination and re-evaluation of taken-for-granted assumptions . . . The male bias in anthropology that I have illustrated here is just as real as the white bias, the middle-class bias, and the academic bias that exist in the discipline" (1975: 49–50). The difficult achievement of political consciousness opened the possibility, but not inevitability or automatic production, of a more adequate science in terms that must be negotiated by the practitioners of the craft. The political consciousness was enabling; it could not substitute for the material, social, semiotic practice of the science. This is a critical point; to stress a political condition of possibility is not to reduce science to another set of social practices. In particular, I wish to argue against the view that Zihlman's development of the woman-the-gatherer model can be *explained* by politics and not science. The relation is more subtle and not caught by simplistic doctrines of cause or ideological category oppositions. Precisely by mis-stating the relations among science, political consciousness, and gender, critics can avoid coming to terms with the scientific issues, including those related to narrative, raised in the work of Zihlman and others.

Also in 1970, Zihlman began teaching a course with her Santa Cruz colleague in cultural anthropology, Nancy Tanner, called "Biological and Cultural Bases of Sex Role Differentiation." These courses were being developed by feminist scientists all over the U.S. in the early 1970s, often as collaborations between biological and social scientists. For Zihlman and Tanner, the course was the beginning of an initially productive, but fatally troubled collaboration that resulted in several publications. These writings have been identified without much differentiation as the core arguments for the woman-the-gatherer reconstruction of early hominid evolution.[5] The working relationship was generative for both women from 1970 until into 1975, and it ended finally by 1977. Their ideas were elaborated in several conference papers in the early 1970s. In 1975, they had a draft of five chapters of a projected joint book. Zihlman withdrew from the project, and Tanner expanded the draft independently into the ten chapters and conclusion published as her important book, *On Becoming Human* (1981).[6] But after the mid-1970s, the kinds of arguments each made were different. Without trying to reconstruct the history of the collaboration and divergence, I will concentrate on the pattern of Zihlman's scientific narratives through the mid-1980s, locating their mutations in the social history of physical anthropology, primate studies, and feminism. My focus is motivated by her social location as a practicing physical anthropologist and her particular accountability to that discursive community. Tanner's constituencies are in many ways quite different from Zihlman's. Because of major delays in volumes resulting from conferences, the dates of the various publications are poor guides to the intellectual and social history of the arguments. This has been a factor in the tangled reception of Zihlman's approaches by younger primate field workers trained in sociobiology and socioecology and by the fraternity in paleoanthropology.

Zihlman's arguments have been overwhelmingly about *early* hominid evolution. The gathering hypothesis was an attempt to revisualize what was new in the adaptive transition from hominoids to hominids, not an attempt to reify the various (and branching rather than linear) stable points over a few million years in hominid evolution into one ideal species type and one sex/gender system. Like other narratives of earliest transitions, however, centering females'/women's extended foraging, sharing, and sexual and reproductive self-determinations constituted a reconstruction of the potent, foundational, originary moments of what it means to be human. One lesson was that interactions with the young of the species and the behavior of the young in general turn out to be more fundamental to species-making stories than determinations of fatherhood.[7] In physical anthropology/paleoanthropology, claiming and naming the oldest fossils, the "first" hominid ancestors, and narrating a compelling account of the earliest specifically "human" adaptations confers, ironically, a "paternal" authority in the story field where science and myth intersect.

Linton's paper piqued Zihlman's acceptance of Washburn's positioning of hunting at the earliest transition from ape to hominid. But what Zihlman shared with Washburn turned out to enable her to do what Linton had not—develop a detailed argument for privileging chimpanzees as models for several key transitional features in early hominid evolution.

Partly because of his long advocacy of a very close African ape-human relationship, Washburn was one of the first to accept the controversial molecular data in the late 1960s, which was interpreted to indicate a very recent human-chimpanzee-gorilla divergence, about five million years ago, rather than the approximately twenty million years often proposed from comparative anatomical and geological work on fossils.[8] The molecular data stimulated Washburn's reactivation in the late 1960s of a much-disputed view of bipedal hominids as made-over knuckle walkers. Zihlman had been friends with Sarich in graduate school. They had taken anatomy together and argued frequently about developments in human evolution. Her dissertation had relied on a chimpanzee model for discussing the evolution of bipedalism and for interpreting the particular anatomical stigmata of the hip joint in chimps, australopithecines, and modern humans. Sarich's arguments that his protein-based phylogeny justified a radical molecular clock were quickly incorporated into Zihlman's justifications for continuing to emphasize chimpanzees in a changed evolutionary narrative. From the data of more recent molecular phylogeny, chimpanzees and humans appear even more closely related than gorillas and chimpanzees. For Zihlman, because of the extremely close genetic relationship implied in the protein and later nucleic acid work, both from nucleotide sequencing and DNA hybridization data, the differences among "us," chimpanzees and humans, must be subtle, with large multiplier effects from behavioral and developmental shifts. If the differences are subtle at the level of code, they are large at the level of semiosis.

This narrative does not lend itself easily to master molecule fantasies about gene programs and genetic determinisms. It does privilege continuities and patterns of dynamic relationships. That makes attention to ontogenetic development in general and to the young members of the species scientifically crucial for a good evolutionary reconstruction (Borchert 1985). If African apes and humans had a quite recent common ancestor, then human origins should be interpreted in terms of a radiation that included the African apes, especially the chimpanzees. That is not to say that either of the two species of modern chimpanzees is the unchanged common ancestor. But, looking for shared or transformed adaptational complexes and ways to move from one pattern to another, developmentally and phylogentically, as populations entered different ecological zones gains power as a scientific narrative strategy. The challenge is to avoid typological thinking while still privileging a strong referential model, rather than the more abstract but very powerful "strategic modeling" advocated by sociobiologists (Tooby and DeVore 1987).

The comparative anatomical and paleontological reasons for emphasizing chimpanzees were complemented by the studies from the 1960s and after of living common chimpanzee social behavior by Jane Goodall and others at Gombe and by Itani Junichiro, Nishida Toshisada, and others in the Mahale Mountains of Tanzania. Jane Goodall narrated her films of chimpanzee behavior at Berkeley in 1964. Chimpanzees emerged from her studies as tool users and makers, skills which required long social learning. Reports came from Gombe about chimpanzees killing and eating small animals; neither bipedalism nor savannah living were required for the initiation of modest hunting and meat-eating. Chimp social lives were stunningly flexible and complex, and chimpanzees seemed to require scientists' attention to individual actors as well as to social structures and biological and ecological parameters. Goodall's long-term studies of mother-infant relationships indicated a complexity and extent of maternal investment that would have been expanded even further in hominid evolution. The Japanese work on unit-group structure, female sexual initiation of copulations, matrifocal social dynamics, and chimp tool use was fundamental to Zihlman's working out her own way of modeling. Flexibility, complexity, individuality, and the patterning of a way of life that suggested how transitions to hominid modes of life could have happened in the move into a new opportunity space or adaptive zone on the savannah were the story elements that made the chimp model irresistible to Zihlman.[9]

In the new ecological zone, a dietary innovation, specifically the making and use of tools by both sexes for gathering calorie-rich plants like large tubers or nuts, could have made the crucial, species-making difference. The shift was not from plants to meat, but from fruits to tubers.[10] Zihlman argued that meat-eating was probably relatively opportunistic and increasing somewhat in frequency and importance during the period that hominids developed large home ranges, extensive sharing, and considerable communicative and technical skill. But meat-eating was not likely to be at the basis of any of those fundamental developments. The continuing mobility of primate/hominid females, as well as males, was central to her narrative.[11] A sharp sexual division of economic labor was not part of the argument; rather both sexes would have become more adept generalist foragers, with at most modest specializations around plant or animal foods. Sharing was not based on a sexual division of labor, with all the attendant tones of the origin of "the family" (Zihlman 1983a). However, the matrifocal structure of social life coupled with the very long period of infant dependency and learning would have made females' gathering and sharing particularly critical to the transition.[12] In the context of a resurgent western feminism that had emphasized versions of maternal thinking and social motherhood since the nineteenth century, the matrifocal group had major symbolic resonances. Females would have chosen to mate with socially skillful males, who were perhaps more willing to share and who could interact non-disruptively. Males would have been ill-advised to compete for mates by agonistic aggression when better social skills could get them further (Zihlman 1976: 27). Mates would not have been the most stable or long-term adult male companions in the social groups; rather, maternal male kin whose ties to mothers and siblings drew them into interaction would have had a larger role, including child care.

The chimpanzee model was complemented by three other elements coming together in the early 1970s, in the context of interpretation of the women's movement from middle-class, white, professional viewpoints within it: female mate choice, "loss" of estrus in hominid females, and pictures of "the human way of life" from studies of surviving human gatherer-hunters. In the late 1960s at nearby Stanford University, Zihlman studied Hugo van Lawick's Gombe chimpanzee film workprints as part of her investigation of chimpanzee locomotor behavior. Working on the film for sequences of feeding and tool use, Randall Morgen offered to show Zihlman "X-rated" footage that would never find its way into a National Geographic television special. Zihlman recalled being struck by the suggestions of female sexual choice in the explicit, close up sequences.[13]Donald Lindburg noted female sexual choice early, and he was heard by his peers from Berkeley.[14] Jane Lancaster (1973, 1979), another colleague from graduate school, also concentrated on female sexual activity, initiations, choice, and self-determination. These were crucial themes in liberal biological, feminist, and political theoretic discourses that intersected in an historical moment in which reproductive politics were extremely controversial.The word "choice" resonated inescapably with the tones of struggles over reproductive "rights." Who has property in the self? It was not simply the personal that was political from the point of view of the women's liberation movement; it was also the biological. Lancaster's theme was the evolution of personal choice. The underlying debate was about what counts as an individual and a citizen in the natural and social body politic.

Zihlman and Lancaster maintained a correspondence throughout the 1970s, building from their feminism in their work in order to write and teach what they insisted were more adequate scientific accounts. Lancaster and Zihlman worked out how to argue that hominid "loss" of estrus was not about sexual "receptivity," constant or otherwise, but it was about enlarged sexual choice. In their accounts, sex was not a source of chaos to be regulated by male action, pair bonding, or culture (mind over body), but rather an indispensable dimension of the western rights-bearing, possessive individual. They did not self-reflexively use the language of liberal theory; rather its logic and metaphors pervaded the narrative, setting up what mattered, what needed explanation, and what could never be acceptable. The discourses of maternalism and of possessive individualism in the woman-the-gatherer models remained in tension with each other in producing a gendered social subject, woman, within evolutionary narratives about females in transition.

The profoundly western aspect of this discourse in its construction of a putatively universal, originary subject cannot be missed. Like most writers within their sister Euro-American feminist discourses of the 1970s, feminist biosocial scientists did not perceive the putative universal salience of the fundamental categories—sex, labor, woman, gender, reproduction, biology, culture, nature—to be problematic.

That tension between maternalism and possessive individualism took a particularly interesting form in Zihlman's early enthusiasm for Robert Trivers's sociobiological theories of parental investment and kin selection. Like the women in primate studies who have been closely identified with sociobiology, especially Sarah Blaffer Hrdy, Zihlman initially adopted sociobiological arguments as soon as they were formulated. They played a significant role in her first versions of the gathering hypothesis. "Trivers' framework is a potentially valuable tool for exploring the implications of high maternal investment (and thus of female choice) to the origin and continuing evolution of the hominid line" (Zihlman 1974: 477). This aspect of the gathering hypothesis has usually been elided in criticisms of it by people who might call themselves sociobiologists, which Zihlman never did. Later, partly in response to the work of her graduate student, Catherine Borchert, Zihlman greatly de-emphasized, but did not eliminate, various components of inclusive fitness arguments in her accounts of the hominoid-hominid transition, in order to foreground cascading effects of developmental processes throughout a population and to demote genetic explanations. But in the early formulations with Nancy Tanner, Zihlman combined the narrative of maternal thinking with that of possessive individualism in the metaphors of reproductive investment and reproductive strategy, in which females as mothers became paradigmatic individuals, rather than resources or plot space for another narrative of subject formation.

Finally, Richard Lee's and his students' and associates' accounts of the Kalahari !Kung people provided the link for the arguments about chimpanzee models and the evolution of sex and gender to normative, indeed species-defining, humans. In the chapter on Sherwood Washburn's program, we saw how the San people came to have a particularly privileged symbolic place in representations of what it meant to be human. In that continuing symbolic context in the 1970s, the reports from the field about women's pivotal economic activities lent authority to emphasizing gathering and sharing in making us human.

Here were the elements of the "woman-the-gatherer" model. Washburn objected that his hunting model had always assumed that gathering and mother-infant relationships had been important; they did not change. The hunting hypothesis was simply intended to emphasize what was *new*. Something new was required to account for change; the logic informed a kind of paternal creation myth. The principle of motion is what was at issue in a highly typical western origin story. But Washburn's account is not convincing. First, placing extensive hunting and a sharp sexual division of economic labor at the origin of human ways of life far transcended the archaeological evidence for hunting tools, indeed by a few million years. But more fundamental was the narrative logic of origins. Zihlman argued that what was *new* was, precisely, extended generalist gathering by both sexes, in which the productive, exchange, and reproductive activities of the females would dynamically propel, but not uniquely cause, a species-making transition. Extended duration of mother-young relations, improved female sexual self-determination, greater complexity of matrifocal groups requiring greater social skill in both cooperation and competition from both sexes, and cultural innovations (tools, social arrangements, and other technologies) enabling these extensions were seen as originary. Zihlman agreed that gathering was a precondition of hunting, but not as an unchanging "matrix" for the generative principle of change. In Zihlman's story logic, both gathering and hunting emerged as repatternings, not opposites, in changed conditions of constraint and opportunity. Narratives of both gathering and hunting ways of life produced genders and citizens.

The narrative difference between Zihlman's and Washburn's accounts turned on how to imagine originary principles—as extensions, complexifications, and repatternings or as seminal generation and heroic action stimulating an otherwise conservative mass. The narrative logics of the "woman the gatherer" and "man the hunter" hypotheses were not symmetrical. The stories were not opposites; they involved different doctrines of cause and origin, not simply different central actors or bits of evidence. Arguments about mating patterns and technology in this context were about more than their manifest content.[15] The presence of the heterosexual nuclear family and the tool-weapon equation in masculinist scientific narratives was about a theory of social reproduction called paternity, the principle of motion in Aristotelian accounts of life.

Mutating Universals: Mosaics of Difference

Zihlman and Tanner's original woman-the-gatherer model was rooted heavily in studies of *Pan troglodytes*, the "common" chimpanzee species found in the Mahale Mountains and the Gombe National Park. Beginning in the mid-1970s, Zihlman began to be more interested in the second chimp species, *Pan paniscus*, the "pygmy" chimpanzee, which was less well known both socially and anatomically.[16] In 1973 Zihlman's friend, anthropologist and anatomical illustrator Douglas Cramer, suggested she look at the extensive Belgian collection of *paniscus* skeletons. Cramer and Zihlman studied this collection in the summer of 1973, and Zihlman has published steadily on the species since then.[17]

Zihlman's attention was caught by the inability to sex a skeleton with the same ease that was possible for common chimps. Further examination of both skeletons and living animals showed them to be generally more bipedal, as well as less sexually dimorphic. Zihlman observed a living pygmy chimpanzee in Susan Savage-Rumbaugh's language research laboratory at the Yerkes Primate Research Center, where studies of communication were underway with the young *paniscus* child, Kanzie, and his adoptive mother Matata, and two common chimps, Sherman and Austin. Field studies reported an array of social behavioral differences, such as more female-female grooming, face to face sexual copulation, and more sexual foreplay, that added intriguing fragments to a case Zihlman was trying to make for the pygmy chimp as the best model among *contemporary* hominoids, a "living link" for imagining the transition from ape to early hominid functional forms and possible behaviors. The pygmy chimp was not constructed as the *ideal* model or as a living *ancestor*, but a particularly suggestive mosaic or mandala for a scientific modeling practice that incorporated as many kinds of data as possible.

The theme of sexual dimorphism emerged centrally in the pygmy chimpanzee research, but Zihlman's interest in the topic dated to the first years after graduate school in the late 1960s, when Loring Brace and M.H. Wolpoff attempted unsuccess-

fully to convince physical anthropologists that much of the variation in australopithecines attributed to species differences was due to sexual dimorphism in a single
species.[18] The implications of such a claim bear directly on modeling constraints for
social behavior, sexual divisions of labor, and mating systems. In particular, the
claims about dimorphism often implicate assertions about the non-innocent category
called "monogamy," the related hoary problem of the origin of "the family," and
explanations of male power and female subordination. Therefore, it is hardly
surprising that several writers from divergent explanatory traditions who work in
the dangerous and productive narrative field structured by their scientific craft and
by feminism have taken a sharp interest in the matter.[19]

Basic to judging intraspecific versus interspecific variation and to reconstructing
selective pressures operative in speciation, sexual dimorphism raises major biological issues across the class of mammals. Darwin's (1871) concept of sexual selection
emerged from his effort to explain sexual dimorphism. Fedigan resists reduction
of questions of sexual dimorphism to sexual selection, and lists the factors that must
minimally be considered in a biologically sound discussion of the matter:

> taxonomic or phylogenetic affiliation; overall body size of the species; group size;
> group composition; breeding system or socionomic sex ratio; intensity of male-
> male competition; aspects of species niche, such as diet and amount of time spent
> on the ground; specialized male predator-protection role; bioenergetic costs of
> pregnancy and lactation; bioenergetic principles and relationships in smaller and
> larger bodies; and, finally, separation of feeding niches between the sexes. (1982:
> 61).

She cautions that sex differences in behavior cannot be deduced from degrees of
sexual dimorphism, despite the common practice of doing so. A rough generalization among primates indicates that very dimorphic species tend to be polygynous
and non-dimorphic species are frequently monogamous. But variability among
primate species makes what seem like analogous patterns granulate into fragments,
whose meaning can only be recovered from an analysis of the whole range of life
history parameters, ecological variables, and evolutionary relationships. Surveying
the class Mammalia, the mammologist Katherine Ralls concluded that it is not
possible to predict sex roles from either social system or degree of sexual dimorphism (1977: 921). Such predictions often rest on restricting attention to the cases
that fit, ignoring the species with similar dimorphic patterns but different behavior.

Discussions of hominid evolution have rarely been notable for inclusion of the
needed complexity in treatment of sexual dimorphism, despite—or probably because of—the particularly heavy symbolic load these particular writings bear. Zihlman's particular contribution to the debates centers around her major emphasis:
even considering "only" the flesh and bones, sexual dimorphism is not one trait; it
is a mosaic of potentially independently varying features in many parts of the body.[20]
For example, it is not really meaningful to ask if common chimpanzees, pygmy
chimpanzees, or modern humans are more or less sexually dimorphic in relation to
each other. Their patterns of dimorphism are distinct, so what is being compared
has to be more carefully specified. Pygmy chimps, humans, and common chimps
all have a similar female to male body weight ratio (females weigh about 80–85%
of what males weigh). But unless the weight ratios of particular tissues are also

measured, the gross ratio could be quite misleading. For example, fat and muscle mass have different functional meanings that might be particularly relevant to reproductive and foraging pattern questions, where debates about sexual dimorphism might be especially acerbic. Most of the sexual dimorphism in weight in humans is due to such body composition differences. Neither body weight nor sexual dimorphism is an "entity" waiting to be measured. Pygmies show no sexual dimorphism in length of limb bones or joint sizes, while common chimps are dimorphic for these measures. What might that mean functionally? Common chimps have high canine teeth sexual dimorphism, pygmies a slight dimorphism, and humans essentially none. Is this difference functionally significant? Without correcting for body size, pygmy chimps do not show cranial capacity sexual dimorphism; common chimps do; and humans are somewhere in between. Should one conclude that female pygmy chimps are relatively more intelligent than the males of their species compared to humans and common chimps? The sorry history of efforts to correlate cranial capacity with mental function within and between human populations would make such a move at the least imprudent.

Zihlman stressed that sexual dimorphism is hard to measure, even in living organisms in sizable populations. In always fragmentary fossil collections, where one rarely has complete individuals or large numbers of individuals, the measures are particularly difficult and often contentious. There is no single pattern of sexual dimorphism among the apes. *Australopithecus*, *Homo*, and *Pan* patterns are distinct. There probably was not a single pattern among the early hominids, and there is excellent reason not to assume a unidirectional development of patterns toward some single endpoint. Whatever the bony and fleshy pattern of dimorphism of ancestral males and females, their behavior would follow no simple curve of correlation. Finally, discussions of sexual difference in paleoanthropology must be contextualized within the extra-ordinary "incitement to discourse" on sex that has characterized the last two hundred years of life sciences. There is no region of the body, living or dead, soft tissue or bony, that has not been interrogated for the secrets of sex and sexual difference. The construction of organic sexual difference has been a major discursive production.[21]

Treating sexual dimorphism as a mosaic of features, whose parts do not follow a single order in relation to each other through time, rather than as a single thing with a linear history, and insisting on interpreting this protean mosaic in a multivariate pattern field, rather than in relation to a stable governor, constitute typical deconstructive and reconstructive interventions by the women whose work I have been examining in "The Politics of Being Female." In dispersing single meanings and subverting stable narratives of sex, they argue for a better science at the level of bones, genes, statistics, or foraging patterns; and simultaneously they open degrees of freedom in their culture's constructions of gender. Again and again, they problematize what sex and sexual difference could possibly be at the level of the potent organic stigmata which have been imagined to ground the whole amazing ediface of sex and gender. Their moves are disarmingly simple, rooted firmly in the details of their material and the debates germane to their local regions of scientific discourse. That is, they practice their science, and thereby are part of a broad cultural reconstruction of what may count as nature. An emergent effect from the process of proliferating and problematizing the variables structured into the semiotic field in which sexual difference is represented in the animal body is the disaggregation of sexual difference itself. Without disappearing, indeed while becoming ever more deeply theorized, ever more deeply embodied, sexual difference becomes less and less dichotomous, less and less able to police the body of nature and appear as the ground for the elaboration of dichotomous sexual difference as gender in the space of representation called culture. As it becomes clear that gender cannot rest on sexual difference, its foundations in power and domination become manifest.

All of these matters came to a sharp point in interpretations of the hominid fossil finds at Hadar, Ethiopia, and Laetoli, Tanzania, which led to the emergence of Lucy and the First Family in the contested narrative of human evolution. We have seen how Owen Lovejoy's interpretation of Lucy brought the old claims about pair bonds, home base, sexual division of labor, and male initiative in hominization back to center stage. Despite the highly unlikely combination of both monogamy and extreme sexual dimorphism in any primate, Lovejoy has tied his scenario to Donald Johanson's and Timothy White's assertions that the Laetoli and Hadar fossils represent a single species—and the oldest and therefore ancestral species—with very great sexual dimorphism. The questions of sexual dimorphism in this case are perhaps most profoundly questions about scientific paternity: the power to be the second Adam, giving true names after the model of the creator. Most comments on these fossils have granted two things: that Lucy and the other material represent a new species at the root of the hominid divergence, *Australopithecus afarensis*, and that the argument for large sexual dimorphism, rather than for two species, is well founded. Working out of her consistent preference for scenarios of hominid evolution with little acute sexual difference, or with sexual difference dispersed into non-isomorphic patterns of relationshps in time and space, Zihlman objected to easy acceptance of the Johanson and White account. Her reasons round out this sketch of the relationship of preferred narratives and scientific argument in Zihlman's version of the paleoanthropology of sex and gender.[22]

First, Zihlman attempts to keep open—or re-open in the face of colleagues' seemingly compliant acceptance of conclusions she regards as poorly founded—the question of the ancestral status of these fossils. Casting doubt on Johanson's and White's identification of primitive features in *afarensis* material, she emphasizes overlap of measurements between Ethiopian and South African *Australopithecus* skulls and pelves. The dates of material from the Hadar Formation might also overlap with those for the southern fossils. How can the ancestors be contemporary with their descendents? Are the Ethiopian fossils signs of a separate species, an original way of life? Second, damping down the newsworthiness, she relativizes Johanson's claim for a single, highly sexually dimorphic species for the variable new fossils by bringing up the earlier and failed attempt by Brace and Wolpoff in the late 1960s and early 1970s. Third, she stresses the generally undecidable nature of the controversy in the face of the still unpublished details of the fossils; independent researchers remain unacceptably dependent on assertions by those who controlled the fossils. "No measurements and little comparative data for the post-cranial fossils have been given which would enable other researchers to assess the degree of dimorphism." Fourth, Zihlman reconstructs facts into dubious opinions. "The interpretation of extreme sexual dimorphism for these fossils has been a mere assertion from the beginning ... and has continued to be so" (1985d: 214). And finally, Zihlman underlines the implications of recognizing that sexual dimorphism is a mosaic of features and not a unidimensional trait. Other australopithecine material, limited as it is for in-depth analysis of sexual dimorphism, suggests both a unique pattern for each species and moderate sex differences.

Should the discipline accept a sharp sexual dimorphism for a putative stem species on the basis of inadequately published fossils, when the pattern would be so odd not only among hominids, but even among hominoids? The sexual dimorphism among *afarensis* individuals would exceed even the degree of difference in the distantly related orangutans, where males may be as much as three times as heavy as females (1985d: 216). The orangutan pattern is pregnant with consequences for the social life of these largely solitary apes, whose forest ecosystem cannot support group foraging for such a large-bodied animal. The orangutan is also perhaps the only primate species in which males really can and do force copulation with unwilling females, arguably justifying in this anomalous instance use of the word *rape* in the biological literature. What examination is now possible of the Hadar and Laetoli fossils indicates perhaps two separate morphological patterns in both cranial and post-cranial anatomy. That is, in addition to size differences within and between sites, two distinct anatomical patterns of upper limb material and distal femur appear to exist. Also, measures of venous sinuses from different localities seem to indicate two different patterns of dimorphism. Two different patterns of sexual dimorphism would be difficult to accommodate in a single species.

The whole issue is hard to address when papers do not give the raw data, and when standards of publication in the field allow conclusions based on one or two kinds of measures, rather than on at least three or four whose meanings would be elucidated within a mosaic conception of sexual dimorphism. Can the species question be settled so that the sexual dimorphism question could be addressed against a stable taxonomic background? Zihlman argues that the Hadar fossil groups *might* turn out to represent one species with very high sexual dimorphism. Or *afarensis* might disappear into prior taxa of early hominids or be dispersed into as yet unnamed species. Zihlman's writing is a classic example of an effort to destabilize a narrative by making its easy acceptance seem odd, irrational, or biased, rather than nearly self-evident on the Word of the Father. Settled facts become distinctly unsettled disputes.

But who is listening? And more broadly, who has listened to Zihlman's scope of arguments about woman the gatherer; about the salience of chimpanzees, especially pygmy chimps, as models for many critical features in hominid evolution; and about sexual dimorphism? Are Zihlman's arguments part of the fray for what will count as better evolutionary science, or are they cultural artifacts of post–1970 U.S. feminism? Can they be both? For whom? Tooby and DeVore (1987) regarded Zihlman as almost certainly wrong, but they did not discuss her work in detail and, from Fedigan's point of view, could reasonably be suspected of dismissive misreading. Or is this a case of complementary biases? Reviews of Zihlman's contribution to *The Pygmy Chimpanzee: Evolutionary Biology and Behavior* quickly dismissed her arguments, even when the reviewer accepted similar reasoning and conclusions from other authors in the same book.[23] Why? Zihlman has given short shrift to potentially important topics, such as female-female competition, male coalitional behavior, or male competition for mates. Her insistence on principles of functional comparative anatomy and intermediate levels of causation, both emphases retained from Washburn's approaches, have set her off from forms of argument that Tooby and DeVore summarized as "strategic modeling." These traits have perhaps led sociobiologically inclined authors, including overtly feminist writers, not to consider seriously many of Zihlman's arguments, including her early use of parental investment principles. I have seen manuscripts in draft from sociobiological writers sympathetic to feminism as one source for their own deconstructive moves in science who nonetheless misread Zihlman to be saying much the same thing as Lovejoy about pair bonding. How could that particular error in reading occur? Does the criticism by sociobiologically and socioecologically inclined primatologists of just about everything and everyone trained at U.C. Berkeley under the influence of the Washburn and Dolhinow tradition in the 1960s and early 1970s make it impossible to read accurately—on both sides of the controversy—in spite of many shared discourses, motivations, and commitments within and across gender?[24]

Zihlman herself attempted to come to terms with "what happened to woman the gatherer?" In addition to writings that have engaged this hypothesis in serious scientific debate and criticism, she cited three suspicious types of reception for the gathering model of early hominization: it has been ignored, dismissed without serious refutation, or co-opted. Examples of cooptation have included 1) simply adding gathering to hunting to get a mixed economy and so still come out with a need for pair bonding, early home bases, and sharp sexual division of labor; and 2) accepting gathering with tools as a major adaptive innovation, but assigning it to males (it now makes them men) and so still coming out with all the ingredients for an Oedipal family and an Oedipal evolutionary narrative.[25] The cooptations consistently "forgot" that the gathering hypothesis was about four things: 1) female mobility; 2) social flexibility for all members of the species; 3) the transformative power of activities of the immature members of the species and of mother-young relationships; and 4) the deconstruction of staples in the narratives of both compulsory heterosexuality and of the dialectic of technological determinism, masculinism, and war: e.g., male-female sexual bonding, male-male agonism, home bases as hearths for nuclear families, and the trope of the tool-weapon.

Washburn minimized the novelty and importance of Zihlman's and Tanner's arguments about gathering with tools and about the causal dimension of female activity in human evolution by claiming that the man-the-hunter hypothesis had simply always assumed gathering was important and that his approach set off what was new and innovative. Alternatively, he criticized the gathering hypothesis for being as biased as the hunting scenario, as if no scientific arguments allowed reasonable choice within the discursive communities that are socially (including technically) empowered to determine what may count as a better account of human evolution.[26] Despite Zihlman's explicit emphasis on the gathering hypothesis as a model for one adaptive complex of a rich temporal hominid evolutionary fabric, critics have read it as a fixed picture for all of human life, as a model to exclude males from human evolution, or as a dichotomous opposite to competing narratives. In short, the scenario has been read as a series of exclusions, fixations, and oppositions; and this reading has been enlisted in subjecting the model to a pattern of exclusion and caricature.

Zihlman suspected that the fate of "woman the gatherer" might be tied all too closely to that of "woman the scientist" (1985b: 15). Both are tropes—figures for sexual difference in large social texts. She phrased the problem for both as one of "citizenship"—in the body politic of nature and of culture. The picture of females and of women delineated in her writing can become an allegory for the relations of the marked sex and gender to the unmarked sex and gender of science. In a sense, science is itself a kind of gender. For the realms of both nature and culture, "science" is the key authorizing activity, the chief sign of rationality and order. Sex is categorically opposed to order; it is what must be ordered; woman remains the sex; woman the scientist becomes the trope figuring bias; man is simply scientist; his gender is unmarked, unremarkable, not a problem, resting easily within the genre (gender) of science. His gender does not seem in danger of becoming the semiotic other to science, namely, a politics.

Keller (1985) has made clear how difficult it has been for the gender woman to be associated with doing science, the exemplar of rational activity, i.e., of being human, for modern westerners. In that context, Zihlman's science cannot be allowed to cohabit with her feminism, which has turned an already marked gender into politics, which is quintessentially the marked "other" to unmarked science. Either feminism or science must be evicted. Where what is perceivable as "general" is still unmarked masculinism, feminism cannot be seen as a realm of reason, only of opinion. If feminism and the marked gender come too close to the preserve of science, the science is polluted, reduced to bias, transmuted from one genre to another. The narratives of hominization are inextricably entwined with the narratives of citizenship, rationality, and gender. One way to account for the odd exclusions, mis-readings, and fixations in too much of the response to the figure of woman the gatherer and to Zihlman's writings is to suggest that she is perceived to have gone too far. Her scientific practice insists too much on what cannot be culturally accommodated under the sign of reason; the reader trips on this disruptive generic retextualization of nature and mis-reads the elementary grammar of sex and gender in her writing. Female, woman, and scientist are finally conflated by elementary errors of reading.

But Zihlman's ideas have also been treated as normal scientific artifacts grounding the ordinary activities of holding conferences, rewriting textbooks, establishing collaborations, and informing ongoing research.[27] What is normal in one discursive community is not normal in another. As a generalization with several exceptions, her writing has had considerably more impact in the primatological and anthropological work of feminist and/or women colleagues than in the fraternity of paleoanthropology. This pattern characterizes a division within academic discourse broadly by the 1980s in the United States, where feminist critical studies have flourished institutionally and theoretically in the academy. In numbers, sophistication, and even material resources, U.S. feminist scholars in most disciplines can and do lead odd double professional lives, partly enmeshed in the "general" (i.e., still male-dominated and male-defined discourse) and partly enmeshed in a very heterogeneous and self-reproducing feminist academic discourse. Even in disciplines where feminist theory seems to have begun fundamentally to restructure the whole discourse, as perhaps in film theory, literature, and cultural anthropology, patterns of gender segregation persist in authority, sites of publication, citation networks, and other apparati of disciplinary production. The segregation says more about social technologies for reproducing *and* reconstructing gender in the everyday processes of determining what may count as knowledge than it does about the merits of work from some impossible objective point of view, scientific or otherwise. Zihlman has benefited from, been hurt by, contributed to, and resisted these structures of production of U.S. academic discourse. This is an ordinary fact of life at this historical moment for people in the institution of the academy in the United States, marked as white, middle class, female, woman, and scientist.

A good example in Zihlman's practice of benefiting from, being marked dangerously by, contributing to, and resisting the relations of gender and science was the day-long symposium on Issues and Controversy in Primate Evolution which she co-organized with Dean Falk at the 1985 Knoxville meetings of the American Association of Physical Anthropologists. With a touch of irony, their public goal was simply to bring in the best people on each topic; the symposium was not explicitly labeled as either "women's" or "feminist." But the subtext was unmistakable; all of the principal speakers were women, representing several explanatory streams and sides of issues.[28] It was not possible for the Knoxville session to go unmarked; it was in fact a kind of scientific celebration of differences among women, as well as of shared concerns and recognized authoritative presence in their sciences. This was the AAPA meeting at which Zihlman delivered the invited keynote address, titled "Sex, Sexes, and Sexism in Human Origins" (Zihlman 1985b).

If Zihlman's scientific narratives have run the risk of being perceived by disciplinary colleagues to be marked indelibly and fatally by her gender and politics, woman the gatherer also runs a perhaps greater danger from feminist points of view. She is in the universalizing humanist lineage of western origin stories, which have been so much a part of colonial discourse. In woman the gatherer, aspects of colonial discourse and Euro-American feminist theory intersect in scientific narrative in knots that can be only partly untied at this historical moment. While continuing to write a universal woman into existence at the origin of the human way of life, Zihlman's own practice has simultaneously been part of the attempt to untie the troubling connections between colonial discourse and U.S. white feminist theory and to problematize the construction of woman by constructing her as a site of internal differences and by deconstructing the boundaries with which sexual difference itself is discursively crafted. Zihlman's illustrations in the *Human Evolution Coloring Book* and her deconstructive approach to sexual dimorphism may be read as subversions of the universalizing tendencies of woman the gatherer. But her 1980s versions of woman the gatherer also directly address the problem of sameness and difference within the gender woman in a context that brings us back to the historical obligations and possibilities signaled by the United Nations Decade for Women.

The direct address recontextualizes the woman-the-gatherer hypothesis and its relation to man-the-hunter narratives in an international, explicitly feminist, anti-colonial frame. In 1987 Zihlman was invited to take part in an American Anthropological Association Project on Gender and the Curriculum. There are significant parallels with Washburn's pedagogic efforts in the period of physical anthropology's reconstructing its object of knowledge called race in the 1950s and 1960s after the UNESCO statements. The project organizers asked Zihlman to write a curriculum guide on the evolutionary history of women, which she provisionally titled "Sex and Gender Two Million B.C."

Zihlman's effort is embedded in a multi-part guide intended for distribution to every anthropology department in the U.S. Each module includes a short orienting essay, suggested course components, an annotated bibliography, and a resource list for classroom use. These are the tools for reconstructing the representation of gender, and gender's representation is a significant tool for its production. Following traditional anthropological divisions, the modules were to be organized by culture region (e.g., Latin America, Native Americans, Southeast Asia, the Middle East, the U.S., etc.) and by subfield. The scholars invited to do each module are feminists who have fundamentally restructured gender as an object of knowledge in anthropology. Of special interest to *Primate Visions*, besides Zihlman, invited authors include Jane Lancaster on biosocial perspectives and Linda and Laurence Fedigan on gender and 'he study of primates. The project director, Sandra Morgen, instructed each contributor on what "cultural diversity" must mean in this curriculum project. It may not be a pluralist or liberal arrangement of many cultures. "I think the project materials will be very useful to teachers in other disciplines and in Women's Studies who seek to do more with cross-cultural materials than 'show and tell' women's lives across the globe."[29]

Rather, the project requires its authors to contextualize and structure their course material and orienting essays to problematize gender as an analytical concept by interrogating the simultaneous relations of gender, race, class, culture, historical context, and the authors' own modes of representation. Diversity as a theme is to be mobilized not to resolve into universals with variation, but to explore how "race, class, ethnicity, sexual orientation, and caste specify women's experiences and the role of gender in cultural systems." The authors were asked specifically to address interconnections structured by colonialism and imperialism and the integration of sexuality with power, class, and race. Recent critical and self-reflexive anthropology was invoked to urge contributors to watch how their own arguments and curricular practices could again reconstruct Third World peoples as "other."

This letter of instruction is a condensed summary of the political agenda of 1980s anti-racist, multicultural U.S. feminist theory, reformulated in the disciplinary language of anthropology. Within this emerging discourse, universalizing projects of all kinds are undergoing severe theoretical, empirical, and political deconstruction. In the feminisms of the 1980s, the historical conditions have been challenged that allowed woman the gatherer to be born, with her animating tensions structured by a maternalist and individualist dialectic that did not finally put into question specific class, race, and national narratives about what it means to be human. Zihlman has realized for years that scientific narratives of human evolution are part of the contest for citizenship in the body politic. That contest has vastly expanded and intensified on a global scale in the last half of the twentieth century; setting the terms of the contest is no longer a western monopoly in life science or in political theory. The challenge facing those who generated the figure of woman the gatherer is to mutate her further to tell better the heteroglossic stories of sexual difference and difference within sex—while remaining responsible scientifically for constructing and reading the data of "Sex and Gender Two Million B.C." In more ways than one, one story is not as good as another.

15

Sarah Blaffer Hrdy: Investment Strategies for the Evolving Portfolio of Primate Females

It was the high drama of their lives, the next episode of the colubine soap opera that got me out of bed in the morning and kept me out under the Indian sun, tramping about their haunts for eleven hours at a stretch. (Hrdy 1977: 76)

[T]he central organizing principle of primate social life is competition between females and especially female lineages. Whereas males compete for transitory status and transient access to females, it is females who tend to play for more enduring stakes. (Hrdy 1981: 128)

Besides, she continued, why call me sexually receptive anyway? That's one of those human words with an opinion written all over it. Call me sexually interested if you will, for I am. . . I'm about as receptive as a lion waiting to be fed! (Josie, the chimpanzee, in Herschberger 1970 [1948]: 9)

[T]he natural body itself became the gold standard of social discourse. (Laqueur 1986: 18)

In primatology as in its other genres, feminist discourse is characterized by its tensions, oppositions, exclusions, complicities with the structures it seeks to deconstruct, and incommensurable languages, as well as by its shared conversations, unexpected alliances, and transformative convergences. Many of the major themes in modern feminism are elaborated in contemporary debates in primatology. Echoing each other across the disciplinary chasms dug to protect the human and natural sciences from each other and across the fortified wall erected to keep genres of science and politics from illicit congress, such monstrously twinned symbolic discourses are highly mediated expressions of pervasive, unresolved social contradictions and struggles (Jameson 1981). That is not the same thing as a reduction of the oddly duplicated arguments to a governing, homogenizing ideology; each discourse has its specific practices for seeking to establish the authority of its accounts of the world. Strathern (1986b) defined feminist discourse as a field structured by its co-textuality, that is, by shared yet power-differentiated and often contentious conversations, but not by agreements and, unhappily, not by equality. Feminism in the academy is a field of discourse characterized by what Annette Kuhn (1982) called "passionate detachment." The discourse is not a series of doctrines, but a web of intersecting and frequently contradictory inquiries and commitments, where gender is inescapably salient and where personal and collective dreams of fundamental change and of bridging the gap between theory and other forms of action remain alive. This is the sense that I wish to invoke for holding together the texts examined in "The Politics of Being Female," as well as my relation to these texts and writers. Here, as in all its feminist senses, gender cannot mean simply the cultural appropriation of biological sexual difference; indeed, sexual difference is itself the more fundamental cultural construction. And even that sense of sexual difference is not enough for feminist theory; gender is woven of asymmetrical and multiply arrayed difference, charged with the currents of power surging through multi-faceted dramatic narratives of domination and struggles for its end.

So, it is not surprising to find that feminist discourse within primatology is heterogeneous and disharmonious, even as each knot in the field is tied to a resonating strand of feminist argument outside primate studies. Jeanne Altmann insisted on muted drama; Linda Fedigan deconstructed narratives empowered by anisogamy and its consequence of escalating sexual opposition; Adrienne Zihlman had no patience for evolutionary plots built on hypertrophied competition or sharp sexual difference. But Sarah Blaffer Hrdy has built her scientific narratives and primatological feminist theory on foundations of popular high drama, originary sexual asymmetry and opposition, and the bedrock importance of competition, especially among females. All four primate women scientists engage in a complicated dance in their textual and professional politics around constructing isomorphisms or noncongruences among the potent entities that emerge as actors in their accounts, called female, woman, and scientist. All four engage in problematizing *gender* by contesting for what can count as *sex*; that is, their reinventions of nature are part of their cultural politics. And all four set on stage a different kind of female primate self or subject to enact the crucial dramas. Jeanne Altmann proposes a monkey ergonomic self; Fedigan's characters are mainly characterized by performance of social roles; Zihlman constructs a liberal, flexible self in the mainstream of Enlightenment doctrine; and Hrdy deploys an investing, strategic self. These primate subjects are not mutually exclusive, but each leads to a different pattern of relations among female, woman, and scientist in the texts of these writers. And each writer finds within the animals that mediate the traffic between nature and culture in their society a kind of sex and sexuality, and a kind of mind and cognition, appropriate to these variant subjects.

Zihlman, Fedigan, and Altmann have placed varying amounts of distance between their explanatory strategies and the complex of doctrines labeled sociobiology since the mid-1970s. Hrdy counts herself an unrepentant sociobiologist and insists that sociobiology must be credited with facilitating female-centered accounts of primate lives. Zihlman put maternalist themes into tension with the narrative resources provided by the liberal theory of possessive individualism. Hrdy locates herself entirely under the sign of liberalism and individualism in a story tightly ruled by the imperatives of reproductive competition under conditions of ultimate ontological scarcity. Ironically, Zihlman reproduced a modernist humanist discourse featuring a troubling universal being, woman the gatherer, who threatened to subdue the heteroglossia of women's power-differentiated lives with the univocal language of sisterhood, while Hrdy delineates a map of proliferating differences written into the primate female body that might indicate directions for a postmodernist, decolonizing bio-politics. But Hrdy's map of differences, her way of narrating that females— and women—differ from each other and are therefore agents, citizens, and subjects in the great dramas of evolution and history, is perhaps more a guide to the cultural logic of late capitalism than to the prefigurative fictions and material practices of international multicultural feminisms. In contrast, Zihlman's universalizing moves might be deconstructed and rewritten into specifications and representations of ways of life necessary to post-colonial women's movements.

Deferring direct discussion of Hrdy's writing, let us broach these weighty matters by beginning with the topic that seems to pervade Hrdy's scientific stories most literally: sexual politics. The salience of sexual politics to major social controversies in the United States in the 1970s and 1980s is undeniable. Sexual politics is a polyvalent term covering a host of life-and-death issues and struggles for meanings in key symbols and practices. A list suggested by the term "sexual politics" defies termination: abortion, sterilization, birth control, population policy, high technology-mediated reproductive practices, wife beating, child abuse, family policy, definition of what counts as a family, the sexual political economy of aging, the science and politics of diet "disorders" and regimes, the shift from nuclear family-based patriarchy in the substitution for the husband by the state in welfare policy, divorce rates and the gender-unequal economic consequences of divorce for men compared to women and children, compulsory heterosexuality, heterodox sexual practices among lesbian feminists, sexual identity politics, lesbian and gay histories and contemporary movements, rape, pornography, transsexuality and other gender reconstructions and reversals, fetal and child purchase through contract with pregnant women ("surrogacy" seems a hopelessly inadequate word), racist sexual exploitation, racist structures of sexual desire and fear, single parenting by men and women, feminization of poverty, poverty in the absence of access to a white male income, women's employment outside the home, unpaid labor in the home, synergisms of race- and sex-stratified labor markets, covertly gendered norms for professional careers, restriction to populations of one sex in health research on non-sex-limited diseases, domestic divisions of labor, class and race divisions among women, productions of popular culture, high theory in the human sciences, technologies of representation, social research methodologies, the ties of masculinism to militarism and especially to nuclear politics, psychoanalytic accounts of gender and culture, gendered effects of agricultural policy and agribusiness, curriculum reform in schools and universities, ties of ecological degradation to misogyny, and on and on. What principle of order could reduce such a list to coherence? It is possible to argue nonfacetiously that every major public issue in the last two decades in the United States has been pervaded by the symbols and stakes of sexual politics. It is in this cultural and political environment that feminism and anti-feminism have emerged in nearly every area of collective and personal life, contesting the constructions and representations of gender.

In her analysis of the ideological development of "second wave" feminism in the United States in the context of the post-war revolution in female (especially mothers' and married women's) labor force participation, controversies about personal and family life, and decline of the cultures of domesticity among white middle- and working-class women, Judith Stacey argued that feminism has often been credited or blamed for transformations, insecurities, and freedoms rooted in the vast rearrangements of "post-industrial" society.[1] Feminism has been a shaper of, but also deeply shaped by, fundamental historical rearrangements of daily life in late industrial society. The ideologies and symbols associated with woman as mother, with woman constituted as object of another's desire and pleasure, and with the female body as the stakes in the contest for honor among men have all been problematized by other cultural discourses on gender and sex. Only some of these challenging moves have been feminist, however defined. But in general, the power of the image of woman as natural mother—a being consumed and fulfilled by dedication to another; a being whose meaning is the species, not the self; a being less than and more than human, but never paradigmatically man—has declined in nearly every discursive arena, from popular culture to legal doctrine to evolutionary theory. Attempts have repeatedly been made since World War II to rehabilitate the "traditional" (i.e., white, bourgeois, western, nineteenth-century) images of the female body organized around the uterus, of social motherhood, and of domesticity; but they have had the feel of a backlash, of a still dangerous but defeated ideology.

At the same time, the languages and issues of reproductive politics have intensified in material and symbolic power. Both symbolically and practically, the fights over reproductive politics are carried out paradigmatically in and on and over the bodies of real women. But they are also carried out in the images and practices of scientific and technological research, science fiction film, metaphoric languages among nuclear weapons researchers, and and neo-liberal and neo-conservative political theory.[2] Reproductive politics provide the figure for the possibility and nature of a future in multinational capitalist and nuclear society. Production is conflated with reproduction. Reproduction has become the prime strategic question, a privileged trope for logics of investment and expansion in late capitalism, and the site of discourse about the limits and promises of the self as individual. Reproductive "strategy" has become the figure for reason itself—the logic of late capitalist survival and expansion, of how to stay in the game in postmodern conditions. Simultaneously, reproductive biotechnology is developed and contested within the large symbolic web of the story of the final removal of making babies from women's bodies, the final appropriation of nature by culture, of woman by man. Symbolically, reproduction displaced to the laboratory and the factory becomes no longer the sign of the power of personal and organic bodies, preeminently the site of sexual politics, but the sign of the conquest of still another "last" frontier in the ideology of masculinist technology and industrial politics. Reproduction, strategic reasoning, and high technology come to inhabit the same sentences in social discourse. This is decidedly not the syntax of maternalism and domesticity.

From western points of view, in sexual reproduction by men and women, the premises of individualism and self-sufficiency break down most dramatically and inescapably. At its simplest, sexual reproduction takes two, no matter how much the theories of masculine potency in western philosophy and medicine attempt to evade the matter.[3] Phrased in the discourse of biology, there is never any reproduction of the individual in sexually reproducing species. Short of cloning, that staple of science fiction, neither parent is continued in the child, who is a randomly reassembled genetic package projected into the next generation. To reproduce does not defeat death any more than killing or other memorable deeds or words. Maternity might be more certain than paternity, but neither secures the self into the future. In short, where there is sex, literal reproduction is a contradiction in terms. The issue from the self is always (an)other. The scandal of sexual difference for the liberal conception of the self is at the heart of the matter. Sexual difference founded on compulsory heterosexuality is itself the key technology for the production and perpetuation of western Man and the assurance of this project as a fantastic lie. In the major western narrative for generating self and other, one is always too few and two are always too many (Haraway 1985). In that awful dialectic lies the plot of the escalation and repressive sublimation of combat as the motor of personal and collective history.

But also at its simplest, so far only women get pregnant. Pregnant women in western cultures are in much more shocking relation than men to doctrines of unencumbered property in the self. In "making babies," female bodies violate western women's liberal singularity during their lifetimes and compromise their claims to full citizenship.[4] For western men in reproduction, setting aside the "problem" of death, the loss of self seems so tiny, the degrees of freedom so many.[5] Ontologically always potentially pregnant, women are both more limited in themselves, with a body that betrays their individuality, and limiting to men's fantastic self-reproductive projects. To achieve themselves, even if the achievement is a history-making fantasy, men must appropriate women. Women are the limiting resource, but not the actors. In postmodern, post-industrial conditions, this continuing narrative of the embattled and calculating mortal individual takes on the added dimension of the breakdown of all coherent subjects and objects, including the western self for both men and women. All subjects and objects seem nothing but strategic assemblages, proximate means to some ultimate game theoretic end achieved by replicating, copying, and simulating—in short, by the means of postmodern reproduction. No wonder cloning is the imaginary figure for the survival of self-identity in cyborg culture.

In this context of the breakdown of ideologies and images of female domesticity and of the intensification of reproductive politics and cultural meanings in postmodern worlds, sociobiology's exuberant emergence and rapid claims to hegemony in evolutionary explanation in the 1970s should come as no surprise. There is a huge literature on controversies generated by sociobiology, especially applied to human beings. I am not here primarily interested in the claims and counterclaims that sociobiology is inherently racist or sexist (or the opposite) or that it is another in a long line of biological determinisms (or the best route to human self-definition and expanded choice). Such arguments have often been rather reductive, caricaturing the discourse and other practices of scientists and critics. Argument at this level reproduces the terms of representation that must be deconstructed. I am here more interested in sociobiology as a postmodern discourse in late capitalism, where versions of feminism readily enter the contest for meanings, at least in retrospect and over the tired bodies of gutsy sociobiological feminists. How have sociobiological feminist arguments, like other western feminisms, enabled deconstruction of masculinist systems of representation, while simultaneously both deepening and problematizing unmarked enabling tropes of western ethnocentrism and neo-imperialism?

Let us characterize, or perhaps caricature for emphasis, central sociobiological images in terms of this discussion. What does the famous "death of the subject" look like in neo-Darwinian evolutionary theory? The ever-granulating "unit of selection" heads the list. No bounded body seems able to resist limitless fragmentation to become at last the luminous unit-agent acting strategically to stay in the game. Who is playing? Has the evolutionary play in the ecological theater become a video game on an automated battlefield? No element of structure and function can unify all the narratives of biological meaning. Species, population, social group, organism, cell, gene: all of these units turn into powder under the explanatory burdens they must bear.[6] No unit, least of all the individual, sexually reproducing organism, is a whole, classically reasonable, potentially rights-bearing subject in the realm of nature. The organism is in constant danger of resolving into nothing but a proximate means for the strategic ends of its own genes. The organism's offspring, its investment, is a congeries of genes that allows calculation of a coefficient of relatedness, but this genetic investment is hardly straightforward reproduction. Inclusive fitness theory demands calculations of coefficients of relatedness and produces a kind of hypothetical or hyperreal individual, put together from its fragments of scattered sameness in all those bodies of others calculating in their own terms how to get more copies of parts of themselves into the game. The imperative is to identify and replicate "same" while holding "other" at bay. But otherness is everywhere, masquerading as same. Altruism must be redefined in terms of the problematic of investment in non-self: how can such a strategy yield a return on the self in the future? Which self? The Darwinian world from the start has been ruled by a reproductive natural economy, but Darwinian reproductive politics have intensified and tightened in the conditions of logical contradiction in which there is no*thing* to reproduce. How can narratives stabilize objects and produce good-enough subjects to get on with the dramas of investment? It is small wonder that female domesticity and selfless maternalism offer few useful images for such a project.

But there is a deep reservoir of universalizing images for the postmodern reproductive politics and reconceptualizations of the relations of mind and body in sociobiology: the female orgasm. Representations of female orgasm may be a map to the politics and epistemologies of sexual pleasure in a world structured by gendered antagonistic difference. In a sociobiological world, sexual difference may no longer be the figure of distinct, hierarchically arrayed, and stably complementary man and woman; but representations of male and female bodies remain ready to figure the strategic calculations of life's unequal investment battles. Neglecting the deep militarization of discourse implied in the ubiquitous battle-strategy-investment metaphors of modern politics and biology, ordering "life's unequal investment battles" around sexual difference has the effect of demoting or erasing other axes of subordination for women, as well as for men. Universalizing representations of the female orgasm are maps of a silent, but nonetheless constitutive, racial discourse.

In particular, the racial and racist nature of sexual politics is distorted by the feminist and anti-feminist discourse of the white middle class that privileges sexual difference as the definitive axis of gender inequality.[7] That is, *sex*, and especially its derivative, *sexual difference*, can be a distorting lens for seeing the asymmetries between and within *genders*, as well as other basic systems of inequality. Because of the history of sexual politics in slavery, lynching, and the contemporary coercive sexualization of non-white women in racist symbology and material life, this issue is particularly sharp around the politics of sexual pleasure and sexual violence in the history of U.S. feminism from the nineteenth century to the present. Ideologies of "social motherhood" and "sexual prudery" were powerful feminist tools in nineteenth-century American struggles for women's control of their bodies, but the ideologies had different resonances and subsequent histories for black and other feminist theorists of color than for white feminists.

Basic economic inequality was a principal reason that black feminists of the post-1960s "second wave" in the U.S. had a different agenda from middle-class white feminists, but it was not the only reason semiotically or materially. Orgasms and female genital anatomy have vastly different racial semiotic fields. Middle-class and white "politics of the female orgasm" can risk privileging women's sexual pleasure in reconstructing notions of agency and property-in-the-self within liberal discourse. Black feminists confront a different history, where black women's putative sexual pleasure connoted closeness to an animal world of insatiable sensuality and black men's sexuality connoted animal aggression and the rape of white women. Hazel Carby showed that black feminists' constructions of black women's respectability were part of an anti-imperialist and anti-racist, as well as feminist, politics in the nineteenth century. In the late twentieth century, anti-racist feminists cannot engage unproblematically in universalizing discourses about sexual pleasure as a sign of female agency without reinscribing feminism within one of the fundamental technologies for enforcing gendered racial inequality. As Carby argued, "A desire for the possibilities of the uncolonized black female body occupies a utopian space. . . . [Nineteenth-century] black feminists expanded the limits of conventional ideologies of womanhood to consider subversive relationships between women, motherhood without wifehood, wifehood as a partnership outside of an economic exchange between men, and men as partners and not patriarchal fathers" (1985: 276). If these texts had been part of the bio-politics of the primate body in the 1970s and 1980s, the discourse about female bodies in sociobiological feminism, as well as in other regions of white middle class feminism, would have been different.

But these texts were not part of the discursive field in which the female orgasm was rediscovered and deployed to signify universal, unmarked woman's natural body and mind within the constraints of hyper-liberalism. In an extension of Thomas Laqueur's discussion of sixteenth- through nineteenth-century medical and political representations of female bodies in "Orgasm, Generation, and the Politics of Reproductive Biology," in the late twentieth century the universalized natural body remains the gold standard of hegemonic social discourse.[8] As a gold *standard*, the natural body is inescapably figured as a convention, i.e., a construction. Neither gold nor bodies enter these equations outside convention. The natural body is a gold standard for power-differentiated social intercourse, for the unequal exchanges of "conversation." Gold is pre-eminently the medium of universal translation, the sign of the promise of a world of frictionless exchange, of final commodification of the body of the world in a hyper-real market ordered by a transparent language, a final common measure. Nineteenth-century Americans used the words *conversation* and *intercourse* interchangeably to signify sexual commerce between men and women. Illicit sex was named *criminal conversation*. Broadly within late twentieth-century scientific discourse, the natural body is conventionally a biotechnological cyborg—an engineered communications device, an information generating and processing system, a technology for recognizing self and non-self (paradigmatically through the immune system), and a strategic assemblage of heterogeneous biotic components held together in a reproductive politics of genetic investment. Genetic currency is golden, a sign of a world always like itself, univocal.[9]

Hrdy has been an active trader in these precious metals, products of the transmutation of representations of genitally organized pleasure into the sign of power and agency. In a process in which Hrdy took significant part, of renegotiating the conventions that set the value of the body for political discourse, orgasmic sexual pleasure became for (unmarked, i.e., white) women what it has been for (unmarked) men before, the sign of the "same," i.e., of the capacity to be (mis)represented as the unmarked, self-identical subject—at least for a few intense seconds—in postmodern conditions. The unmarked category is the category present to itself, the category of identity, of the "same," of gold, versus the marked category of otherness, of value defined by another, of lack of power to name, of base metal. Orgasm becomes the sign of mind, the point of consciousness, of self-presence, that holds it all together well enough to enable the subject to make moves in the game, instead of being the (marked) board on which the game is played.

Since the eighteenth century, liberal theory has required the body to be the bearer of the rational subject.[10] If sociobiology is a hyper-liberalism, the organs of its hyper-bodies must be signs for the subject constituted through strategic reason. Has the "mentula muliebris," the little mind of women, a common name for the clitoris in sixteenth-century learned texts, like the phallic "mentula," or little mind of men, become a late twentieth-century guarantor of the status of (a desiring and investing) subject, of representability, and of agency in life's dramatic hyper-narratives? What kind of feminism could this be?

Laqueur argues that before the latter part of the eighteenth century in Europe, most medical writers assumed orgasmic female sexual pleasure was essential for conception. To simplify a complex story, it was only with the constitution of sexual difference in a (re)productionist frame that female orgasms came to seem either non-existent or pathological from the point of view of western medicine.[11] Sexual difference was constituted discursively through the nineteenth-century biological reorganization of the female body around the ovaries and uterus and of the male body around a spermatic economy that linked phallus and mind through the commerce between nervous and reproductive energy. Prostitutes, non-white women, sick women—these might have orgasms and large clitorises, but "civilization" itself seemed to require the little mind of women to disappear. The "problem" of the clitoral *versus* vaginal orgasm would have been incomprehensible to a Renaissance doctor. But by the late nineteenth century, surgeons removed the clitoris from some of their female patients as part of reconstituting them as properly feminine, unambiguously different from the male, which seemed to be almost another species—or better, masculine and feminine connoted the odd taxonomic and linguistic inversion of two genera in a single species. Female and male structures before the late eighteenth century were almost universally regarded as homologous; the female was a kind of male turned inside herself. She even appeared to some writers to have two organs homologous to the male penis, but this apparent duplicity caused no evident trouble in the text of the body or the text of the medical writer. The female was a *lesser* human, for example, less hot, less spirited; but she was not a *different* human.

Changing the meaning of and practices for establishing homologies, biological discourse about reproduction and sexual difference was part of a great redeployment of male and female in cultural and political space. Hierarchical homologies were abandoned for a different discursive order, comparative functional anatomy and its many bio-political offspring. Incommensurability replaced hierarchy as the principle of relationship between the sexes. The sexes became "opposite" or "complementary," rather than more or less. And new ways of interpreting the body were new ways of representing and constituting social realities.

In particular, European-derived feminist and anti-feminist debates proliferating from the late eighteenth century located themselves on the terrain of the meaning of sexual difference. The history of modern feminism would be incomprehensible without the history of modern reproductive biology and clinical gynecology—as a moral discourse about social order and as a social technology. Evolutionary discussions were part of this larger discursive frame. Again to oversimplify, in anti-feminist discourse wherever the boundaries of old hierarchies were threatened by new Enlightenment liberal doctrines of universal man, biological sexual and racial difference reimposed "natural" limits.[12] The body as bearer of the rational subject in liberal theory was "neutral," sexed but not gendered, and as such could be used to threaten gender hierarchy (Wollstonecraft 1792). Liberal theory was a resource for feminists, but only at the price of renouncing anything specific about women's voice and position and carefully avoiding the problem of difference, for example, race, among women. The "neutral" body was always the unmarked masculine. As species Man, women were silenced as such. But for anti-feminists, as females biologically and functionally, women were imagined to be without capacity for full citizenship. Their specificity fitted them for the home, for domestic space, not for the competitive world of business, scholarship, and politics.[13] Yet even feminists drawing on liberal theory in these debates could not and did not dispense with the contradictory resources of the emerging discourses on functional sexual difference.

Nineteenth-century feminist doctrines of social motherhood, a major argument for the vote, depended on the discursive terrain of the new reproductive biology. In these arguments, as mothers women were especially fitted for citizenship. They would heal the body politic from the wounds inflicted by militarist and competitive men. They would be mothers of the republic, acting in a public space, as they were mothers in the domestic space. The doctrine of separate spheres could be and was deployed by those opposing and those advancing (some) women's rights. From the late eighteenth to the late twentieth century, to claim the right to public speech as women, feminists had few alternatives to maternalist discourse. The "passionless" woman became an historical figure. One casualty of this map of discursive possibility was the meaning of female genital sexual pleasure. It might still occur, and the sexologists from Havelock-Ellis on tried manfully to re-establish non-pathological meanings for female sexual pleasure. But what did female orgasm *mean*—and in what regions of the body did it occur—for "good" women, that is? While they *were* The Sex, they could not *have* sex—only babies, in the literal form of children or the symbolic form of an enfantilized public world crying for politics as mothering.

Like their human cousins, primate females seem to have been born into the post-eighteenth-century liberal world of primatology without orgasms and as natural altruistic mothers. This is no condition for a good strategic investor, that late twentieth-century figure of the ideal citizen, for whom a self-possessed *mentula* is absolutely essential. How did female monkeys and apes get their orgasms back, and what does the story have to do with the white, middle class branch of the Women's Liberation Movement and one of its daughters, sociobiological feminism?

Let us begin by going back to the chimpanzee, Josie, whom we left in a cage at Robert Yerkes's Orange Park facility in the late 1930s. Complaining about Yerkes's confusion about sexual, nutritional, and social hungers among male and female chimpanzees, and especially about his theories of the biological origins of prostitution and female natural subordination, Josie was given speech in Ruth Herschberger's prefigurative feminist text, *Adam's Rib*, in the 1940s. Herschberger wanted to rewrite more than Josie's lines in Yerkes's papers; she was interested in a thorough scientific critique of sexist biology of the female body, most certainly including a rediscovery of the physiological and evolutionary point of having a clitoris. In 1944 Herschberger wrote Yerkes to ask him if he had seen any signs of orgasm in female chimpanzees. Yerkes expressed interest in her query and referred it to his appropriate colleague in reproductive physiology, William Young. Herschberger followed with a detailed request for specific observations about clitoral erection. Yerkes again referred her to an appropriate colleague, this time Dr. Blandau of the University of Rochester School of Medicine and Dentistry.

And there, as far as I can find, the matter rested until the same period that gave rise to the Women's Liberation Movement. Twentieth-century sex advice manuals stressed women's sexual satisfaction and even orgasm before the second wave of feminism, and Herschberger's little book is a useful guide to the terms of the discussion between the two world wars. In general, the therapeutics of marriage contained the bio-politics of women's always-on-the-verge-of-pathological sexual pleasure or sexual quiescence ("frigidity"). If before the late eighteenth century, female orgasm was assumed to be necessary to conception, before the late twentieth century, once its existence was re-admitted to polite society, its only hope for normality lay in heterosexuality, itself one of the great constructions of the last two hundred years. In the shadow of the normal has lurked the specter of the pathological—women's sexual pleasure for their own ends, as a sign of their existence as ends and not as functions, no longer as mothers, wives, or even free lovers under the reign of gender, i.e., male domination. The concept of existing as ends is incompatible with the binary division into normal and pathological; the binarism is about functions, means, not ends. In patriarchal ideology, Woman is contained by her functions, not achieved through and for her own ends.

By means of the link through orgasm of self-possession and existence as ends rather than means, in a typical universalizing discourse, women as females could be semiotically reconstructed under the sign of reason and citizenship. As long as "the family" or the "pair bond" contained the meaning of women's sexuality, women could not be social subjects, ends-in-themselves, in the hegemonic narratives of liberal theory and bio-politics. In effect, without getting their orgasms back on their own terms, late twentieth-century western middle-class women could not have minds. This point was implicit in the extraordinary attention that the politics of women's orgasms got in the popular media and in polemics of the Women's Liberation Movement internationally throughout the 1970s. Deep controversies about female sexual desire continue to rend the western women's movements and "general" society in the 1980s.[14] In the early years of the Women's Liberation Movement, orgasms on one's own terms signified property in the self as no other bodily sign could. Indeed, masturbation came to promise the best kind of orgasm for women, while lesbian sado-masochism could be a utopian sign of freedom from the taxonomy of functionality and normality. In the curious logic of signs, non-reproductive, non-heterosexual, female-controlled, women's orgasmic practices could point to a possible world without gender, a science fiction world where the rule of the normal was broken for women, and not merely once again for men. Sexual difference and the whole apparatus of liberal functionalism as a category of analysis were at stake in these semiotic contests. Male orgasm had signified self-containment and self-transcendence simultaneously, property in the self and transcendence of the body through reason and desire, autonomy and ecstasy. These were the symbolic prizes sought by those who had been semiotically contained and fixed for another in the notion of the normal, healthy, and functional. No longer pinned in the crack between the normal and the pathological, multiply orgasmic, unmarked, universal females might find themselves possessed of reason, desire, citizenship, and individuality. This is quite a performance to ask of the mentula.

It is in this historical context that I wish to read Hrdy's sociobiological feminism.[15] Interesting contradictions leap out at once. Sociobiology is narratively a hyper-functionalism and a hyper-liberalism. The promise of the self-contained and self-transcendent subject, which has historically been the fantastic longing embedded in liberalism, seems even more fraudulent in the postmodern landscape of disaggregating subjects and units powdered into impossible fragments in a vast simulated world ruled by strategic maneuvers. Both evoked and blocked earlier within bourgeois liberalism, female individualism in the postmodern landscape threatens to be a pyrrhic victory. To escape the rule of biological sexual difference, with its patriarchal moral discourse on the normal and the pathological, only to play the board game of reproductive politics, as replication of the self within sociobiological narratives, seems a parody of feminist critiques of gender. The tie between recasting females as strategic reasoners (proximate ends in themselves) and the fierce reproductive teleology of the ultimate game takes the bite out of sociobiology's deconstructions of the social functionalist dramas of maternalism. Sociobiological females/women are cast again as mothers with a vengeance, this time in the problematic guise of the rational genetic investor, a kind of genetic receptacle, holding company, or trust, not so much for the spermatic word of the male as for the contentious ultimate elements of code that assemble to make up the postmodern organism itself. There seems nothing that can elude the bottom line of genetic investment logic, no matter how much free play is allowed in the intermediate accounts. In sociobiological stories, the poor hand of the ideology of sexual difference seems to be called in against social functionalism only to reappear in spades in the escalating sexual combat rooted in anisogamy. To locate the reconstruction of the biopolitics of being female in sociobiology is fraught with irony. But the irony and the contradictions cannot evict this rich contemporary discourse from the capacious and contentious house of feminism. Let us look for hints of a radical future even for hyper-liberal feminism.[16]

Hrdy recounted the history of the debate about primate female orgasms subsequent to Herschberger's 1940s queries in her popular feminist sociobiological book, *The Woman That Never Evolved*.[17] In the context of 1960s laboratory-based studies of endocrine physiology and sexual behavior, the psychiatrists Doris Zumpe and Richard Michael described a "clutch reflex" in rhesus monkey females at the time of their male partners' ejaculations. The authors suggested these spasmodic arm movements might be signs of female orgasm, or as they put it, "an external expression of consummatory sexual behavior."[18] Perhaps rhesus females not only had orgasms, but even had the most fantastic kind in the heterosexual imagination— simultaneous with a male partner. Prompted by Masters and Johnson's (1966) famous studies, in an experimental design verging on a caricature of rape, the anthropologist Frances Burton "subjected three [restrained] rhesus females to five minutes of clitoral stimulation mechanically applied by the experimenter, four minutes of rest, and five more minutes of vaginal stimulation."[19] The point was to modify Masters and Johnson's recording techniques to study females who could give no verbal reports about their states of excitement. Removing vestigial doubt about the matter, Masters and Johnson "proved" that women have orgasms (and that there is no difference between clitoral and vaginal orgasms) by a method that did not rely on verbal report by women either. These privileged orgasms took on the authority of natural science. They wrote, rather than spoke, their truth. Women's orgasmic bodies themselves became inscription devices, coupled to other recording instruments, tracing the trajectory of excitement and its resolution. These recordings put in writing, seemingly without the polluting intervention of an interpreting subject, graphic evidence of women's multiple peaks of sexual climax. Orgasms were recorded by a kind of automatic writing technology that could also be coupled to the body of the nonhuman primate, joining females in a sisterhood of officially recorded sexual pleasure across species and across the boundary of nature and culture. However, what the pleasure *meant* was not resolved by documentary evidence of its presence. All stages that Masters and Johnson found in women, except the anatomically obscured final stage (orgasm itself), were clearly visible in these rhesus females. Burton concluded her subjects were orgasmic under these laboratory conditions. In view of the repeated copulations reported from field observations of several species, likely building up sexual excitement, female orgasms in the wild seemed a reasonable conclusion.

Beginning about the same time as Burton, during her dissertation research on the ontogeny of communication, for which she took data on communication before and after copulation, Suzanne Chevalier-Skolnikoff, Washburn's student, reported extensive observations of likely female orgasms in captive colonies of stumptail macaques at Stanford, resulting from both heterosexual and homosexual mounting. Chevalier-Skolnikoff speculated that female orgasm was widespread among mammals and, echoing Renaissance physicians, was "an essential ingredient of fertile coitus in at least some mammalian species."[20] She argued that female orgasms likely functioned as a motivational mechanism in all female mammals. But, although she gave a model for the evolution of female orgasm, she did not have access to the sociobiological narrative to develop this germ of a transformative story. Chevalier-Skolnikoff recalled that no one seemed to be paying much attention to homosexual behavior among nonhuman primates then; and her observations, which were incidental to her main study focus on communication, bore on both homosexual and female orgasmic sexuality in the stumptail subjects. Lucille Neuman invited her to give a talk on these matters at a symposium Neuman was organizing at the Association for the American Advancement of Science meetings in Philadelphia in 1971; Jane Lancaster, also in Chevalier-Skolnikoff's cohort at Berkeley, asked her to speak in a symposium on female sex roles at meetings of the International Primatological Society. The feminist biosocial network provided one ready audience for these observations of active female sexuality. Over a short period, Chevalier-Skolnikoff gave four different papers, publishing three of them.

Both the genesis and reactions to these papers took her by surprise. She recounted that in the original symposia where she read papers, no one said a word during the discussions, but that she would then receive 600–800 requests for copies, whereas she might have received 10–15 for a rather successful conference paper before.[21] She was invited to be on the editorial board of the *Archives of Sexual Behavior*, but was rebuffed by the Stanford feminist social anthropologist, Michelle Rosaldo, when she suggested putting a paper on female orgasm into what became one of the foundational U.S. texts in feminist anthropology.[22] At a 1972 presentation before a Los Angeles group supporting transsexual operations, Chevalier-Skolnikoff was cheered by an audience standing and waving banners as she walked up to the podium.

Although women's groups did not seem to receive the sex papers with the same enthusiasm, Sarah Blaffer Hrdy, a Harvard anthropology graduate student who did not yet define herself as a feminist, heard and absorbed Chevalier-Skolnikoff's 1971 presentation at the American Anthropological Association meetings in New York on the stumptail female sexual response and its implications for female mammalian sexuality. Hrdy approached Chevalier-Skolnikoff later, saying she had been impressed and amazed, but did not yet have an explanatory framework for fitting it in. Chevalier-Skolnikoff's first evolutionary model of female orgasms bore the same relation to Hrdy's subsequent formulations that Slocum's critique of man the hunter and first elaborations of a gathering hypothesis had to Zihlman and Tanner's thesis. Intellectually formed in Irven DeVore's undergraduate primate behavior course, for which Robert Trivers was her teaching assistant, and then by debates in DeVore's graduate primate seminars, in which Trivers was a central presence, and by E.O. Wilson's ideas and personal support in graduate school, Hrdy interpreted her dissertation research data *Langurs of Abu* in terms of sociobiological approaches to male and female strategies of reproduction. [Figure 15.1] Active female sexuality was already part of that story; but in *The Woman That Never Evolved* and subsequent publications, the full implications of female sexual assertiveness, symbolized and facilitated by the female orgasm, emerged.[23]

In *The Woman That Never Evolved*, which Harvard University Press advertised in the *New York Review of Books* with a drawing of a "women's culture" stitchery that read "women's place is in the jungle," Hrdy argued against sociobiologist-anthropologist Donald Symons's view that female orgasms were a by-product of a male adaptation. [Figure 15.2] Symons put it plainly: "orgasm may be possible for female mammals because it is adaptive for males."[24] Males have nipples because mammals are built from a bipotential plan, but nipples are adaptively meaningless except in females; ditto in reverse for the *mentula muliebris*. Milk and semen travel in different channels and to different ends. Hrdy salvaged the heterodox ideas of the feminist psychiatrist, Mary Jane Sherfy, who had argued that human females' recently authorized capacity for multiple orgasms showed them to be virtually sexually insatiable. Sherfy had linked this wonderful sexuality to a past matriarchal stage in human development. Hrdy had a much more respectable evolutionary narrative for restoring rationality to female genital sexual pleasure. "In strictly anatomical terms, then, the human clitoris is the core of the problem" (1981: 166). Lying in the way of an easy argument for the functional significance of female orgasm were troubling facts. There was no way to argue that female orgasm was essential to conception, as there once had been. Intercourse had, to say the least, been shown to be quite possible without female pleasure at all, much less orgasm. Even among orgasmic females, orgasm seemed occasional rather than entirely reliable (in contrast to males?). But still, "[n]o function other than sexual stimulation of the female has ever been assigned to the clitoris." And true to a strong biological tendency broadly, and a sociobiological imperative, an adaptive function should be found for anything as constant as that little phallus in female mammals. Hrdy struck the right tone of mild sarcasm toward dissenters: "Are we to assume, then, that this organ is irrelevant—a pudendal equivalent of the intestinal appendix?" (1981: 167). Since the proximate end was

Figure 15.1 Sarah Hrdy watching a young langur monkey on Mount Abu. Published with permission of Sarah Blaffer Hrdy/ Anthrophoto.

female pleasure, Hrdy had to find the ultimate explanation in the gold coin of reproductive advantage.

So how could the occasional, or the insatiable multiple, orgasm enhance the reproductive chances of females so endowed? It wouldn't make them better mothers, but it might make them safer. And above all, orgasms enhanced females' tool kit as good genetic strategists: it made them better able to lie, a mark of postmodern intelligence in primate studies as elsewhere. Hrdy reasoned,

Figure 15.2   Advertisement for *The Woman That
Never Evolved*. Drawing by Sarah Landry. Published
with permission of Harvard University Press. Compare this image to the Coelho drawing of the heads
of female primates in Figure 11.2.

> If we recognize that a female's reproductive success can depend in critical ways on
> the tolerance of nearby males, on male willingness to assist an infant, or at least to
> leave it alone, the selective importance of an active, promiscuous sexuality becomes
> readily apparent. Female primates influence males by consorting with them,
> thereby manipulating the information available to males about possible paternity.
> To the extent that her subsequent offspring benefit, the female has benefitted from
> her seeming nymphomania. (1981: 174)

This is not a story about happily complementary heterosexuality. It is certainly not
a tale about human female sexuality taking on its meaning from the origin of
monogamy, enabled by constant female "receptivity" and the "loss" of estrus, called
by Hrdy "situation-dependent, concealed ovulation."[25] Hrdy's is a narrative about
root conflict of interest that cuts across sexual difference, in which sexual pleasure
and concealed ovulation can give the added competitive edge to some females, in
competition with other females, in their struggle to turn males, pursuing their own
reproductive ends, into a resource or at least less of an enemy.

Hrdy believed keeping males uncertain about paternity was one advantage retained by females in an evolutionary game heavily weighted toward male physical
power.[26] How did this situation come about? I have read the female orgasm here
as the privileged sign of female agency, where mind and body come together in
female possession of the phallus. A more sober commentator would protest that
orgasm is a sideshow for sociobiologists, with only a few lines in the great epic text
of genetic investment strategies. What is the sociobiological first book of Genesis
that recounts original hostile difference? How did it come about that non-identity
must mean competitive difference? And how did that original difference translate
into cosmic limits on female power and agency?

"Part of the answer, remote in time and nonnegotiable—but not particularly
satisfying—is *anisogamy*, from the Greek *aniso* (meaning unequal) plus *gametes* (eggs

and sperm): gametes differing in size" (Hrdy 1981: 21). Whatever the merits of the tale, whose doubtfulness Hrdy signals even while invoking it in order to get her own narrative going, it is always satisfying to start an important origin story with a Greek etymology. In the primeval soup, by chance some protocells were bigger than others. These richer cells became coveted resources for any cell that could commandeer them to its own energetic replicative ends. "Competition among the small cells for access to the largest ones favored smaller, faster, and more maneuver-able cells, analogous to sperm. The hostages we might as well call ova. . . . The ground rules for the evolution of two very different creatures—males and females—were laid down at this early date." One group began specializing in amassing resources into themselves, while the other specialized in "competing among them-selves for access to these stockpiling organisms" (1981: 21). When these early strate-gists took the further fateful chance step of fully combining, inherently antagonistic sexual reproduction was born out of a prior asymmetrical, but rather innocent, economic exploitation. This mode of reproduction, with its radical legacy of exces-sive heterogeneity, of untamable difference across and between the generations, quickly became an inescapable necessity.

But while eggs remained necessarily vastly larger than sperm, the bearers of the gametes, males and females, took the opposite course. In very few species of pri-mates are females bigger than males. Why? Darwin's answer was sexual selection, intra-sexual competition for access to the means of reproduction, i.e., the "opposite" sex. Since females, from large eggs, gestation, and lactation through lifelong energy-demanding social entanglements with their offspring, put more into reproduction, they become the so-called "investing sex." And the investing sex becomes the limiting resource: the sign and embodiment of what is most desired and always scarce, never really one's own, the needed and hated female as prize. The limiting resource always runs the risk of being nothing but the prize, not a player in its own right—no, not *its*, *her* own right. The limiting resource is *logically* gender female; this is the "syntac-tic" basis for the equation of nature and female in western-gendered narratives. Nature in a productionist paradigm is limiting resource for humanist projects; nature is female, the limiting resource for the reproduction of man, loved and hated and needed, but held in check as agent in her own right. Among primates, the intensity of male competition leads to bigger, stronger, savvier males. And in gen-eral, the greater the inequality of reproductive success among males of the species, the larger the size difference between males and females will be. Inequalities and antagonistic differences cascade synergistically in these stories. Females are not passive in all of this, but their goose was cooked in the primeval Greek alphabet soup.

Anisogamy and sexual selection converge with a third story element that knots behavioral ecology into the weave of sociobiological narrative. Given that females, not males, are the limiting resource, their ways of getting food will have the more fundamental consequences for the possibilities of social life in the species. Among primates, why do some females congregate with other females and males in large groups, while others space themselves singly or in still other social arrangements? Hrdy uses Richard Wrangham's female-centered model for the social systems of primates to advance her account.[27] If competition for scarce resources drives the whole system, competition for food will determine much else. Females will distribute themselves in space as a function of their inherited physiologies and the neighborhood's ecological opportunities and constraints for making a living. A very large-bodied fruit eater will have a very different range of distributive possibilities compared to a smaller-bodied leaf eater, and so on. Males will space themselves in relation to females and to their own competitive relations with each other. These competitive relations can include all kinds of coalitions and forms of cooperation, which are all proximate matters in the great ultimate game of genetic investment according to the principles of methodological individualism. Where they can, females will form social groups with related females, to take advantage of various kinds of help. In this sense, "sisterhood" is powerful; it is a question of coefficients of relatedness and the terms of economic survival. In large social groups, lineages of females will compete with each other. Ranks of females within lineages and ranks among different lineages will concern females greatly and will have large consequences for male possibilities.

Competition among assertive, dominance-oriented females is an absolutely central principle of primate social life, built into their natural status as limiting resources whose eating habits are the pivot of sexual politics (Hrdy 1981: 122–27). Female genetic investment strategies could not have produced the mother-for-the-species of the social functionalists; complementary sexual difference in the organic liberal order gives way to the full array of sexualized agonistic strategic possibilities in hyper-liberalism. Females remain committed to reproduction, but not within a maternalist discourse. Rather, females are redeployed semiotically within a strategic investment discourse, where mind, sex, and economy collapse into a single, highly problematic figure—the hyper-real unit of selection.

This narrative of female economic agency, whose foundational female spatial "deployment" and competitive relations with each other set the constraints for shifting male "strategies," converges with the story of female sexual agency through such means as orgasmic sexual assertiveness and concealed ovulation to produce the sociobiological account of females as full citizens of the evolutionary hyper-liberal state of nature. Here is where the story element of females as the site of internal differences is foregrounded. The bottom line of the story is that females evolve too. That is, females, as well as males, differ among themselves in parameters that matter to differential survival and fecundity, and so are subject to natural selection. The point of this story is that females are dynamic actors of self- and species-forming dimensions. Hrdy has contributed particularly fully to elaborating these arguments. Female reproductive fitness can vary as a function of at least five categories that she has characterized: female choice of mate, female elucidation of male support or protection, competition with other females for resources, cooperation with other females, and female ergonomic efficiency. In each category, females are actors, albeit, like males, heavily constrained ones. One female is not like another; their differences matter to the most basic of life's dramas; they are not the backdrop to someone else's action; their action is differentiated and interesting, full of conflict and ultimate risks. Females are heroes in the narrative of nature, not its plot space or its prize. This is the universal account Hrdy is determined to give.

Hrdy's universalized, yet inherently differentiated and heterogeneous females cannot be dismissed from the permanently contentious field of feminist discourse important to deconstructing and reconstructing what it means to be gendered in late capitalism, if not in the primeval soup. Within the discourse of primate studies, with its practices for determining strong and weak accounts, Hrdy's writing faces different standards than I have measured it by in this chapter. And yet, I believe the criteria explored here can interact productively with more orthodox forms of biological criticism. I do not think deconstructive analysis of scientific narrative is hostile to biological science in its own terms, even if it is far from the same thing. Sociobiologists should be especially, indeed, professionally, sensitive to accounts of narrative strategies and models for moves in textual games. They have been among the best teachers of the analysis of narrative strategy and of how to imagine entering every high-stakes game. But, while shifting the ground from under some accounts, like functionalist maternalism, sociobiologists have been much less good at recognizing how to destabilize and defamiliarize their own ways of constraining any possible account. They have perhaps lacked a sense of *irony* about their own narrative resources. Their repertoire of tropes has excluded that corrosive, but salutary figure for any writer entering her craft centuries after the genesis of a large cultural text on the problem of self and other. Like it or not, sociobiologists write intertextually within the whole historically dynamic fabric of western accounts of development, change, individualism, mind, body, liberalism, difference, race, nature, and sex. A careful dissection of the logic and history of the dialectic between same and different in evolutionary biology's stories, and of those stories' constitutive silences that construct what can be marked and unmarked in contending models of change, might well be required in the training of a biologist.

For those who locate themselves outside the territory of primate sciences, it would be too easy to argue that Hrdy's origin stories once again reify gender outside of history in another ethnophilosophical naturalizing narrative about beginnings and ends, this time in an effort to account for limits to female bonding while still foregrounding female agency within a frame of liberal individualism. But that argument fails to take seriously the craft of constrained story-telling intrinsic to biological sciences and simply assumes there is some safer place for narrative, called "inside" history, "outside" nature. There is no disinterested map for these fictional, potent boundaries. And in any case, boundary crossing and redrawing is a major feminist pleasure that might make some modest difference in a struggle for differentiated meanings, material abundance, and comprehensive equality. Hrdy's naturalized postmodern females emerge full of resources for exploring many interesting topics: female complicity and collusion in male dominance, female agency of many kinds, the potency of sexual discourse in narratives of agency and of mind, the limits of functionalist principles of complementarity and difference, the power of narratives about competition, the logical errors involved in leaving the male an unmarked category in evolutionary models, ways of describing females as sites of internal difference that allow meditation on both bonding and competition, and the deep saliency of reproductive politics semiotically and materially in contemporary U.S. and multinational society.

The silences that structure these naturalized postmodern females also are instructive about unintended but nonetheless endemic replications of colonial or neo-imperialist discourse in white, middle-class feminism. While uncovering the logical error of the unmarked male universal, Hrdy's sociobiological construction of females as agents and sites of difference leaves intact an ethnocentric logic of sexual politics and a deep (re)productionist ethnophilosophy translated into a technostrategic language of universal investment games.[28] Female "soul-making" in natural history occurs on the ground of *constitutive* silence about other differences that make sexual narratives so exclusively productive in western colonizing and neo-imperialist cultures. And, in the end, Hrdy's narrative logic can only do one thing with difference—categorically. Difference is the motor of antagonism, of root, inescapable, final hostility among whatever units can provisionally stabilize themselves within sociobiology's versions of the death of the subject, no matter how many proximate causes and intermediate levels allow something else to have narrative space for a time.[29] Here is the stable, lethal, univocal inheritance from the complex history of the principles and stories of liberal possessive individualism passed onto the mutant hyper-liberal children of the late twentieth century. In sociobiological hyper-liberalism, reproductive bio-politics are finally about war. Check mate.

16

REPRISE: SCIENCE FICTION, FICTIONS OF SCIENCE, AND PRIMATOLOGY

Reading Primatology as Science Fiction: The Second Foundation and Stanford's Second Primate Project, 1983–1984

> However, because the genetic interests of individuals are not identical (unless they are clones), conflicts of interest perpetually endanger the survival of cooperative relationships. (Smuts, Cheney, Seyfarth, Wrangham, and Struhsaker 1987: 297)

"Unless they are clones": Surely, this is an innocent parenthetical exception, nothing but a punctuated precaution for hominid zoologists professionally alert to the array of modes of reproduction and replication in the living world, where clones appear naturally in many species, for example, among the colonial insects. Even in the stodgy, conservative primate order, itself a kind of right-wing reaction to the publicly visible, widespread, and baroque practices among fungi and invertebrates, identical twins and kinky replicative habits occur, if infrequently and generally only in tropical forests. Or in laboratories. Laboratories are the material and mythic space of modern science, and the naturalistic field is one of the laboratories of modern primatology. Indeed, the field has been primate science's privileged semiotic center.

The primate field, naturalistic and textual, has been a site for elaborating and contesting the bio-politics of difference and identity for members of industrial and post-industrial cultures. Cloning is simultaneously a literal natural and a cultural technology, a science fiction staple, and a mythic figure for the repetition of the same, for a stable identity and a safe route through time seemingly outside human reach. Evolutionary biology's bottom line on difference is succinctly stated in the quotation opening this conclusion: in the end, non-identity is antagonistic; it always threatens "the survival of cooperative relationships."In the end, only the sign of the Same, of the replication of the one identical to itself, seems to promise peace. Can patriarchal monotheistic cultures ever allow another primal story?

Using Isaac Asimov's imagination of the *Second Foundation* to set the stage, I would like to begin the conclusion to *Primate Visions* with a return to its recurring themes of repetition, identity, cooperation, whole, difference, change, conflict, fragment, reproduction, sex, and mind. Running through the weave of these themes has been the thread of preoccupation with biological and political questions of survival, catastrophe, and extinction. Explicit in the opening quotation above, questions of difference are questions about survival, for both fragments and wholes. Primatology has been a rich cultural fabric for exploring these matters. "Teddy Bear Patriarchy," "The Bio-politics of a Multicultural Field," "Mothering as a Scientist for National Geographic," "Remodeling the Human Way of Life," and "The Politics of Being Female" have all turned repeatedly on narratives of the bio-politics of difference and identity in large dramas of twentieth-century history, reaching from pre-World War II African colonialism through post-nuclear and post-colonial struggles over race and gender. Questions about the nature of war, technology, power, and community echo through the primate literature. Given meaning through readings of the bodies and lives of our primate kin, who were semiotically placed in allochronic time and allotopic space, reinvented origins have been figures for reinvented possible futures. Primatology is a First World survival literature in the conditions of twentieth-century global history.

In Asimov's *Second Foundation* (1964 [1953]), the Seldon Plan for speeding up the return of collective advanced galactic civilization has reached a critical point. Foreseeing the decay of the present Empire, before his death Harry Seldon invented the science capable of predicting social patterns from human interactions in vast masses, a discourse he called Psychohistory. Seldon predicted and manipulated one gametic fragment for the new order and planted the second essential germ cell in the interstices of the fragmenting old Empire. Located "at opposite ends of the galaxy," the first fragment represented science and technology, and the other nurtured advanced mental powers. But the galaxy's shape makes the meaning of their relative location hopelessly ambiguous; the two foundations might be in the same place, yet unknown to each other. They might be mirror-image clones, more than haploid fragments. They turn out to relate as center and periphery, nucleus and margin. The Second Foundation finally controlled the meanings and fate of the First Foundation. These spatial ambiguities about the relation of fragments that might be clones, gametes, or parts of the same cell can be metaphors in narratives of the relations of variant explanatory frameworks in scientific repetition, fertilization, or succession. In the *Second Foundation*, a sterile mutant, the Mule, appears by chance in the story. He is the unique event that the Psychohistorians could not have predicted, and this mutation threatens to undermine the Seldon Plan. The Mule has tremendous mental powers for controlling others' minds, and he puts his power to work conquering the First Foundation and searching for the Second Foundation to add it to his upstart and monstrous empire based on violence and conquest.

Ultimately, the psychohistorians of the Second Foundation overcome the Mule's power, restoring the hegemony of their mental talents needed to knit together a cooperative new civilization.

Asimov's story provides a loose-fitting but still suggestive way to read the Center for Advanced Study in the Behavioral Sciences' second Primate Project in 1983–84, in comparison and contrast with its twin, complement, and predecessor, the first Primate Project in 1962–63. Both projects took place at the prestigious Center located near Stanford University; the Center may be imagined to be a kind of real-time Institute for Psychohistory, where accounts of the foundations of social and cognitive life are regularly reinvented by selected cultural authorities. Throughout *Primate Visions*, science fiction has provided one of the lenses for reading primatological texts. Mixing, juxtaposing, and reversing reading conventions appropriate to each genre can yield fruitful ways of understanding the production of origin narratives in a society that privileges science and technology in its constructions of what may count as nature and for regulating the traffic between what it divides as nature and culture.

The field-defining, synthetic books produced from each project's year of study, writing, and seminars are maps to changing explanatory frameworks for understanding the relations of parts to wholes and sameness to difference in post-war primatology, as well as for understanding networks of competition, cooperation, and professional reproduction among primatologists. These books, *Primate Behavior: Field Studies of Monkeys and Apes* (DeVore 1965a) and *Primate Societies* (Smuts et al 1987), mark critical reinventions of what may count scientifically as primate society. On one level, the second Primate Project was a deliberate repetition of the first, the next generation, a reproduction, a kind of duplicated cultural genetic region, with mutations coding for a novel but affiliated end product, whose substitutions and homologies can be identified, and whose function remains the recognition of difference between self and non-self, human and animal. The second primate year also dramatized the marginalization of the major paradigms and the social networks of the first project. The second project was simultaneously a nucleus directing translations of the primate story, a germinal fragment of a whole, a highly mutated clone, and the successor. *Primate Societies* is located at the opposite end of the galaxy of post-war primate field studies from *Primate Behavior*. But the opposition is based on an identity and repetition. The texts occupy the same field; they are in the same place. And for each, what counts as the core and motor of primate social life is at stake. The dynamics of cooperation and competition are endlessly elaborated in a repeating but differentiated primatological survival literature.

Both primate years at the Palo Alto Center and the resulting books owed many of their conditions of existence to the same powerful paternal figure in the biomedically oriented behavioral sciences, especially the endocrinological and neurosciences and experimental psychiatry, David Hamburg. Hamburg and Sherwood Washburn at the University of California at Berkeley collaborated to organize the first primate project. Washburn's favored former student, Irven DeVore, played a large role in planning the project year; and he edited the resulting volume, which synthesized and exhibited the dominant frameworks for most United States primate field anthropology for many years. At the time of the Primate Project in 1962–63, Hamburg was the head of the Psychiatry Department at Stanford University's School of Medicine, where he was responsible for the department's redirection to a much more experimental approach. Washburn was then entering the high plateau of the success that his "plan" would achieve for bringing together primate field studies, functional comparative anatomy, and studies of fossils. Stress was the multivalent concept for bringing together body, mind, evolution, and health in the Washburn-Hamburg vision. "Stress" was a widespread, complex element of post-Hiroshima American Cold War discourse on the relation of human beings to their technological products and animal inheritance. Stress was about cultural perceptions of the traffic between nature and human society and about the connections between body and mind. Stress was part of a discourse about the prospects of survival for nuclear humanity that also faced deep ecological crisis. The terms of possible human community were at the heart of this post-World War II, universalizing, evolutionary narrative about the origins and biology of conflict and cooperation. The solution was the concept of the "social group" as the principal primate adaptation and the "sharing way of life" as its progressive hominid variant. Difference, signified especially by race in a period of decolonization and civil rights struggles, was contained by functionalism and liberalism within an ideology privileging a science-based cultural and political order, reaffirmed in Hamburg's 1983 inaugural lecture as president of the Carnegie Corporation. From the global and local social struggles of the 1960s through the reactionary Reagan years, Hamburg kept the faith that the sharing way of life would be recuperated through international science.

Twenty years after the first primate project, Hamburg, then president of the National Institute of Medicine, invited Barbara Smuts, his former graduate student at Stanford, who was also affiliated with Irven DeVore and his student network at Harvard, to organize the second primate year. [Figure 16.1] Smuts' webs of connections to people from Gombe, Stanford, Harvard, Cambridge (England), and

Figure 16.1   Barbara Smuts with a group of baboons at Gilgil. Barbara Smuts/Anthrophoto. Published with permission.

several other sites where primatologists evolved or immigrated made her the ideal person to undertake the task. As in the first year, the core of the project was a small group resident at the Center for Advanced Study in the Behavioral Sciences. They utilized their professional networks to solicit papers broadly. *Primate Societies* had forty-six contributors, about equal to the total number of primatologists who had done field studies by the mid-1960s.[1] Twenty contributors to *Primate Societies* were women, including one of the two Japanese participants (Hiraiwa-Hasegawa Mariko). Substantial gender equality in authority and authorship, relative ethnic homogeneity, and extensive collective interactions pervade the text of *Primate Societies*.[2] Of the thirty-one people who came together to produce *Primate Behavior*, four were women (Frances Reynolds; Jane Goodall, who contributed a paper but was not present; Jane Lancaster; and Phyllis Jay). The center of gravity of the first project was Berkeley; the second project was enmeshed in an inter-institutional and inter-site web exemplified by Smuts's many connections. Prominently missing in 1982–83 was anyone from the Washburn network, or indeed from the recent or present primate people at the University of California at Berkeley.[3] The co-editors who spent the year together—Smuts, Cheney, Seyfarth, Wrangham, and Struhsaker—were multiply linked through Robert Hinde at Cambridge, Gombe before 1975, Marler's associates at the Rockefeller University, Stanford, Harvard, and the University of Michigan. The University of California at Davis and the University of Chicago were other well-represented centers of primate work in the volume.

At opposite ends of Darwinism's galaxy, primatology's Second Foundation operated with different enabling explanatory constraints for understanding primate social life compared to the first Primate Project. But like the First Foundation, the links of concepts of mind, body, and community were embedded in a larger discourse on the nature and meaning of difference and on the prospects for primate survival, including implicitly human survival, in the late twentieth century. One of the dialects or codes for that persisting larger social discourse was evolutionary theory, especially contests for the mantle of Darwin. There was one key contrast in this context between the books from the two primate years that bears directly on the bio-politics of difference in the First and Second Primate Foundations. The contrast centers around the treatment of difference, variation, parts, fragments, and wholes. Ironically, the focus on cooperation, complementary differentiation of parts in a social whole, and the social group as the primate order's defining adaptation produced a universalizing and essentializing discourse that finally sharpened narratives of antagonistic difference and preoccupations with dominance and competition. Equally ironically, the commitment in *Primate Societies* to individual selection,[4] inclusive fitness doctrines, socioecological analysis, strategic modeling, and similar explanatory resources, ordered in the last instance around the principle of antagonistic rather than complementary difference, produced opposite effects. The evolutionary arguments demanded extensive attention to several factors: situational specificity; extraordinary flexibility at all levels of analysis; alertness to myriad forms of coalition, reciprocity, and cooperation; emphasis on animals' social intelligence and generally rich mental and emotional capacities; major interest in previously relatively invisible kinds of individuals, like the aged or juvenile females; and a sense of the politics of conservation more tempered by awareness of the power-differentiated historical positions and non-harmonious interests of all the those with stakes, human and animal.[5]

In addition, the foregrounded *anthropo*logical, rather than zoological, referent of *Primate Behavior* had a paradoxical effect of narrowing the sense of possible continuities between human and animal, while constraining vision of the specificity and multiplicities in the animals' ways of life. As paradoxical, the greater zoological and ecological emphasis of *Primate Societies* seems to permit a richer map of connections between human and animal and a more diverse tool kit of available narratives for the animals. There are many reasons for these contrasts between the two books that are not linked to the explanatory strategies and variant developments of Darwinism, not least the accumulated data from twenty more years of field and laboratory studies, more than a decade of highly visible contestation over biological versions of sex and gender in primate studies, and painfully sharpened conservation dilemmas.

However, at root *Primate Societies* displays a methodological and explanatory commitment to specificity and non-reductive difference that exceeds its bottom-line equation of non-identity ("unless they are clones") with antagonistic opposition. In evolutionary discourse, and indeed much more broadly, reproductive bio-politics are the paradigmatic, iconic condensation of the whole set of narratives about same and different, self and other, one and many. The bio-politics of *Primate Societies* are about situational specificity; intrinsic explanatory and generic heterogeneity; and the construction, as natural-technical objects of knowledge, of multiple centers of agency and power in always permeable and conditional social wholes. The world of *Primate Societies* is capable of producing surprises, unexpected and promising ways of narrating the meanings of difference and sameness. Ruled by an orthodox reductionism to antagonistic difference and methodological individualism "in the last instance," the discourse of *Primate Societies* repeatedly privileges multiplicity, difference ordered by an exuberant array of possibilities, and above all, specificity. The textual richness of *Primate Societies*—and of the primate and primatological practices that enable the text—is vastly in excess of the its explicit law. Here is the interesting aspect of the Second Primate Foundation from the point of view of *Primate Visions*.

In this sense, I read *Primate Societies* as an exemplar of a widespread groping in 1980s western bio-political and other cultural discourse for ways to narrate difference that are as deeply enmeshed in feminism, anti-colonialism, and searches for non-antagonistic and non-organicist forms of individual and collective life, as by the hyper-real worlds of late capitalism, neo-imperialism, and the technocratic actualization of masculinist nuclear fantasies. The persistent binarism between antagonistic *versus* complementary or organicist ("cooperative") difference, coded in primate evolutionary biology in terms of the opposition between group selectionism or genic/individual selection, is what is cracking apart in these hydra-headed, medusoid gropings in the Primate Order.

Let me illustrate this way of reading this recent, well-authorized textbook in primate studies by briefly characterizing fragments of the writing of its first editor, Barbara Smuts. In her book based on her thesis research on baboons at Gilgil, Kenya, *Sex and Friendship in Baboons*, Smuts (1985) adopted many of the same writing strategies as those analyzed above in the section on Jeanne Altmann, "The Time-Energy Budgets of Dual Career Mothering." The generic heterogeneity in the abrupt juxtapositions of quantitative and highly allegorical and narrative accounts, as well as in the iconography of the photographs, tables, and figures, constantly forced the reader to shift reading conventions. [Figure 16.2] The text's different

Figure 16.2   Baboon friends. Barbara Smuts/Anthrophoto. Published with permission. Smuts narrates, "Such physical intimacy is rare among most male-female pairs, but common among Friends" (1985: 60). In a non-narrative mode, Smuts makes a similar point through measures like the "comparison of the number of Friend dyads in which $\%A_f$ − $\%L_f$ was greater than the same index for Non-Friend males versus the number in which the reverse was true. Sign test: $x = 33$, $z = 3.96$, $N = 40$, $p > .001$" (1985: 77). (A_f is approaches by the female; L_f is leaves by the female.) Neither the photographic image, the narrative prose, nor the statistical statement constitutes a "value-free," intrinsically "objective," non-language-mediated "fact." All three together structure a miniature scientific discursive field.

generic moves did not resolve into a single story. Smuts foregrounded the word "friendship" because she needed its polyvalent connotative loading to represent the animals as she scientifically experienced/constructed them as objects of knowledge. Her account of reproductive politics in the baboons de-centered sex and centered social intelligence and, above all, the agency of the animals. Her principal implicit ideological object of interest was heterosexual friendship. She used all these elements to suggest a story for the transition from animal to human that depended not on the division of labor and economic exchange, but on social, communicative commerce.

In *Primate Societies*, Smuts (1987a, 1987b) deepened her thematic commitment to multi-contextuality, biological anti-essentialism, multi-species perspectivity, and constant renegotiation of the bio-politics of difference as they are written into primate bodies. In "Sexual Competition and Mate Choice," she highlighted variability, complexity, and flexibility, while demoting and contextualizing the explanatory status of dominance. Her systematic attention to female choice led to an extended treatment of the complexities of the bases for female mate selections, for which individuals and histories, in principle not exhaustively knowable to an observer, counted for a great deal. "Gender, Aggression, and Influence" extended the same thematics. From one point of view, Smuts simply erred in using the word "gender" in her title instead of "sex." Without explicit discussion of the many debates about the culturally specific and contested meanings of sex and gender, she glided within the essay from the term "behavioral sex differences" to "gender." She did not take account of a large body of critical theory that maintains that gender is not about differences, but is about a relationship of power. The concept of sex differences, behavioral or otherwise, reduces an analytic about power to a positivist discourse about roles, properties, or other pre-existent observables.

But from another point of view, Smuts's "mistake" was the result of her destabilization of essentialism, and it may be productive within the terrain of feminist deconstructions of gender. By the time she was through reconstructing biological sex difference, there was no more given biological resource waiting for cultural reformation and appropriation into gender. The entire essay may be read as an argument against biological essentialism in relation to sex. In particular, Smuts makes the concept of "inherent" sex differences impossible to use in discussing differential reproductive strategies within the narratives usually called sociobiological. "Sex" became in Smuts's text a signifier for a dynamic, context-dependent (thus obviously constrained, sensitive to inequalities, and not utopian) array of possibilities. Overall, Smuts's text worked to shift attention away from *intrinsic properties* of individuals and toward constitutive social interactions and contexts with complex dimensions in time and space. The destabilization of intrinsic difference had especially intriguing effects in a narrative explicitly ruled by the premises of methodological individualism in evolutionary theory broadly and in sociobiology particularly. For all their constant strategic thinking, the "individuals" in Smuts's paradoxical text do not pass muster as good methodological individualists. Their boundaries are too permeable and webbed with others'. Internally and externally, these individuals are continuously reconstituted in intersecting, partially incongruous, unfinishable patterns. When biology is practiced as a radically situational discourse and animals are experienced/constructed as active, non-unitary subjects in complex relation to each other and to writers and observers, the gaps between discourses on nature and culture seem very narrow indeed.

In the Second Foundation's concluding chapter on the "Future of Primate Research," the rough analogy to Seldon's Psychohistorians and the problem of the dangerous mutant mental power of the Mule seems unavoidable (Cheney et al 1987). The relation of cognitive science and complex social behavior was the primatologists' penultimate topic, just before the concluding essay on conservation and primate survival. The topics and their order—mind and survival—are unsurprising in a discourse that constantly appealed to models of strategic reasoning, originary asymmetries, and evolutionary stable strategies at the heart of evolutionary biology. In evolutionary theory staying in the game is fundamentally a question of reproductive politics. Reproductive politics and communications technologies lie very near each other in this discourse. They are both aspects of strategic reasoning in relation to survival, and they are both emblematic of the breakdown of the hermetically sealed individual. Strategic reasoning is social intelligence; both are part of the technology of communication that has been progressively constructed as a central object of knowledge in twentieth-century life, human, and physical sciences.

In the relation between cognitive science and complex social behavior, communication is the luminous object of attention. And communication is where machine, animal, and human boundaries broke down dramatically in post-World War II popular and scientific discourses. Linguistics, machine communications sciences, social theory, neurobiology, and semiology all inter-digitate and sometimes conflate in contemporary cognitive sciences. Cheney, Seyfarth, Smuts, and Wrangham make the pregnant continuities, communicative commerce, and reproductive politics among animal, human, and machine explicit in their characterization of a research strategy for joining evolution, development, complex social behavior, and cognitive science.

> One research strategy, pursued in a variety of forms, has been to investigate "almost minds," such as the minds of children or the "minds" of computer programs, to see what makes them different and what would be needed to elevate them to fully human status. As the chapters in this volume illustrate, nonhuman primates provide an extraordinary diversity of "almost minds" that, in their social interactions with one another, promise to provide unique insights into the study of intelligence. ... In their natural habitats, however, primates are uniquely poised to reveal how, in the first instance, some minds gained an advantage over others. (Cheney et al 1987: 494).

In the first instance, in the beginning, there was difference, and so began the struggle of some minds to gain an advantage over others. This is a fragment of strategic narrative, oedipal narrative, and modern technological narrative, where survival—possible futures—is at stake in a techno-fetal world of 'almost minds.'[6] Do "almost minds" have "half-lives"?

Children, AI computer programs, and nonhuman primates: all here embody "almost minds." Who or what has "fully human status"? As if the answer were self-evident, the adult human scientists who wrote "Future of Primate Research" did not ask that question. And yet, primatology has persistently been about just what "fully human status" will be allowed to mean. The authors quietly embodied the maturations of the "almost minds" that they signaled: adult to child, human to nonhuman primate, scientist to machine artificial intelligence. What is the end, or telos, of this discourse on approximation, reproduction, and communication, in which the boundaries among and within machines, animals, and humans are exceedingly permeable? Where will this evolutionary, developmental, and historical communicative commerce take us in the techno-bio-politics of difference?

To address this question, we must move from reading primatology as science fiction to the next logical step—reading science fiction as primatology. Let me turn to Octavia Butler's novel, *Dawn*, the first in her trilogy on Xenogenesis (1987, 1988), as if it were a report from the primate field in the allotopic space of earth after a nuclear holocaust. Let us look at *Dawn* as the first chapter for the text that might issue from the next primate year, the Third Foundation for the third planet from the sun at the Center for Advanced Study in the Behavioral Sciences.

Reading Science Fiction as Primatology: Xenogenesis and Feminism

> Lilith: "It won't be a daughter. It will be a thing—not human. It's inside me, and it isn't human."

Ooloi: "The differences will be hidden until metamorphosis."

"I had gone back to school," [Lilith] said. "I was majoring in anthropology." She laughed bitterly. "I suppose I could think of this as fieldwork—but how the hell do I get out of the field?"
(Butler 1987: 262–3, 91)

Throughout *Primate Visions*, I have read both popular and technical discourses on monkeys and apes "out of context" (Strathern 1987). My hope has been that the always oblique and sometimes perverse focusing would facilitate revisionings of fundamental, persistent western narratives about difference, especially racial and sexual difference; about reproduction, especially in terms of the multiplicities of generators and offspring; and about survival, especially survival imagined in the boundary conditions of both the origins and ends of history, as told within western traditions of that complex genre. *Primate Visions* is replete with representations of representations, deliberately mixing genres and contexts to play with scientific and popular accounts in ways that their "original" authors would rarely authorize (Rabinow 1986: 250). But *Primate Visions* is not innocent of the intent to have effects on the authorized primate texts in both mass cultural and scientific productions, in order to shift reading and writing practices in this fascinating and important cultural field of meanings for industrial and post-industrial people.

Primate Visions does not work by prohibiting origin stories, or biological explanations of what some would insist must be exclusively cultural matters, or any other of the enabling devices among primate discourses' apparatuses of bodily production. I am not interested in policing the boundaries between nature and culture—quite the opposite, I am edified by the traffic. Indeed, I have always preferred the prospect of pregnancy with the embryo of another species; and I read this "gender"-transgressing desire in primatology's text, from the Teddy Bear Patriarchs' labor to be the father of the game, through *Primate Societies'* developmental-evolutionary narrative fragment about a heterogeneous sibling group of "almost minds." Gender is kind, syntax, relation, genre; gender is not the transubstantiation of biological sexual difference. The argument in *Primate Visions* works by telling and retelling stories in the attempt to shift the webs of intertextuality and to facilitate perhaps new possibilities for the meanings of difference, reproduction, and survival for specifically located members of the primate order—on both sides of the bio-political and cultural divide between human and animal.

Tucked in the margins and endnotes of "Teddy Bear Patriarchy" was a little white girl in Brightest Africa in the early 1920s. Little Alice Hastings Bradley was brought there by Carl Akeley, the father of the game, on his scientific hunt for gorilla, in the hope that her golden-haired presence would transform the ethic of hunting into the ethic of conservation and survival, as "man" and his surrogates, sucked into decadence, stood at the brink of extinction. The gorilla taken during that "last" hunt turned into the Giant of Karisimbi, potent and alone in his reproduction of the true image of man. After death, that gorilla became a clone of the father of the game, whose own life ended at the scene of his dreams. Duplicitous, the little girl turned into James Tiptree, Jr., and Racoona Sheldon, a man and a mother, the female author who could not be read as a woman and who wrote science fiction stories that interrogated the conditions of communication and reproduction of self and other in alien and home worlds. But Tiptree's gender, species, and genre transfigurations were only beginning to germinate in the child placed in the world still authored by the father of the game and the law of the father.[7]

But in a post-colonial world of the politics of being female, the earlier margins of possibility can become the main story. Not in the margins of the opening chapter, but at the culmination of *Primate Visions*, Octavia Butler's speculative/science fiction is preoccupied with forced reproduction, unequal power, the ownership of self by another, the siblingship of humans with aliens, and the failure of siblingship within species. Butler's is a fiction predicated on the natural status of adoption and the unnatural violence of kin. Like Tiptree—and like modern primatologists—Butler explores the interdigitations of human, machine, nonhuman animal or alien, and their mutants in relation to the intimacies of bodily exchange and mental communication. She interrogates kind, genre, and gender in a post-nuclear, post-slavery survival literature. Her fiction, especially in Xenogenesis, is about the monstrous fear and hope that the child will not, after all, be like the parent. There is never just one parent. Monsters share more than the word's root with the verb "to demonstrate"; monsters signify. Butler's fiction is about resistance to the imperative to recreate the sacred image of the same (Butler 1978). Butler is like "Doris Lessing, Marge Piercy, Joanna Russ, Ursula LeGuin, Maragret Atwood, and Christa Wolf, [for whom] reinscribing the narrative of catastrophe engages them in the invention of an alternate fictional world in which the other (gender, race, species) is no longer subordinated to the same" (Brewer 1987: 46).

But unlike Lessing, Piercy, Russ, LeGuin, Atwood, Wolf, or Tiptree, Butler's uses of the conventions of science fiction to fashion speculative pasts and futures for the species seem deeply informed by Afro-American perspectives with strong tones of womanism or feminism.[8] Butler's gender, kind, and genre germinations and transgressions begin with two protean, parental figures: the body-changing Doro, originally from the ancient Kush people of East Africa, who, after being clothed in many bodies, belongs to no people, including humanity as a whole; and the Wild Seed woman, Anywanyu, taken by Doro to colonial New England from West Africa during the slave trade. The story begins not with the white girl child brought into Africa, but with the black woman taken out, who seeds the diaspora that stands as a figure of the history and possible future of a very polymorphous species (Butler 1977, 1980). This is survival fiction more than salvation history. Catastrophe, survival, and metamorphosis are Butler's constant themes. From the perspective of an ontology based on mutation, metamorphosis, and diaspora, restoring an original sacred image can be a bad joke. Origins are precisely that to which Butler's people do not have access. But patterns are another matter.

At the end of *Dawn*, Butler has Lilith—whose name recalls her original unfaithful double, the repudiated wife of Adam—pregnant with the child of five progenitors, who come from two species, at least three genders, two sexes, and an indeterminate number of races. Adam's rib would be poor starting material to mold this new mother of humanity or her offspring. Preoccupied with marked bodies, Butler writes not of Cain or Ham, but of Lilith, the woman of color whose confrontations with the terms of selfhood, survival, and reproduction in the face of repeated ultimate catastrophe presage an ironic salvation history, with a salutary twist on the promise of a woman who will crush the head of the serpent. Butler's salvation history is not utopian, but remains deeply furrowed by the contradictions and questions of power within all communication. Butler's fiction is about miscegenation, not reproduction of the One. Butler's communities are assembled out of the genocides of history, not rooted in the fantasies of natural roots and recoverable origins. Hers is survival fiction. Most of the action of *Dawn* takes place on the Oankali ship, itself a living part of their embodied culture of "exchange." The image of deracinated fragments of humanity packed into the body of the aliens' ship inescapably evokes the reader's memories of the terrible middle passage of the Atlantic slave trade that brought Lilith's ancestors to a "New World," where a "gene trade" was also enforced. Implicated in these histories, Butler's narrative has the possibility—indeed, the necessity—of figuring something other than the Second Coming of the sacred image. Some other order of difference must be possible in Xenogenesis that could never be born in the Oedipal family narrative.

In the story, Lilith is a young American black woman rescued with a motley assortment of remnants of humanity from an earth in the grip of nuclear war. Like all the surviving humans, Lilith has lost everything. Her son and her second generation, Nigerian-American husband had died in an accident before the war. She had gone back to school, vaguely thinking she might become an anthropologist. But nuclear catastrophe, even more radically and comprehensively than the slave trade and history's other great genocides, ripped all rational and natural connections with past and future from her and everyone else. Except for intermittent periods of questioning, the human remnant is kept in suspended animation for 250 years by the Oankali, the alien species that originally believed humanity was intent on committing suicide and so would be far too dangerous to try to save. Without human sensory organs, the Oankali are primatoid Medusa figures, their heads and bodies covered with multi-talented tentacles like a terran marine invertebrate's. These humanoid serpent people speak to the woman and urge her to touch them in an intimacy that would lead humanity to a monstrous metamorphosis. Multiply stripped, Lilith fights for survival, agency, and choice on the shifting boundaries that shape the possibility of meaning.

The Oankali do not rescue human beings only to return them unchanged to a restored earth. Their own origins lost to them through an infinitely long series of mergings and exchanges reaching deep into time and space, the Oankali *are* gene traders. Their essence is embodied commerce, conversation, communication—with a vengeance. Their nature is always to be midwife to themselves as other. Their bodies themselves are genetic technologies, driven to exchange, replication, dangerous intimacy across the boundaries of self and other, and the power of images. Not unlike us. But unlike us, the hydra-headed Oankali do not build non-living technologies to mediate their self-formations and reformations. Rather, they are complexly webbed into a universe of living machines, all of which are partners in their apparatus of bodily production, including the ship on which the action of *Dawn* takes place. The resting humans sleep in tamed carnivorous plant-like pods, while the Oankali do what they can to heal the ruined earth. Much is lost forever, but the fragile layer of life able to sustain other life is restored, making earth ready for recolonization by large animals. The Oankali are intensely interested in humans as potential exchange partners partly because humans are built from such beautiful and dangerous genetic structures. The Oankali believe humans to be fatally, but reparably, flawed by their genetic nature as simultaneously intelligent and hierarchical. Instead, the aliens live in the post modern geometries of vast webs and networks, in which the nodal points of individuals are still intensely important. These webs are hardly innocent of power and violence; hierarchy is not power's only shape—for aliens, primates, or humans.

The Oankali make "prints" of all their refugees, and they can print out replicas of the humans from these mental-organic-technical images. The replicas allow a great deal of gene trading. The Oankali are also fascinated with Lilith's "talent" for cancer, which killed several of her relatives, but which in Oankali "hands" would become a technology for regeneration and metamorphoses. But the Oankali want more from humanity; they want a full trade, which will require the intimacies of sexual mingling and embodied pregnancy in a shared colonial venture in, of all places, the Amazon valley. Human individuality will be challenged by more than the Oankali communication technology that translates other beings into themselves as signs, images, and memories. Pregnancy raises the tricky question of consent, property in the self, and the humans' love of themselves as the sacred image, the sign of the same. The Oankali intend to return to earth as trading partners with humanity's remnants. In difference is the irretrievable loss of the illusion of the one.

Lilith is chosen to train and lead the first party of awakened humans. She will be a kind of midwife/mother for these radically atomized peoples' emergence from their cocoons. Their task will be to form a community. But first Lilith is paired in an Oankali family with the just pre-metamorphic youngster, Nikanj, an ooloi. She is to learn from Nikanj, who alters her mind and body subtly so that she can live more freely among the Oankali; and she is to protect it during its metamorphosis, from which they both emerge deeply bonded to each other. Endowed with a second pair of arms, an adult ooloi is the third gender of the Oankali, a neuter being who uses its special appendages to mediate and engineer the gene trading of the species and of each family. Each child among the Oankali has a male and female parent, usually sister and brother to each other, and an ooloi from another group, race, or moitie. One translation in Oankali languages for ooloi is "treasured strangers." The ooloi will be the mediators among the four other parents of the planned cross-species children. Heterosexuality remains unquestioned, if more complexly mediated. The different social subjects, the different genders that could emerge from another embodiment of resistance to compulsory heterosexual reproductive politics, do not inhabit this *Dawn*. In this critical sense, *Dawn* fails in its promise to tell another story, about another birth, a xenogenesis. Too much of the sacred image of the same is left intact.

Even so, the treasured strangers give intense pleasure across the boundaries of group, sex, gender, and species. It is a fatal pleasure that marks Lilith for the other awakened humans, even though she has not yet consented to a pregnancy. Faced with her bodily and mental alterations and her bonding with Nikanj, the other humans do not trust that she is still human, whether or not she bears a human-alien child. Neither does Lilith. Worrying that she is a Judas-goat, she undertakes to train the humans with the intention that they will survive and run as soon as they return to earth, keeping their humanity as people before them kept theirs. In the training period, each female human pairs with a male human, and then each pair, willing or not, is adopted by an adult ooloi. Lilith loses her Chinese-American lover, Joseph, who is murdered by the suspicious and enraged humans. At the end, the first group of humans, estranged from their ooloi and hoping to escape, leave for earth. Whether they can be fertile without their ooloi is doubtful. Perhaps it is not only the individual of a sexually reproducing species who always has more than one parent; the species too might require multiple mediation of its reproductive bio-politics. Lilith finds she must remain behind to train another group, her return to earth indefinitely deferred. Nikanj has made her pregnant by Joseph's sperm and the genes of its own mates. [Figure 16.3] Lilith has not consented, and the first book of Xenogenesis leaves her with the ooloi's uncomprehending comfort that "the differences will be hidden until metamorphosis."Lilith remains unreconciled: "But they won't be human. That's what matters. You can't understand, but that is what matters." The treasured stranger responds, "The child inside you matters" (Butler 1987: 263). Butler does not resolve this dilemma.

In the narrative of *Primate Visions*, the terms for gestating the germ of future worlds constitute a defining dilemma of reproductive politics. The contending shapes of sameness and difference in any possible future are at stake in the primate

Figure 16.3   Jacket illustration for the second novel, *Adulthood Rites* (1988) in Octavia Butler's Xenogenesis series. Copyright Wayne Barlowe. Published with permission. Mediating the contact of egg and sperm, this medusoid, alien-human, poly-racial hybrid figure of uncertain gender represents one of Lilith's children born from her unfree "exchange" with the Oankali, the alien species introduced in the first book of the series, *Dawn*. Barlowe's polyvalent illustration contrasts sharply with that by another artist for the cover of *Dawn*, in which the Afro-American woman, Lilith, was pictured as an ivory white brunette mediating the awakening of an ivory white blond woman aboard the Oankali ship. Illustrating the workings of the unmarked category, "white," *Dawn*'s cover art has allowed several readers whom I know to read the book without noticing either the textual cues indicating that Lilith is black or the multi-racialism pervading Xenogenesis.

order's unfinished narrative of traffic across the specific cultural and political boundaries that separate and link animal, human, and machine in a contemporary global world where survival is at stake. Finally, this contested world is the primate field, where, with or without our consent, we are located. "She laughed bitterly. 'I suppose I could think of this as fieldwork—but how the hell do I get out of the field?' "

MIRA'S MORNING SONG

BY

RAYNA RAPP, FOR MIREILLE RAPP-HOOPER

(to the tune of "You are my Sunshine")

You're not a tarsir
You're not a lemur
You're not a monkey or a shrew
You're not an orang or a baboon
You're a hominid and we love you

You are not simian
For you are sapient
And someday you'll rise to your feet
You'll learn to walk and you'll learn to talk for
You're a hominid and we think you're sweet

You are not gibbon
Nor gorilla
Although from apes you once did come
Your little vision is stereoscopic
And you have an opposable thumb

You are not Dryo
You are not Rama
You are not gracile nor robust
You are not habile nor neanderthal
You are sapiens, sapiens we trust

This is the story
My darling Mira
That science tells us of our worth
Welcome to culture
My dearest daughter
It's the greatest show on earth

NOTES

1 The Persistence of Vision

1 John Varley's science fiction short story, "The Persistence of Vision," is part of the inspiration for *Primate Visions*. In the story, Varley constructs a utopian community designed and built by the deaf-blind. He then explores these people's technologies and other mediations of communication and their relations to sighted children and visitors (Varley 1978). The interrogation of the limits and violence of vision is part of the politics of learning to revision.

2 Foucault (1973); Albury (1977); Canguilhem (1978); Figlio (1976).

3 See also Latour (1983, 1987, 1988); Bijker et al (1987); Callon and Latour (1981); Knorr-Cetina (1983); Knorr-Cetina and Mulkay (1983); Traweek (1988).

4 Hartsock (1983b); Harding (1986); Rose (1983).

5 Young (1977, 1985a); Yoxen (1981, 1983); Figlio (1977).

6 Fee (1986); Gould (1981); Hammonds (1986); Hubbard, Henifin and Fried (1982); Gilman (1985); Lowe and Hubbard (1983); Keller (1985).

7 I owe this analysis of Linnaeus as the Eye/I of God to Camille Limoges, Université de Québec à Montréal.

2 Primate Colonies and the Extraction of Value

1 Yerkes and Yerkes (1929); Zuckerman (1978); Schultz (1971); Clarke (1985).

2 Calmette (1924), Honoré (1927), Delormé (1929).

3 R.M. Yerkes Papers (RMY), folder 194:n.d., about 1924.

4 Köhler (1925); Rothmann (1912).

5 Derscheid (1927), M.J. Akeley (1929a).

6 Kohts (1923, 1935); Hahn (1988: 173–78).

7 Corner (1958); Aberle and Corner (1953).

8 van Wagenen and Aberle (1931), van Wagenen (1972).

9 Ruch (1941), Schultz (1971).

10 Zuckerman (1932); Lancaster and Lee (1965).

11 Hamilton (1914); Kempf (1917). G.S. Miller (1931), a mammalogist at the U. S. National Museum, relied on such a perspective, as well as on the new sciences of endocrinology and behavior, in producing theories on the origin of rape, prostitution, and marriage.

12 Zuckerman (1932); Hediger (1950); Benchley (1942); Marais (1926, 1939, 1969).

3 Teddy Bear Patriarchy

1 H.F. Osborn (1908, in Kennedy 1968: 347).

2 Edgar Rice Burroughs, in Porges (1975: 129).

3 A plaque at the Deauvereaux or Hotel Colorado in Glenwood Springs inscribes one version of the origin of the Teddy Bear, emblem of Theodore Roosevelt: T.R. returned empty-handed from a hunting trip to the hotel, and so a hotel maid created a little stuffed bear and gave it to him. Word spread, and the Bear was soon manufactured in Germany. It is a pleasure to compose a chapter in feminist theory on the subject of stuffed animals.

4 Believing "man" arose in Asia, H. F. Osborn presided over Museum expeditions into the Gobi desert in the 1920s in an attempt to prove this position. However, Africa still had special meaning as the core of primitive nature, and so as origin in the sense of potential restoration, a reservoir of original conditions where "true primitives" survived. Africa was not established as the scene of the original emergence of our species until well after the 1930s. Pietz (1983) theorizes Africa as the locus for the inscription of capitalist desire in history.

5 Feminist theory emphasizes the body as generative political construction (Hartsock 1983a; Moraga 1983; de Lauretis 1984, 1987; Martin 1987; Moraga and Anzaldua 1981; Hartouni 1987; Spillers 1988). See also Social Research, Winter 1974, essays from the New School for Social Research, "Conference on the Meaning of Citizenship."

6 Visual communion, a form of erotic fusion in themes of heroic action, especially death, infuses modern scientific ideologies. Its role in masculinist epistemology in science, with its politics of rebirth, is fundamental. Feminist theory so far has paid more attention to gendered subject/object splitting and not enough to love in specular domination's construction of nature and her sisters (Merchant 1980; Keller 1985; Keller and Grontkowski 1983; Sofoulis 1988).

7 William Pietz's 1983 UCSC slide lecture on the Chicago Field Museum analyzed museums as scenes of ritual transformation.

8 See also McCullough (1981); Cutright (1956). On travel and the modern Western self, especially the penetration of Brazil, see Defert (1982). Travel as science and as heroic quest interdigitate.

9 Women had all the frightening babies, a detail basic to their immigrant life in a racist society (Gordon 1976; Reed 1978; McCann 1987). Roosevelt's 1905 speech popularized the term "race suicide."

10 Akeley to Osborn, 29 March 1911, in Kennedy (1968: 186).

11 On principles of composition: Clark (1936); The Mentor, January 1926; Natural History, January 1936. Lowe (1982) illuminates the production of the transcendental subject from the structured relations of human eye/subject/technical apparatus.

12 For a genealogy of darkest Africa, see Brantlinger (1985).

13 Malvina Hoffman's bronzes of African men and women in this hall and her heads of Africans at

the entrance to the hall testify to a crafted human beauty, not a story of natural primitives. On Osborn's failed effort to enlist Hoffman in his projects, see Porter (n.d.) and Taylor (1979).

14 James Clifford's sharp eye supplied this perception. See Landau (1981, 1984) on evolutionary texts as narrative.

15 Akeley (1923: 211). The jealous mistress trope is a ubiquitous element of the heterosexist gender anxieties in male scientists' writing about their endeavors (Keller 1985).

16 Nesbit (1926); Guggisberg (1977).

17 Akeley (1923: 226). Bradley (1922) wrote the white woman's account of this trip. The white child, daughter of Mary Hastings and Herbert Bradley, became James Tiptree, Jr., the science fiction writer. Introducing *Warm Worlds and Otherwise* (Tiptree 1975), Robert Silverberg used Tiptree, whom he later learned was Dr. Alice B. Sheldon, as an example of fine masculine writing that must have been produced by a "real" man. Sheldon earned her doctorate in experimental psychology at age 52 and then began a career as a science fiction writer. Silverberg compared Tiptree to Hemingway—citing "that prevailing masculinity about both of them" (Silverberg, in Tiptree 1975: xv). Tiptree's fiction drew deeply from her travels with her naturalist parents to Africa and Indonesia. Writing as Racoona Sheldon when she wished a female-identified persona, Tiptree kept her "real" gender identity obscure until 1976, near the time of publication of her first novel, *Up the Walls of the World* (1978), which explores an alien species in which males mother the young. Tiptree's fiction and her publishing practices both interrogate gender. A man and a mother, a scientist and a writer of science fiction, a woman and a masculine author, Tiptree is an oxymoronic figure reconstructing social subjectivities out of a childhood colonial past and into a post-colonial world of other possibilities. In ill health, Tiptree committed suicide with her aged husband in 1987 in their Virginia home.

18 Reserving it for internal use only, the Museum refused permission to publish this photograph. Is it still so sensitive after 68 years?

19 Frank Chapman of the Department of Mammalogy and Ornithology was working on North American bird habitat groups, installed for the public in a large hall in 1903. In the 1880s, British Museum workers innovated methods for mounting birds, including making extremely lifelike vegetation. The American Museum founded its own department of taxidermy in 1885 and hired two London taxidermists, the brother and sister Henry Minturn and Mrs. E. S. Mogridge, to teach how to mount the groups. Joel Asaph Abel, Head of Mammalogy and Ornithology, hired Frank Chapman in 1887. Chapman, a major figure in the history of American ornithology, influenced the start of field primatology in the 1930s. American Museum bird groups from about 1886 were very popular. "Wealthy sportsmen, in particular, began to give to the museum." This turning point in fortunes was critical to the U.S. conservation movement. Department of Mammalogy and Ornithology scientists significantly enhanced the scientific reputation of the American Museum in the late 1800s (Kennedy 1968: 97–104; Chapman 1929, 1933; Chapman and Palmer 1933; pamphlet of Chicago Field Columbia Museum, 1902, "The Four Seasons"; "The Work of Carl E. Akeley in the Field Museum of Natural History," Chicago: Field Museum, 1927).

20 M. J. Akeley (1929b: 127–30, 1940: 115).

21 Akeley (1923: 223–4). Akeley recognized the utility of his camera's telephoto feature to anthropologists for making "motion pictures of natives of uncivilized countries without their knowledge" (Akeley 1923: 166).

22 October 1923, prospectus, AMNH; Johnson (1936); M. J. Akeley (1929b: 129); July 26, 1923, Akeley memorandum on Martin Johnson Film Expedition, microfilm 1114a and 1114b. The Johnsons' films were *Simba*, made on the Eastman-Pomeroy expedition, and *Trailing African Wild Animals*. *Cannibal of the South Seas* was earlier. See Osa Johnson's (1940) thriller about their lives.

23 October 1923, prospectus to the AMNH, microfilm 1114a.

24 October 1923, Osborn endorsement, AMNH microfilm 1114a.

25 Martin Johnson, July 26, 1923, prospectus draft, microfilm 1114a. The expectation that a film (*Simba*) made in the 1920s would be the last wildlife extravaganza was a wonderful statement of the belief that nature existed in essentially one form and could be captured in one vision, if only the visualizing technology were adequate.

26 The principal sources for this section are correspondence, annual reports, photographic archives,

and artifacts in the AMNH; Akeley (1923); M. J. Akeley (1940); Akeley and Akeley (1922); Mary Jobe and Carl Akeley's articles in *The World's Work;* and Delia Akeley's adventure book (1930). Delia is Delia Denning/Akeley/Howe. See N.Y. Times, 23 May 1970, 23. The buoyant racism in the books and articles of this contemporary of Margaret Mead makes Mary Jobe and Carl look cautious. Olds (1985: 71–154) provides a biographic portrait of Delia Akeley. Not sharing the elite social origins of most women explorers, Delia was born about 1875, on a farm near a small Wisconsin town, the youngest of nine children of devout Catholic Irish immigrant parents. She ran away from home at 13 and married a barber a year later. Nothing is known about the end of that marriage. Probably meeting him on hunting trips with her husband in Wisconsin, she married Carl Akeley in 1902, when he was still a taxidermist-sculptor at the Milwaukee Public Museum. Without hint of irony, Olds comments on the 1905–06 Akeley trip in Kenya: "The indispensable 'boys' took the place of horses, mules, or donkeys, because the tsetse fly made use of beasts of burden impossible" (Olds 1985: 87). In the project of recovering great white foremothers, Olds writes in 1985 in the same colonialist tones that permeate Akeley's work 60 years earlier. Olds's book is appropriately endorsed on the back cover by a NASA administrator. Olds makes a convincing case for the official scientific community's covering up Delia's role in Carl Akeley's explorations in favor of the story of Mary Jobe (Olds 1985: 150). Delia's bull elephant kill from 1906 is mounted in the Chicago Field Museum. She collected 19 mammalian species listed in the Field Museum catalogue, in addition to a large bird collection. Six weeks after her divorce from Carl in 1923, Delia was commissioned by the Brooklyn Museum of Arts and Sciences to lead an expedition to East and Central Africa. The museum director reported that it would be a "one-woman expedition"; i.e., "her sole companions on trips into the interior will be natives selected and trained by her" (Olds 1985: 114). To be with "natives" was to be "alone" epistemically. This scientific expedition was the first such venture led by a woman. Including a Dutch heiress, Alexine Tinne in 1862, an American feminist, May French Seldon, in 1891, and the British intellectual, Mary Kingsley, who traded throughout the Congo Free State in the 1890s, the theme of adventurer-white women "alone" in the "interior" of Africa does not begin with Jane Goodall and the National Geographic sagas. But the later coding of the woman as scientist is different. Contrast the popular reporting of the Goodall story in the 1960s with the 1923 headline in the *New York World:* "Woman to Forget Marital Woe by Fighting African Jungle Beasts." For the world in which Delia and Mary Jobe worked, see Rossiter (1982).

27   From English extraction on both sides, Mary Jobe Akeley (1878–1966) was born and went to college in Ohio. Her father's family had been in America since colonial times. Mary Jobe studied English and history for two years in graduate school at Bryn Mawr, earned a Master's degree at Columbia in 1909 and was on the Hunter College faculty until 1916. She owned and ran a summer camp for upper class girls in Mystic, Connecticut, from 1916–30, where Martin and Osa Johnson talked of their adventures. Married to Carl Akeley in 1924, she led her own expeditions in 1935 and 1947. Her wildlife photography dates from about 1914 (McKay 1980).

28   Engraved on a plaque at the entrance to Earth History Hall, AMNH.

29   H. F. Osborn, 54th Annual Report to the Trustees, p. 2, AMNH.

30   Carnegie (1889); Domhoff (1967); Kolko (1977); Weinstein (1969); Wiebe (1966); Hofstadter (1955); Starr (1982); Oleson and Voss (1979); Nielson (1972).

31   Latour (1988); Latour and Woolgar (1979); Knorr-Cetina and Mulkay (1983).

32   On decadence and the crisis of white manhood: F. Scott Fitgerald, *The Great Gatsby* (1925); Henry Adams, *The Education of Henry Adams* (privately printed 1907); Ernest Hemingway, *Green Hills of Africa* (1935). On the history of conservation: Nash (1977, 1982); Hays (1959). On eugenics, race doctrines, and immigration: Higham (1975); Haller (1971); Chase (1977); Ludmerer (1972); Pickens (1968); Gould (1981); Chorover (1979); Cravens (1978); Kevles (1985). On sexuality, hygiene, decadence, birth control, and sex research in the early 1900s in life and social sciences: Rosenberg (1982); McCann (1987); Sayers (1982).

33   AMNH: Osborn, "The American Museum and Citizenship," 53rd Annual Report, 1922, p. 2. For the sweep of his work, see Osborn (1930).

34   Osborn, "Citizenship," p. 2.

35   Osborn, 53rd Annual Report, 1921, pp. 31–32. Ethel Tobach helped me find material on AMNH

social networks, eugenics, racism, and sexism. Galton Society organizing meetings were in Osborn's home.

4 A Pilot Plant for Human Engineering

1   Yerkes (1925: Chpt. 13); Yerkes (1943: Section III).

2   Foucault (1973); Jacob (1974). Lamenting the capitalist logic of managers, but revering patriarchy, Lasch (1977: Chpt. 2) gives another view of the central place of personality sciences and study of social relations in capitalist management of the family in this period.

3   "[Sex as a political issue] was at the pivot of the two axes along which developed the entire political technology of life[:] the disciplines of the body [and] the regulation of populations. . . . [T]he mechanisms of power are addressed to the body, to life, to what causes it to proliferate, to what reinforces the species, its stamina, its ability to dominate, or its capacity for being used" (Foucault 1978: 145, 147). See also Baritz (1960); Barnes and Shapin (1979); Barker-Benfield (1972).

4   Noble (1977: 264); Braverman (1974).

5   Morawski (1986); Cross (1982); Angell (1907); Domhoff (1967); Mills (1956).

6   Robert M. Yerkes Papers (RMY): *Testament*, iv. The *Testament* is a 400+ page unpublished manuscript. A full biography of Yerkes is planned by James Reed of Rutgers. For biographies by Yerkes's contemporaries, see Hilgard (1965), Boring (1956), Carmichael (1957). Yerkes (1932a) wrote an earlier autobiographical statement. Excerpts from *Testament* dealing with primates were posthumously published (Yerkes 1963). According to Yerkes's daughter, Roberta Yerkes Blanshard, an editor at Yale Press at the time *Testament* was rejected for publication: "The MS did not seem to us to present a balanced account of RMY's life work; it overemphasized the phase of thought that he was going through at the time of writing it, his rethinking of his growth away from religion. He was doing his own thinking, but the results were not new; many had traveled that road before him. The original contributions to his life were correspondingly underemphasized" (personal communication, 23 July 1978). But the historian is particularly interested in what *Testament* shows to be shared. The itinerary from religion to science within a strong ideology of service is repeated and multiply varied in America in the early twentieth century. The story is a New Testament, where the mythic elements of salvation history are reworked in a scientific primal story. Yerkes's Laboratories of Comparative Psychobiology and his ideal of human engineering as the scientific production of cooperation cannot properly be understood apart from the tradition of Christian utopian literature and its search for the organon (logic, tool) of community. The mythic structure of Yerkes's *Testament*, as well as of his published work like *Almost Human* and *Chimpanzees: A Laboratory Colony*, is not idiosyncratic in primatology, where the fundamental text has remained a tale of origins, language, and tools in the historical crafting of nature, family, work, and community.

7   The classic and still indispensable history of the great chain of being in western thought is Lovejoy (1960) [1936].

8   RMY, correspondence folder 18: Yerkes to Angell, 12 May 1924; "Memorandum for President James R. Angell, Subject: Proposed Academic Center or Institute for Psychology," 12 February 1924.

9   RMY, correspondence folder 18: Raymond Pearl to Yerkes, 2 June 1913, folder 28; *Testament:* 221–27; Yerkes to Angell, 12 May 1923. "There is one portion on the research program of this [Migration] Committee which might logically be located at Yale if our plans were realized. It has to do with the internationalizing or universalizing of methods of mental measurement." Brigham was kept by Princeton when Yale could not make a sufficiently attractive offer. Yerkes to Angell, 6 June 1924.

10 Yerkes (1916, 1925, 1932b, 1943).

11 RMY: Annual Reports (1924–29) of the Institute of Psychology; of the Anthropoid Experiment Station of the Laboratories of Comparative Psychobiology (1930–35), later the Yale Laboratories of Primate Biology (1935–42); and the Yerkes-Angell correspondence. The Florida center housed about 30–40 animals at any one time. Many chimpanzees were gifts from the Abreu estate in Cuba, a private collection which formed the basis for *Almost Human*. Others had been collected by Nissen from French West Africa. The goal was eventually to have all laboratory-bred animals. The Yerkes

laboratory became a Regional Primate Research Center of the National Institutes of Health at Emory University, Atlanta, scene of Frederick Wiseman's documentary film, *Primate* (1974).

12   RMY, folder 626: R.M. Yerkes, "Plan for the Development of the Department of Comparative Psycho-Biology," 28 March 1929, p. 1.

13   *Human Biology and Racial Welfare,* a textbook for students in the life sciences, enunciated the ideology particularly clearly (Cowdry 1930). Yerkes's chapter, "Mental Evolution in the Primates," was organized around the chain of being and recapitulation doctrines. The book's preface announced that the text was for "readers of mature years occupied both in science and in business" (vi). Edwin Embree, former Rockefeller Foundation officer in charge of the 10-year grant to Yerkes for an ape breeding station, wrote the "Introduction." It was an apology for scientific management and human engineering in the fulfillment of human destiny to dominate nature through knowledge and action. In the early Depression's gnawing doubt about the beneficence of science, biology was claimed to provide the missing sense of purpose and unity—through the sciences of functional adjustment. In the section entitled Man as a Physiological Unit, Clark Wissler contributed "The Integration of the Sexes—Marriage." The culminating section, The Future, included papers on the biology of human populations (Raymond Pearl), mingling of races (Charles Davenport), purposive improvement of the human race (E. G. Conklin), and intentional shaping of human opinion (H. A. Overstreet). Other contributors included John Dewey, W. B. Cannon, C. S. Sherrington, and W. M. Wheeler. These were the senior men who laid the foundations for the 1930s life sciences of integration. Teleologically, I read the book as a culmination of evolutionary naturalism and a transition to the variation of organicism-functionalism relocated in the hierarchy of systems sciences by 1940 (Cross and Albury 1987; Haraway 1981–82; Curti 1943: Young 1981; Burnham 1968a, 1968b).

14   RMY: Yerkes, "Plan for the Development of the Department of Comparative Psycho-Biology," p. 2.

15   Yerkes and Yerkes (1934: vii, 331). RMY, folder 626: Plans and Budget of the Laboratories of Comparative Psychobiology, 1928–34. On Fulton, see Hutchinson (1960). The multiple meanings of physiological dominance are key themes in primate neurobiology. Yerkes's collaborations in the 1930s included work on reproductive physiology with Edgar Allen and J. H. Elder and on the nervous system with Fulton and C. F. Jacobsen (RMY, correspondence folders 180–81: John Fulton, 1929–55). The 1930s also produced important papers on experimental sociology and animal models for psychopathology (NRC-NAS archives: Conference on Experimental Neuroses and Allied Problems under the Auspices of the Inter-Divisional Committee on Borderland Problems of the Life Sciences, April, 1937).

16   RMY, correspondence folder 19: Angell, 1930–49; folder 665: Report of the Committee on Appraisal and Recommendation, 1935, and Administrative Committee, Minutes of Meetings, 1939–42; Annual Reports of the Anthropoid Experiment Station of Yale University. On Weaver's policy see Kohler (1976) and Abir-Am (1982). Ironically, the many science promotion committees Yerkes worked on partly provided a model for Weaver's activist management and promotion that worked to exclude the Yale psychobiologist.

17   For thesis that biology was constituted around life and organization, see Albury (1977); Cross (1977); Figlio (1977, 1976); Foucault (1971, 1973); Jacob (1974).

18   Rubin (1975); Rich (1980).

19   Yerkes (1916, 1925: 95, 1927a, 1927b, 1928, 1930, 1935–36); Yerkes and Yerkes (1929).

20   Relating to dissimilar doctrines of adaptation, the action patterns of psychobiology differ from those of Konrad Lorenz's ethology except that both required clinical observation skills (Klopfer 1974; Lorenz 1957, 1971; Richards 1974; Thorpe 1979).

21   Yerkes (1927b: 153, 1930: 134). Yerkes worked passionately to protect anthropoids both in captivity and in natural habitats. He greatly admired Carl Akeley's efforts to save gorillas in Parc Albert in the Belgian Congo.

22   Yerkes and Yerkes (1929: 549); Yerkes (1930: 129).

23   Yerkes (1935–36). This lecture must be seen in light of Yerkes's extreme funding troubles with Warren Weaver at the time. Strident emphasis on "culture methods" and physical-chemical language were unusual.

24 Yerkes (1937); Yerkes and Yerkes (1935: 1027–31).

25 Dawkins (1976); Tooby and DeVore (1987).

26 RMY: Regulations printed in Yerkes 1932b; Annual Reports of Anthropoid Experiment Station of Yale University, 22 May 1929 to 30 June, 1930, "Scientific Activities," pp. 12–13.

27 These beliefs were so strong that, in a book which continually pointed out the scant or invented observations that the incautious had used to make assertions in the past, the authors classified social type before any systematic field observations had been made. Yerkes relied on Mme. Abreu of Havana. He interviewed her and observed her colony in 1924. "It would seem from all that has been learned about chimpanzees by Mme. Abreu and all that the literature affords, that this [a father, mother, and baby] is a typical family. If it were resident in the wilds of Africa, there would probably be other children in the group, for it is well established that family relations are long maintained" (Yerkes 1925: 138).

28 Crawford (1939); Pauly (1976); Maslow (1936a, 1936b, 1936c, 1940, 1943, 1954, 1966). M.P. Crawford was an investigator in Yerkes's laboratories; Abraham Maslow was Harry Harlow's graduate student.

29 May and Doob (1937); Adorno (1950); Fromm (1941); Yerkes and Yerkes (1935–6: 1018). Yerkes's physical anthropologist friend, Earnest Hooton (1942), developed the theme of the communist howlers for a popular audience in a book which used anthropomorphic categories of domineering, individualist, reactionary, communist, proletarian, etc., as the primary story telling device.

30 RMY, box 7: Herschberger to Yerkes, 2 March 1944. Herschberger was interested in encouraging Yerkes's associates to pay more attention to the female orgasm in chimpanzees. She objected strongly to Yerkes's terminology of "the naturally more dominant" and "naturally more subordinate" animals, and she hypothesized that females might be more assertive over food during estrus because they were hungry—not for status but for food. Yerkes justified his "natural" terminology as more convenient and unlikely to be misunderstood. Indeed. Yerkes to Herschberger, 30 March 44, Box 7.

31 Yerkes (1943: 69); Freedman (1974).

5 The Semiotics of the Naturalistic Field

1 A consequence of the 1937 Asiatic Primate Expedition, the Cayo Santiago rhesus colony has been a major primate behavior research resource since the 1950s. The colony was tied to Columbia's associated School of Tropical Medicine in the University of Puerto Rico. A convergence of interests from reproductive physiology, naturalistic behavior studies, and investigation of infectious diseases justified the Markle Foundation 3-year grant to establish the colony. A partial Indian embargo on the export of rhesus monkeys was an additional stimulus. Robert M. Yerkes Papers (RMY), folder 84: Carpenter, "Macaques on Santiago Island, Puerto Rico"; Carpenter-Yerkes correspondence, 1937–38; C.R. Carpenter Papers (CRC): Carpenter field notebooks, "The Markle-Columbia Primate Expedition"; Carpenter (1942, 1946); Windle (1980).

2 CRC: Carpenter field notebooks, "Markle-Columbia Primate Expedition."

3 CRC: Carpenter field notes, "Santiago Island," June–July 1939, Feb.–May 1940, May–June 1940. Carpenter (1964: 276–341).

4 CRC: "Notes on Santiago Island Colony," May–June 1940, p.97.

5 CRC: "Notes on Santiago Island Colony," May–June 1940, pp. 120, 122, 139, 159.

6 Carpenter 1929, 1933a, 1933b. The doctoral work was funded from Stone's grant from the Committee for Research in Problems of Sex, chaired by Yerkes from 1922–47.

7 RMY, folder 82: Carpenter to Yerkes, 18 Feb. 1931; National Research Council-National Academy of Sciences (NAS): letters of recommendation from Stone, Miles, Terman, and McDougall, National Research Fellowship records.

8 RMY, folder 82: Yerkes to Carpenter, 10 March 1931; Yerkes to National Research Fellowship Board, 28 March 1931; Yerkes to Walter Miles, Yale Institute of Human Relations, 10 March 1931.

9 RMY: Yerkes to Thomas Barbour, Museum of Comparative Zoology and Institute for Research in Tropical America, Barro Colorado Island Biological Laboratory, 14 Nov. 1931.

10 Chapman to Yerkes, 1 Feb. 1931. Barro Colorado Island Biological Laboratories were part of the Institute for Research in Tropical America, set up under National Research Council auspices in 1923. NAS: Institute for Research in Tropical America file; Barbour (1945); Chapman (1929); Fairchild (1924).

11 Older work on pigeons and domestic fowl was a rich source for Carpenter's work. The bird literature affected Carpenter's primate studies in three interlocked areas in the 1930s; territoriality, population biology, and the animal sociology of cooperation and competition—all part of a general physiological functional approach to organismic wholes. Carpenter incorporated all three into his howler monkey study (Howard 1920; Hutchinson 1978). The space of territory was both physical and psychological. Behavior, not physical boundaries, marked off territory. Behavior differentiating space—such as bird song, scent-marking, aggressive display, or periodic travel to particular places—was a fundamental mechanism of integration of animal societies. Mapping space by mapping behavior became an operation appropriate to social physiology. See Mayr (1975: 365–396) for connections between ornithology and primate studies. Unlike mammology, with its deeper ties to medical physiology, or physical anthropology, with its links to racialism and reconstructions of narratives of human origins, ornithology has an extra-ordinary tradition of interest in organisms in their natural habitats. Important figures include John Emlen, Nicholas Collias, Frank Chapman, J.H. Crook, Robert Hinde, Peter Marler, and many others. Emlen, a graduate student of Cornell's Arthur Allen worked on communication and territory. George Schaller, who studied the mountain gorilla in the early 1960s, and several other later primate field biologists were Emlen's graduate students at Wisconsin. At Cornell, Allen (1962) was the first university professor of ornithology in the United States. A conduit for European ethology into primatology, Collias collaborated with an important figure in primate population biology and resource conservation, Charles Southwick (Collias 1950; Collias and Southwick 1952). The history of ornithology, with its late university professionalization among the biologies and long connections with museums and field expeditions, illuminates the transformation of biological research from "workshop" to "industrial" conditions. Primatology shares some of the same dates in this process. C.O. Whitman, the first director of the Marine Biological Laboratory of Woods Hole and chair of Chicago's Department of Zoology, and his student Wallace Craig studied pigeons in ways that lessened the distance usually thought to exist between American comparative psychology and Continental ethology before the 1950s. Whitman, cited favorably by Konrad Lorenz in the 1930s, carefully studied motor action patterns of pigeons as morphologically differentiated forms useful in making taxonomic judgments (Whitman 1919; Craig 1908, 1918). Margaret Nice, an American student of sparrows, called Carpenter's attention in 1935 to Lorenz's 1931 and 1932 papers (CRC: Nice to Carpenter, 16 March 1935, 29 March 1935; Nice (1937); Chapman and Palmer (1933). Ernst Mayr felt that Carpenter, formed by comparative psychology, never understood the modern evolutionary synthesis and Darwinian approaches to territoriality (Washburn to Haraway, 4 July 1979). Carpenter was a mélange of neo-Lamarckian and neo-Darwinian strands in the 1930s.

12 Raymond Pearl led in the study of human and animal populations as physiological objects, for which variations in growth form were in principle subject to medical management (Pearl, in Cowdry 1930). Pearl, with L. J. Reed, rediscovered the famous logistic equation to describe population growth (Hutchinson 1978: 1–40; Kingsland 1985). An employee of the Metropolitan Life Insurance Company, Alfred J. Lotka (1925), worked for a year in the mid-1920s at The Johns Hopkins University School of Hygiene and Public Health in association with Pearl. Lotka illuminates discursive interconnections of analytic techniques shared by the insurance industry and biological theory. Political and natural economy have a long common history; economics and bionomics are both management sciences based on production systems, for which rate equations are fundamental representations and technologies.

13 Mitmann (1988); Caron (1977); Allee (1931, 1942).

14 Although the term ecosystem was coined in 1935, a cybernetic systems concept of the basic unit in ecology did not take form until after World War II (McIntosh 1977; Shelford 1913; Clements and Shelford 1939). For the transformation of ecology from a science based on organic communities to one based on cybernetic systems, see Farley (1977), Taylor (1987), Hutchinson (1948), Odum (1977).

15 Allen (1975); Parascondola (1971); Russett (1966); Haraway (1976).

16 Carpenter came from a poor family in Cherryville, North Carolina. His parents wrote a moving letter to Yerkes, thanking him for his many kindnesses and expressing their worry about their son heading for a foreign and strange place so far from home. RMY, folder 82C: E. and G. L. Carpenter to Yerkes, 22 Dec. 1931.

17 RMY, folder 82: Carpenter to Yerkes 31 Dec. 1931.

18 From the beginning Carpenter photographed his howlers, following the lead of Chapman, who had introduced photography in his bird work in 1908. RMY: Carpenter to Yerkes, 5 Feb. 1932; Mayr (1975: 391–92).

19 RMY: Yerkes to Carpenter, 18 Jan. 1932.

20 RMY: Carpenter to Yerkes, 26 March 1932.

21 RMY: Carpenter to Thomas Barbour, 9 Feb. 1933; CRC: Carpenter field notes, "Field Studies at Camp La Vaca, the Clark Expedition," pp. 47–48, 54–56, 58–61.

22 Such as f−y, M−f, y−y, M−y, etc.; f=female; y=young; M=Male. The capital category is unconscious. Later, he used the formula $[N (N−1)/2]$ for the maximum number of possible pair-wise relations that would constitute a social group.

23 Similarities between Harold Coolidge and Frank Chapman illuminate aspects of primatology. Without doctorates, both were consummate specialists in a zoological field. Chapman's institutional base was ornithology at the American Museum of Natural History; Coolidge's was the mammalogy section of Harvard's Museum of Comparative Zoology. Both were instrumental in national and international wildlife conservation. Legally enforced conservation was a mixed blessing to anatomists and physiologists, making collecting more difficult, if favoring the survival of something to collect. In the 1930s, scientists and travelers were only beginning to adjust to the individual consequences of rational conservation. Coolidge selected Asia for the primate expedition because the London Convention on Africa, effective in January 1936, strictly protected the gorilla and partly protected the chimpanzee:

> It is true that specimens can be obtained on a scientific permit, but this new law is so recent that it will be some years before what one might call a liberal scientific permit would be issued. In Asia the Dutch have already closed down on the gibbon and the orang. The British are planning a Pan-Asiatic Convention similar to the African Convention. . . . When this happens, it will be too late to obtain a sufficient series of one species to make much needed studies of variation. It will be also extremely difficult to obtain reproductive material and especially fetuses, which are at present extremely uncommon in the collections. . . . In years to come we can continue our field studies on the gibbon, but in order to obtain collections to correlate with these field studies it will be necessary to have a scientific permit, and such a permit would not be extensive enough to justify the expense of taking a party of primate specialists into the field. (RMY: Coolidge to Yerkes, 6 June 1936)

Coolidge urged Yerkes's confidentiality on these reasons for the expedition. The public press announcements stated "Party to Trace Human Traits in Asiatic Apes" (*New York Herald Tribune*, 16 Dec. 1936, p. 25); "Studying the Key Ape: The Gibbon in Siam May Unlock Secrets of Man's Past" (*New York Times* 2 Dec. 1936).

24 Committee for Research in Problems of Sex (CRPS): 1925–23, Grantee: C. Wissler. Beatrice Blackwood (1935), a protegée of Bronislaw Malinowski, like most anthropologists of her generation, sought a group "uncontaminated by civilization."

25 CRPS: 1925–33, Grantee: C. Wissler, "Report on Anthropological Studies Supported by the CRPS," pp. 5–7.

26 Carpenter's gibbon work continued to emphasize the theoretical framework of organicism joining the biological and social sciences in the 1930s. Within this framework, Lawrence Henderson (1935) extended his laboratory physiology into sociology, via a reading of Vilfredo Pareto. Heir to this link, the social systems theorist Talcott Parsons (1970) knew his debt.

27 The journal was founded in 1937. Drawing his first sociograms while still in Berlin in 1923, Moreno ran an Institute of Sociometry from New York City. Sociometry was prominent in the study of democratic and authoritarian personalities in the late 1930s.

28 RMY, folder 84: Carpenter to Yerkes, 20 March 1939; Yerkes to Smith, 23 March 1939; Smith to Yerkes, 27 March 1939; Smith to Carpenter, 27 March 1939; Yerkes to Carpenter, 9 May 1939. Smith and Engle did not intend to recommend Carpenter's research proposal on maternal drives and endocrines to the Committee for Research in Problems of Sex (Smith to Yerkes, 27 March 1939).

29 In World War II Carpenter was an Air Force technical adviser on training films for jungle warfare. In 1944 he chaired the American Psychological Association's committee on audiovisual aids. He established the Psychological Film Registry at Pennsylvania State and directed the evaluation of films produced by the military for use in psychological instruction. Carpenter directed the Instructional Film Research Program at Pennsylvania State, and worked to use television for national and international "integration." Carpenter remained a reserve officer in the Air Force, with responsibilities in training-technology. He served as special consultant on educational programs in Guam and Samoa, participant in numerous international conferences on problems of integration through new communications technologies, consultant on legislation on multimedia approaches to primary and secondary education, consultant to the General Electric Company and Raytheon EDEX Educational Systems, director of the Washington Center for Applied Educational Research, and Ford Foundation adviser to the Indian government on development of a communications system.

Educational technology for Carpenter remained tied to a medical, psychological, and psychiatric approach to human cooperation. In the Air Force, he helped organize the German Youth Reeducation Program and was the Director of the Research Section of the Biarritz American University in France. He interpreted his work as part of rebuilding democracy through judicious application of educational control as a psychiatric therapy to produce adjustment after pathology. Carpenter echoed a persistent subtheme in post-war life science, including primatology, i.e., psychiatry addressed to mass social therapeutics, especially in managing aggression and stress. C. B. Chisholm, a Canadian psychiatrist and former army officer who took a leading role in the World Health Organization, identified educators and psychiatrists as key partners in a new mass preventive psychiatry addressed to a whole society, with the same logic exercised previously on small groups. As usual, "the family" was both a major concrete focus and an ideological model for therapeutics and theory. Educators and psychiatrists were to assume the burdens of parenthood in helping the world citizens adjust to the post-war realities of global integration and interdependence. Local prejudices would have to give place to a more inclusive identity prescribed by the experts. Therapeutic attitudes should replace religious outlooks in a psychiatrically ordered program of education. Maturity meant the ability to work in organizations under rational authority and the willingness to show tolerance, adaptation, and compromise (Lasch 1977: Chpt. 5).

30 Interview with S.A. Altmann, April 1, 1982.

31 In the late 1950s and early 1960s, Altmann's approach stood in especially sharp contrast with the anthropological social functionalist frames of Sherwood Washburn and his students.

32 Yoxen (1977); Heims (1980); Kevles (1978).

33 Waddington (1977, 1970). Waddington was most attracted to topological models translating concepts of boundaries and probabilities of trajectories into visible-tangible landscapes. This work was continuous with his pre-war organismic embryology (Needham 1936; Haraway 1976).

34 Murrell (1965). Murrell, an officer in the British Admiralty, described the founding of the interdisciplinary Ergonomics Research Society in his office in 1949, in order to continue the progress made in the war in studying human performance as an engineering systems design problem. In the United States, the Office of Scientific Research and Development played a similar role to the British Admiralty and Royal Air Force in facilitating the science of cybernetics—the study of self-regulating, automated devices ordered by the control of rates and directions of flow of information. Feedback control is the most familiar strategy in cybernetic systems. The classic description of the emergence of cybernetics from problems in gunnery control and the breakdown of human operators is Wiener (1948). See also Yoxen (1977: Part III; Heims 1980: Chpt. 9). E.O. Wilson pioneered in the application of ergonomics to study of the division of labor in insect colonies (Haraway 1981–82; Murrell 1965). "It is helpful to think of a colony of social insects as operating somewhat like a factory constructed inside a fortress . . . [T]he colony must send foragers out to gather food while converting the secured food inside the nest into virgin queens and males as rapidly and efficiently as possible" (Wilson 1968: 42).

35 Stress medicine after the war may be approached through *Life Stress and Bodily Disease: Proceedings of the Association for Research in Nervous and Mental Disease*, XXIX (New York: Afner, 1968, reprint of 1950 ed.).

36 Theories of textuality in the semiologies of the human sciences made parallel moves in the post-war period. The objects of work and author no longer grounded theories of meaning.

37 The fundamental technical point in the Shannon-Weaver equation (H = -Spi log pi) is strict focus on the statistical nature of information. Weaver (in Shannon and Weaver 1949: 104–05):

> [F]rom the point of view of engineering, a communication system must face the problem of handling any message that the source can produce. If it is not possible or practicable to design a system which can handle everything perfectly, then the system should be designed to handle well the jobs it is most likely to be asked to do, and should resign itself to be less efficient for the rare task. This sort of consideration leads at once to the necessity of characterizing the statistical nature of the whole ensemble of messages which a given kind of source can and will produce. And information, as used in communication theory, does just this.

Warren Weaver, director of the Division of Natural Sciences of the Rockefeller Foundation from 1932 to 1955, changed the rules for funding biological research, such that Yerkes's psychobiology lost favor. Molecular biology was one spectacular result; comprehensive cybernetic theories of behavior and evolution were another (Kohler 1976; Abir-Am 1982).

38 von Foerster (1950–55). The conference proceedings were published, beginning with the sixth meeting. Under the leadership of the neurophysiologist-psychiatrist Warren McCulloch, these conferences brought together an extraordinary collection of notables in information and systems theories, many of whom had been active in pre-war work that came to fruition under the intense demands of wartime science. Besides McCulloch, participants included anthropologists Margaret Mead and Gregory Bateson; social psychologists Alex Bavelas and Kurt Lewin; engineer Julian Bigelow; psychiatrist Henry Brosin and neurologist-psychoanalyst Lawrence Kubie; mathematicians Norbert Wiener, John von Neumann, and Leonard Savage; logician Walter Pitts; physiologists Rafael Lorente de No, Ralph Gerard, and Arturo S. Rosenblueth; comparative psychologist and specialist on insect societies, Theodore Schneirla; experimental psychologist Heinrich Kluver; philosopher F. S. C. Northrop; and ecologist G. Evelyn Hutchinson. Networks were rich; for example, Walter Pitts had studied with Rudolph Carnap, Charles Morris's colleague at Chicago. Invited guests such as Claude Shannon from the Bell Telephone Laboratories; neuropsychiatrist David McKenzie Rioch; physicist Donald M. MacKay; and primate comparative psychologist from Robert Yerkes's Orange Park laboratories, Herbert G. Birch, actively participated in the meetings. Most of these people did not produce militaristic models and oppressive applications from cybernetics. Mead, Bateson, Hutchinson and Schneirla are all correctly considered humanists; Heims (1980) documents Wiener's rejection of militarism. My point is that recently constituted natural-technical objects, whether or not a particular scientist develops "humanist" meanings, has the shape of a communication-command system. See Ashby (1961); Lilienfeld (1978).

39 The earliest uses of the term "sociobiology" were embedded in contexts rich with reference to interdisciplinary work and application to human life. Comparative psychologist John Paul Scott in 1946 used the term at a conference on genetics and social behavior; in 1948 Charles Hockett, a linguist seen by Altmann in the tradition of Charles Morris and a major resource for Altmann's use of cybernetic communication theories, proposed the term. Scott (1950: 1004) suggested the term for the "interdisciplinary science which lies between the fields of biology (particularly ecology and physiology) and psychology and sociology." Schneirla was an active participant in the 1948 conference to which Scott's 1950 foreword refers. The search for an adequate theoretical frame was a major theme. From 1956–1964 Scott and others constituted the Section on Animal Behavior and Sociobiology of the Ecological Society of America. This section became the Animal Behavior Society in 1964. In 1965, Altmann's professional title at the Yerkes Primate Regional Research Center in Atlanta was "Sociobiologist." In a letter to Stuart Altmann (20 Oct. 1981, courtesy of E.O. Wilson), Wilson recalled himself and his student in Puerto Rico, when Altmann was doing his dissertation research in January 1956, discussing the need for a "deep comparative sociobiology." Wilson wrote that he was motivated by the recent successes of the reductionist program in molecular biology. "I believe that we succeeded, in part because of the information-analysis techniques that were avail-

able—again, as you say—thanks to Ma Bell's research arm." In an interview with me, Wilson (19 March 1982) recalled with pleasure his own promotion to full professor in the early 1960s ahead of James Watson. Wilson saw the theoretical power and analytical program of sociobiology both as the legitimate successor to the evolutionary biology of the previous generation, represented by Ernst Mayr at Harvard, and the answer to Watson's contempt for evolutionary explanations. The term "sociobiology" also has many other genealogies and meanings.

40 Altmann fundamentally disagrees with my interpretation of the importance of military and industrial themes in the inheritance of post-war social biological theory and practice (Altmann to Haraway, 12 Oct. 1981; interview 1 April 1982).

41 CRC: Altmann to Carpenter, 5 Feb. 1956, 5 June 1956; Morrison to Carpenter, 23 Nov. 1955; Rioch to Carpenter et al, 6 Jan. 1956; "Agreement between the National Institute of Neurological Diseases and Blindness and the University of Puerto Rico," 16 April 1956;Conference on the Procurement and Production of Rhesus Monkeys, Institute of Animal Resources, National Research Council, June and Sept. 1955.

42 CRC: Rioch to Carpenter, 6 Jan. 1956.

43 CRC: Altmann to Carpenter, 22 March 1956, with protocol of Altmann's research questions. The 1962 publication of his dissertation research appeared in a special issue of *The Annals of the New York Academy of Sciences,* a landmark in the rebirth of primate social-behavioral biology after the war.

44 CRC: Altmann protocol, Altmann to Carpenter, 22 March 1956.

45 Miller (1951); Miller, Wiener and Stevens (1946); Edwards (1988).

46 The basic design features he explored were: channel of communication, directed transmission, presence of metacommunication, multiple coding potential of the system, interchangeability of group members, semanticity of signals, and arbitrary denotation (Altmann 1967: 325–62).

47 The Bermuda Biological Station provided support facilities. The Public Health Service granted $130,000 for a 1970–72 investigation of "Telemetric Control of Free-Ranging Gibbons" (CRC: grant application). The Rockland State Hospital experiments were part of the historical research base for Marge Piercy's feminist science fiction novel, *Woman on the Edge of Time* (1976), in which Consuelo, a Chicana mental hospital patient, rebels against coercive experimental brain surgery and behavior control.

48 CRC: "Telemetric Control of Free-Ranging Gibbons," p. 18.

49 In the 1960s, the popular conservative writer Robert Ardrey (1961, 1966, 1970) extensively used Carpenter's work on territory, aggression, and population equilibrium. Ardrey and Carpenter conducted a long, mutually enthusiastic correspondence (CRC: Robert Ardrey file, 1964–70). Ardrey referred a New York labor arbitration official to Carpenter for expert advice on applying "territoriality" to settling labor disputes (Carpenter to Donald Strauss, American Arbitration Association, 2 Nov. 1966; Strauss to Ardrey, 29 Sept. 1966; Strauss to Carpenter, 25 Oct. 1966). Carpenter referred worried white parents during the civil rights struggle to Ardrey (CRC: Primates, General Correspondance File, Dorothy Mayhue to Carpenter, 16 March 1970; Carpenter to Mayhue, 20 April 1970). Carpenter tried to get favorable reviews in magazines to promote *The Social Contract. Life* serialized the book, hoping that its controversial content would help the financially troubled magazine. *Life*'s special interest was population control (Ardrey to Carpenter, 21 March 1970). The BBC was enthusiastic about making thirteen 50-minute features written by Ardrey on the origin of violence (Ardrey to Carpenter, 23 July 1970; Carpenter to Ardrey, 23 Oct. 1970; Carpenter to Ardrey, 9 Sept. 1970). After finishing *The Territorial Imperative* in 1966, Ardrey wrote the film script for *Khartoum,* a $7 million United Artists extravaganza. *Territorial Imperative* was especially popular in secondary schools (Ardrey to Carpenter, 27 Jan. 1967). Carpenter's approval of Ardrey continued only as long as the latter kept his place. "Finally I must report that I had a two hour conversation with Robert Ardrey in Atlanta recently and found him the same egocentric authority that he was previously. I think he is making a mistake by trying to become the authority rather than the scribe and interpreter" (Carpenter to Alan Elms, Psychology Department, University of California at Davis, 11 Jan. 1973).

50 CRC: "Telemetric Control," pp. 22, 23, 27.

51 CRC: Bermuda Notebook, "Suggestions for the Gibbon Colony."

6 Reinstituting Primatology after World War II

1 Latour and Woolgar (1979: 43–53); Derrida (1976); Haraway (1988).

2 For his own presentation of the work and meanings of the Yerkes Primate Research Center, see Bourne (1971).

3 Dickson (1984); Kevles (1978).

4 C.R. Carpenter Papers: Weiss, "Nonhuman Primate Committee Report to the Institute of Animal Resources," 22 Sept. 1955, p. 2. Weiss, an organicist developmental biologist, reminds us that epistemological commitment to anti-reductionist explanatory strategies hardly ensures a democratic science politics (Haraway 1976).

5 U.S. Department of Health, Education, and Welfare, National Institutes of Health Primate Research Centers: A Major Scientific Resource, Dec. 1971; *Regional Primate Research Centers: The Creation of a Program*, 1968, 67 pp; Leo Whitehair and William Gay, "The Seven NIH Primate Research Centers," *Lab Animal*, 1982.

6 *The Economist*, 9 Feb. 1980, p. 92.

7 Goodwin, in Bermant and Lindberg (1975: 9). Ruch's (1941) 4090 bibliographic entries were the start of the Primate Information Center (PIC) data base, supported by the NIH at the Washington Regional Primate Research Center.

8 Napier and Barnicot (1963); Buettner-Janusch (1962).

9 Conferees included Adolph Schultz, who led in founding *Folia Primatologica, Bibliotheca Primatologica*, and the *Handbook of Primates*, and Hans Kummer (1968, 1971) from Zurich. Zurich has been a major center of modern primatology (Schultz 1969). Also present were Helmut Hofer at Giessen and Frankfurt; François Bourlière, from the Physiology Laboratory of the Faculté de Medicine in Paris; Bourlière's student, J.J. Petter; Louis Leakey from East Africa; John Napier, at London, who, with P.H. Napier, his wife, has been influential in primate physical anthropology (Napier and Napier 1967); Jane Goodall (set up at Gombe by Leakey and just beginning doctoral studies with Robert Hinde of Cambridge, later sponsor of several important primate theses); C.R. Carpenter; Cynthia Booth (also set up by Louis Leakey); Sherwood Washburn's student, Phyllis Jay; Charles Southwick, who like George Schaller and several later primate field workers, completed his doctoral thesis with John Emlen at the University of Wisconsin; Yale zoologists, Richard Andrew and Alison Bishop (later Alison Jolly); Stuart Altmann, just finishing his doctoral study of rhesus monkeys on Cayo Santiago; and K.R.L. Hall, chair of psychology at Bristol University. Hall observed baboons and patas monkeys while a professor of psychology at the University of Cape Town in South Africa. He was a major influence on his student, Stephen Gartlan. Bristol students drew especially from ornithology and socioecology. J.R.L. Crook was officially Gartlan's sponsor after Hall's early death from infection with a virus contracted from a vervet monkey bite. The list of conferees also included Adrian Kortlandt, from Amsterdam, who studied chimpanzee "protohominid behavior"—throwing, clubbing, stabbing, and other components of supposed hominid hunting behavior—first through questionnaires sent to zoo keepers and then in 1960 in a paw paw plantation in East Africa (Kortlandt 1962; Reynolds 1967: 117–22). Kortlandt came out of the same bird-oriented context that produced Niko Tinbergen (Thorpe 1979). Also present was Utrecht's J.A.H. van Hooff, a former ethology doctoral student of Desmond Morris who got his Ph.D. with Niko Tinbergen at Oxford. Van Hooff in turn would sponsor the doctoral work of Frans de Waal (1982). This list implies a web of concerns from anthropology, medicine, ethology, zoology, and paleontology. Lineages traced from institutional nodes in this web would lead to most practicing Euro-American primatologists.

10 Irven DeVore*, Mizuhara Hiroki, K.R.L. Hall, Phyllis Jay*, David Hamburg, Frances and Vernon Reynolds, Jane Lancaster*, John Ellefson*, Paul Simonds*, Richard Lee*, William Mason (then a colleague of Irwin Bernstein at the Yerkes Laboratories of Primate Biology in Florida; Arthur Riopelle (also from the Orange Park, Florida, labs), Donald Sade* (also a Frank Beach student), Charles Southwick, Harry Harlow, Peter Marler (Thomas Struhsaker's mentor), George Schaller, and others. Bernstein represents a strand of comparative psychology in these networks, emphasizing ties between hormones and behavior, especially in studies of sex and aggression; * designates

Washburn students (DeVore 1965a). People important to the U.S. Primate Research Centers are prominent on this list.

11 Kortlandt, with an extreme case of hunting monomania, actually proposed at the 1962 London meeting that a semi-free ranging colony of chimpanzees be set up in Kivu Lake and be trained "to adopt a proto-paleolithic way of life" that might revive dormant behavior patterns inherited by chimpanzees from the common ancestor with "man." "Perhaps this crucial test will result in the emergence of a chimpanzee tribe having a self-perpetuating, sub- or semi-australopithecine type of hunting culture" (Kortlandt and Kooij 1963: 86). Kortlandt believed that chimpanzees had devolved from a more hominized evolutionary past and that their bad contemporary throwing aim, happily evident in the shit pies which missed most zoo watchers, was a regression from truer past goal-directed behavior. South Africa's Raymond Dart (1953) also weighed in at the London meeting with his inimitable prose on the hominid lust for flesh. Whenever I worry that the adherents of my favorite hypotheses might be excessively ideological in rejecting offensive theories, I return to these kinds of prose, abundant in the primate literature of the 1960s and hardly absent in the 1980s, and take note of the great professional restraint of even the most partisan of my congenerics!

12 Notable long-term field primate research sites are the macaques of Japan, under the sponsorship of the Japan Monkey Center and the Primate Research Institute (Itani 1954); baboons of Amboseli National Park in Kenya, where Jeanne and Stuart Altmann and Glenn Hausfater have been the central organizers (after initial study in the 1960s, research has been continuous since 1970); the rhesus of Cayo Santiago in Puerto Rico, where Donald Sade and his students were critical in years following Altmann's thesis work (observers from several institutions have been active on Cayo Santiago); the howlers of Barro Colorado, under the administration of the Smithsonian Institution (Collias and Southwick 1952; Milton 1980); the chimpanzees of the Gombe National Park in Tanzania, begun by Jane Goodall in 1960 (Goodall et al 1979; Hamburg and McCown 1979; Goodall 1986); the chimpanzees of the Mahale Mountains in Tanzania, where Japanese workers have been since 1965 (Nishida 1972, 1979); red colobus and other species of Uganda's Kibale Forest (Struhsaker 1981); Ethiopian geladas and other cercopithicines (R.I.M. Dunbar and E.P. Dunbar, Bristol University people); baboons on the former Cole ranch near Gilgil, Kenya (organized by Robert Harding and Shirley Strum and continued by Strum, studies by Nancy Nicolson, Barbara Smuts and others, 3 troops transplanted by Strum [1987a, 1987b] to another site in 1984); several forest species studied from about 1973 in Manu National Park, Peru (Terborgh 1983); several rain forest species near the C.N.R.S. Laboratory of Equatorial Primatology and Ecology in Gabon (Gautier, Gautier-Hion, Pierre Charles-Dominique, others); gorillas of Rwanda's Parc des Volcans, begun by Dian Fossey in 1967; orangutans of Central Borneo's Tanjung Puting Reserve, initiated by Biruté Galdikas and Ron Brindamour in 1971; the Japanese pygmy chimpanzee project (Kano 1979; Kuroda 1982); S.U.N.Y. Lomako Forest Pygmy Chimpanzee Research Project (begins 1979, anatomy: Randall Sussman; social behavior: Alison Badrian, Nancy Handler; diet: Noel Badrian, Richard Malenky); the Smithsonian Biological Program in Ceylon (begins 1967, director was John Eisenberg, also joint grant to Suzanne Ripley and Eisenberg, joined by A. Hladik and C.M. Hladik, R. Rudran [Ph.D. U. of Ceylon], N. Muckenfuss, W. Dittus); Sussman (1979: 484n109).

13 Rowell (1966). I believe that a unifying theme in the primatology done by women has been their high likelihood of being skeptical of generalizations and their strong preference for explanations full of specificity, diversity, complexity, and contextuality. In the 1960s, consider Jane Goodall, Thelma Rowell, Alison Jolly, Phyllis Jay, and Suzanne Ripley. Jolly considers that Goodall and Thelma Rowell, both Robert Hinde students at Cambridge, had a major impact on Hinde's approach to primates, bringing him to focus on the individuality of monkeys and apes in a very different way than he had learned in the Tinbergen-Lorenz variants of ethology (Jolly, interview 1982). Hinde reported a similar opinion in discussing the impact of Rowell and Yvette Spencer-Booth in the early days of his rhesus colony at Cambridge's Sub-Department of Animal Behaviour at Madingley (interview 1982; Hinde and Spencer-Booth 1967; Spencer-Booth and Hinde 1971; Hinde and Rowell 1962). Jolly defined "parsimony" in scientific method in primate studies in relation to her expectation of complexity and individuality: "Instead of assuming that all behavior is produced by the simplest possible mechanism consistent with the observations, it is more cautious to assume that behavior is produced by complex mechanisms, as complex as those that we would expect for ourselves, except where one can definitely prove that a factor present in us is not present in the [other] primate" (Jolly 1972: 264). Referring to watching other primates as "the privilege to observe,"

Jolly believed that traditional, learned values of patience often made women in primatology in the 1960s better field observers. She saw the origins of post-World War II field primatology as part of the revolt against behaviorism's straightjackets (Jolly, interview 1982).

14 Bolwig (1952–53, 1959).

15 "The chimpanzees we watched had a lovely life" (Reynolds 1967: 127; see also Reynolds and Reynolds 1965). This was the dominant note of the chimp reports of the 1960s; their somber side was scientifically narrated in the harsher political times of the late 1970s. In a nice turn on the more usual 1960s orthodoxies about the terrors of the city for stressed man, Vernon Reynolds discounted hunting in hominization and regarded the modern city, similar to the exciting worlds of chimp flexible society, as a return to peace-generating conditions of the early human past, before the awful world of settled agriculture: "Now in his city, man has achieved a remarkable combination: he can be, and is, a good human citizen and a good instinctive ape" (Reynolds 1967: 274).

16 In the late 1960s Thelma Rowell inherited the study facilities used for birds by Peter Marler at UC Berkeley, who was moving to the Rockefeller University. Rowell had set up Robert Hinde's rhesus research facility at Madingley about the time she finished her ethological dissertation on maternal behavior in golden hamsters (Rowell, interview 21 July 1977). Her savvy in cage design permitting captive primate behavior unusual depth and complexity is a recurrent theme from my interviews.

17 Gartlan 1964, interview 12 Aug. 1982; Hall 1962, 1965; Hall and DeVore 1965; Rowell 1966, 1972. Rowell (1974) summarized her harsh criticisms of the dominance approach, which was then most closely associated with Irven DeVore and other Washburn network people, but had a much longer history in animal behavior studies (Haraway 1978a). Rowell's work drew from and converged with that of comparative psychologist Irwin Bernstein, who did field studies and worked primarily with captive animals at the Yerkes Regional Primate Research Center, first at Orange Park and then associated with Emory University. Bernstein (1968, 1976, 1981) has been a major critic of logical and experimental fallacies in dominance explanations. Bernstein and Rowell brought together experimental and observational studies on captive and free monkey populations, tying together the physiology of hormones and behavioral studies of social role and structure from a functional comparative psychological point of view. Another major stream of criticism of the notion that macaque society is organized by the linear dominance hierarchy of adult males came from the work of Donald Sade (1965, 1967) on Cayo Santiago and the workers from the Japan Monkey Center, beginning in the 1950s (Imanishi 1960, 1963; Altmann 1965b). Criticism of male dominance hierarchy explanations did not derive from feminist sources for many of its chief opponents, although that was a factor for some. For history of the debates, see Fedigan (1983) and Bernstein (1981). For an account of Shirley Strum's effort to eliminate dominance hierarchies for describing baboon males, and her subsequent decision to build them back in minimally, in order to show their insignificance for her explanations of what matters to baboons, see Strum's description of a controversy she had with other participants (including the people from the Amboseli baboon studies, Jeanne and Stuart Altmann and Glenn Hausfater) at her Wenner-Gren conference in 1978 on methods and models in baboon studies (Strum 1987b: 157–67). Strum was a controversial innovator in arguing that explanations of baboon society required an account of the animals' complex emotional and cognitive capacities evidenced in their pervasive strategic social behavior and relationships. Social strategies, not male dominance hierarchies, gave shape to primate life (Strum 1978, 1982; 1983a, 1983b; Strum and Mitchell 1987; Western and Strum 1983). The implications for theories of animal consciousness are both obvious and complex.

18 Alison Jolly to Donna Haraway, 10 Dec. 1982; Struhsaker 1967; Altmann and Altmann 1970.

19 Taub (1983); Tiger and Fowler (1978); Wasser (1983b); Small (1984); Fedigan (1982).

20 Harlow and Mears (1979); Knutsen (1973); Reite and Caine (1983); Hausfater and Hrdy (1984); Cohen, Malpass and Klein (1980); Coelho, Hamburg, and Adams (1974); McGuire (1974).

21 Other projects for rehabilitating young apes to the wild include:
    1) In Kuching, Sarawak, Indonesia, in the 1950s and 1960s, Tom and Barbara Harrison reared baby orangutans, confiscated from the illicit export trade by the Forest Department (Reynolds 1967; Harrison 1963; IPPL Newsletter 10[3], Dec. 1983, p. 19). Barbara Harrison also started a rehabilitation center for orangutans in Bako, Sarawak, later moved to Sepilok, Sabah, East Malaysia.
    2) Biruté Galdikas, originally with her husband Ron Brindamour, started a "halfway house" for

young orangs at their camp in the Tanjung Puting Reserve, Borneo, Indonesia, beginning in 1971 (Galdikas 1979). They named their first site Camp Leakey, after Louis Leakey, who helped obtain funding. Their rehabilitants were confiscated from illegal traders by the Nature Protection and Wildlife Management arm of the Indonesian Forestry Service. Other Orangutan Rehabilitation Centers were established in Malaysian Borneo and Indonesian North Sumatra (Galdikas 1975). Indonesian and U.S. university students, officers of the Indonesian Forestry Department, trained Dayak tree climbers, and local staff all became part of the project (Galdikas 1980).

3) The Bohorok Center in Gunung Leuser National Park, North Sumatra, Indonesia, was established in June 1973, sponsored by the World Wildlife Fund and funded by the Frankfurt Zoological Society. Swiss zoologists Monika Borner and Regina Frey pioneered the station, followed by Conrad and Rosalind Aveling from England. In 1979 the station came under management by the Indonesian government, which pays a field manager and staff (Djojosudharmo 1984). The manager in 1984, Suharto Djojosudharmo, pointed out that the majority of rehabilitants no longer come from the pet trade, but from timber and oil concessions, plantations, or shifting cultivation areas. About 90 apes have been released over 10 years. The main functions of the centers are conservation education, aiding enforcement of wildlife protective legislation, and if possible, curbing deforestation. Tourism and research are secondary goals.

4) Both Jane Goodall and Dian Fossey attempted to deal with orphan infants who had been kept as pets or captured for export.

5) To foster breeding in captivity, chimpanzees whose behavior was deeply troubled by cruel lab cage confinement have been rehabilitated in the sense of being removed from extreme isolation and restriction and integrated into captive social groups (Linda Koebner, "Surrogate Human," *Science* 82 [July–August], pp. 33–39).

22   Hahn (1988); Brewer (1978); Carter (1982, 1988).

23   In a story announced on the cover of *Psychology Today* (Nov. 1975), "The Sexual Coming of Age of My Chimpanzee Daughter Lucy," Dr. Maurice Temerlin discussed his preoccupation with the incest taboo in relation to this ape daughter, raised by himself and his wife Jane, with their son Steven. The article skirted the line of titillating the reader with the theme of father-daughter sex. The father/scientist frankly admitted he loved Lucy with a freedom and lack of guilt he could not express with his human son. The article was ambiguous whether it was the sex or species of the son that blocked paternal spontaneity. Lucy, an Ameslan-using chimp, at adolescence retreated from her loving father, turning instead to almost any other human male of a proper age. "Because Lucy had not seen another chimpanzee while growing up, I had expected that her sexual interests would be directed toward me when she matured. After all, I was the male with whom she had had the most close and enduring relationship. I was entirely wrong" (Temerlin 1975: 61). A modern parent, Temerlin provided his maturing child with a copy of *Playgirl*. Lucy enthusiastically masturbated to the human male centerfold. The lesson: "Lucy has helped me become more understanding of myself, for she rejects neither her own animal nature nor her acquired humanness" (1975: 103). Temerlin worried about finding a proper chimp mate for his mature animal daughter. Carter's versions of Lucy differ dramatically from Temerlin's.

24   Linden (1986); Naureckas (1986: 18).

25   Famous home-reared apes have included the baby gorilla, Goma, raised by Dr. Ernst Lang, director of the Basel Zoo, and his wife (Lang 1962); the infant chimpanzee, Viki, from the Yerkes Primate Laboratories, raised at home by Cathy and Keith Hayes (Hayes 1951); chimpanzee Gua, raised for nine months at home by W.N. and L.A. Kellogg (Kellogg and Kellogg 1933); chimp Lucy (Temerlin 1975; Carter 1982, 1988); gorilla Toto, who grew up after 1932 with Mrs. Maria Hoyt and a keeper named Tomas (Hoyt 1941; Hahn 1988: 41–53); chimpanzee Christine, who was raised in the 1950s on a Pennsylvania farm by Mrs. Lilo Hess (Hess 1954); chimp Meshie (Raven 1933); gorilla Froma raised by the Basel zoo director and his wife, Jorg and Mrs. Hess (Reynolds 1967: 163–93; *Gorilla* [film] 1981).

The debates about whether or not the apes really communicate in a human-like way with a real human language have been sharp. The debates have been incorporated into popular films (*Quest for Fire, Greystoke, Project X*), novels reconstructing human evolution (Auel's [1980] *Clan of the Cave Bear*), and science fiction (Piercy's [1976] *Woman on the Edge of Time*, McIntyre's [1983] *Superluminal*, Brin's [1983] *Startide Rising*). The language-teaching efforts and debates can be followed in Gardner and Gardner (1980); Fouts and Budd (1979); Tanner (1981); Premack and Premack (1972); Rumbaugh (1977); Savage-Rumbaugh, Rumbaugh, and Boysen (1978); Savage-Rumbaugh (1986); Patterson and Linden (1981); Terrace (1979); Linden (1976, 1986); Sebeok and Sebeok (1980). Results with the pygmy chimpanzee, Kanzi, studied by Sue Savage-Rumbaugh and her colleagues at the Yerkes Primate Research Center in Atlanta, Georgia, were discussed in the *New York Times*, 24 and 25 June 1985. Headlines were: "Pygmy Chimp Readily Learns Language Skill" and "Kanzie the Chimp: A Life in Science." The narrative has Kanzie and Mulika living a charmed life, even getting to select TV programs. His favorites were reported to be human wrestling and the National Geographic Jane Goodall specials. One of the most fascinating kinds of evidence for a shared approach to language among apes and people has been demonstrations of chimpanzee and gorilla lying. The ingenuity of people's efforts to prove simian deception is exceeded only by the apes' stratagems to deceive (Patterson 1978: 459).

7 Apes in Eden, Apes in Space

1 After taking over Gulf Oil, Chevron sponsored the National Geographic specials. On Gulf's troubles in the 1970s, see Moskowitz, Katz and Levering (1980: 506–10).

2 Keller (1985); Sofoulis (1988).

3 Here, as elsewhere, it is crucial to remember that the "Jane Goodall" produced by the media is a representation within a particular context, not the human being herself. Similarly, the representations which Goodall "authored"—her books, articles, interviews, speeches, field notes, letters, etc.—are also not unobscured windows onto the person and her work. Finally, *Primate Visions* is a visualizing technology—a way to see some things and not others—not the resolution of other representations into the true and the false.

4 Flo received probably the first obituary for a wild chimpanzee in a major newspaper, in Sept. 1972, in the Sunday *New York Times*.

5 *Time*, 10 Feb. 1961: 58. The caption under HAM's photograph read "from Chop Chop Chang to No. 65 to a pioneering role." For HAM's flight and the Holloman chimps' training see Weaver (1961) and *Life Magazine*, 10 Feb. 1961. *Life* headlined, "From Jungles to the Lab: The Astrochimps." All were all captured from Africa; that means many other chimps died in the "harvest" of babies. The astrochimps were chosen over other chimps for, among other things, "high IQ." Kratochvil (1968) stated that in 1963 the Holloman colony numbered over 60 chimps in aerospace medical research, especially on high-risk environments: effects of decompression to near vacuum, deep-sea diving experiments, retinal burns for the Defense Atomic Support Agency, and evaluation of drug toxicity. In the late 1960s Holloman was ready to use its colony of over 160 chimps for infectious disease research.

6 *Time*, 8 Dec. 1961: 50; *Newsweek*, 5 March 1962: 19.

7 In her readings of *2001* and of other science fiction films, Sofia (1984; Sofoulis 1988), developed a theory of "extraterrestrialism" and argued that every technology is a reproductive technology. Bryan's (1987) authorized, celebratory history of the National Geographic Society posited a similar relation between spaces of the biological body and outer space that I read ironically in juxtaposing HAM and David Greybeard. The penultimate chapter in Bryan's book is "Inner Space," introduced with the epigraph, "The stuff of stars has come alive" (1987: 454). The chapter explores the National Geographic's voyages into deep body space, complete with the color-enhanced scanning electron micrographs of killer T cells, macrophages, and other Body Wars inner wonders. The paired final chapter is titled "Space," headed by the phrase, "The choice is the universe—or nothing" (1987: 352). Both the bodily and space images in Bryan's account are coded in ET terms, and the jungle's complementary, mutually constitutive "natural" body is absent. Linden (1976: 275–388) makes this postmodern connection between the allotopic spaces of "nature" and "space" explicit in his final chapter on the ape language experiments, called "Washoe and the Moon Shot: Dionysus and Apollo." Linden sees human language as the world of surrogates and displacements; it is the world of technological rationality and the moon shot. Washoe's mastery of American Sign Language promises a bridge back to Eden, away from the edge of the abyss of final separation from the natural body of earth. Washoe's *language*

promises "man" a "return to *biology*" (287, my emphasis). My argument is that all the participants in the boundary-eroding discourses of body, jungle, and space in nuclear, late-capitalist culture are part of a postmodern cyborg world of communication, technology, war, and reproduction.

8   Virilio and Lotringer (1983); Gray (1988).

9   Goodall (1965a, 1968b).

10 Lorenz (1957); Thorpe (1979); Richards (1974).

11 Two streams of communications analysis developed separately. Located in the U.S., one emphasized the design features and statistics of signal exchange. The other, beginning in the U.K. and Europe, conducted morphological and functional studies of ethograms. Both approaches constructed signal catalogues, but used them differently (Altmann 1965a; Marler 1976).

12 The *Reader's Digest* (Oct. 1985, from *Life*, July 1985) version organized the tale around Koko's mourning and subsequent choice of a new kitten after All Ball was run over by a car. Finding the kitten dead, Penny gently told Koko. The bereaved gorilla "began to hoot the same soft distress cry she had made as an infant when Patterson left her for the night. For two months Koko's behavior was subdued" (202, 203).

13 Not all popular post-war narratives of touch with apes have been mediated by women, but the exceptions support the argument that the politics and codes of gender are fundamental to the meanings of the primate stories and to their role in political struggles in European and Euro-American culture over "human nature." The National Geographic special, *Gorilla*, visits John Aspinall's 55 acre, originally private gorilla collection in Howletts Zoo Park in Kent, England. Aspinall stresses the values of wildness and natural group life, as well as close emotional bonds between staff and animals. But the emotional bonds are rigorously disciplined by gender. Aspinall says that the ultimate honor is to be accepted by gorilla mothers and their young, and he is shown enjoying the kiss of a large male, who proceeds to mouth the man's eyelashes in a loving touch. But the element of challenge and danger provides the spice for the most valued touch—the wrestling bout with the large male silverback. Aspinall (1976) waxes eloquent over his desire to go in with the big males, to confront them as males on the edge of love and danger. A little probing in the book reveals Aspinall's old colonial connections and values from British India. His politics remain survivalist and primitivist, complete with the claim that male domination in apes and humans is simply natural. Aspinall is an Akeley for our time, who lives out in a private enclosure the colonial dream no longer to be encountered and wrestled with in the wild.

The second exception to human female mediation of the ape's touch is David Attenborough's *The Primates* (1981), in his PBS *Life on Earth* series (Attenborough 1979). Pointed out to me by Katie King, Attenborough adopts a unique method of mediating the relation of viewer and gorilla: he is literally in the middle. You must go through him to the animals. The mode of mediation in the Goodall, Galdikas, and Fossey films and photographs leads the viewer to identify with the woman and with the animal in a saving fusion after the ordeal of seeking communication. But in Attenborough's film, the viewer cannot identify with Attenborough, who whispers the audience into the film, constantly turning his head from viewer to animal scene and back, actively drawing back the curtain like a theater master. It is not the drama of touch that fills the screen; it is Attenborough, the master of ceremonies. Attenborough reveals his virtuosity in an orgy of touch with a blackback male. The explicit commentary is about the gorilla's peaceful nature and the bum rap the animal has taken as the symbolic embodiment of aggression. Even so, we are cautioned not to forget the silverback is the lord of the forest and immensely strong. This is the theater of male exhibition. Since the visual rhetoric forbade the viewer's identification, making the audience conscious of its status as voyeur, the action is Attenborough's.

14 In *Gorillas in the Mist*, the 1988 film with Sigourney Weaver as Dian Fossey, Campbell was Fossey's lover. But it was only Digit's touch that mattered. Campbell could only be spectator to the drama between the woman and the animal, whose story did not fully conform to Oedipal or pre-Oedipal rules, even when their union was legitimized in a kind of "marriage" in Karisoke's gorilla graveyard, when the Christian African field assistant joined the separate circles of stones around Fossey's and Digit's grave mounds. One in life, they were joined in death.

15 "White" is a designation of a political space, not a biological "race" within the pre-World War II chain of being framework, nor a "population" defined within the modern evolutionary synthesis.

"White" is obviously not a marker of actual skin color. That point makes it easier to remember how the Irish moved from being perceived as colored in the early nineteenth century in the United States to quite white in Boston's school busing struggles in the 1970s, or how U.S. Jews have been ascribed white status more or less stably after World War II, while Arabs continue to be written as colored in the daily news. This book attempts to use categories like "Euro-American" to shake the taken-for-granted quality of the unmarked category and to insist on the adjectival status of all human beings—we are all modified by history; none of us is outside history, either as origin or goal.

16 Two recent fictions in which women of color appear as scientists confirm by contrast the analysis of the triply coded web of gender, science, and race in the National Geographic narratives: Toni CadeBambara's *Salt Eaters* (1980) and Marge Piercy's *Woman on the Edge of Time* (1976). Both lie at the juncture of literature, political allegory, and science fiction; the National Geographic films and articles ride the line between science and science fiction.

Velma, the central figure in *Salt Eaters*, is a computer scientist at a moment of crisis for the fictional black community of Claybourne, Georgia. The coming apocalypse around which the novel is organized is ambiguously the end of the world, an explosion at a nearby chemical plant, carnival, the re-enactment of a slave insurrection, the moment of revolution in the present and future, and a thunderstorm like the Old Testament's deluge. Velma, herself cracked and threatened by the tensions in her personal and political life, had attempted suicide and is healed by a fabled wise woman of the district, Minnie Ransom. The healing in the Southwest Community Infirmary is witnessed by a male medical doctor, whose experiences in the novel lead him back into the service of the oppressed. Velma's existence is constantly between two poles, bridging contradictions between traditional healing arts and science, the men and the women, and the community and the world. Her blackness is central to the story of the coming together of the world's colored peoples, embodied in the Seven Sisters, a traveling theater group of women of color. The white, male dominant, multinational corporate devastation of the planet is explicitly reversed and displaced in the recoding of nature and science. Velma's blackness, her married status, her maternal status, her gender, and her relation to scientific and traditional practices are all required to mediate the healing of the planet and the human relation to it. Velma can represent humanity, which excludes the category white, because she is structured to mediate all the contending forces that threaten to rend the fabric of the whole. In the oppressed coming to heal themselves and the planet there is hope of global survival. The National Geographic narrative too is about the question of survival, of humanity and of nature, but the required healing touch is enacted in the lonely drama of the white woman scientist. See Hull (1985).

Piercy's *Woman on the Edge of Time* narrates the present and future world of a Chicana, Consuelo or Connie, who in the present is forcibly committed to a mental hospital and subjected to experimental brain surgery. In the future, to which she has access by her sensitivity as a "receiver," Connie sees her alternative possibility as a scientist in the person of Luciente. Luciente appears to be a man to Connie at first, but her womanhood soon becomes apparent. The combination of the brownness, gender, and relation to science of Connie/Luciente is the organizing code of the novel. As in the National Geographic world, the "past" is the contested zone in *Woman on the Edge of Time*. The "past" is the time of origins, when possibilities are set. But the past of the possible future in Piercy's novel is the actual present for Connie, site of her struggles in the mental hospital, and by extension the historical struggles of all oppressed people, embodied in the "crazy" brown woman confined and subjected by "science."

17 Gilman (1985) demonstrated how the genitals of Khoisan women (represented by Cuvier's publication of the dissection of the body of a woman known as Saartjie Baartman, the "Hottentot Venus" in 1817) were used to illustrate the essential pathological nature of black women's sex. Sex = genitals = female body = woman, especially woman of color, who stands as the generic category, the vanishing point of convergence for this family of equations. There is here no concept of gender, the space opened up to contest biological essentialism in both feminist and social science discourse in the 1970s. The genitals of the Hottentot Venus were compared to those of the European prostitute, another degenerate female form. Drawing on a commonplace of early European travel literature, Buffon stated that African people's apelike sexual appetite led black women to copulate with apes (Gilman 1985: 212). All women, as "the sex" in nineteenth century biologizing discourse, partook in the animality envisioned; but black women were ontologically the essence of animality and abnormality. Black woman and prostitute were joined

by other sisters with sick reproductive behavior in the nineteenth century, e.g., the lesbian, the overly intellectual middle class woman, the woman who liked sex too much, the neuraesthenic, and the hysteric, who like all the others displayed on her body the stigmata of masculinist terror of its own creature/monster, woman. Appearing as substantial actors in popular scientific story fields, these assertions have been banished to the margins of official post-World War II biology and medicine, but they remain as the restless undead even there, vampirizing truths created in the light of reason.

18 The 1932 *Tarzan, the Ape Man*, with Johnny Weissmuller and MaureenO'Sullivan playing Tarzan and Jane, makes this race and animality coding starkly evident in the undisguised racism of that period. By contrast, the National Geographic's approach seems almost subtle and progressive. In the dramatic scenes of Tarzan, Jane and her party have been taken captive by a village of pygmies, who put them in a pit for a King Kong figure to bash to death and probably eat, while the "natives" gleefully dance and shoot arrows at the hapless captives. The white ape-man, Tarzan, of course, arrives to rescue Jane, et al, from the savage beast and the savage primitives, who echo each other visually. An analysis of all the scenes in which the pygmies appear confirms that the entire "pygmy" village is made up of achondroplastic black male dwarfs. Hollywood anticipated the most extravagant casting of Federico Fellini in this early sound film. Some of the men are white dwarfs in black face. The black dwarfs' savagery and evil are marked in what are supposed to be read as their monstrous bodies. The all-male village seems a kind of unconscious extra, letting slip the implicit masculinist reproductive politics that oddly echoes the male cyborg fetuses in extra-terrestrial space in late twentieth-century science fiction films.

19 I am indebted in this section to Pauly (1979).

20 Grosvenor, quoted in Pauly(1979: 526–27); Bloom (1987).

21 Mostly accompanied by compelling cover photographs, the primate articles in the *National Geographic* have been: Goodall (1963b, 1965b, 1979); Fossey (1970, 1971, 1981); Galdikas (1975, 1980); Patterson (1978); Vessels (1985); Strum (1975, 1987a); Riopelle (1967, 1970). Goodall (1967b) was also a NGS publication.

22 Strum interview, 29 June 1983.

23 The group had been named by R.S.O. Harding, who as a University of California Berkeley graduate student initiated research on this troop on the Cole ranch in Kenya in 1970. Harding had just finished reading Tom Wolfe's *Pump House Gang* (1969), and the baboons hung out around an old ranch pumphouse (interview, 29 Oct. 1982).

24 I have examined three drafts of the Pumphouse Gang article, kindly lent me by Shirley Strum. The first draft opened with, "Ray and I joined the group at the same time." That draft explored the problem of an outsider joining an established group. The bond between Strum and Ray was forged on the basis of shared initial fear, ignorance, and exclusion. The transient males have to integrate themselves into a new troop frequently, and to do so they require the friendship of adult females. The "dailiness" of baboon life emerging in Strum's account became the basis of Cynthia Moss's (1982 [1975]) chapter on "Baboons" in her popular presentation of behavior studies on East African mammals. Moss contrasted her narrative of daily life with the drama of male aggression in Ardrey (1966). Moss emphasized that primates are not particularly more social or more complex than the other social mammals. She doubly displaced the primate drama of the social contract and the original state of nature, first, by rejecting the heroic and masculinist story and second, by rejecting the special place of primates in nature. Imaginations of non-hierarchical difference and the valuing of "dailiness" have been major themes in recent feminist writing, and without naming itself feminist, Moss's book fits easily. Zebras and elephants in Moss's version are simply wonderful (Moss 1988).

25 Scott Gilbert, a developmental biologist and historian of science at Swarthmore College, sent me the card, with the inscription, "feminist critique of primate research." Gilbert teaches feminist issues in the sciences at Swarthmore (The Biology and Gender Study Group 1988).

26 Carpenter's films include: *Characteristics of Gibbon Behavior* (1942), *Behavioral Characteristics of the Rhesus Monkey* (1947), *Social Behavior of Rhesus Monkeys* (1947), *Howler Monkeys of Barro Colorado Island* (1960), *Behavior of the Macaques of Japan* (1969), *Activity Characteristics of Gibbons*, in three parts, *Ecology and Maintenance Behavior, Locomotion,* and *Social Behavior* (1974).

27 See Schneider (1984) for a critique of the concept of kinship in human anthropology. His arguments

apply to primate dominance and functional specialization. The natural-technical object of knowledge in both conceptual sets is "The Family" and all its variants.

28 The sources for this brief history, ending with the kidnapping of 1975 are Goodall's popular and professional publications, interviews with researchers who worked at Gombe or supervised Gombe research (Harold Bauer, Robert Hinde, William McGrew, Peter Marler, Nancy Nicolson, Joan Silk, Barbara Smuts, Richard Wrangham, Adrienne Zihlman), Stanford University catalogues and descriptions of Gombe West, reconstructions of networks of primate research projects and people based at universities from which the major Gombe chimpanzee researchers came (especially Cambridge, Stanford, Edinburgh), Cole (1975), and Hamburg and McCown (1979). No matter the data base, this history is not the truth to quell the lie of the popular representations. Rather, this section is a node in a discursive field of representations.

29 Tanganyika gained independence from Britain in 1961 and joined with newly independent Zanzibar in 1964 to form Tanzania. Kenya achieved independence in 1963. The first African nation to be decolonized officially after World War II was Libya in 1951.

30 Goodall (1962, 1963a).

31 Stanford Human Biology undergraduate Gombe students who went on to get Ph.D.s in primate studies elsewhere include Curt Busse (U.C. Davis, 1981, on infanticide and paternal care in chacma baboons), Anthony Collins (Edinburgh, 1981, on social behavior and mating patterns in adult yellow baboons), Nancy Nicolson (Harvard, 1982, on weaning and the development of independence in olive baboons), Craig Packer (Sussex, 1977, on inter-troop transfer and inbreeding avoidance in *Papio anubis*), David Riss, and Joan Silk (U.C. Davis, 1981, on influence of kinship and rank on competition and cooperation in female macaques). Primatology students accommodated in Hamburg's Neuro- and Behavioral Sciences doctoral program included Harold Bauer (Stanford, 1976, on "Ethological Aspects of Gombe Chimpanzee Aggregations with Implications for Hominization"), Anne Pusey (Stanford, 1978, on "The Physical and Social Development of Wild Adolescent Chimpanzees"), and Barbara Smuts (Stanford, 1982, on "Special Relationships between Adult Male and Female Olive Baboons"). In addition, Patrick McGinnis (Cambridge Ph.D. 1973, "Patterns of Sexual Behaviour in a Community of Free-Living Chimpanzees") and Richard Wrangham (Cambridge Ph.D. 1975, "The Behavioural Ecology of Chimpanzees in Gombe National Park, Tanzania") were intermeshed with the Hamburg Stanford web. Hamburg's own work was distinct from that of the primatologists.

32 Other Cambridge doctoral students who came to Gombe included Laurie Baldwin (no degree), Richard Barnes (work interrupted by the kidnapping, went on to do Ph.D. on elephant behavior, also directed the mountain gorilla Karisoke Research Center), David Bygott (Ph.D. 1974), Timothy Clutton-Brock (Ph.D. 1972 for work on red colobus, later supervised Craig Packer's dissertation at Sussex, tying primate work to the game theoretical evolutionary theory of John Maynard Smith), Anthony Collins (Ph.D. 1981 for work on yellow baboons), Hetty Plooij, Margaretha Thorndahl, and Richard Wrangham (Ph.D. 1974).

33 Thelma Rowell's strong opposition to sociobiology in those years is cited by both sides of the issue as a divider between her network at Berkeley and that at U.C. Davis, associated with Peter Rodman and W.J. Hamilton. The Davis group, even before Hrdy's arrival, had a strong female-centered cohort, in both theory and practice. Joan Silk and Amy Samuels went from Davis to work under Jeanne Altmann's influence at Amboseli. See Small (1984). People in the network include Meredith Small, Linda Scott (who worked at Gilgil), Joan Silk, Amy Samuels, John Matani, Bruce Wheatley, C. Clark-Wheatley, and others. Despite the number of science doctorates among Asian-Americans and the prominence of Japanese nationals in primatology, Matani is the only Asian-American primatologist I have found.

34 Goodall (1968a, 1986); Wrangham (1975, 1979b); Hamburg and McCown (1979); interviews with Joan Silk, Richard Wrangham, Nancy Nicolson, Barbara Smuts.

35 I compiled an extensive list of European and Euro-American assistants and students at Gombe from 1966–1975, including many who did not proceed to doctoral work. Of the 58 people on the list, 35 (60%) are women. Of 22 assistants and students from Gombe whom I have identified as having earned doctorates in some field of behavioral biology, 11 are women. So among those who asked Jane Goodall to come to Gombe to watch primates who did not seek higher degrees, women

outnumbered men 24:12. If the list were complete, the difference might be larger. It was easier to identify people who earned higher degrees.

36 That project was congenial both to Hinde's and Hamburg's interests. Hinde's laboratory had emphasized mother-infant relations and infant development from the time of Hinde and Rowell's early papers on infant rhesus, related to Hinde's interest in John Bowlby's work on attachment and loss in human infant development. Goodall expected chimp research to give insight into methods by which "we" raise our children, the problem of human aggression, non-verbal communication, the difficulty of human adolescence, and human mental illness (Goodall 1971: 290–92). The Goodall-Hamburg way of asking questions about evolution, with its strong social and personality functionalism, contrasted sharply with the cluster of "neo-Darwinian" approaches (sociobiology, investment theory, cost/benefit analysis, reproductive strategies, behavioral ecology, socio-ecology) adopted by most of the doctoral students after the early 1970s.

37 For a summary of her life-long interest in the topic in the context of chimpanzee-human comparison on "the family," see Goodall (1984).

38 Thelma Rowell consistently and ironically commented on the evident pleasure primates take in each other's company. At least on a phenomenological level, sociality in most primates is prior to individualism. It took all the resources of late-capitalist economic theory to make the autonomous, competitive individual fill the primate scientist's field of vision. The temptation to note that "masculine" westerners (male or female) are more likely to need to explain the group, while taking the smallest "autonomous" individual as a non-problematic starting point in their narratives, is strengthened by the pattern of western liberal political theory (Macphearson 1962). The masculinism in that theoretical tradition has been systematically examined in feminist political theory (Hartsock 1983a). Savvy about the form and logical oddity of the narratives, but silent about the questions of capitalist masculinism, Latour and Strum (1986) wittily dissected evolutionary origin narratives. Sensitive to gender in the pattern, the mathematical biologist and feminist theorist, Evelyn Fox Keller (1986 and forthcoming), outlined the gratuitous mathematical and logical moves in population genetic, evolutionary, and ecological theories that underpin the scientific hegemony of methodological individualism in late twentieth-century biology.

39 Goodall (1973); Bygott (1974).

40 Bygott (1979); Wrangham (1979a, 1979b).

41 Pusey (1979: 479); Smuts, interview, 18 March 1982.

42 Smuts interview, 18 March 1982; Wrangham interview, 13 August 1982.

43 Smuts went beyond Trivers to reformulate the issues in her later baboon work. Smuts noted that she and Nancy Nicolson collected virtually identical sets of data as Jeanne Altmann without having met or spoken to one another. When they came together, they were stunned by the degree of overlap among them. "Fascinated, not scared—[there was] enough of our own so that was not a problem" (Smuts, interview, 18 March 1982). Douglas Shapiro, at La Pergera doing alternate service to the military as an M.D., taught Smuts how to watch monkeys. She worked in the field with Jill Stein.

44 Wrangham (1979a, 1979b, 1980); Wittenberger (1980); Wasser (1983a).

45 A "reasonable length of time" might not always mean the same thing for men and women workers, as a function of differences in career planning and expectations (interviews).

46 Wrangham interview, 13 Aug. 1982.

47 Katie King suggested reading this film in terms of sf "first contact" conventions. It was also her insight that Goodall was constructed as the complete single woman and girl guide in this first film.

8 Remodeling the Human Way of Life

1 Mayr (1950, 1982a).

2 Landau (1981, 1984); Beer (1983); Haraway (1986).

3 Thanks to Sarah Williams (1987) for calling my attention to this portrait and for her related analysis.

4   Dart's small-brained child was originally spurned by the paleontological community as an unlikely ancestor. Found between 1908 and 1915 in England, the Piltdown skull's large brain and ape-like jaw fitted better with prevailing expectations; uncovering of Piltdown as a deliberate fraud occurred in the 1950s (Weiner et al 1953). Henry Fairfield Osborn of the American Museum of Natural History regarded Asia as the birthplace of human ancestors. He helped launch the "Missing Link Expedition" to Mongolia in 1921, called by Osborn the "paleontological Garden of Eden" (Boaz 1982: 240; Andrews 1926). But this Central Asiatic Expedition, perhaps the largest until the emergence of the international paleoanthropology team expeditions of the 1960s, found no hominid fossils. Boaz described 26 scientific personnel, a large supporting staff, numerous motor cars and camels, private donations over $600,000, and logistical support by Standard Oil and U.S. Rubber, among others.

5   Director of animation was Ann Smelzer. Lucy's portrayal was the work of Diana Namkoon. The most wonderful special effects appeared in the simulation of the formation of the rift systems of East Africa. These fabulous reconstructions were oddly juxtaposed to the staid, nineteenth-century genre of the academic seminar for the verbal lecture on geology.

6   The organizing interpretation of Lucy in the film is that of Donald Johanson of the Cleveland Museum of Natural History and his close allies and colleagues. A popular book presentation of the same arguments is Johanson and Edey (1981), excerpted by *Readers' Digest*, Sept. 1981. Johanson has been a master with the press since the finds of 1974. Actual scientific consensus about Lucy is another matter (Lewin 1987a).

7   Films: *The Hunters* (1957); *N'ai: The Story of a !Kung Woman* (1980); *The Making of Mankind: III. A Human Way of Life* (1982). North American books: Lee and DeVore (1968, 1976); Shostack (1981); Lee (1979, 1984). Japanese monograph: Tanaka (1980).

8   Lee (1985: 19); see also Pratt (1985, 1986).

9   I am indebted to the Swarthmore College developmental biologist, Scott Gilbert, for his interpretation of the Matternes painting in terms of sex roles and Eden. Matternes' technique of reconstruction should be a study in itself, as he works up from the bones to life in a way reminiscent of Carl Akeley's invention of modern taxidermic technique.

10 Stocking (1968); Boas (1925); Cravens (1978).

11 Ernst Mayr summarized the uneven "mosaic evolution" of the synthesis (Mayr and Provine 1982). See also Haldane (1932); Mayr (1942, 1982b); Provine (1971). Many naturalists, especially ornithologists, were writing in populational terms in the 1910s, while Mendelian geneticists were working with typological notions of "the wild type." The pieces of the synthesis were in place between 1910 and 1930; they came together between 1937 and 1947 in the reconciliation of experimentalist and naturalist research traditions, along with the modeling of theoretical population genetics and population ecology. The major themes of the evolutionary synthesis were selectionism and adaptationism, although themes of non-adaptive evolution were developed in early versions, compared to a hardened adaptationist program and narrower genetic models in later versions (Simpson 1944, 1953; Gould 1980). Washburn, deeply indebted to Simpson, adopted a particularly firm version of selectionism/adaptationism in his 1950s program for the "new physical anthropology."

12 Montagu (1952, 1965); Greene (1981: 162–79); Kaye (1986); Huxley (1942); Dobzhansky (1937, 1962).

13 Hutchinson (1978); Foucault (1978); Kingsland (1985); Jacob (1973).

14 UNESCO (1952: 103); see also Lewontin, Rose, and Kamin (1984).

15 Wilson (1975); Dawkins (1976); Maynard-Smith (1964).

16 The only woman among Hooton's former students of that period, Alice Brues (Ph.D. 1940, the same year as Washburn's), was not one of the experts consulted. Brues (1954) also published within the context of the modern evolutionary synthesis.

17 Spencer (1981: 363; 1982).

18 Armelagos et al (1982: 309–13); on Hooton, see Shapiro (1981); Brace (1982); Howells (1954).

19 Washburn Papers: Julian Steward to Sherwood Washburn (7 April 1959); Washburn to Steward (27 April 1959).

20 Brace (1982: 21–22); Coon (1939, 1962, 1965).

21 Fleagle and Jungers (1982: 196–8, 209).

22 Washburn, interview 21 June 1977; Washburn to Haraway (24 Jan. 1981).

23 Schultz (1969, 1971).

24 For his career account, see Washburn (1983); for his history human evolution and primate studies, see Washburn (1977, 1978, 1982).

25 Washburn (1946a, 1946b); Washburn and Detwiler (1943).

26 Kevles (1985: 355); Dunn and Dobzhansky (1946).

27 Tax interview, 14 Nov. 1985; Sol Tax to D. Haraway (15 Feb. 1985).

28 Howells (1962); Jepson et al (1949); Boaz (1982); Politzer (1981); Fleagle and Jungers (1982).

29 Washburn (1952). The Cold Spring Harbor, "New Physical Anthropology," and Wenner-Gren papers were reprinted numerous times in volumes intended for classroom use. The pedagogical context of the papers helped form the structure of assumptions for the next two decades' students.

30 Tax interview, 14 Nov. 1985.

31 WP: informal report on Viking Fund Grant, Washburn to Fejos n.d.; Washburn (1983); Baker and Eveleth (1982: 38–39).

32 Sarich (1971); Sarich and Wilson (1967).

33 Gilmore (1981: 388–89); Ribnick (1982).

34 WP: Report on the Evolution of Human Behavior, n.d., p.1; Grant Proposal to the Ford Foundation, p.2. Out of interest in Washburn's lectures during 1957–58, a seminar formed, including Warren Kinsey (zoo-based research on the use of the primate hand), John Frisch (a jesuit whose fluency in Japanese connected the Washburn group to Japanese primatology), Virginia Avis (zoo-based study of brachiation in relation to human origins), G. Cole (an archaeology student who surveyed possible baboon research sites in East Africa), Jack Prost (behavioral study of primate joints), Phyllis Jay (preparing for monkey field work in India), and Irven DeVore. At Berkeley, Washburn's early students included Theodore Grand (initially slated to study factors in the balance of the head in primates, actually studied functional anatomy of the chimpanzee shoulder), and Ralph Holloway (beginning with studies of the vertebral column dissected according to motions, finishing with a doctoral study of the primate brain).

35 WP: Grant Proposal to the Ford Foundation, pp. 3, 5.

36 WP: "Evolution of Human Behavior," 7 Jan. 1958.

37 DeVore interview, 18 March 1982; Tax interview, 14 Nov. 1985.

38 DeVore listed the observational studies that he used to prepare for field work, including Harlow, at the "leading edge" of his mother-infant attachment studies in the late 1950s; S. Altmann, who was preparing to do his Cayo Santiago rhesus study; Emlen and Schaller (Schaller 1963), involved in a field study of the mountain gorilla; Imanishi (1960) and Itani (1954), completing their first trip to the U.S.; Kummer (1968), whose M.A. thesis was available and whom DeVore saw in Switzerland on his way to Kenya; Scott (1950), whose sheep study was one the few published field studies of any mammal; Bartholomew (1952) on elephant seals; Darling (1937) on the red deer; and Carpenter (1964), who was rather bitter about the deafening silence that had met his 1930s path-breaking studies. For his early work, see DeVore (1962, 1963, 1965b); Hall and DeVore (1965). For DeVore's role in canonizing male dominance as a principal organizer of primate society, see DeVore (1968); Eimerl and DeVore (1965).

39 Ribnick (1982); Gilmore (1981); Malinowski (1927, 1944); Radcliffe-Brown (1937, 1952).

40 Dollard (1964); Cross (1982); Morawski (1986).

41 Jay (1963a); DeVore (1963)

42 Jay (1962, 1963b, 1965, 1968); Dolhinow (1972); Dolhinow and Bishop (1972).

43 WG: Fejos to Washburn, 3 Nov. 1958; Washburn to Fejos, 2 Dec. 1958; Conference Report, 22–30 June 1959; DeVore to Fejos, 4 Sept. 1962; Washburn and DeVore (1961).

44   Hamburg (1953); Lederberg (1983).

45   Hamburg (1961, 1963, 1972). See also Hinkle (1973/74).

46   *Stanford Univ. Bul.* 1971–72: 499.

47   Coelho et al (1974); Washburn et al (1974).

48   WP: Hamburg, Evolution of Emotion, AAAS paper, 1966.

49   On masculinism, science, and nuclearism, see Easlea (1983); Cohn (1987); Sofia (1984); Sofoulis (1988).

50   Schneider (1984: 165–74, and personal communications). Interviews: J. Altmann, 2 April 1982; S. Altmann, 1 April 1982; Struhsaker, 11 Aug. 1982; Marler, 12 Aug. 1982.

51   Lee and DeVore (1976); Lee (1979, 1984)

52   Kaye (1986); Gould (1986).

53   Haraway (1986); Hrdy (1981, 1986); Smuts (1983, 1985).

54   Rubin (1975); Harding (1983, 1986).

9   Harry Harlow and the Technology of Love

1   Harlow and Mears (1979: 5). Sources for this chapter include a biographical memorial (Suomi and LeRoy 1982); a collected, popular volume of Harlow papers covering 30 years of research (Harlow and Mears 1979); a close reading of a paper at the margin between professional and popular publishing, site of inter-translation of the cultures and ideologies of laboratory and mass industrial culture (Harlow, Harlow, and Suomi 1971); an analysis of a program for a 1960 popular television series (*Mother Love*); an interview with Stephen Suomi, former Harlow graduate student and collaborator (Madison, WI, October, 1978); and the Robert Yerkes and C.R. Carpenter archives. No attempt is made to produce an internal intellectual history of Harlow's comparative psychology; the focus is the boundary of translation and traffic between the laboratory and other areas of 1950–1975, U.S. middle class, white culture.

2   This founding story of the research for which Harlow's lab is remembered is repeated freqently. See also Harlow (1977) and Suomi and LeRoy (1982: 323). The stratosphere (or above) is the type location for man's second birth in the mythologies of phallocratic, high-technological societies (Sofoulis 1988). Sofoulis characterizes the mythologies of technology in science fiction film in ways useful to understanding the iconography, artifacts, and architecture of psychological research laboratories. The masculinist research style was often explicit. In a letter from Harry Harlow to Sherwood Washburn, 21 May 1958, the psychologist wrote, "Experimental social behavior is a real tough game, but I am sure at least a number of problems are soluble." Washburn had written Harlow with a number of suggestions for surrogate mother research, including non-primate animate pseudo-mothers.

3   For a list of Harlow graduate students, see Suomi and LeRoy (1982).

4   Margaret K. Harlow is the wife/colleague memorialized in histories of the lab as the designer of the "nuclear family apparatus" and co-source of the concept of "affectional systems." Harlow first married Clara Mears, then Margaret Kuenne, then after Margaret's death, again Clara Mears. Both women were his professional colleagues. Other women important to the work of the Harlow lab include the secretary for Harlow's last 20 years at the Wisconsin Psychology Primate Laboratory, Helen LeRoy, and Helen Lauersdorf, editorial secretary for many years, credited by Harlow with having written or rewritten many papers officially authored by staff or graduate students. Of Harlow's 36 graduate students, probably 5 were women, 3 of those after 1970. Contrast this figure with Sherwood Washburn's and Robert Hinde's primate-centered Ph.D.s for the post-war period. The oral culture of the Harlow lab includes repeated stories of animal technicians letting isolated experimental animals out when scientists were not present, disrupting protocols.

5   Memmi (1967); Hartsock (1983b); de Beauvoir (1954); Benjamin (1980).

6   "Our own data show without question that the monkeys we raise in the laboratory are healthier, less disease-prone, more fearless, brighter, and presumably happier than any monkeys ever raised in a

wild or feral state" (Harlow and Mears 1979: 5). Not all the monkeys, if the papers on the effects of normal semi-isolation rearing, not to mention various experimental series, are to believed. But, the monkeys raised in the nuclear family apparatus were especially happy and particularly superior to those in a state of nature.

7 *Mother Love* script, 1960; Harlow and Mears (1979: 9). For a discussion of conditions of U.S. women in the period from 1950 to 1970, see Chafe (1972).

8 Mulvey (1975); Kuhn (1982: 61).

9 In the TV program, the emphasis was on the structuring of vision in experiments. White-coated men gather around a view window to watch an "open field test" conducted in a box ingeniously designed by one of them (Dr. Robert Zimmermann, a 1958 Harlow Ph.D. student). "Diabolical fear objects," like mechanical drummer teddy bears, are released into the field, and baby monkeys are shown clutching their surrogate mothers for comfort. The narration works by a question and answer format, through which the scientist's expertise is conveyed verbally and shown in the experimental apparatus. The apparatus for the production of primate love was quite literal and conscious in this laboratory (Suomi and Harlow 1969).

10 It is ironic that Harry Harlow's first graduate student at Wisconsin was Abraham Maslow, who did his doctoral work on "The Role of Dominance in Social Behavior of Primates," finishing in 1934. Maslow subsequently became famous for his theories of human motivation in "self actualization"; he was a father of the human potential movements that swept U.S. society from the 1960s (Maslow 1936a, 1936b, 1936c, 1940, 1943, 1954, 1966).

11 The contest between the cloth and wire surrogate mothers for the affections of infant rhesus monkeys deprived of more orthodox maternal figures was featured in teaching films for the next generation of psychologists (H.F. Harlow and R. Zimmermann, Nature and Developpment of Affection, 1959) and in popular films reaching broad television audiences (National Geographic Society, *Monkeys, Apes, and Man*, 1971; CBS Conquest Series, *Mother Love*, 1960).

12 Ehrenreich and English (1979); Ehrenreich (1983); Lasch (1977)

13 Bowlby (1969, 1973); Riley (1983).

14 Alison Jolly (interview, 10 August 1982) told of her, she believed typical, reaction to a Harlow lecture, in 1961–62 in the Cornell Messenger Lecture Series, as a humanizing message. In Robert Hinde's laboratories at Cambridge, infant-mother separations were done by removing the mother only, leaving the infant in a group of known animals, and for short periods only. Hinde people argued that this was more like the human situation for which a model was sought. At the Yerkes Laboratories for Primate Biology, Davenport supervised experiments of up to two years of isolation for infant chimpanzees. In 1982, Jolly believed that the Harlow infant isolation and mother surrogate experiments retrospectively showed the limits of ethically permissible animal experimentation; at some point, knowledge just is not worth the cost in animal suffering and human arrogance. Establishing that point is a political process, in which knowledge-power relations are constituted. The failure of cost/benefit ethics is only heightened by the argument that the similarity of the animal makes "it" relevant to the human need. Jane Goodall resisted editors' efforts in her first professional paper to substitute "it" for her use of "he" and "she" to refer to chimpanzees (Goodall 1986: 90; Fedigan and Fedigan n.d.: 16). If "we," animal and human, are so similar to each other, then perhaps there is not enough difference to permit morally the objectification of animals as models for people. Perhaps they are not models, but independent beings, indeed, selves. In liberal societies, such independence and selfhood imply "rights." For an excellent ethnographic treatment of re-searchers and animal rights activists' conflicting discourses on animal liberation, see Sperling (1988).

15 Harlow (1963); Harlow and Mears (1979: 8).

16 The theme recurs in science fiction (Fairbairns 1979; Piercy 1976).

17 Harlow (1951: 183–238); doctoral theses list and Harlow cumulative bibliography (Suomi and LeRoy 1982: 341–42). The logic of using monkeys as models for humans turned on the ability of the laboratory to multiply, accelerate, magnify, and condense. The rhesus monkey was a good model only partly because of similarity to humans; its differences were also crucial, especially the differences in rates of maturation. The rhesus matured several times faster than the human child. If a set of developmental comparisons can be constructed between two species, then the "faster" species can

be used to multiply and condense tests interesting for the "slower" species. Other rate differences were also basic, like reproductive times. A rhesus breeding lab can produce many more subjects per unit time than a human family, even in the fabled American Republic of the late-eighteenth century when white women did their racial duty so prolifically. The alternative to breeding is cropping, but here variables get out of control, like disease in the animals and politics in the animal trade system. The point of the laboratory is to make as many interesting things as possible happen in as short a time and small a space as is convenient to people. Bruno Latour called the laboratory a machine for making many more (interesting) mistakes than would otherwise be practical or possible. An interesting mistake is not all that easy to arrange. Such an apparatus is worth a great deal, financially and socially. Harlow was implicitly aware of this point in the sociology of science (Harlow and Mears 1979: 4; Suomi and Harlow 1969; Latour 1983).

18 Expansion of the laboratory program was based on two wider social facts: growth of federal interest in mental health research and radiation research in the context of nuclear development and the space race. Mental health, family system, and nuclear-related research were all part of the relations of production and reproduction in Cold War America, mediated by the monkey body and mind. Beginning in the mid–1950s, Harlow's publications initiated what became a central focus of the laboratory, i.e., research in psychiatry and mental health (e.g., Harlow 1952; the series on affectional systems, such as Harlow 1962; Harlow and Harlow 1965; the series on mother-infant separations and depression, such as Seay and Harlow 1966; McKinney, Suomi and Harlow 1971; Mason and Sponholz 1963). David Hamburg at the Stanford's Department of Psychiatry encouraged experimental animal research on emotional development and mental health in the same period. Also in the 1950s publications appeared on effects on learning of high altitude cosmic rays (Harlow, Schrier and Simons 1956); whole body X-irradiation (Harlow and Moon 1956); and on "brainwashing" (Farber, Harlow, and West 1957). Rhesus monkeys had become all-purpose primate material, and Harlow never missed an opportunity to be of service to the ideological, technological, and clinical state apparatus for the production of the primate body, including a 1960s series of talks for the Voice of America Forum Lectures on animal study. Harlow benefited from the establishment of the federally funded Regional Primate Research Centers; the Wisconsin Primate Center was always separate from the university psychology Primate Laboratory, but Harlow headed both in the late 1960s. Harlow was active in the organization and promotion of lab and field primate behavioral and social research in the post-war period, serving on national committees, reviewing proposals, etc.

19 The overtly misogynist jokes in the Harlow corpus are numerous, e.g., the exuberantly told story of the orangutan, Maggie, at the Villas Park Zoo, who used to beat her "husband," Jiggs, but became docile after she had been hit in the head by the zoo keeper who was afraid she would attack a human child: "Apparently, after all those years, she had finally found the man of her dreams, a man who understood the psychology of women" (Harlow and Mears 1979: 87). Similar humor appears in Harlow and Mears (1979: 9, 23, 106, 108, 124, 130, 132–3, 135, 261; this is not an exhaustive list, merely a tiring one).

20 Up to 1982, he was the first and only primatologist to receive this award, the highest U.S. official scientific award (Suomi and LeRoy 1982: 323).

21 Rubin (1975); Wittig (1981); Rich (1980).

22 Suomi and LeRoy (1982: 325). Suomi was quoted in the Milwaukee Journal for 6 Aug. 1981 saying that the wells of despair were a scientific failure because individual differences in infant responses could not be eliminated. "Suomi also stated that 'pitting' was 'unnecessarily harsh' as well as 'unpleasant' and 'distasteful,' and that participating in such experiments gave him 'nightmares' " (quoted in "Suomi Denounces Harlow Techniques," International Primate Protection League Newsletter 8,3 [September 1981]: 13). Suomi, who worked with the chambers when he was under Harlow's direction, continued, "One of the first things we did when Harry left in the early 1970s was get those things out of here." Suomi defended use of the monkeys for psychopathology experiments in principle, and equally cruel apparatus was introduced into the Wisconsin laboratory after Harlow left at retirement in 1974 for the University of Arizona. The productive search for the human model continues in Suomi's timely interest in monkeys in relation to child abuse (Suomi and Ripp 1983). In a perfect, if unintended, self parody of his professionally tendered economic principles, so useful in Pinochet's Chile, famous worldwide for its record on humane treatment, Milton

Friedman informed the IPPL that the record of the National Institutes of Health's funding of cruel and useless experiments in the Wisconsin Primate Laboratory was yet more reason for his opposition to such needless federal waste; the NIH should be abolished! (IPPL Newsletter, 8,3 [Sept. 1981]: 13.) On grounds of cruel and useless professional activity, Friedman might check out the parable of the mote in one's neighbor's eye.

10   The Bio-politics of a Multicultural Field

1   Asquith (1981: 211); Itani (1954, 1975). I have followed the Japanese convention of placing the family name first and personal name second.

2   Nishida (1968, 1979); Hiraiwa-Hasegawa et al (1983).

3   Kuroda (1980); Kano (1979, 1982); Kano and Mulavwa (1984); Nishida (1972).

4   S.A. Altmann interview, 1 April 1982.

5   Frisch (1959); Kitahara-Frisch resides in Japan. Altmann (1965b).

6   See Traweek (1988) for cultural similarities and differences among United States, European, and Japanese high energy physicists. She traces the threads of elite international scientific cultures, particular national variants, and gender-specific scientific practices and imaginations woven into the fabric of field theories and built into the hardware of test devices.

7   Doi (1986: 23–33, 76–86).

8   Asquith (1981, 1983, 1984, 1986a, 1986b).

9   Nakane (1982) provides one Japanese woman's account of becoming an anthropologist. I have not found a comparable account by a woman anthropologist studying primates.

10   Just as the Christian ideal of stewardship is complex, historically dynamic, and able to accommodate mutually contradictory forms of social relationship with nature, so too Japanese cultural reworkings of Buddhism, Confucianism, and western science are multi-faceted and historically mobile. For some sense of how modern cultural reinvention works in Japanese popular culture, see Buruma (1985). However, Buruma's "Orientalist" treatment reinscribes for an American audience stereo-typed Japanese exoticisms (Kondo 1984). In a structuralist analysis, Ohnuki-Tierney (1987) examines Japanese concepts of self and other by inquiring into the meanings of monkeys and the "special status" people who produce the popular monkey performances. In Japan monkeys have been mediators between deities and people, scapegoats, and clowns linked to social criticism. On "cultural reinvention" and ethnographic narratives of the self, see Clifford (1988).

11   Hern's (1971, orig. 1899) retelling of Japanese ghost stories evokes the regions outside the world of ordinary humans, a kind of naturalistic and demon-filled wilderness, but not one closely resembling the modern western idea of wilderness (where the deer and the antelope play, etc.). From its origin, this western wilderness is permeated with nostalgia and a sense of the impending disappearance of nature. That is, this concept of wilderness is implicit in European and Euro-American culture only after the "penetration" and "discovery" of colonized others, including the landscape with its plants and animals, were underway. However, one should not assume that Japanese constructions of nature are necessarily more compatible with conservation or broadly non-exploitative practices necessary for species and habitat survival, nationally or globally, compared to western constructions. Both inside and outside Japan, the Japanese are enmeshed in terrible environmental problems. My point is that the different ways of loving, knowing, and otherwise constructing "nature" have different strengths or hazards for building non-exploitative conservation and survival practices.

12   Altmann (1962); Itani (1963); Imanishi (1960); Kawamura (1958).

13   Haraway (1981–82); Keller (1985).

14   DeVore (1962, 1963, 1965a), interview, 18 March 1982; Smuts interview, 18 March 1982.

15   Imanishi (1963: 69); Asquith (1981, 1984).

16   Kawamura (1963); Itani (1963).

17   Japan has been a colonial power, and I do not intend to romanticize Japanese treatment of its

colonial "others," for example, the Koreans. But Japan's indigenous species have not been enlisted in its cultural symbology marking insider-outsider.

18   Tournier (1972); Akeley (1923); Fossey (1983); Mowat (1987).

19   Asquith, personal communication, 28 Aug. 1985.

20   Hrdy (1986); Hrdy and Williams (1983); Haraway (1978a, 1978b, 1983a); Rowell (1984); Small (1984).

21   Altmann (1967); Losos (1985); Clutton-Brock (1983).

22   A U.C. Berkeley student, Donald Sade, reported the same structure among the rhesus monkeys of Cayo Santiago (Sade 1965, 1967). Sade discounted influence from the women's movement in primatology, and he was not comfortably part of the Washburn network (Losos 1985: 16–19). Losos's citation analysis showed that members of the Washburn network who continued to emphasize the male dominance hierarchy as the organizing axis of society did not often cite Sade on matrilineal rhesus social organization. Losos argued that male dominance hierarchy explanations, never hegemonic, were undermined by socioecology through the late 1960s and early 1970s independently of feminism.

23   For a film stressing that "the young revolutionize the culture" (soundtrack), see National Geographic's *Monkeys, Apes, and Man* (1971). The remark comes right after the filmic narration of Gray and Eileen Eaton's research with a *Macaca fuscata* colony at Oregon Regional Primate Research Center. The soundtrack calls the Oregon Japanese monkeys "ambassadors from some foreign nation—the animal kingdom." The rich ambiguities implicit in the words ambassador, colony, and nation resonate with the move from human nation to animal species. The next filmic scene is the monkey's "home"—Japan. The first stop, in a snowstorm, shows the adaptation of the monkey citizens in the extreme northern part of their distribution. The viewer learns that this simian is not a tropical creature. The point silently joins the resonances around ambassadors, nations, and colonies. Then we go south, to Koshima Island, to see monkeys swimming happily in the ocean, washing sand out of wheat scattered on the beach, and scrubbing sweet potatoes. "The young revolutionize culture." Commenting on the island culture's similarity to "a small nation," the narration stresses the "primitive inventor" in the "culture of Koshima." On screen, monkeys walk upright, carrying their food to wash it before eating.

From "culture" in Japan, the film goes to West Africa, to narrate the lives of chimpanzees tied to the trees, living in a "carefree style" in the tropics (soundtrack). The codes of race and colonization in the drama of "civilization" in the animal kingdom could hardly be more explicit. Here, the codes override the usual placement of monkeys and apes, in which the apes are ascribed the greater social and behavioral complexity and form the link between animal and human. The film moves out of the rain forest into the open, where progress and civilization can begin, here represented in Adriaan Kortlandt's studies of chimpanzees' understanding of death and of their use of clubs to beat up a threatening stuffed leopard.

24   Landau (1981, 1984).

25   The *Konjiki* and the *Nihongi*, the earliest written historical chronicles, from about 720 C.E., were written in Chinese. They are full of accounts of important roles of women and female deities in Japan as a nation, e.g., the story of the Empress Suiko (692–628 B.C.E.). Her title was posthumously bestowed because the writing of history was considered to be an important innovation of her reign. The close assocation of women with writing at the origin of Japan as a national and linguistic community contrasts markedly with the symbolic function of Eve's eating from the tree of knowledge of good and evil for the western and Islamic peoples of the Book.

26   Haraway (1979, 1985).

27   S.A. Altmann, speech to the American Society of Primatologists, 3 June 1980.

28   Kawai (1969: 293), in Asquith (1981: 344).

29   Kawai (1965); Asquith (1981: 344n).

30   Chodorow (1978); Irigaray (1985); Gilligan (1982); Keller (1983, 1985); Merchant (1980).

31   *IPPL Newsletter* 14(2), July 1987: 10.

32   *IPPL Newsletter* 5(1): 3.

33 The *IPPL Newsletter* from 1977–83, especially the ten year memorial issue, vol. 10, no. 3, 1983, details the history of the export ban and IPPL's role in the struggle. Shirley McGreal founded the controversial conservation organization in 1973. The IPPL has a highly motivated intelligence and action network in about two dozen countries. Respected primatologists represent the organization in their countries and/or on the advisory board: Suzuki Akira, Japan; Frances Burton, Canada; William McGrew, Scotland; Zakir Husain, Bangladesh; Qazi Javed Iqbal, Pakistan; S.M. Mohnot, Central and West India; Jane Goodall, Tanzania; Edward Brewer, The Gambia; Vernon Reynolds, U.K.; Geza Teleki, U.S.; Dao Van Tien, Vietnam. Other activities include Sumit Hemasol's Thai-language conservation magazine for young people. Sumit Hemasol, formerly a reporter for the *Bankok Post,* exposed illegal hunting activities of highly placed Thai military officers in an incident that contributed to the fall of the military dictatorship in 1974. The democractic regime that followed instituted a ban on all primate exports. IPPL has uncovered multiple outrages in the legal international primate trade, in conditions of experimentation and housing, and in the intricate multinational illegal smuggling networks purveying protected primates. The IPPL's use of the Freedom of Information Act has been inspired. Its style of action tends to be direct and dramatic, e.g., organizing teams of Thai university students in "Project Bangkok Airport" in 1975 to log wildlife shipments to document cruel and illegal activities, thereby generating publicity that helped lead to a ban on exports. IPPL has been a thorn in the side of U.S. embassies and the State Department, which prefers to see the pesky group as fringe extremists. Its newsletter's rhetorical strategy moves from meticulous documentation, with facsimile reproduction of condemning evidence, to pleas for direct letter campaigns like Amnesty International's, to delicate line drawings of various species by a Thai art student (Kamol Komolphalin), to heart-wrenching descriptions and photographs of the murder of the gorilla Digit by poachers fighting Dian Fossey in Rwanda, to centerfolds of primate babies of the year. IPPL is a consummate narrator in a field of master story-tellers.

34 McGreal (1981). The large majority of monkeys imported to the United States were not used for military research, but in the medical, pharmaceutical, cosmetic, pet, and zoo trades. Over 150,000 rhesus monkeys per year were imported into the U.S. in the mid–1950s, about 120,000 of whom were killed in the production of vaccines, before later techniques reduced the number of "sacrificed" monkeys. From 60,000 to 80,000/year were imported at the height of the polio campaign. By the imposition of the Indian and Bangladesh embargoes in 1978, about 1125 monkeys (125 green monkeys from Africa and 1000 rhesus) a year were needed for oral polio vaccine production, a number quickly supplied from U.S. breeding sources (Conference on Procurement and Production of Rhesus Monkeys, Institute of Animal Resources, National Research Council, 7 June and 22 Sept. 1955; Bermant and Lindberg [1975]; *Oregon Journal*, 22 March 1982). The 20,000 rhesus per year coming to the U.S. from India in 1977 were worth $17 million retail in 1980 dollars (*The Economist*, 9 Feb. 1980). Exporting primates to the vast western biomedical industry in order to earn foreign exchange is, along with tourism and timber concessions to multinational corporations, economically important for primate-habitat national governments. The economic and political dimensions of agriculture, both subsistence activities and agribusiness, are also fundamental for primatology.

35 Southwick and Siddiqi (1985). Doctoral-level collaborators in this long project and related ones in Nepal: M. Rafig Siddiqi, M. Azher Beg, M.L. Roonwal, R.P. Mukherjee, M. Babar Mirza, S.M. Alam, T.N. Ananthakrishnan, K.K. Tiwari, R.K. Lahiri, J.A. Kahn, John Oppenheimer, Jane Teas, Thomas Richie, Henry Taylor, Bunny M. Marriott. The Indian field work was sponsored by U.S. Public Health Service grants (Southwick et al 1980; Siddiqi and Southwick 1977; Southwick and Beg 1961).

36 Emlen interview, 6 April 1982.

37 Collias and Southwick (1952); Southwick (1963).

38 E.g., Mohnot and Roonwal (1977); Mohnot (1974). From the early 1960s, the Japanese collaborated with Indian scientists to study langurs (Sugiyama, in Altmann, 1967; Sugiyama et al 1965; Miyade 1964). Several students of Sherwood Washburn and Phyllis Dolhinow studied Indian primates. Hrdy (1977) observed Indian langurs in a series of several-month stays, partly reflecting difficulty in obtaining permissions for long-term research by foreigners. Several U.S. universities (Johns Hopkins, U.C. Berkeley, Yale) had primate projects in India.

39 For an example of modest change, see the publication lists from the Kibale Forest Project in Uganda,

under the directorship of Thomas Struhsaker of the New York Zoological Society, Rockefeller University, and Makerere University. By 1982, three young Ugandans, including one woman who did a primate-focused thesis, had completed Masters of Science degrees in ecology, behavioral ecology, or conservation at Kibale through Makerere University (Baranga 1978).

40 Before having those results, in 1978 when the trade embargo was instigated Southwick appeared to believe the rhesus population could sustain cropping for biomedical export—at three times the existing export of 20,000/year, according to an IPPL response to Southwick and Siddiqi (1985), which did not cite the IPPL's role in the embargo or mention the issue of nuclear military research (*IPPL Newsletter* 12[1]: 12, 1985). But see Southwick and Siddiqi's August 1985 letter in *Natural History* acknowledging the IPPL role, but referring only to their demonstration that the rhesus were being used "for inhumane laboratory experiments" (p.4).

41 Du Chaillu (1861); Morell (1986); Shoumatoff (1986); McGuire (1987).

42 Schaller (1964: 9, 1963).

43 Schaller (1964: 270). Schaller, a major conservationist, later studied Indian wildlife (1967), social carnivores (1972), and collaborated with the Chinese on the giant panda (Schaller et al 1984).

44 Mburanumve, 9 Nov. 1960, quoted in Schaller (1964: 274).

45 Fossey (1976) earned her doctorate under Robert Hinde's sponsorship at Cambridge. For the mountain gorilla field studies by the generation that both worked with and succeeded Fossey, see Fossey (1983); Harcourt (1979); Stewart (1981); Harcourt and Fossey (1981); Veit (1983). The Mountain Gorilla Project was established in 1979, after severe poaching episodes, by the British Fauna and Flora Preservation Society. Together with other conservation groups, the Project provides equipment, expertise, and funds to help Rwanda in gorilla conservation. Most of the poachers come from the Zaire side of the park, but agricultural pressures are mostly from the Ugandan and Rwandan areas of the gorilla range in the Virunga Volcanoes. Veit (1983) estimated there were about 1000 living mountain gorillas.

46 Fossey (1970, 1971, 1981).

47 Weaver talked about her feelings about Fossey, the gorillas, and her role in the film to several interviewers, including Christopher Durang (1988: 34–43).

48 For the premiere of the movie (based on Isak Dinesen's autobiography), *Out of Africa*, in Nairobi on 31 Jan. 1986, to benefit the African Wildlife Foundation, an all-white swing band played for white actors (none of the big names), the film's co-producer, and Kenyan Vice-President Mwai Kibaki, while Mr. Kenya 1985, wearing an authentic Masai warrior headdress, auctioned off Meryl Streep's boots and Robert Redford's hat (*Vanity Fair*, May 1986: 38).

49 Jolly was particularly drawn to Hutchinson's way of thinking about evolution and species diversity in "Hommage to Santa Rosalia" (Hutchinson 1959). Hutchinson served on Jolly's thesis committee, typical of his support of heterodox women graduate students whose interests did not fit easily into Yale's official program. The theme of powerful senior male mentors for women in primatology emerged repeatedly, e.g., David Hamburg's concern for Barbara Smuts, Louis Leakey's importance for Jane Goodall, Sherwood Washburn's energetic sponsorship of some of his students, and E.O. Wilson's encouragment of Sarah Blaffer Hrdy.

50 Jolly interview, August 1982.

51 Zuckerman (1932, 1933); Haraway (1978b). Also aggressively defensive, Carpenter commented at the New York Academy of Sciences session that included papers by Stuart Altmann, J.J. Petter, Phyllis Jay, Alison Bishop (Jolly), Jane Goodall, and Cynthia Booth. Every paper but Booth's was a report on doctoral research. Carpenter was comfortably superior in tone to Goodall, Bishop, Jay, and Booth. He chided Bishop for her title and for "neglecting due proportionate emphases on more elusive processes." Goodall had an "excellent specialized paper," but had to learn to interpret her observations. Booth's paper was a veritable object lesson in the sad mistakes deriving from inadequate facilities and training. She needed to learn how to be objective. By contrast, "the science of primatology is advanced significantly by Petter's sensitive basic observations of lemurs." And although Carpenter did "not see in Altmann's list [of items in a taxonomy of behavior treated statistically] any theme or rationale," and chided Altmann for working more independently than necessary (i.e., not using Carpenter's previous Cayo Santiago report sufficiently), Altmann's was still

a report "ranking along with Schaller's magnificent study of gorillas" (Carpenter 1962: 488–96). Carpenter managed a doyen's tone of superiority all around, but only the women were systematically enfantilized. The topics of Goodall's reports at London and New York on chimpanzee nest building and tool use in termiting were soon to become world famous. Carpenter later tried to get Jane Goodall to affiliate with Georgia, allying her Gombe chimpanzee studies with Carpenter's efforts to build a primate doctoral program (Carpenter papers). She cast her lot with Stanford's David Hamburg.

52   Jolly interview 1982; Jolly (1980: 3–9).

53   Jolly interview, August 1982. See also Alison Richard (1978, 1985). Most of her lemur research was done on Malagasy animals before 1975. She went to Madagascar in 1970 to get comparative data from prosimian leaf eaters on the ranges of flexibility in a species' social behavior, in light of extensive data from the previous ten years on ranges of social behavior within species for Old World monkeys. This variation undermined the prior tendency to search for species-specific social behavior patterns. Richard approached her work in the context of reframing hypotheses in socioecology following the limited success after Crook and Gartlan's (1966) pioneering efforts. Focused on primates as mammals and not as human relatives or surrogates, her history of primatology explored why the field has not been more part of ecology until recently. Richard is another node in networks of women primatologists. Mary Pearl (1982) was her graduate student at Yale. Richard's (1981) history of views and critique of sociobiology is also a map of her explanatory networks
In her 1982 Harvard primate behavior and ecology course, Barbara Smuts taught with Richard's fieldstudy-oriented textbook, *Primates in Nature*, when it was still in manuscript. Smuts's other texts were Jeanne Altmann's *Baboon Mothers and Infants* (1980), Sarah Hrdy's *The Woman That Never Evolved* (1981), and Martin Daly's and Margo Wilson's sociobiological textbook, *Sex, Evolution and Behavior* (1978). Journal articles drew from Jane Goodall, Shirley Strum, Katie Milton, Richard Wrangham, Timothy Clutton-Brock and Paul Harvey's *Readings in Sociobiology* (1979, W.D. Hamilton and Robert Trivers articles), Caroline Tutin, and C. Bachmann and Hans Kummer. While the women in primatology certainly do not neglect research by their male colleagues, the evidence is strong for their acute awareness of each other, both cooperatively and competitively (interviews with Smuts, Hrdy, Richard, Jolly, J. Altmann, Zihlman, Morbeck, Strum, Nicolson; Smuts's course outline). Hrdy names Alison Jolly as her major role model, who had in her turn seen Jane Goodall in that role. Smuts' course outline documented 1980s webs: women who draw from each other, but who emphasize divergent explanatory traditions (Smuts, Richard, Hrdy, Altmann, Strum, Milton); socioecologists and sociobiologists, especially from King's College, Cambridge, U.C. Davis, and Harvard (Smuts, Daly and Wilson, Trivers, Clutton-Brock, Harvey, Wrangham, Hamilton); Gombe researchers (Goodall, Smuts, Wrangham, Tutin); Gilgil baboon researchers (Smuts, Strum); international baboon researchers (Smuts, Kummer, Strum, Altmann); researchers focusing frequently on female behavior (Wrangham, Smuts, Altmann, Hrdy, Tutin); researchers focusing intently on male behavior (Strum, Smuts, Clutton-Brock, Trivers, Bachmann and Kummer).

54   Interview, 10 Aug. 1982. Jolly (1972 [1985], 1980, 1985, 1987); Bruner, Jolly, and Sylva (1976); Jolly, Oberle, and Albignac (1984).

55   Gould (1981); Gilman (1985); Carby (1985).

56   Interview, August 1982. Besides Jolly's first intended meaning that nuclear war threatens life on earth for all species, disarmament of non-nuclear weapons is also critical for conservation. Westing (1980), coining the term "ecocide," detailed the effects of non-nuclear war in Indochina. In the 1980s, bombing in El Salvador and war-related forest destruction in Honduras and Nicaragua are stark reminders that the issue is all too current.

57   Quoted in Jolly (1987: 179); interview, 15 Jan. 1988.

58   Wright (1988) gives a sense of the collective and individual research tied to conservation in 1980s Madagascar. Her studies led the Malagasy government to make an important piece of rain forest into a national park to preserve three rare species of bamboo-eating lemurs. Since Wright is located among those biologists-naturalists willing to order their professional lives around bio-political issues with a different chronology from optimal academic publishing strategies, it seems bio-poetic justice that she has encountered a lemur species previously unknown to scientists.

11   Women's Place is in the Jungle

1   In the *New York Evening Express,* 5 March 1867 (Harris 1973: 190–91).

2   For specific analyses, see Haraway (1981, 1983a).

3   Aristotle (1984); Delaney (1986); Horowitz (1976); Lange (1983); O'Brien (1981).

4   Zihlman (1983a); Laporte and Zihlman (1983).

5   The following publications give a rudimentary hint of the international, multiethnic, and multicultural women's movements and radically self-redefining consciousnesses and of their partial expression in the documents and meetings in the context of the UN Decade for Women: the Indian feminist journal *Manushi,* from Delhi; *Migration World Magazine* XIV, no. 1/2 (1986), special issue, The Spirit of Nairobi and the U.N. Decade for Women; Morgan (1984); Fraser (1987); Seager and Olson (1986); publications from the International Labor Organization in Geneva (e.g., *Women and Rural Development: Critical Issues,* 1981; *Women at Work,* twice yearly); several issues of the *Multinational Monitor;* the publications from ISIS in Geneva (a twice yearly thematic journal); publications of The Minority Rights Group (e.g., *Arab Women,* no. 27, 1983; *Western Europe's Migrant Workers,* no 28, 1982); *Connexions: An International Women's Quarterly; International Journal of Women's Studies; Women's Review of Books; Women's Studies International Forum* 8, no. 2 (1985), special issue, "The UN Decade for Women: An International Evaluation"; United Nations publications from the Decade for Women (1975, 1980, 1985); publications of The International Women's Tribune Center, including registers of women's organizations and guides to the Decade; publications and papers from the African Training and Research Center for Women in Addis Ababa and from the Association of African Women for Research and Development; and New Internationalist Cooperative (1985).

6   For histories and critical discussion of the concept of gender, see Scott (1986); Haraway (1988); de Lauretis (1987).

7   Smith-Rosenberg and Rosenberg (1984); Schiebinger (1986); Jordanova (1980, 1986); Spencer (1852); Blackwell (1875); Geddes and Thomson (1890); Sayers (1982); Hubbard (1982). For histories of feminist concepts of gender, see Scott (1986); Haraway (forthcoming b). For a review of the history and philosophy of women in science, see Schiebinger (1987).

8   Jones (1978, 1980, 1981, 1983). In a radically different historical and cultural context, African women have also begun to gain the advanced degrees to author the natural history, conservation, and primate literature, e.g., Baranga (1978). In a non-degree context, in the 1980s two young Kenyan women in their early twenties, Norah Njiraini and Soila Saiyielel, became assistants in the long-term Amboseli Elephant Research Project, with Cynthia Moss, Joyce Poole, and Sandy Andelman.

> They have been trained to carry out the full range of field work. . . . Norah often went out by herself in a four-wheel-drive vehicle over rough terrain, collecting data on musth males for Joyce. She got charged by huge bulls, got stuck in the mud, had punctures and breakdowns, but handled it all with great calm and a growing self-confidence that was rewarding to watch. Soila has learned to identify all the adult females and most of the calves faster than anyone else I know and while Sandy was away she did all the radio-tracking of two females who had been fitted with radio collars. . . . What makes me happiest is that they love their work and they love the elephants and are as taken up with their lives as any of us have been. (Moss 1988: 312)

This picture of women's cross-cultural scientific collaboration and mentoring seems part of the construction of a complex and historically new mutation in possibilities for love and knowledge of nature. Moss's narrative contradicts my informants' repeated assertion that East African women could not possibly want to be, nor would they be allowed to be by the men of their cultures, field assistants. Implicit in this assertion is the tendency of western women and men to assume that men of other cultures are more sexist and women less self-defining than those from European and Euro-American groups.

9   Few women worked on primates in the field or lab before World War II. A major exception was Gertrude van Wagenen at Yale, who had unusual success with a breeding laboratory colony of rhesus. Van Wagenen's research was in reproductive physiology in the frame of the National

Research Council's Committee for Research in Problems of Sex. An early field report by a woman was Nolte (1954).

10 Hahn (1988) is a good place to look at the para-scientific cultures of women and apes in the twentieth century.

11 To compare Euro-American, U.C. Berkeley-trained women primate scientists' consciousness of female animals just before and just after the eruption of the Women's Liberation Movement, see Jay (1963c) and Lancaster (1973), both pioneering papers.

12 Compare Reed (1975) and Skybreak (1984) for different political moments in popular marxist struggles for beliefs, ideologies, and scientific doctrines on women's evolution and the origin of sexual inequality. They make intriguingly different uses of Engels. Reed was an anthropologist and a member of the elder generation of American socialists; Skybreak is a pseudonym for a writer who is both sympathetic to the (Maoist) Revolutionary Communist Party and well-informed about current debates in ecology and evolution. Skybreak also has a sense of humor.

13 Hausfater and Hrdy (1984); Wasser (1983b); Small (1984); Smuts et al (1987).

12 Time-Energy Budgets of Dual Career Mothering

1 To name a few others from the 1960s: Thelma Rowell, Stephen Gartlan, Thomas Struhsaker, Donald Sade, Stuart Altmann.

2 From 1957 to 1970, Altmann did special tutoring in mathematics, including teaching minimally brain-damaged children and adults with limited literacy. She also designed and implemented an experimental mathematics teaching program for Decatur City Schools in Georgia. She had a wide range of experience and a deep commitment to helping people think mathematically, rather than mystify mathematics as an arcane set of techniques accessible only to "geniuses."

3 Interviews: Joan Silk, Nancy Nicolson, Barbara Smuts, Patricia Whitten, Dorothy Cheney, Robert Seyfarth, Jeanne Altmann, Stuart Altmann, Peter Marler, Robert Hinde, Naomi Bishop, Suzanne Ripley.

4 Whiting and Whiting (1975). Just as C.R. Carpenter's approaches to primate field studies must be seen in the context of social and human sciences, such as semiotics, and Stuart Altmann's rhesus study on Cayo Santiago owed many of its central practices to George Miller's communications research, Jeanne Altmann's interventions in primate biology and evolution were rooted in quantitative social sciences. It is simply not true that all the moves toward quantitative "reduction" in natural sciences follow a one-way flow along the ideological positivist hierarchy from physics through chemistry to biology and, always weakened and polluted, finally across the River Styx between nature and culture to behavioral and social sciences, and goddesses-forbid, beyond to the human sciences. Modern semiology ought to give the lie to illusions that technical analysis in human sciences follows positivist dicta. Modern mathematics or theoretical physics ought to give the complementary lie to illusions that fully mathematized discourses are not based upon interpretive practice. Technical (including but not restricted to mathematical forms of technical analysis) and interpretive are not opposites. Discussions of the problems of "reduction" ought to foreground this point.

5 Jeanne Altmann also emphasized that Stuart Altmann served as her mentor in field primatology in many respects. He had earlier independently begun interactions with the Whitings, just as he had been alert to ideas for field primatology in George Miller's approaches in language and communication analysis. The mutual criticism, methodological exchanges, and general collaboration between Jeanne and Stuart Altmann over many years has been a critical dimension of their professional work.

6 The moves work as follows. Identification: as a woman/female, I would identify with other women/ females and therefore claim privileged access to certain kinds of knowledge. A kind of being grounds a kind of knowledge; i.e., ontology grounds epistemology. "Nature" in the last instance grounds identification. Deconstruction: I would try to show how woman/female gets constructed and sustained in discourse, including science and politics. My unpacking of semiotic and other practices necessary to construct these objects aims to destabilize and problematize any moves I or you might make to claim authority based on "natural" identification. In my sense, deconstruction does not

forbid identification; rather, it displays what kind of move it is and forces responsibility for making it if one chooses to do so. The responsibility is never authorized or justified by "nature" or the "body," which are precisely what get constructed by moves in the same political-semiotic family with identification. Identification without acknowledging itself claims to come to rest outside constructive practices; deconstruction is forced to remain inside fields of shiftable meanings. Making meanings shift is a material political practice.

7   J. Altmann, interview, 2 April 1982. In my reading of Altmann's practice, sometimes she judged it better not to eliminate a word, but rather to force a category, e.g., "dominance," into such rich relation with other matters that they structure the narrative foreground. "Dominance" is too useful to give up, and in any case disliking a concept is not grounds for declaring it scientifically meaningless (It's me, not Altmann, who dislikes the category dominance). Unlike "rape," "dominance" has been too solidly built into durable pictures of primate life for the great majority of accounts to be able to dispense with it completely as an analytical device. It still remains not to make the essentializing move of mistaking the category for the world, which happily does not come pre-delineated and pre-packaged into units for which the only problem is deciding which names to assign. Labels are endlessly intriguing. Primatologist Jane Teas described a session at a scientific meeting on baboon strangers immigrating into a new group (personal communication). There was a subtle, maybe unconscious, contest between men and women for the terms "ease of entry" versus "penetration." My own use above of the metaphors "immigration" and "strangers" shifts the verbal play that surrounds any category formation from the sexual connotations that Teas reported on to the terrain of travel and relocation, with its tones of exclusion, choice, coercion, pain, opportunity, and danger.

8   King (1987a, 1988); Haraway (1988). I have adapted the concept of the apparatus of bodily production from King's notion of the apparatus of literary production in the construction of the poem as an object of value.

9   My own writing has been far from free of these problems, e.g., in "The Contest for Primate Nature" about the infanticide debate and women's narrative practices in primate studies (Haraway 1983a). One effect can be—and sometimes was—that readers did not take the scientific work of the women I was writing about seriously or did not regard my treatment as equally serious as that in other papers about male scientists. Focusing on sex and violence gets attention for both men and women, but the attention is more likely to undermine than to enhance women's authority. Women rarely control the field structure of interpretation.

10 Altmann's account is akin to the German women's collective, Frauenformen, which developed the practice of "memory work" to excavate and construct past "experience." The stories produced by memory work were for collective criticism. They were not read as autobiography, coming together in a coherent self, but as defamiliarized fragments, belonging to strangers. The stories were clues not to a progressive historical narrative of the self, but to the contradictions of the sexualization of the female body (Haug et al 1987).

11 In a popular immunology textbook the premium placed on this matter was highlighted by the label for the variable gene regions implicated in generation of diversity in somatic recombination theories: G.O.D., the generator of diversity (Golub 1987). "Master" theories, a kind of scientific creationism, abound in modern biology, perhaps most exuberantly in master molecule versions of molecular genetics. This is the mythic-scientific discourse that produced the broadly popularized notions of the Central Dogma (information flows in only one direction—from the stable source of all difference outward to its mediators and ephemeral embodiments: DNA-RNA-protein), the secret of life, etc. Male action in primatology, action differentiated by the geometry of the dominance hierarchy, has played similar roles in evolutionary biology. Modern evolutionary theory ascribes generation of difference to genetic mutation or rearrangements and restriction of difference to selection. Thus, in an orthodox sense, that males are more highly selected means they are the source of restriction of variation, not its generation. In another sense, the exclusive preoccupation with male action in terms of their differential reproductive success is also about "generation of diversity" because, from this "female-blind" perspective, genetic variations among females are relatively inconsequential since they are roughly equally passed on to the next generation. Variations among males, though "generated" in the genetic system and not through natural selection, are the ones that count in terms of determining evolutionary change. The males, or their variations, are the dynamic actors in the evolutionary play.

12 I am assuming for discussion that good evidence would be available that more dominant males would leave more kids because they got more sex with females at the right times. I read the record on this point to be rather messy. This is one of the issues that has divided workers at Amboseli, like the Altmanns and Glenn Hausfater, so sharply from Shirley Strum at Gilgil (Strum 1987b: 160–62). Hausfater's data (1975) showed advantages of high rank for males in mating, but also made plain frequent male changes in rank and female rank stability.

13 In addition to giving me copies of their grant applications from the early 1960s through the early 1980s, Stuart and Jeanne Altmann prepared a manuscript summarizing "Research on Baboons in Amboseli, 1963–78" for the controversial, contentious, and finally unsuccessful conference Shirley Strum organized, Baboon Field Research: Myths and Models, in 1978. Strum invited Bruno Latour, a major innovator in the social studies of science, to serve as a stimulator of reflection on how scientific categories are put together at the same time as the conference participants worked to develop shared frameworks for getting comparable data from the disparate baboon research sites and projects (Latour and Woolgar 1979; Latour 1978; Latour and Strum 1986). Not knowing until just before the conference that Latour would be present, with part of his agenda being to study their interactions, several conference participants objected, arguing his observing-interacting presence could interfere with their deliberations. The distress was so high that before discussion could begin, Latour had to sign a pledge not to publish anything from his observations of the scientists (Latour, personal communication; interviews: Strum, Hausfater, J. Altmann, S. Altmann, Osmundsen). It would be safe to say the primatologists were not habituated to being observed.

14 Draper (1976); Cheney (1978); Seyfarth (1976); Berman (1978); Hrdy (1977); L.T. Nash (1978); Struhsaker (1971).

13 Publishing Strategies as Model Interventions

1 In "Gender and the Study of Primates," in manuscript in 1988 for the AAA Gender and Curriculum Project, comparing primate studies and ethnography Linda and Larry Fedigan stressed western women's contributions to primatology in terms of "1) special respect for individual differences and proper attention paid to gaining insight from the exceptional case; 2) a belief that the complexity of nature exceeds our own imaginative possibilities and that reductionistic solutions demonstrate insufficient humility in the face of such complexity; 3) a reluctance to impose an a priori or premature theoretical design on the material, but rather a desire to listen to the material . . .; and 4) the ability to persist under difficult circumstances, particularly lack of recognition and respect from colleagues." They insisted that these traits had nothing to do with the stereotype of women with cute, furry animals and that the traits were not biological capacities exclusive to women, but were historical positionings. Fedigan's summaries of other women primatologists' books stressed these authors' representations of animals as actors, not resources. Positioned in a curriculum reform project, Fedigan's list of traits and ways of characterizing the women's writing should be read as oppositional narratives about women and science, at least as much as descriptions of women's scientific practice.

2 Originally, the monkeys were a personal gift to Emlen, who worked for a more institutionally secure arrangement. Involving many of the senior actors in the field, such as C.S. Southwick, C.R. Carpenter, and Emlen, the story of the troop's search for a home is a fascinating window onto themes of international primate politics, scientific hopes, and amateur and professional lovers of animals in the late 1960s and 1970s. The habitat for the Arashiyama monkeys had become too crowded, and the Japanese anthropologists did not want to see the redundant monkeys subjected to physiological research. The Japanese at one point corresponded with Emlen in the wonderful reversed east-west cultural idiom of his being a "Columbus looking for a home for the monkeys in the New World." Emlen spoke about "wanting to seed the monkeys" (Emlen interview). One possible home in America was a casualty to nuclear politics; the Savannah River atomic research site of the University of Georgia included a promising semi-natural island habitat. But federal regulations prohibited foreigners at a nuclear site, an impossible situation for U.S.-Japan primate research cooperation. Coming from several universities, especially Texas, Wisconsin, and Oregon, student workers at the Texas sites were a close group (interviews: Emlen, Stephenson; Clark and Mano 1975). Claud and Sharon Bramblett found the LaMoca site through an undergraduate student of

Claud Bramblett's, who was Edward Dryden's daughter, and the second site through a former M.A. student, Lou Griffin. "Through a combination of good luck and good will, the entire Arashiyama enterprise has been a model of East-West cooperation" (Fedigan to Haraway, 5 April 1988; Linda Fedigan, "The History of the Arashiyama West Japanese Macaques in Texas," in manuscript, 1987). In August 1987, Pamela Asquith and Linda Fedigan organized a Wenner-Gren conference of all presently active Arashiyama researchers from Japan and North America.

3   Fedigan to Haraway, 13 Sept. 1982.

4   Fedigan (1982: iv, 156–74). Fausto-Sterling (1985: 209–13) argues that feminist approaches to biology turn on the issue of simplicity versus complexity in structuring knowledge and in views of the object to be known. Frequently, she argues, the advocates of anti-essentialism, flexibility, and situational meanings are women, and often women in complex relation to feminism. This emphasis is among the strategies available to "subjugated knowledges" (Foucault 1980). Fausto-Sterling centers the explicit role of feminism in what she calls good science because of feminism's insights on subject/object separation, the value of experience in rational explanation, the presense of misogyny in ordinary scientific literature and practice, and the complexity of subject positions for those denied unitary (if still fictional) identities. Insisting on contextuality at all nodes in the knowledge-producing field is a way to destabilize and resist. It is a way to construct, but without the kind of authority represented as hegemonic and oppressive. Privileging "contextuality" is not a function of female/women's nature, but of possibilities in a particular field of knowledge-power. Similarly, feminists must deconstruct what counts as "experience" in order to foreground multiplicities, contradictions, and constructions of such potent entities.

5   In discussing female sexual choice, Fedigan explicitly made visible female sexual choice of female partners, as part of a pattern of female sexual initiative and not a function of the absence of available male partners (Fedigan 1982: 284; see also Wolfe 1984). This matter is often absent in the rigorously heterosexual texts of authors who are otherwise intent on emphasizing female sexual and other social agency (Strum 1987b).

6   Jill Morawski's (1987) history of sex and gender differences research in psychology attempts to show how gender categories are generated in particular research practices at particular social moments. Her goal is not to criticize faulty scientific ideas, but to show precisely how scientific practices are constructed to foster certain interests and confirm certain realities. The obscuring effects of categories of sex and gender are not a function of their being "faulty" scientific ideas resulting from conspiratorial sexist bias, but their overdetermined power to regenerate the conditions that made them emerge as analytic tools. Science and scientists actively "do" gender in many ways, including in sex and gender differences research in primatology (West and Zimmermann 1987).

7   As I read the record of published and circulating pre-publication manuscripts in primate studies through the late 1970s and certainly by 1980, several primatologists, among whom women were prominent but not unique, insisted on epistemologically reconstructing their animal subjects, both male and female, as far more complicated social actors than they had been seen to be in most earlier accounts. I think many of these writers acted out of consciousness of sexism, perhaps especially in reaction to the structural location of females, not to mention women, as resources for male action in biological discourse and in the organization of scientific practice. This structure was paradigmatically clear in much discussion of dominance interactions and of priority of access to estrus females. But response to sexism is not an exhaustive explanation of the growing stress on nonhuman primate subjects as agents and savvy social actors in precisely the same period in which the same people carried out more quantitatively sophisticated and problem-oriented field studies (e.g., J. Altmann 1980; Cheney 1977, 1978; Fedigan 1976, 1982; Hinde 1983; Seyfarth 1976; Smuts 1982, 1985; Strum 1978, 1982, 1983a, 1983b, 1987b). In addition to her 1978 Wenner-Gren paper for her controversial conference, I had copies of the Strum manuscripts emphasizing complex social agency around 1980 and Fedigan's in 1979. The ideas in the Strum papers had begun to be articulated in field work, discussions, and drafts in the middle and late 1970s. Timothy Ransom was Strum's companion in her first field work. Bonnie and Timothy Ransom (1971) had discussed "special relationships" among baboons in their report of work at Gombe. Ransom had the benefit of both Thelma Rowell's and Jane Goodall's sharp sensitivity to the developed sociality and individuality of the animals. Thelma Rowell, Alison Jolly, and Jane Goodall stressed from the 1960s the theme of the fundamental importance for primatology of the complex social and intellectual capacities among

the monkeys and apes. But the theme was also important, in different historical idioms, to Robert Yerkes in the 1920s and in the modern period, for example, to Irwin Bernstein, Robert Hinde, Robert Seyfarth, and, especially, to Frans de Waal.

The theme of social agency is fundamental to western origin stories of the self and the individual; whether or not animals in general and females in particular can be paradigmatic social agents and social subjects has been more problematic. Negotiating those issues is negotiating the traffic between the western categories of nature and culture. In my opinion, because the writers who centered the capacities of their animals as social actors, especially but not only if the animals were females, often perceived themselves to be heretics and to be taking risks with terminology (such as calling animal associates "friends"), a perception which I share in cautiously contextualized ways, they could too easily see writers with whom they might be in a range of competitive and cooperative interactions, age/gender relations, and mentor-mentored relations to be taking sole credit for their discoveries and emphases and not citing each other adequately.

Strum (1987b: 197–204) takes the risk to reveal some of the complex and painful aspects of these common matters in print. I originally read her bibliography in the light of the history of strained relations among baboon workers. In a list labeled "technical" that emphasizes accounts of the baboons as intelligent social actors, Strum did not cite people I expected, for example, J. Altmann, Ransom, Kummer, Smuts, Bernstein, Cheney, Wasser, and Seyfarth. No Japanese studies of the cercopithicines are cited, although they emphasize social complexity, long-term observations, and the importance of individuals. (Intelligent social agency and social complexity are not the same thing. Strum's core idea is that the baboons are socially smart—they are real agents. Biological discourse traditionally equivocates on this touchy issue. Putting social intelligence at the center of explanations of animal behavior remains a radical gesture in western traditions.) Fedigan, Rowell, Jolly, and de Waal are cited; Harding and Ransom appear briefly in the text and Kummer appears in a footnote but not in the references in a popular book with a short bibliography. What should the standards of citation in a popular and deeply personal book be?

One answer is that the author has rather few degrees of freedom in deciding the question in a popular book. Strum wrote a popular book when she did in order to finance the transplantation of endangered troops from Gilgil to a safer spot. "Pure" science funding bodies would not pay for the transfers, but were interested in backing subsequent research on the consequences for social interactions, reproduction, etc. Conservation organizations would not fund the transfer because baboons as such were not endangered, just the animals at Gilgil. The publisher's advance paid for the transfer. Demanding that a bibliography in a popular book be very short, the publisher warned Strum that pages of text would have to be traded for pages of citations. Popular works often cite only studies by the author. Strum's bibliography was a streamlined effort to name general works and give some sense of the literature most directly relevant to topics actually discussed in the book. Ransom's absence from the list was a proofreading error. (Strum to Haraway, 28 Sept. 1988). But, allowing for the constraints, errors, and purposes, I still read the bibliography as in part a public sign, conscious or otherwise, of the dynamics in ordinary scientific work of deep feelings about priority, appropriation, and exclusion.

For anger at Strum's practice in this context, see Stewart (1988). It is possible to read Stewart's hostile review as a reaction to her experience with being the object of Dian Fossey's pain and anger at the mountain gorilla research site in Rwanda. Stewart's work brought her into complex conflict with Dian Fossey, a founder of a research site who could not finally accept the next generation of field workers, approve their approach to conservation politics, their different emotional relation to the animals, or feel at home in the intense publication-centered peer competition in primatology (Mowat 1987; Fossey 1983; Stewart 1981). The situation at Gilgil was different from that at Karisoke in all those variables, but subsequent competitive relations make it easy to read them together.

In *Almost Human,* Strum lists Rowell's 1966 paper on forest baboons, but not her fuller 1972 textbook, a risk-taking early critique of dominance hierarchy explanations, nor Bernstein's, nor subsequent studies. Baboon field workers with whom she has had sharp priority, professional-personal, or theoretical controversies do not appear by name, e.g., S. or J. Altmann, Hausfater, Smuts, and Nicolson.

Emphasizing strategies much more effective for the animals than aggression and dominance in obtaining desirable goods, Strum's view of strategic social manipulation as social intelligence, elaborating a version of reciprocal altruism (full of risk-taking initiative) for the males and kin selection-based (and more emotionally conservative) social intelligence for the females, is closer to

Frans de Waal's picture of chimpanzee society than to other versions of social agency and intelligence, e.g., Jolly's or Smuts's (Western and Strum 1983; de Waal 1982; Strum 1987b: 139–43). At the time she wrote *Almost Human*, Strum had thirteen years of data on females, mostly not published, since she had written the papers on the male social strategies first. Strum suspected sex (or gender?) differences in expression of social intelligence among baboons, but not in capacity. "Females were certainly as smart as males and as socially aware. Peggy and others proved this to me. Yet did they, surrounded by family, really need this smartness as much as males did? Males had plenty of opportunities and a great need to exercise their social skills; how many chances did females have?" (Strum 1987b: 141). Compare this statement with Darwin's in *The Descent of Man* (1981 [1871]) on the evolution through sexual selection of different male and female mental traits. For Darwin, females simply were not as smart as males, just more coy, nurturant, etc. For Strum, females had unused intelligence, just awaiting the evolution of equal opportunity and a crisis in family support systems. In Foucault's terms, these contrasting positions rely on a common episteme spanning at least a hundred years.

Analysis of Smuts (1982, 1985) convinces me that Strum's prior as well as simultaneous work at Gilgil, while listed, is not foregrounded in just those places where the innovative aspects of seeing the animals as friends and intelligent social actors are most stressed. Jeanne Altmann (1980) discussed special relationships between male and female baboons extensively. I am convinced that such emphases were complexly independent and interdependent for these writers and others, but that the material social practices of scientific priority and competition, in founding and nurturing research sites and in publication, overdetermine exclusions and splits among the women every bit as much as among the men or between men and women. A striking example of competitive exclusion of Strum's work by a cohort of workers that is powerful in primate studies is the *Science* article by Cheney, Seyfarth, and Smuts (1986) on nonhuman primate social relations and social cognition. Here the bibliography is supposed to be unambiguously technical, and the genre excludes the conventions of deeply personal narrative prominent in *Almost Human*. In several places, the text produces the effect of a history of the ideas, as well as references to the leading current reviews. Whatever one's view of dates, priority, interdependence, and dependence among Strum, Smuts, Cheney, Seyfarth, etc. on constructing nonhuman primates as savvy social actors at the heart of evolutionary theory, the complete absence from this published technical paper of any reference to Strum except for a bit of her older work on baboon predation behavior is stunning. At least three papers published before the *Science* article were directly relevant (Strum 1982, 1983a, 1983b). In addition, Ransom and Rowell are invisible. Overwhelmingly, Cheney, Seyfarth, and Smuts cite their cohorts and networks, and from my point of view, produce a disturbing historical narrative.

For a judgment that Strum's theoretical ideas and behavior in the field (her challenge of dominance explanations, pioneering explanations based on complex social agency, and transplant of the endangered Gilgil troops) were indeed heterodox, original, fruitful, and rejected from bias or adopted without proper recognition of their source, see Rowell's (1988) favorable review of *Almost Human*. In "Gender and the Study of Primates" (in manuscript), without denying several men's contributions as well, Fedigan assesses positively Strum's role—along with that of Jeanne Altmann, Barbara Smuts, and Thelma Rowell—in effectively displacing reductionist male-centered dominance explanations with "revisionist" views of baboons as complex social actors. Fedigan also comments on the rejection of Strum's ideas by her peers. Rowell herself has often been in a heterodox position in primatology— without the power (gender, professional position, particular networks, mentors, etc.) that can magically transform heterodoxy or "difficult personalities" into the founding ideas and authorities that everyone must cite, whether they were known from the published written record first or from conference papers, conversations, or similar mechanisms.

8   Goodall (1986); de Waal (1982); Smuts (1985); Fossey (1983).

9   For opposing positions, see Dawkins (1976) and Lewontin (1978; 1979).

10 None of the sociobiology literature in primate studies seriously examines the hypotheses about non-adaptive processes. Tooby and DeVore (1987) briefly comment that the non-adaptive hypotheses of Lewontin and others are virtually untestable in principle. But that charge has been brought against evolutionary biology as a whole, and natural selection/adaptation arguments do not have an easy time of it either on that score. Tooby and DeVore, and sociobiologists widely, use strategic modeling to test conceptually, but the relation to an empirical field of the kind of modeling they do

is extremely controversial among biologists. Simulation and representation exist in tense relation in modern and postmodern biological practices. I would suggest another reason why Tooby, DeVore, and other strategic reasoners resist non-adaptive possibilities for the key narrative fragments in the big stories: Adaptationism lends itself inescapably to stories. Non-adaptationist (in contrast to anti-adaptive) processes are hard to narrativize. How do you get a good story out of genetic drift, loose selective constraints, or topological transformations entailed by slightly altered developmental trajectories? It is possible, but not nearly as easy, especially for scientists formed as people of the Book of Nature with a long history of cultural stories about adaptation, design, providence, sex, technology, and war. For a good feminist-scientific critique of sociobiologiocal adaptationism, particularly human sociobiology, see Fausto-Sterling (1985: 174–204). For the best overall critical evaluation of human sociobiology, see Kitcher (1985).

11   The famous writer in several genres, socialist, and sister of J.B.S. Haldane, Naomi Mitchison published her first science fiction novel, *Memoirs of a Space Woman*, at the age of 63 in 1962. It was reissued in 1985 by The Women's Press as one of its first titles in a project to publish science fiction by women and about women. The Women's Press aimed to present provocative feminist imaginings of possible futures that offered alternative visions of science, technology, and gender. Mitchison imagined a xenobiologist whose task is to establish communication with exotic species on space ventures. The fundamental restriction on the biologist is that she must not conduct invasive, interfering experiments. Emotional, erotic, and scientific entanglements abound. The xenobiologist had somehow to negotiate the terms of communication that could subvert the always lurking imperialist project embedded in exploration narratives, including both western scientific and feminist quest literature. Reading *Memoirs* with Fedigan's *Primate Paradigms* is instructive for each. Mitchison is also an adopted full member of the Bakgatla tribe in Botswana, with whom she has been concerned especially with women's affairs.

12   Sally Linton Slocum was a charter member of the American Primatological Association in 1977.

13   Fedigan (1986: 58); Zihlman (1981, 1982b, 1983a, 1984, 1985a, 1985d).

14   The Paleonanthropology of Sex and Gender

1   This chapter draws on formal and informal interviews with Zihlman; on her 1984 manuscript, "Color Me Anthropologist: A Personal History"; and on an interview with Nancy Tanner.

2   Distressed by her first experience of apartheid as a graduate student and consistently opposing racism and apartheid in the U.S., South Africa, and elsewhere, Zihlman has nonetheless continued research, travel, and publication in South Africa, associating herself with South African white opponents of apartheid who are also important in physical anthropology. This position has put her into significant opposition within parts of her U.S. feminist community, including students and faculty in the University of California Santa Cruz Women's Studies Program. This struggle is a good and painful example of the difficulties of alliance and opposition within principled communities that remain heterogeneous in ways liberal pluralism cannot and should not resolve.

3   The weave of scientific careers with heterosexual and white middle-class privilege and with the political and social—called personal and private—obligations of women in the domestic division of labor is complex. This is the weave that constitutes a specific, historical, gendered position, not gender in some universal sense. It was decidedly not normal in a statistical sense in the U.S. population, but it was normative in a moral and political sense for women in Zihlman's world. Recalling women role models in science simultaneously in terms of their success as mothers, wives, and scientists had little to do with Zihlman's own choices; she did not have children, nor was she particularly anxious to marry. But joining the maternal and the scientific had a major semiotic function for women like Zihlman. The maternal and the scientific are culturally set up as opposites; "marrying" them can be very significant.

4   However, aside from structural sexism that cannot be uniquely the responsibility of particular men, women's experience of explicit, personal sexism has not been irrelevant to the intersection of gender and science in primatology. My informants have described incidents they named as sexual harrassment, including direct acts to damage their careers after sexual invitations from male colleagues were turned down; infantilization and disrespect of a specifically gendered kind from

senior men; plagiarism, and more mildly, failure to cite, that would seem unlikely to be practiced on a male colleague; lack of trust from male supervisors when a woman student married, including taking away a desirable research topic and assigning it to a male student; and direct threats to give low grades on required courses by a male teaching assistant who felt a woman student had no right to displace men from positions in medical school. Benign sexism was sometimes described as even more hurtful than hostile antagonism to women in science. For example, women and men both described the kind of asymmetrical, emotionally and professionally complex complicity that can develop between senior men and young women and that prevents the junior women from learning independence in a highly competitive field, while their male peers relate to their adviser in a constrained agonistic mode appropriate to successful career patterns in the masculine-normative structures of science.

5   Tanner and Zihlman (1976); Zihlman and Tanner (1978); Zihlman (1978a, 1978b, 1981); Tanner (1981, 1987).

6   In spring 1987, Tanner gave a course at the University of California Santa Cruz in which she included in the xeroxed reader two versions of her most recent formulation of the gathering hypothesis. One was the published version (Tanner 1987); the other was her typescript, on which the deletions made by the editors for the published version were highlighted. Deleted were criticisms of the hunting model and especially of Tooby and DeVore (1987), appearing in the same volume (Kinzey 1987). In view of the history of the controversy over and dismissal of the gathering hypothesis as simplistic, biased, or not up-to-date on the best recent paradigms (behavioral ecology or inclusive fitness arguments), these deletions appear dubious.

7   For appreciation of the critical importance of play for primates, see Lancaster (1972); Poirier (1972); Bruner et al. (1976).

8   Lewin (1987a, 1987b); Sarich and Wilson (1967); Sarich and Cronin (1976); Goodman and Cronin (1982).

9   Tanner and Zihlman (1976); Zihlman and Tanner (1978); Zihlman (1978a, 1978b).

10 Zihlman (1983b). Beginning in the late 1970s, placing fossil material in the integrated context of climate, vegetation, molecular data, and plate tectonics, Zihlman collaborated with geologist and paleontologist Léo Laporte to reconstruct the geological-ecological zone in which Miocene East African hominoid and later Plio-Pleistocene hominid evolution occurred. Their conclusion was that the savannah-mosaic environment was in no sense a spare, dessicated, tree-less world, but neither was it a forest garden. Mobility to take advantage of rich but widely spaced resources would be essential for early hominids (Laporte and Zihlman 1983). Drawing from several theoretical discourses, the gathering hypothesis has been rooted in wide-ranging and data-rich sources—including social and organismic behavioral, anatomical, molecular, geological, and ecological research. Fedigan's conclusion that this hypothesis has been perhaps the most data-rich among contemporary reconstructions of hominid evolution seems reasonable. That does not make the hypothesis correct, but it does problematize its quick dismissal, reductions to previous formulations, and patent distortions and oversimplifications by critics, as well as by those who want an uncritical female-centered model.

11 The basic hominid adaptation to bipedalism remained central to Zihlman's thinking about mobility. It made no biological sense to her to envision females at any point as sedentary. In hers and Douglas Cramer's article on "Human Locomotion" in *Natural History*, Zihlman wanted to illustrate the argument about this root human adaptation with an action photograph of a modern San gathering woman—pregnant, with a young child, loaded with a large burden in a carrying bag, and striding energetically along at a considerable distance from her camp. The basic argument was about motion, a species-defining interrelation among erect posture, ability to cover long distances, throwing, and carrying (Zihlman and Cramer 1976; Zihlman interview). The magazine editors, without consultation, substituted two photographs of a modern American white male football player, one inscribed with the pelvic and leg bones and the other with the whole-body musculature poised for receiving a pass.

12 This section might have been called "The Reproduction of Primate Mothering," under the sign of the book by Nancy Chodorow (1978), *The Reproduction of Mothering*. Chodorow's arguments, also presented in the influential *Woman, Culture, and Society* (Rosaldo and Lamphere 1974), greatly

influenced white, middle-class, liberal and socialist feminism in the late 1970s and early 1980s. See also Ruddick (1980). Chodorow was Zihlman's colleague at Santa Cruz, a University of California campus with early established, intellectually and politically active women's studies and feminist theory programs drawing from and challenging several disciplines. Zihlman's doctoral student, Catherine Borchert, was a graduate student in the interdisciplinary History of Consciousness Board at the University of California Santa Cruz, which had about thirty students in the early 1980s writing Ph.D. dissertations in some area of feminist critical studies. Tying together questions in the history of science, primate studies of development and communication, models of evolution, and medical issues in human development, Borchert simultaneously was enrolled as a medical student at the University of California San Francisco. As an historian of biology at UCSC with an appointment in feminist theory, I was co-adviser with Zihlman of Borchert's dissertation. These patterns of colleagueship are not maps of agreements nor of influence, but of location in a shared historical moment.

13 Zihlman (interview, 19 July 1983; "Color Me Anthropologist" ms.; 1985b).

14 Lindburg (1967, 1971, 1975).

15 Zihlman repeatedly stressed that "mating system" or "pair bond" could not be a synonym for "marriage." This last is an institution, in which such matters as labor, sex, and reproduction are first constituted as categories (e.g., functions) and then variously disaggregated in relation to each other. Marriage is a socially heterogeneous construction, not simply the human version, with its variations, of a mating system. Only in some forms of marriage do sex, labor, and children flow in the same directions, and even there control patterns for the flows would be variable. Mating is a word for a biological parameter of reproductive "strategies" in an evolutionary framework. Marriage is a word for a multitude of entailed obligations, constitutions of social subjects, signifying practices, and possibilities, where reproduction in the biological sense is not necessarily of central concern. Marriage is not reproductive strategy. I would argue that marriage is also not the cultural appropriation of the biological as resource. Marriage and mating are two different kinds of categories in two different kinds of discourses that need not be related as culture to nature. Arguments labeled as biological reductionism or biological determinism are always particularly easy to make if categories like marriage and mating system are imagined to be related to each other as culture to nature, where the productivist logic of resource appropriation is built in, both for those who maintain the hegemony of culture in human life and for those who prefer a reductionist or continuity account. "Decolonizing" narratives of what it means to be human and animal require, for example, that an object of knowledge like a mating system is not the animal version of human marriage, nor is it the cake beneath the icing of culture, nor is it in any of a series of similar relationships insistently replicated in western stories. Category construction is a basic representational practice which reproduces the world in its image; representation is an active technology, not a passive reflection. This argument also destabilizes any argument taking the form that some form of marriage is universal in human society. The category, marriage, loses its transparent, unmarked quality.

16 Coolidge (1933); Nishida (1972); Kano (1979); Sussman (1984).

17 Zihlman and Cramer (1978); Cramer and Zihlman (1978); Zihlman (1979, 1984); Zihlman, Cronin, Cramer, and Sarich (1978).

18 Brace (1973); Wolpoff (1971); Boaz (1982: 249–50).

19 Kleiman (1977a); Ralls (1976); Fedigan (1982: 51–67); Hrdy and Bennett (1981); Leibowitz (1975); Wright (1984). A related issue in discussions of monogamy and of sexual dimorphism is paternal, maternal, and biparental care of the young. As many as half the species of New World primates have extensive biparental systems, in which the father often becomes the primary infant caretaker, leaving the mother to forage more efficiently and thereby bear the heavy costs of lactation and gestation in particular ecological conditions, in which food-patch size and distribution make multi-female groups, with the advantages for mothers of female-bonded support systems, uneconomical and therefore make monogamous pairing with high paternal investment a good reproductive strategy for both sexes (Wright 1984; Wrangham 1980). The point of this example is to emphasize the kind of taxonomic, biological, and ecological analysis that should be necessary to make any claims about mating systems, parental investment levels, and sexual dimorphism among primates. The New World primates, unfortunately little discussed in this book, provide rich material for

learning this caution. Patricia Wright's work has been significant in theorizing the issues. To get an initial sense of recent field work on New World primates, see Terborgh (1983); Goldizen (1987); Robinson, Wright, and Kinzey (1987); Crockett and Eisenberg (1987); Robinson and Janson (1987); Kleiman (1977b); and Coimbra-Filho and Mittermeier (1981). The strong tropical ecological, socio-ecological, and behavioral ecological core of this work stands out, perhaps partly because the New World primates have been less entrapped in debates about human evolution than the great apes or savannah or forest-edge Old World species, and even more deeply entrapped in the bio-politics of the destruction of tropical forests. But the history-laden, global and local political ecology of conservation inheres in the fabric of all primatology, inescapably engaging the most recalcitrant "pure scientist" by the 1980s.

20 Zihlman (1976: 24, 1983c, 1985d).

21 Schiebinger (1986); Jordanova (1980); Sayers (1982); Fee (1979); Longino and Doell (1983); Long (1988).

22 Zihlman (1985d); Johanson (1980); Johanson and White (1979).

23 Sussman (1984); Corruccini (1986); Dahl (1986).

24 A mid–1980s reminder of the continuing social and intellectual separation of these two broadly defined and slightly overlapping groups came at the Philadelphia meetings of the American Anthropology Association on December 4, 1986, in the day-long, double session on primates. The afternoon session on Primate Reproductive Strategies was organized by Joan Silk and Sarah Blaffer Hrdy and included papers by a multiply linked network of field workers, including besides Hrdy and Silk, Jeanne Altmann, Dorothy Cheney, Alison Jolly, Patricia Wright, Janette Wallis, and Peter Rodman. Discussants were Robert Boyd (Silk's collaborator and husband) and Robert Seyfarth (Cheney's collaborator and husband). The morning session on Primate Behavior was organized by Warren Kinsey and included papers by the Berkeley-characterized network of Diahan Farley, Phyllis Dolhinow and Frieda Richenbach, Mark Taff, Linda Wolfe, Carolyn Ehart, Susan Cachel, and Connie Anderson, with Susan Sperling Landes as discussant. Landes had been a primate behavior graduate student in anthropology at Berkeley, but ultimately did her field work and dissertation on the animal rights movement (Sperling 1988). Both morning and afternoon groups had a lot to say about both behavior and reproduction; the titles of the sessions seemed signs for other differentiations by institutions, field sites, and explanatory and narrative patterns.

25 For Zihlman's assessment of the fate of her model, see 1982b and 1985b; for examples of cooptation, see Issac (1978) and Lovejoy (1981).

26 Washburn to Haraway, 5 Dec. 1978; Washburn and Ranieri (1981).

27 Zihlman's recent collaborators include Mary Ellen Morbeck and Jane Goodall. Morbeck, whose Berkeley Ph.D. was in 1972 for a study of the forelimb of the Miocene Hominoidae, ran a very successful Wenner-Gren conference drawing from fossil and modern primate studies and tying together functional morphology, behavior, and ecological parameters (Morbeck, Preuschoft, and Gomberg 1979). Morbeck and J. de Rousseau organized a conference for Wenner-Gren in 1987 on Primate Life History and Evolution, where Morbeck and Zihlman reported on their collaboration on post-mortem anatomical studies of the Gombe chimpanzees (Martin and Harvey 1987). Goodall sends Zihlman the skeletons her team recovers from deceased chimps; Zihlman's laboratory in Santa Cruz has a cabinet full of the bones of the famous TV personalities, Flo, Flint, and many others. Walking by the labeled drawers evokes an odd consciousness of the juxtaposition of popular and technical discourses in primatology. The known extensive life histories of these animals greatly enhance the interpretive power of the post-mortem functional anatomy and will permit greater confidence in extrapolations to life from other anatomical material where less or nothing is known from direct observation before death.

28 Speakers included Joyce Sirianni on prenatal development and sexual dimorphism, Elizabeth Watts on developmental dimorphisms and primate evolution, Dana Olson on female choice in patas monkey matings, Antoinette Brown on the role of females in primate provisioning, Patricia Wright on commonalities among monogamous primates, Jane Buikstra (with co-authors) on diet and demographic change, Patricia Shipman on the taphonomic reconstruction of early hominid life styles, Adrienne Zihlman on sex and the single species problem in australopithecines, Este Armstrong on limbic system evolution and the human revolution, Christine de Lacoste (with co-author)

on sex differences in the development of human fetal morphological asymmetries, and Dean Falk on the role of sex in the evolution of intelligence. Jane Lancaster had to cancel, but she had intended to speak on sex differences in primates' attachment to home range and natal group. Meredith Small read Sarah Hrdy's paper on parental investment in sons and daughters. Linda Fedigan was invited, but had to be in the field. Mary Ellen Morbeck was uncomfortable with the gender agenda and declined, an act similar to Barbara Smuts's friendly refusal to be part of *Female Primates* if all the authors had to be women (Smuts, interview, 18 March 1982; Small 1984). The strong interweaving of developmental and evolutionary themes in this symposium should not be missed. Phylogeny and ontogeny have been variously twinned repeatedly in the reconstructions of life's narrative. The structure of narrative itself, with its tropes, temporal flows and plots, encourages these linkages in any historical genre. Taphonomy is the science of the reconstruction of extinct life ways from material remains of activity, such as marks on bones or stone-bone assemblages. A kind of fossil version of archaeology, the small field of taphonomy, in sharp contrast to paleoanthropology, is dominated by women scientists perceived to speak with authority (Behrensmeyer 1976).

29   Morgen to participants, 22 May 1987.

15   The Evolving Portfolio of Primate Females

1   Stacey (1987); see also Rosenfelt and Stacey (1987); Eisenstein (1984).

2   Sofoulis (1984, 1988); Cohn (1987a, 1987b); Petchesky (1987); Edwards (1987, 1988)

3   Keller shows the functions and contradictions of the peculiar language for discussing sexual reproduction "as an autonomous function of the individual organism." "I suggest that it provides crucial support for the central project of evolutionary theory—namely that of locating causal efficacy in the intrinsic properties of the individual organism" (Keller 1986, 1987b). The logical and linguistic requirements for discussing sexual reproduction should have serious consequences for the premises of methodological individualism in evolutionary theory. Her paper has direct relevance to the debates about units of selection in current evolutionary biology, where possibilities for confusion abound about exactly what level (genes, genotype, organism, population, species) pertains in precise definitions of fitness and differential survival (Brandon and Burian 1984; Sober and Lewontin 1982; Dawkins 1976). Keller concludes, "For sexually reproducing organisms, fitness is in general not an individual property but a composite of the entire interbreeding population—including, but certainly not determined by, genic, genotypic, and mating pair contributions. To the extent that the advent of sex undermines the reproductive autonomy of the individual organism, it simultaneously undermines the possibility of locating the causal efficacy of evolutionary change in individual properties" (1987b:85). This position is quite different from what sociobiologists have dismissed as group selectionism, which rested on methodological individualism (where the group was the bounded whole and functionalist complementary role theory served as the main social analytic), just as much as currently more popular genic selection arguments do (Wynne-Edwards 1962; Williams 1966).

4   See Hartouni (1987). Strathern (1980 and in press) has shown why it is necessary repeatedly to encumber these sentences on reproductive politics, languages, and facts with the annoying qualifier "western." The qualification is insufficient and far too reductive and general, only hinting at the complex specific analysis required to fill out the arguments that are barely suggested in this introduction to Hrdy and white, middle-class, sociobiological feminism. But perhaps the qualifier's annoying quality will serve to remind both reader and writer of "our" relentless ethnocentric insistence on self-evident common sense. The "common sense" that women make babies should have the same deconstructed status as the question of "primitives' " knowledge of the true principle of paternity (Delaney 1986). Strathern (in press) goes to great lengths to show precisely how, in Melanesia, women do not make babies (David Schneider, personal communication, December 30, 1987). The western productionist narrative of the world and everything in it is what is at stake here. In the context of 1980s U.S. feminism and anti-feminism and biomedical-biotechnical politics, I am tempted to answer every question about where babies come from with a shaggy dog story of infinite regress, refusing every move of appropriation, until finally the stork is seen for the great gift-giver it is and children can be represented as something other than someone's alienable, inalienable, or

alienated property. Sociobiology provides one of the most radical representations of children as investment opportunity in the great species commodity futures market and breeding stock exchange.

5   O'Brien (1981); de Beauvoir (1954).

6   Dawkins (1976); Latour and Strum (1986).

7   Davis (1982); Mohanty (1984); Hooks (1981); Giddings (1985). On the history of race concepts in science that affect constructions of sexual difference, see Stepan (1982); Hammonds (1986); Gould (198); Fausto-Sterling (1988).

8   Laqueur (1986); see also Douglas (1970); Jordanova (1980).

9   Thus, the multi-billion dollar project to sequence the human genome is the paradigmatic biotechnical undertaking in the late 1980s. A puzzle in the reporting on the genome project in the science news press and general professional magazines like *Science* is the absence of discussion about whose genome will be sequenced. In this great project literally, in writing, to standardize the human genetic legacy, the bio-politics of standardization are left unvoiced. Discussion seems to take place within a biological world where genes are fixed and there could be a type genome for the species, rather than addresses for gene locations. Of course, every molecular biologist knows that genes vary and that a species is made up of populations with differing gene frequencies. For many important genes, mapping has been able to reveal relatively variable and constant regions and to allow a molecular comparative analysis of great resolution. But still, public writing about the genome project is all in the language of "the human genome," in the singular number and eternal time. The human genome project seems to function semiotically as a promise of a gold standard and a library-bank. The complete nucleic acid linear sequence of the genetic code becomes a sign of the full library of the natural human body, the source, and the ultimate repository of information, on which all bodily conversation must rely. The genome project operates under the sign of totality, of fullness. The genome project would produce the library of human nature as a master dictionary. Every book in this strange library will be a dictionary in the ur-language of living bodies, the original tongue from which all merely historical words, merely proximate meanings, must be traced. This library is the figure of a postmodern apparatus of bodily production. These bodies are vast writing technologies. The genome project's particular transfiguration of the much older European dream of the perfect library and discovering the original tongue of man gives a nice turn to the history of philology and hermeneutics. The code is indeed the message. Contemporary disputes over whether or not the human genome can be patented and whether access to the dictionary can be sold give the final twist to the trope of bodily conversation, reproductive politics, and genetic investment as commerce. Would it be criminal conversation "to access" the genome outside the law? A kind of computer crime of illicit access to the ur-program?

10 For sexual difference in the history of philosophy of mind and doctrines of rationality, see Lloyd (1984); MacKinnon (1982).

11 Canguilhem (1978) remains essential for understanding the construction of the modern concepts of "normal" and "pathological," without which the whole architecture of sexual and racial difference could not have been built or sustained.

12 Laqueur (1986: 18); Sayers (1982).

13 This doctrine never kept slave women out of the fields nor other working women out of factories when they were required. Anti-feminist and racist positions, which were still formulated inside the discursive parameters of liberal theory, allowed powerful concepts of specific difference to be articulated cogently, but specific difference in the context of fundamental power inequality was (and remains) a ticket to continued blatant oppression and exclusion. Liberalism has been an extremely complex inheritance for feminists. In particular, liberalism's theory of sameness and difference does not resolve the problem of inequality and power. Wendy Brown, in her review of *Feminism Unmodified*, summarized MacKinnon's (1987) analysis of the comprehensive trap of liberal theory: the naturalization and neutralization of gender inequality as sex and/or gender difference; the constant requirement of sameness for equality; the depoliticization, often through the elaboration of life and social scientific discourse, of socially constructed inequalities by such categories as interests, preferences, choices, and accidents of fate; and the tacit claim to normatively neutral (i.e., not gendered, not marked by race or class, simply human) standards, discourses, and juridical perspectives that are actually rooted in the hegemony of particular classes, genders, and races.

These are the aspects of liberal theory that constantly lead to re-formation of the binary dualisms that mystify women's situations: public/private, action/speech, domination/difference, consent/coercion, political/natural. The deep historical ties of liberal theory and the construction of the biologically marked bodies of race, sex, and class can be followed in Young (1970, 1981, 1985b); Figlio (1976, 1977); Cross and Albury (1987); Cross (1982); Haraway (1978, 1978b, 1979); Jordanova (1986); Eisenstein (1981); Sayers (1982); Davis (1982); Gates (1985); and Spivak (1985).

14 Haug (1986); Haug et al (1987); Snitow, Stansell, and Thompson (1983); Vance (1984); Moraga (1983); Carby (1985); MacKinnon (1987).

15 Hrdy has been important in networks of women and feminist primatologists, including Patricia Whitten, Nancy Nicolson, and Barbara Smuts, who were all at Harvard together in various capacities in the late 1970s and early 1980s. Not unlike Suzanne Chevalier-Skolnikoff, Hrdy had the money and social position derived partly from marriage and class origin to allow her to publish and so prosper professionally, while having children, without a ladder faculty position and without much support from her former adviser, for several years after finishing her doctorate. She had research associate titles at the Peabody Museum at Harvard from 1979–85. Beginning to name and construct herself as a feminist while writing *The Woman That Never Evolved*, she has been articulate about anti-woman structures at Harvard, including in her program. In 1984 Hrdy was hired in her first ladder appointment, as a full professor in the Anthropology Department at the University of California at Davis, where she entered into networks of women and men in primate studies who were sympathetic to female-centered accounts and to versions of feminism (Small 1984). Hrdy has participated in conferences on feminism and science and represented a feminist perspective in several contexts in her field (Bleier 1986; Hrdy and Williams 1983).

16 Eisenstein (1981) discusses the latent radical implications of liberal feminism.

17 Hrdy (1981: 166–72); see also Lloyd, in manuscript. By 1985, Hrdy's book had appeared in Japanese, Italian, and French editions, and had gone through five paperback editions with Harvard University Press. The book was well received by its reviewer for the radical feminist monthly, *Off Our Backs* (Henry 1982). Science confirmed orgasms for females and women scientifically in the 1960s; but by the 1980s, nonhuman primates were modeling PMS, premenstrual syndrome, instead (Eckholm 1985; Hausfater and Skoblick 1985). Although sexual liberation is not women's liberation, something was lost between between the 1960s and 1980s!

18 Zumpe and Michael (1968), quoted in Hrdy (1981: 169).

19 Hrdy (1981: 169); Burton (1971, 1977).

20 Chevalier-Skolnikoff (1971b: 8; see also 1971a, 1974, 1975, 1976).

21 Interview, 7 February 1983.

22 Rosaldo and Lamphere (1974). It would be interesting to know why Rosaldo, who died tragically in 1983, did not want this paper. The likely explanation seems resistance to biological explanations of the cultural domain. In contrast, in the other pivotal collection of early U.S. feminist anthropology, with its roots in marxian narratives that permitted a careful evolutionary story, Sally Linton Slocum's and Leila Leibowitz's papers were both included and much discussed by the book's audience (Reiter 1975). Neither Chevalier-Skolnikoff nor Hrdy seem to have known of Ruth Herschberger's speculations and efforts to get data in the 1930s. The main focus of Chevalier-Skolnikoff's writing has been on Piagetian models for theorizing the evolution of primate intelligence and social organization (Chevalier-Skolnikoff 1977). Frank Beach had been important to Chevalier-Skolnikoff at Berkeley, and Beach responded enthusiastically to her sexual behavior papers (Beach 1965).

23 Hrdy (1977, 1979, 1981, 1986); Hrdy and Williams (1983).

24 Symons (1979), quoted in Hrdy (1981: 165).

25 Hrdy (1981: 187, 1983); Hrdy and Bennett (1981). See Zihlman (1985c) for a serious spoof and analysis on the problem of "lost" estrus.

26 Hrdy has not argued that females are naturally subordinate to males, only that they do not have the same possibilities for the same moves in the reproductive struggle. Indeed, females compete and struggle for all they are worth, but are limited by the realities of power and resources. No naturally cooperative, globally altruistic female could have evolved. Non-kin-based sisterhood is not natural, no matter how effective it might have been in equalizing sexual politics. Inherent female

competition with each other, especially competition among female genetically related lineages (called "kin" in sociobiological terms), blocked female bonding as a sex against male strategies, such as infanticide, that disadvantaged females as a group. Solidarity could always be undercut by a cheater that got ahead reproductively by currying favor with (or deceiving) one of the oppressors, limiting severely the possibility of the evolution of comprehensive, non-competitive female cooperation. Cooperation between females and males was limited by their genetic difference, and so non-identical reproductive interests. Hrdy regarded feminist solidarity as a human achievement, fragile and precious, that could only happen in a species that could act to counter natural conditions favoring competition. In that sense, Hrdy is ironically close to important socialist feminist arguments that stressed the achieved quality of a feminist standpoint, rooted in an analysis of women's material conditions but not naturally part of women's consciousness (Hartsock 1983b). In this section, I have not scrupulously observed the niceties of the difference between human and nonhuman situations in Hrdy's arguments because I wish to highlight a narrative logic that joins the regions of nature and culture in a particular political setting. In any case, Hrdy has a strong sympathy for human sociobiology applied to analysis of child abuse, sex ratios, differential treatment of male and female children, and several other topics. The infanticide debates took place in a period in which child abuse was becoming a public issue. Hrdy's arguments are not simplistic, but they are firmly anchored in the narrative of genetic investment. For an account of Hrdy's location in debates about whether male killing of infants among non-human primates is adaptive or the result of pathological stress—a contest between reproductive strategy, sociobiological arguments and evolutionary structural functionalism emphasizing other narratives of adaptation—see Haraway (1983a), Hrdy (1977), Bogess (1979), and Hausfater and Hrdy (1984). For a view of Hrdy's critique of Lovejoy and of similar stories of the origin of the family, see Haraway (1986).

27 Wrangham (1979a, 1979b, 1980).

28 Spivak (1985, 1986); Cohn (1987); Strathern (forthcoming).

29 For feminist efforts to theorize "difference" differently, see Trinh T. Minh-ha (1988); Anzaldua (1987); de Lauretis (1987). Trinh Minh-ha's critique of the Hegelian logic of appropriation and escalating domination is suggested in her typographical convention for a possible subjectivity: the inappropriate/d other.

16 Science Fiction, Fictions of Science and Primatology

1 The fifty or so field primatologists active by 1965 came from about 10 nations. Those on fellowship at the Center were DeVore, K.R.L. Hall, Phyllis Jay (who also had a role in the planning), Hiroki Mizuhara, Vernon Reynolds, and George Schaller. The stable study group also contained Hamburg, Washburn, and Frances Reynolds. During the project year, a conference was held at the Center, involving about 18 additional people. Also, Peter Marler, William Mason, Adriaan Kortlandt, and J.P. Scott came for short periods. Washburn and Hamburg wrote the synthesizing penultimate chapter on "Implications of Primate Research," in which the social group as the principal primate adaptation was the organizing concept. This piece set the logic of future field studies, while Schaller wrote the concluding guide to field procedures. The recommended norm was to conduct a preliminary ecological survey, to follow by a detailed, non-interventionist observation of the social life of a selected group; producing a "species repertoire" based on quantitative data, including a good population census; and then to conduct intensive studies into some particular aspect of behavior, using experimental as well as observational methods. United States scientists far outnumbered others. With some exceptions, the organizing idea of the project was the characterization of species-specific ways of social life in ecological context. Differences in social behavior within species in various habitats were accommodated by explanations like referring them to differential effects on a common behavioral plan of crowding and stress (DeVore 1965a; Jay 1965a).

2 In *Primate Societies*, fifteen chapters had only men authors; twelve had only women authors; thirteen were written jointly by both available genders. Six of the chapters were authored by married couples; one was jointly authored by women colleagues and five by men colleagues.

3 One former Washburn student from Chicago, Warren Kinzey, then at the City University of New York, was present at the Center in 1983–84; but neither Kinzey nor DeVore's students indicate

continuity with the 1960s and early 1970s Washburn-affiliated approaches. By contrast, two former students of Peter Marler at Berkeley in the 1960s in zoology, Thomas Struhsaker and John Eisenberg, were important to the later project, with its strong emphasis on socioecology, behavioral ecology, and conservation. Several past or present Marler associates (six) from the Rockefeller University, including two of the editors (Dorothy Cheney and Robert Seyfarth), were authors in the 1987 volume. At least six authors had histories at Gombe, and nine were linked through Cambridge University. Studies of New World primate species were much in evidence, in sharp contrast to the essays in *Primate Behavior*. A kind of symbolic paternal figure whose Cambridge students were really the center of the action in the 1982–83 year, Robert Hinde played a role in the volume similar to Washburn's in the 1965 book.

4 Genic selection plays little role; the individual organism remains the practical unit of selection throughout the books' many narratives.

5 For example, Mittermeier and Cheney (1987: 489) emphasize that the "lesser developed countries" already commit a greater percentage of their resources to conservation than richer countries, which benefit most from those resources, in the short run, in strengthening their national scientific and medical establishments and advancing the professional careers of their citizens.

6 Surviving the holocaust means surviving not only the Oedipal narrative, but, ultimately, "the symbolic drive toward atomisation, a cultural Symbolic unconscious whose drives are harnessed in the service of conflict, inequality, alienation, and violence. . . . Moreover, the narrative of difference in feminist writing requires that we reflect on our [whose?] global self-destructive adherence to the Oedipal narrative of rivalry and conflict that not only denies woman a place in its economy but also, according to the same dictates, programs the Symbolic order's drive to extinction as well. Surviving fictions engages nothing less than our imaginative capacity of surviving the nuclear symbolic in its narrative dimensions." (Brewer 1987: 48,50)

  Lurking just underneath the discussion of difference throughout this book has been feminist theory's relations to psychoanalytic theory, especially Lacanian versions. It is too late to force this potent monster up from the depths, breaking the surface tension of my discussions of difference, to join the monkeys and apes in the upper stories of the primate text. Let the beast continue to inhabit the fluid regions that threaten to flood the primal scenes where "almost minds" communicate.

7 Born 1915, died 1987, Dr. Alice Hastings Bradley Sheldon's generic literary personae were James Tiptree, Jr. (1973, 1975, 1978a, 1978b, 1981, 1985) and Racoona Sheldon (1985). Alice—named by mother, father, husband, the publishing industry, and a scientific academic credential—wrote constantly about alien conversations. The revelation of Tiptree's "true" gender, her identification as "Alice" in 1977 after the death of her mother, was said to have caused a crushing depression (*Washington Post*, 19 May 1987, A1, 14). Her literary disguise was uncovered by an investigative fan, Jeffrey D. Smith. Sheldon and her husband helped in the founding of the C.I.A. after World War II. She then did photo-intelligence work, compiling dossiers on people, until the mid–1960s. When she left the C.I.A., she used its techniques to establish a whole new identity. "I was somebody else" (Aldiss 1986: 365–6). The persona of James Tiptree, Jr., was not her first exploration of the modes of construction and deconstruction of identities; "Alice" was truly a twentieth century Milton's daughter. Neither Tiptree nor Octavia Butler could ever start writing from the beginning.

8 For introductions to Butler, see Crossley (1988); Mixon (1979); Salvaggio (1986); Williams (1986); Zaki (1988).

SOURCES

Manuscript Collections and Private Papers Consulted
(abbreviations used in notes indicated in parentheses)

American Museum of Natural History (AMNH), Photographic Collection and Archives, Department of Library Services, New York, NY

Clarence Ray Carpenter Papers (CRC), Manuscripts and Archives, The Pennsylvania State University

Robert M. Yerkes Papers (RMY), Manusripts and Archives, Sterling Memorial Library, Yale University, New Haven, CN

Committee for Research in Problems of Sex (CRPS), National Research Council of the National Academy of Sciences (NRC-NAS), Washington, D.C.

National Research Council of the National Academy of Sciences (NRC-NAS), Washington, D.C.

Rockefeller Foundation Archives (RF), Tarrytown, NY

Department of Anthropology, University of Chicago, Chicago, IL

Sherwood L. Washburn private papers (WP), Berkeley, CA

Wenner-Gren Foundation for Anthropological Research (WG), New York, NY

Interviews

Over a period of several years, the following people graciously consented to formal interviews. In addition, several of these people and other actors in primate studies have corresponded with me and allowed me to see letters, grant proposals, unpublished drafts of papers, course outlines, and other private papers. I am deeply grateful for their generous help. Their participation does not indicate their agreement with my arguments or with my interpretations of their comments and published and unpublished materials.

Jeanne Altmann, 2 April 1982

Stuart A. Altmann, 1 April 1982

Harold Bauer, 9 August 1982

Irwin Bernstein, 10 August 1982

Naomi Bishop, 7 March 1982

Dorothy Cheney, 13 August 1982

Suzanne Chevalier-Skolnikoff, 7 February 1983

Irven DeVore, 18 March 1982

Phyllis Dolhinow, August 1981

John Emlen, 5 and 6 April 1982

Stephen Gartlan, 12 August 1982

Robert S.O. Harding, 29 October 1982

Glenn Hausfater, 11 August 1982

Robert Hinde, 12 August 1982

Sarah Blaffer Hrdy, 1 November 1984

Alison Jolly, 10 August 1982, 15 January 1988

Barbara King, 9 August 1982

Richard Lee, 14 October 1985

Donald Lindburg, 10 August 1982

Wayne McGuire, 9 August 1982

Peter Marler, 12 August 1982

Nancy Nicolson, 12 August 1982

Lita Osmundsen, 3 March, 1982, 22 March 1982

Mary Pearl, 9 August 1982

Suzanne Ripley, 12 March 1982

Thelma Rowell, 21 July 1977

Robert Seyfarth, 13 August 1982

Joan Silk, 13 August 1982

Barbara Smuts, 18 March 1982

Gordon Stephenson, 6 April 1982

Thomas Struhsaker, 11 August 1982

Shirley Strum, 1 June 1981, 29 June 1983

Stephen Suomi, 25 October 1978

Nancy Tanner, 9 March 1983

Sol Tax, 14 November 1985

Russell Tuttle, 15 November 1985

Sherwood Washburn, 21 June 1977

E.O. Wilson, 19 March 1982

Patricia Whitten, August 1982

Richard Wrangham, 13 August 1982

Adrienne Zihlman, 19 July 1983

Bibliography

A

Aberle, Sophie and George W. Corner. 1953. *Twenty-five Years of Sex Research: History of the National Research Council Committee for Research in Problems of Sex, 1922–47.* Philadelphia: W. B. Saunders.

Abir-Am, Pnina. 1982. The discourse of physical power and biological knowledge in the 1930s: A reappraisal of the Rockefeller Foundation's 'policy' in molecular biology. *Social Studies of Science* 12:341–82.

Adorno, Theodor. 1950. *The Authoritarian Personality.* New York: Harper & Row.

Akeley, Carl E. 1923. *In Brightest Africa.* New York: Doubleday, Page & Co.

Akeley, Delia. 1930. *Jungle Portraits.* New York: Macmillan.

Akeley, Mary Jobe. 1929a. Africa's great National Park. The formal inauguration of Parc Albert at Brussels. *Natural History* 29:638–50.

Akeley, Mary Jobe. 1929b. *Carl Akeley's Africa.* New York: Dodd & Mead.

Akeley, Mary Jobe. 1940. *The Wilderness Lives Again. Carl Akeley and the Great Adventure.* New York: Dodd & Mead.

Akeley, Carl E. and Mary Jobe Akeley. 1922. *Lions, Gorillas, and their Neighbors.* New York: Dodd & Mead.

Albury, William R. 1977. Experiment and explanation of Bichat and Magendie. *Studies in the History of Biology* 1:47–131.

Aldiss, Brian W. 1986. *Trillion Year Spree.* New York: Avon.

Allee, W.C. 1931. *Animal Cooperation: A Study in General Sociology.* Chicago: University of Chicago Press.

Allee, W.C. 1942. Social dominance and subordination among vertebrates. *Biological Symposium* 8:139–62.

Allee, W.C. 1951. *Cooperation among Animals.* New York: Schuman.

Allee, W.C., A.E. Emerson, O. Park, T. Park, and K.P. Schmidt. 1949. *Principles of Animal Ecology.* Philadelphia: Saunders.

Allen, Arthur A. 1962. Cornell's Laboratory of Ornithology. *Living Bird* 1:7–36.

Allen, Edgar, ed. 1932. *Sex and Internal Secretions, a Survey of Recent Research.* Baltimore: Williams & Wilkins.

Allen, Garland E. 1975. *Life Science in the Twentieth Century.* New York: Wiley.

Altmann, Jeanne. 1974. Observational study of behavior: Sampling methods. *Behaviour* 49:227–67.

Altmann, Jeanne. 1980. *Baboon Mothers and Infants.* Cambridge, MA:Harvard University Press.

Altmann, Stuart A. 1962. A field study of the sociobiology of rhesus monkeys, *Macaca mulatta. Annals of the New York Academy of Sciences* 102(2):338–435.

Altmann, Stuart A. 1965a. Sociobiology of rhesus monkeys, II. Stochastics of social communication. *Journal of Theoretical Biology* 8:490–522.

Altmann, Stuart A., ed. 1965b. *Japanese Monkeys: A Collection of Translations.* Selected by Kinji Imanishi. Atlanta: Yerkes Primate Research Center.

Altmann, Stuart A., ed. 1967. *Social Communication among Primates.* Chicago: University of Chicago Press.

Altmann, Stuart A. and Jeanne Altmann. 1970. *Baboon Ecology: African Field Research.* Chicago: University of Chicago Press.

Andrews, R.C. 1926. *On the Trail of Ancient Man.* New York: Putnam.

Angell, James Rowland. 1907. The province of a functional psychology. *Psychological Review* XIV: 61–91.

Anzaldua, Gloria. 1987. *Borderlands/La Frontera: The New Mestiza.* San Francisco: Spinsters/Aunt Lute.

Ardrey, Robert. 1961. *African Genesis.* London: Collins.

Ardrey, Robert. 1966. *The Territorial Imperative*. New York: Atheneum.

Ardrey, Robert. 1970. *The Social Contract*. New York: Atheneum.

Ardrey, Robert. 1976. *The Hunting Hypothesis*. New York: Atheneum.

Aristotle. 1984. *Generation of Animals*. In Jonathan Barnes, ed. *The Complete Works of Aristotle*, 2 vols. Princeton: Princeton University Press.

Ariyoshi, Sawako. 1980. *The River Ki*. Translated by Mildred Tahata. Tokyo and New York: Kodansha, Ltd.

Armelagos, George J., David S. Carlson, and Dennis P. Van Gerven. 1982. The theoretical foundations and development of skeletal biology. In Spencer 1982:305–28.

Ashby, W. Ross. 1961. *An Introduction to Cybernetics*. London: Chapman & Hall.

Asimov, Isaac. 1964 [1953]. *The Second Foundation*. New York: Avon.

Aspinall, John. 1976. *The Best of Friends*. New York: Harper & Row.

Asquith, Pamela. 1981. Some aspects of anthropomorphism in the terminology and philosophy underlying Western and Japanese studies of the social behaviour of non-human primates. Ph.D. thesis, Oxford University.

Asquith, Pamela. 1983. The Monkey Memorial Service of Japanese primatologists. *RAIN*, no. 54(February):3–4.

Asquith, Pamela. 1984. Bases for differences in Japanese and Western primatology. Paper delivered at the 12th Meeting of CAPA/AAPC, University of Alberta, November 15–18.

Asquith, Pamela. 1986a. Imanishi's impact in Japan. *Nature* 323:675–76.

Asquith, Pamela. 1986b. Anthropomorphism and the Japanese and Western traditions in primatology. In J. Else and P. Lee, eds. *Primate Ontogeny, Cognition and Behavior: Developments in Field and Laboratory Research*. New York: Academic Press, pp. 61–71.

Attenborough, David. 1979. *Life on Earth*. Boston: Little, Brown & Co.

Auel, Jean. 1980. *Clan of the Cave Bear*. New York: Crown Books.

B

Baker, Thelma and Phyllis Eveleth. 1982. The effects of funding patterns on the development of physical anthropology. In Spencer 1982:31–48.

Bambara, Toni Cade. 1980. *The Salt Eaters*. New York: Random House.

Baranga, Deborah. 1978. The role of nutritive value in the food preferences of the Red Colobus and Black-and-White Colobus in the Kibale Forest, Uganda. M.Sc. thesis, Makerere University, Uganda.

Barbour, Thomas. 1945. *A Naturalist in Cuba*. Boston: Little, Brown.

Baritz, Leon. 1960. *The Servants of Power: A History of the Use of Social Science in American Industry*. Middletown, CN: Wesleyan University Press.

Barker-Benfield, G.J. 1972. The spermatic economy: A nineteenth-century view of sexuality. *Feminist Studies* 1:45–74.

Barnes, Barry and Steven Shapin, eds. 1979. *Natural Order Historical Studies of Scientific Culture*. London: Sage.

Bartholomew, G.A. 1952. Reproductive and social behavior of the northern elephant seal. *University of California Publications in Zoology* 47(15):369–472.

Bartholomew, G.A. and J.B. Birdsell. 1953. Ecology and the protohominids. *American Anthropologist* 55:481–98.

Baty, Paige. 1985. I, Vampire, and the Woolf-(Wo)Man. Unpublished paper, University of California Santa Cruz.

Beach, Frank A., ed. 1965. *Sex and Behavior*. New York: Wiley, 2nd ed., 1974.

de Beauvoir, Simone. 1954. *The Second Sex*. Translated and edited by H.M. Parshley. New York: Vintage.

Beer, Gillian. 1983. *Darwin's Plots: Evolutionary Narrative in Darwin, George Eliot and Nineteenth-Century Fiction*. London: Routledge & Kegan Paul.

Behrensmeyer, A. Kay. 1976. Taphonomy and paleoecology in the hominid fossil record. *Yearbook of Physical Anthropology 1975* 19:36–50.

Benchley, Belle. 1942. *My Friends the Apes*. Boston: Little, Brown.

Benedict, Ruth. 1934. *Patterns of Culture*. New York: Houghton-Mifflin.

Benjamin, Jessica. 1980. The bonds of love: Rational violence and erotic domination. *Feminist Studies* 6(1): 144–74.

Berman, Carol M. 1978. Social relationships among free-ranging infant rhesus monkeys. Ph.D. thesis, Cambridge University.

Bermant, Gordon and Donald Lindberg, eds. 1975. *Primate Utilization and Conservation*. New York: Wiley-Interscience.

Bernstein, Irwin S. 1968. Primate status hierarchies. *American Zoologist* 8:741 (abstract).

Bernstein, Irwin S. 1976. Dominance, aggression, and reproduction in primate societies. *Journal of Theoretical Biology* 60:459–72.

Bernstein, Irwin S. 1981. Dominance: The baby and the bathwater. *Behavioral and Brain Science* 4:419–58.

Bijker, Wiebe E., Thomas P. Hughes, and Trevor Pinch, eds. 1987. *The Social Construction of Technological Systems. New Directions in the Sociology and History of Technology*. Cambridge, MA: M.I.T.Press.

Bingham, H.C. 1932. Gorillas in a native habitat. *Carnegie Institution Publications*, no. 426:1–66.

Biology and Gender Study Group. 1988. The importance of feminist critique of contemporary cell biology. *Hypatia* 3(1):61–76.

Blackwell, Antoinette Brown. 1875. *The Sexes throughout Nature*. New York: Putnam & Sons.

Blackwood, Beatrice. 1935. *Both Sides of Burka Passage*. London: Oxford University Press.

Bleier, Ruth. 1984. *Science and Gender: A Critique of Biology and Its Theories on Women*. New York: Pergamon.

Bleier, Ruth, ed. 1986. *Feminist Approaches to Science*. New York: Pergamon.

Bloom, Lisa. 1987. The North Pole: A semiology of nationalist and masculinist discourse. Ph.D. qualifying essay, University of California Santa Cruz.

Boas, Franz. 1925. What is race? *The Nation* 120:89–91.

Boaz, Neil. 1982. American research on Austraopithecines and early *Homo*, 1925–80. In Spencer 1982:239–60.

Bogess, Jane. 1979. Troop male membership changes and infant killing in langurs (*Presbytis entellus*). *Folia Primatologica* 32:65–107.

Bolwig, Niels. 1952–53. Field and laboratory studies on an African monkey, *Cercopithecus ascanius schmiditi Matschie. Proceedings of the Zoological Society of London* 122(II):297–394.

Bolwig, Niels. 1959. A study of the behavior of the chacma baboon, *Papio ursinus. Behaviour* 14:136–63.

Boorstin, Daniel. 1974. *The Americans: The Democratic Experience*. New York: Vintage.

Borchert, Catherine. 1985. A critique of neo-Darwinism and its implications for the evolution of language. Ph.D. thesis, University of California Santa Cruz.

Boring, E. G. 1956. Robert Means Yerkes. *Yearbook of the American Philosophical Society* 20:133–40.

Bourne, Geoffrey. 1971. *The Ape People*. New York: Signet.

Bowden, D. 1966. Primate behavioral research in the USSR: the Sukhumi Medico-Biological Station. *Folia Primatologica* 4(4):346–60.

Bowlby, John. 1969. *Attachment and Loss, Vol. 1, Attachment*. New York: Basic Books.

Bowlby, John. 1973. *Attachment and Loss, Vol. 2, Separation*. London: Hogarth.

Brace, C. Loring. 1973. Sexual dimorphism in human evolution. *Yearbook of Physical Anthropology* 16:31–49.

Brace, C. Loring. 1982. The roots of the race concept in American physical anthropology. In Spencer 1982:12–29.

Bradley, Mary Hastings. 1922. *On the Gorilla Trail*. New York: Appleton.

Brandon, R. and Richard Burian. 1984. *Genes, Organisms, and Populations*. Cambridge, MA: M.I.T. Press.

Brantlinger, Patrick. 1985. Victorians and Africans: The geneology of the myth of the Dark Continent. *Critical Inquiry* 12(1):166–203.

Braverman, Harry. 1974. *Labor and Monopoly Capital*. New York: Monthly Review Press.

Brehm, Alfred E. 1922. *Tierleben. Allegemeine Kunde des Teirreichs*. Leipzig: Bibliographisches Institut.

Brewer, Maria Minch. 1987. Surviving fictions: Gender and difference in postmodern and postnuclear narrative. *Discourse* 9:37–52.

Brewer, Stella. 1978. *The Chimpanzees of Mount Asserik*. New York: Knopf.

Brin, David. 1983. *Startide Rising*. New York: Bantam.

Brown, Wendy. 1989. Review of MacKinnon, *Feminism Unmodified*. *Political Theory*, in press.

Brues, Alice. 1954. Selection and polymorphism in the ABO blood groups. *American Journal of Physical Anthropology* 12:559–97.

Bruner, S., Alison Jolly, and K. Sylva, eds. 1976. *Play: Its Role in Development and Evolution*. New York: Basic Books.

Bryan, C.D.B. 1987. *The National Geographic Society: 100 Years of Adventure and Discovery*. New York: Harry N. Abrams, Inc.

Buettner-Janusch, John, ed. 1962. *The Relatives of Man: Modern Studies of the Relation of Evolution of Nonhuman Primates to Human Evolution. Annals of the New York Academy of Sciences* 102(2):181–514.

Burnham, John C. 1968a. The new psychology: from narcissism to social control. In John C. Burnham, R. Braemner, and D. Brody, eds.*Change and Continuity in Twentieth Century America, the 1920s*. Ohio: Ohio State University Press.

Burnham, John C. 1968b. Historical background for the study of personality. In E. Borgatta and W. Lambert, eds. *Handbook of Personality Study and Research*. Chicago: Rand McNally.

Burroughs, Edgar Rice. 1976 [1912]. *Tarzan of the Apes*. New York: Ballantine.

Burton, Frances D. 1971. Sexual climax in female *Macaca mulatta*. *Proceedings of the Third International Congress of Primatology* 3:180–91.

Burton, Frances D. 1977. Ethology and the development of sex and gender identity in nonhuman primates. *Acta Biotheoretica* 26:1–18.

Buruma, Ian. 1985. *Behind the Mask: On Sexual Demons, Sacred Mothers, Transvestites, Gangsters, Drifters and Other Japanese Cultural Heroes*. New York: Pantheon.

Butler, Octavia. 1977. *Mind of My Mind*. New York: Doubleday.

Butler, Octavia. 1978. *Survivor*. New York: Doubleday.

Butler, Octavia. 1980. *Wild Seed*. New York: Doubleday.

Butler, Octavia. 1987. *Dawn*. The Xenogenesis Trilogy, vol. 1. New York: Warner Books.

Butler, Octavia. 1988. *Adulthood Rites*. The Xenogenesis Trilogy, vol. 2. New York: Warner Books.

Bygott, J.D. 1974. Agonistic behaviour and dominance in wild chimpanzees. Ph.D. thesis, Cambridge University.

Bygott, J.D. 1979. Agonistic behavior, dominance and social structure in wild chimpanzees of the Gombe National Park. In Hamburg and McCown 1979:405–28.

C

Callon, Michel and Bruno Latour. 1981. Unscrewing the big Leviathan, or how do actors microstructure reality? In Karin D. Knorr-Cetina and A. Cicourel, eds. *Advances in Social Theory and Methodology: Toward an Integration of Micro and Macro Sociologies*. London: Routledge & Kegan Paul.

Calmette, A. 1924. La Laboratoire Pasteur de Kindia. *La Nature*, no. 2638 (October 25):257–62.

Campbell, Bernard, ed. 1972. *Sexual Selection and the Descent of Man, 1871–1971*. Chicago: Aldine.

Canguilhem, Georges. 1978. *On the Normal and the Pathological*. Translated by Carolyn R. Fawcett. Dordrecht: Reidel.

Carby, Hazel V. 1985. "On the threshold of a woman's era": Lynching, empire, and sexuality in Black feminist theory. *Critical Inquiry* 12(1):262–77.

Carmichael, Leonard. 1957. Robert Means Yerkes, 1876–1956. *Psychological Review* 64: 1–7.

Carnegie, Andrew. 1889. The gospel of Wealth. *North American Review*, vol. 149–50.

Caron, Joseph A. 1977. La théorie de la cooperation animale dans l'écologie de W. C. Allee: Analyse du double registre d'un discours. M.Sc. thesis, Université de Montréal.

Carpenter, Clarence Ray. 1929. A study of sex behavior of the common pigeon with special emphasis on the monogamic tendencies of mated pairs. M. A. thesis, Duke University.

Carpenter, Clarence Ray. 1933a. Psychobiological studies of social behavior in Aves: I. The effect of complete and incomplete gonadectomy on primary sexual activity of the male pigeon. *Journal of Comparative Psychology* 16:25–27.

Carpenter, Clarence Ray. 1933b. Psychobiological studies of social behavior in Aves: II. The effect of complete and incomplete gonadectomy on secondary sexual activity with histological studies. *Journal of Comparative Psychology* 16:59–97.

Carpenter, Clarence Ray. 1934. A field study of the behavior and social relations of howling monkeys. *Comparative Psychology Monographs* 10(2):1–168.

Carpenter, Clarence Ray. 1940. A field study in Siam of the behavior and social relations of the gibbon (*Hylobates lar*). *Comparative Psychology Monographs* 16(5):1–212.

Carpenter, Clarence Ray. 1942. Sexual behavior of free ranging rhesus monkeys. *Journal of Comparative Psychology* 33(1): 113–62.

Carpenter, Clarence Ray. 1945. Concepts and problems of primate sociometry. *Sociometry* 8:56–61.

Carpenter, Clarence Ray. 1946. Rhesus monkeys for American laboratories. *Science* 91:284–86.

Carpenter, Clarence Ray. 1962. Concluding remarks. *Annals of the New York Academy of Sciences* 102(2):488–96.

Carpenter, Clarence Ray. 1964. *Naturalistic Behavior of Nonhuman Primates*. University Park: Pennsylvania State University Press.

Carroll, Lewis. 1971 [1876]. The hunting of the Snark. In *Alice in Wonderland*. New York: Norton Critical Edition.

Carter, Janis. 1981, April. A Journey to freedom. *Smithsonian* 12(1):90–101.

Carter, Janis. 1982. Lucy, me and the chimps. *Reader's Digest* (August):99–103.

Carter, Janis. 1988, June. Freed from keepers and cages, chimpanzees come of age on baboon island. *Smithsonian* 19(3):36–49.

Carthy, J.D. and F.J. Ebling. 1964. *The Natural History of Aggression*. London and New York: Published for the Institute of Biology by Academic Press.

Cartmill, Matt, David Pilbeam, and Glynn Isaac. 1986. One hundred years of paleoanthropology. *American Scientist* 74:410–20.

Cassin, René. 1968. Looking back on the Universal Declaration of 1948. *Review of Contemporary Law*, 15:13–26.

Chafe, William. 1972. *The American Woman, her Changing Social, Economic, and Political Roles, 1920–70*. New York: Oxford University Press.

Chapman, Frank M. 1929. *My Tropical Air Castle*. New York: Appleton-Century.

Chapman, Frank M. 1933. *Autobiography of a Bird Lover*. New York: Appleton-Century.

Chapman, Frank M. and T. S. Palmer, eds. 1933. *Fifty Years of Progress in American Ornithology, 1883–1933*. Lancaster, PA: American Ornithologists Union.

Chase, Allan. 1977. *The Legacy of Malthus: The Social Costs of the New Scientific Racism*. New York: Knopf.

Cheney, Dorothy L. 1977. Social development of immature male and female baboons. Ph.D. thesis. University of Cambridge.

Cheney, Dorothy L. 1978. Interaction of immature male and female baboons with adult females. *Animal Behaviour* 26:389–408.

Cheney, Dorothy L. and Robert M. Seyfarth. 1985. The social and non-social world of non-human primates. In Robert A. Hinde, A. Perret-Clermont and Joan Stephenson-Hinde, eds. *Social Relationships and Cognitive Development*. Oxford: Oxford University Press.

Cheney, Dorothy L., Robert M. Seyfarth, and Barbara B. Smuts. 1986. Social relationships and social cognition in nonhuman primates. *Science* 234 (12 Dec. 1986): 1361–66.

Cheney, Dorothy L., Robert M. Seyfarth, Barbara B. Smuts, and Richard W. Wrangham. 1987. Future of primate research. In Smuts, et al 1987:491–98.

Chevalier-Skolnikoff, Suzanne. 1971a. The ontogeny of communication in *Macaca speciosa*. Ph.D. thesis, University of California Berkeley.

Chevalier-Skolnikoff, Suzanne. 1971b. The female sexual response in stumptail monkeys (*Macaca speciosa*), and its broad implications for female mammalian sexuality. Paper presented at the American Anthropological Association meetings, New York City.

Chevalier-Skolnikoff, Suzanne. 1974. Male-female, female-female, and male-male sexual behavior in the stumptail monkey, with special attention to the female orgasm. *Archives of Sexual Behavior* 3(2):95–116.

Chevalier-Skolnikoff, Suzanne. 1975. Heterosexual copulatory patterns in stumptail macaques, *Macaca arctoides*, and in other macaque species. *Archives of Sexual Behavior* 4(2):199–220.

Chevalier-Skolnikoff, Suzanne. 1976. Homosexual behavior in a laboratory group of stumptail monkeys (*Macaca arctoides*): Form, contents and possible social functions. *Archives of Sexual Behavior* 5(6):511–27.

Chevalier-Skolnikoff, Suzanne. 1977. A Piagetian model for describing and comparing socialization in monkey, ape, and human infants. In Suzanne Chevalier-Skolnikoff and Frank E. Poirier, eds. *Primate Bio-Social Development: Biological, Social, and Ecological Determinants*. New York: Garland Press, pp. 159–87.

Child, Charles Manning. 1924. *Physiological Foundations of Behavior*. New York: Holt.

Child, Charles Manning. 1928. Biological foundations of social integration. *Publications of the American Sociological Society* 221:26–42.

Child, Charles Manning. 1940. Social integration as a biological process. *American Naturalist* 74:389–97.

Chodorow, Nancy. 1978. *The Reproduction of Mothering*. Los Angeles: University of California Press.

Chorover, Stephen L. 1979. *From Genesis to Genocide*. Cambridge, MA: M.I.T. Press.

Clark, James. 1936. The image of Africa. In *The Complete Book of Africa Hall*. New York: American Museum of Natural History.

Clark, Timothy and Tetsuo Mano. 1975. Transplantation and adaptation of Arashiyama A troop of Japanese macaques to a Texas brushland habitat. In S. Kondo, M. Kawai, and A. Ehara, eds. *Contemporary Primatology*. Basel: Karger, pp. 358–61.

Clarke, Adele. 1985. Emergence of the reproductive research enterprise: A sociology of biological, medical, and agricultural science in the U.S., 1910–1940. Ph.D. thesis, University of California San Francisco.

Clements, F[rederick] E. and V[ictor] E. Shelford. 1939. *Bioecology*. New York: Wiley.

Clifford, James. 1988. *The Predicament of Culture*. Cambridge, MA: Harvard University Press.

Clifford, James and George E. Marcus, eds. 1986. *Writing Culture:The Poetics and Politics of Ethnography*. Berkeley: University of California Press.

Clutton-Brock, Timothy H., ed. 1977. *Primate Ecology: Studies of Feeding and Ranging Behaviour in Lemurs, Monkeys and Apes*. London: Academic Press.

Clutton-Brock, Timothy H. 1983. Behavioural ecology and the female. *Nature* 306:716.

Clutton-Brock, Timothy H. and P.H. Harvey, eds. 1979. *Readings in Sociobiology*. San Francisco: W.H. Freeman.

Coelho, George V., David A. Hamburg and John E. Adams, eds. 1974. *Coping and Adaptation*. New York: Basic Books.

Cohen, Mark Nathan, Roy S. Malpass, and Harold Klein, eds. 1980. *Biosocial Mechanisms of Population Regulation*. New Haven: Yale University Press.

Cohn, Carol. 1987. Sex and death in the rational world of defense intellectuals. *Signs* 12(4):687–718.

Coimbra-Filho, A.F. and Russell A. Mittermeier, eds. 1981. *Ecology and Behavior of Neotropical Primates*, vol. 1. Rio de Janeiro: Academia Brasileira de Ciencias.

Cole, Sonia. 1975. *Leakey's Luck*. New York: Harcourt Brace Jovanovich.

Collias, Nicholas E. 1950. Social life and the individual among vertebrate animals. *Annals of the New York Academy of Sciences* 51:1076–92.

Collias, Nicholas E. and Charles H. Southwick. 1952. A field study of population density and social organization in howling monkeys. *Proceedings of the American Philosophical Society* 96:143–56.

Coolidge, Harold. 1933. *Pan paniscus*. Pygmy chimpanzee from south of the Congo River. *American Journal of Physical Anthropology* 18:1–59.

Coon, Carleton S. 1939. *The Races of Europe*. New York: Macmillan.

Coon, Carleton S., ed. 1948. *A Reader in General Anthropology*. New York: Holt.

Coon, Carleton S. 1962. *The Origin of Races*. New York: Knopf.

Coon, Carleton S. 1965. *The Living Races of Man*. New York: Knopf.

Corner, George. 1958. *Anatomist at Large, an Autobiography and Selected Essays*. New York: Basic Books.

Corruccini, Robert. 1986. Relative growth, dietary adaptation, and a distinctive ape species. *Reviews in Anthropology* 13(1):25–28.

Cowdry, E.V., ed. 1930. *Human Biology and Racial Welfare*. New York: Hoeber.

Craig, Wallace. 1908. The voice of pigeons regarded as a means of social control. *American Journal of Sociology* 19: 29–80.

Craig, Wallace. 1918. Appetites and aversions as constituents of instincts. *Biological Bulletin* 34: 91–107.

Cramer, Douglas L. and Adrienne L. Zihlman. 1978. Sexual dimorphism in the pygmay chimpanzee, *Pan paniscus*. In D.J. Chivers and K. A. Joysey, eds. *Recent Advances in Primatology*, vol. 3, *Evolution*. London: Academic Press.

Cravens, Hamilton. 1978. *The Triumph of Evolution: American Scientists and the Heredity Environment Controversy, 1900–41*. Philadelphia: University of Pennsylvania Press.

Crawford, M. P. 1939. The social psychology of vertebrates. *Psychological Bulletin* 36:407–46.

Crockett, Carolyn and John F. Eisenberg. 1987. Howlers: Variations in group size and demography. In Smuts et al. 1987:54–68.

Crook, John Hurrell and Stephen Gartlan. 1966. On the evolution of primate societies. *Nature* 210:1200–03.

Cross, Stephen J. 1977. John Hunter's theories of life, organization and disease: A conceptual study. M.A. thesis, Johns Hopkins University.

Cross, Stephen J. 1982. Origins of the Institue of Human Relations at Yale, 1922–29: From psychobiology and psychiatry to a unified science of human behavior. Abstract of a paper from the Joint Atlantic Seminar in the History of Biology.

Cross, Stephen J. and William R. Albury. 1987. Walter B. Cannon, L. J. Henderson, and the organic analogy. *Osiris*, 2nd series, 3:165–92.

Crossley, Robert. 1988. Introduction. In Octavia E. Butler. *Kindred*. Boston: Beacon, pp. ix–xxvii.

Curti, Merl. 1943. *The Growth of American Thought*. New York: Harper.

Cutright, Paul Russell. 1956. *Theodore Roosevelt the Naturalist*. New York: Harper & Row.

D

Dahl, Jeremy. 1986. *Pan paniscus*, a catalyst for the study of apes and humans. *American Journal of Primatology* 10:97–99.

Dahlberg, Frances, ed. 1981. *Woman the Gatherer*. New Haven: Yale University Press.

Daly, Martin and Margo Wilson. 1978. *Sex, Evolution, and Behavior*. North Scituate, MA: Duxbury Press.

Darling, F.E. 1937. *A Herd of Red Deer*. London: Oxford University Press.

Dart, Raymond A. 1953. The predatory transition from ape to man. *International Anthrop. Linguistics Review* 1:201–17.

Darwin, Charles. 1859. *On the Origin of Species by Means of Natural Selection*. London: John Murray.

Darwin, Charles. 1981 [1871]. *The Descent of Man and Selection in Relation to Sex*. Princeton: Princeton University Press.

Davis, Angela. 1982. *Women, Race, and Class*. London: Women's Press.

Dawkins. Richard. 1976. *The Selfish Gene* London: Oxford University Press.

de Lauretis, Teresa. 1980. "Signs of Wa/onder." In Teresa de Lauretis, A. Huyssen, and K. Woodward, eds., *The Technological Imagination*. Madison, WI: Coda Press.

de Lauretis, Teresa. 1984. *Alice Doesn't: Feminism, Semiotics and Cinema*. Bloomington, IN: Indiana University Press.

de Lauretis, Teresa. 1987. *Technologies of Gender: Essays on Theory, Film, and Fiction*. Bloomington: Indiana University Press.

Defert, Daniel. 1982. The collection of the world: Accounts of voyages from the sixteenth to the eighteenth centuries. *Dialectical Anthropology* 7:11–20.

Delaney, Carol. 1986. The meaning of paternity and the virgin birth debate. *Man* (n.s.)21:494–513.

Delgado, José. 1967. Social rank and radio-stimulated aggressiveness in monkeys. *The Journal of Nervous and Mental Diseases* 144(5):383–90.

Delormé, M. 1929. "Pastoria" (Institut Pasteur de Kindia): Son histoire et son role aujourd'hui et demain. *La Guinée militaire* supplement 8–9:1–16.

Derscheid, J.M. 1927. Notes sur les gorilles des volcans du Kivu (Parc National Albert). *Ann. Soc. zool. Belgiq.* 58:149–59.

Derrida, Jacques. 1976. *Of Grammatology*. Baltimore: Johns Hopkins University Press.

Derrida, Jacques. 1985. Racism's last word. *Critical Inquiry* 12(1):290–99.

DeVore, Irven. 1962. The social behavior and organization of baboon troops. Ph.D. thesis, University of Chicago.

DeVore, Irven. 1963. Mother-infant relations in free ranging baboons. In Harriet L. Rheingold, ed., *Maternal Behavior in Mammals*. New York: John Wiley & Sons, pp. 305–35.

DeVore, Irven, ed. 1965a. *Primate Behavior: Field Studies of Monkeys and Apes*. New York: Holt, Rinehart & Winston.

DeVore, Irven. 1965b. Male dominance and mating behavior in baboons. In Frank A. Beach, ed. *Sex and Behavior*. New York: Krieger, pp. 266–89.

DeVore, Irven. 1968. Social behavior, animal: Primate behavior. In *International Encyclopedia of the Social Sciences*. New York: Macmillan, pp. 351–60.

Dickson, David. 1984. *The New Politics of Science*. New York: Pantheon.

Dixon, Vernon. 1976. World views and research methodology. In L.M. King, V. Dixon, and W.W. Nobles, eds. *African Philosophy: Assumptions and Paradigms for Research on Black Persons*. Los Angeles: Fanon Center, Charles R. Drew Postgraduate School.

Djojosudharmo, Suharto. 1984, April. The Bohorok Orang-utan Rehabilitation Center. IPPL *Newsletter* 11(1):13–15.

Dobzhansky, Theodosius. 1937. *Genetics and the Origin of Species.* New York: Columbia University Press.

Dobzhansky, Theodosius. 1944. On species and races of living and fossil man. *American Journal of Physical Anthropology* 2:251–65.

Dobzhansky, Theodosius. 1962. *Mankind Evolving: The Evolution of the Human Species*. New Haven: Yale University Press.

Doi, Takeo. 1986. *The Anatomy of Self*. Translated by Mark Harbison. Tokyo: Kodansha International, Ltd.

Dolhinow, Phyllis, ed. 1972. *Primate Patterns*. New York: Holt, Rinehart & Winston.

Dolhinow, Phyllis and Naomi Bishop. 1972. The development of motor skills and social relationships among primates through play. In Dolhinow 1972:312–37.

Dollard, John. 1964. Yale's Institute of Human Relations: What was it? *Ventures* 4:32–40.

Domhoff, G. William. 1967. *Who Rules America?* New Jersey: Prentice-Hall.

Douglas, Mary. 1966. *Purity and Danger: An Analysis of the Concepts of Pollution and Taboo*. London: Routledge & Kegan Paul.

Douglas, Mary. 1970. *Natural Symbols*. London: Cresset.

Douglas-Hamilton, Iain and Oria Douglas-Hamilton. 1975. *Life among the Elephants*. New York: Viking.

Draper, Patricia. 1976. Social and economic constraints on child life among the !Kung. In Lee and DeVore 1976.

Du Chaillu, Paul. 1861. *Explorations and adventures in Equatorial Africa; with accounts of the manners and customs of the people, and the chase of the gorilla, crocodile, leopard, elephant, hippopotamus, and other animals.* London: Murray.

Dunn, L.C. and Theodosius Dobzhansky. 1946. *Heredity, Race, and Society*. New York: New American Library.

Durang, Christopher. 1988, July. Dream Weaver. *Interview* 18:34–43.

E

Easlea, Brian. 1983. *Fathering the Unthinkable: Masculinity, Scientists and the Nuclear Arms Race*. London: Pluto Press.

Eckholm, Erik. 1985, June 4. Premenstrual problems seem to beset baboons. New York *Times*, C2.

Eco, Umberto. 1983. *The Name of the Rose*. Translated by William Weaver. New York: Harcourt Brace Jovanovich. (*Il nome della rosa*, Milano: Bompiani, 1980).

Edwards, Paul. 1987. The army and the microworld: Computers and the militarized politics of gender. Manuscript accepted by *Signs*.

Edwards, Paul. 1988. The Closed World: Computers and the Politics of Discourse. Ph.D. thesis, University of California Santa Cruz.

Ehrenreich, Barbara. 1983. *The Hearts of Men*. Garden City: Anchor Press.

Ehrenreich, Barbara and Deirdre English. 1979. *For Her Own Good: 150 Years of the Experts' Advice to Women*. Garden City, NY: Doubleday.

Eimerl, Sarel and Irven DeVore. 1965. *The Primates*. Life Nature Library. New York: Time, Inc.

Eisenstein, Zillah. 1981. *The Radical Future of Liberal Feminism*. New York: Longman.

Eisenstein, Zillah. 1984. *Feminism and Sexual Equality: Crisis in Liberal America*. New York: Monthly Review Press.

Elgin, Suzette Haden. 1984. *Native Tongue*. New York: Daw.

Emerson, Alfred Earl. 1939. Social coordination and the superorganism. *American Midland Naturalist* 21:182–209

Emerson, Alfred Earl. 1954. Dynamic homeostasis: A unifying principle in organic, social, and ethical evolution. *Scientific Monthly* 78:67–85.

Erikson, G.E. 1981. Adolph Hans Schultz, 1891–1976. *American Journal of Physical Anthropology* 56:365–71.

Evans, Mary Alice and Howard Ensign Evans. *William Morton Wheeler, Biologist*. Cambridge, MA: Harvard University Press.

F

Fabian, Johannes. 1983. *Time and the Other: How Anthropology Makes Its Object*. New York: Columbia University Press.

Fairbairns, Zoe. 1979. *Benefits*. New York: Avon.

Fairchild, David. 1924. Barro Colorado Island Laboratory. *Journal of Heredity* 15: 99–112.

Farber, I.E., Harry Harlow, and L.J. West. 1957. Brainwashing, conditioning, and DDD (debility, dependency, and dread). *Sociometry* 20:271–285.

Farley, Michael. 1977. Formations et transformations de la synthese écologique aux Etats-Unis, 1949–1971. M. Sc. thesis, L'Institut d'histoire et de sociopolitique des sciences, Université de Montréal.

Fausto-Sterling, Anne. 1985. *Myths of Gender: Biological Theories about Women and Men*. New York: Basic Books.

Fausto-Sterling, Anne. 1988. In search of Sarah Bartmann. Unpublished manuscript.

Fedigan, Linda Marie. 1976. A study of roles in the Arashiyama West troop of Japanese macaques (*Macaca fuscata*). *Contributions to Primatology* 9. Basel: Karger.

Fedigan, Linda Marie. 1982. *Primate Paradigms: Sex Roles and Social Bonds*. Montreal: Eden Press.

Fedigan, Linda Marie. 1983. Dominance and reproductive success in primates. *Yearbook of Physical Anthropology* 26:91–129.

Fedigan, Linda Marie. 1984. Sex ratios and sex differences in primatology. *American Journal of Primatology* 7:305–08.

Fedigan, Linda Marie. 1986. The changing role of women in models of human evolution. *Annual Review of Anthropology* 15:25–66.

Fedigan, Linda Marie and Laurence Fedigan. n.d. Gender and the study of primates. American Anthropological Association: Project on Gender Curriculum. In manuscript.

Fee, Elizabeth. 1979. Nineteenth-century craniology: The study of the female skull. *Bulletin of the History of Medicine* 53(3):415–33.

Fee, Elizabeth. 1986. Critiques of modern science: The relationship of feminism to other radical epistemologies. In Bleier 1986:42–56.

Figlio, Karl. 1976. The metaphor of organization: An historiographical perspective on the bio-medical sciences of the early 19th century. *History of Science* 14: 17–53.

Figlio, Karl. 1977. The historiography of scientific medicine: An invitation to the human sciences. *Comparative Studies in Society and History* 19: 262–86.

Fisher, Elizabeth. 1979. *Woman's Creation: Sexual Evolution and the Shaping of Society*. New York: McGraw-Hill.

Fisher, Helen. 1982. *The Sex Contract: The Evolution of Human Behavior*. New York: Morrow.

Fleagle, John G. and William L. Jungers. 1982. Fifty years of primate phylogeny. In Spencer 1982: 187–230.

von Foerster, Heinz, ed. 1950–55. *Cybernetics: Circular Causal and Feedback Mechanisms in Biological and Social Systems*. Nos. 6–10. New York: Josiah Macy Jr. Foundation.

Fossey, Dian. 1970, January. Making friends with mountain gorillas. *National Geographic Magazine* 137:48–68.

Fossey, Dian. 1971. More years with mountain gorillas. *National Geographic* 140:574–85.

Fossey, Dian. 1976. The behaviour of the mountain gorilla. Ph.D. thesis, University of Cambridge.

Fossey, Dian. 1981. The imperiled mountain gorilla. *National Geographic* 159:501–23.

Fossey, Dian. 1983. *Gorillas in the Mist*. Boston: Houghton-Mifflin.

Fossey, Dian. 1986. His name was Digit. *International Primate Protection League Newsletter* 13(1):10–15.

Foucault, Michel. 1971. *The Order of Things*. New York: Pantheon.

Foucault, Michel. 1973. *The Birth of the Clinic: An Archaeology of Medical Perception*. Translated by A. M. Sheridan Smith. New York: Pantheon.

Foucault, Michel. 1978. *The History of Sexuality. Vol. 1. An Introduction*. Translated by Robert Hurley. New York: Pantheon.

Foucault, Michel. 1979. *Discipline and Punish*. New York: Pantheon.

Foucault, Michel. 1980. *Power/Knowledge: Selected Interviews and Other Writings*. Colin Gordon, ed. New York: Pantheon.

Fouts, Roger and R.L. Budd. 1979. Artificial and human language acquisition in the chimpanzee. In Hamburg and McCown 1979:374–92.

Fraser, Arvonne S. 1987. *The UN Decade for Women: Documents and Dialogue*. Boulder, CO: Westview Press.

Freedman, Estelle. 1974. The new woman: Changing views of women in the 1920s. *Journal of American History* 61: 372–93.

Frisch, Jean (Kitahara). 1959. Research on primate behavior in Japan. *American Anthropologist* 61:584–96.

Fromm, Erik. 1941. *Escape from Freedom*. New York: Holt, Rinehart & Winston.

G

Galdikas, Biruté. 1975, October. Orangutans, Indonesia's 'people of the forest'. *National Geographic* 148(4):444–73.

Galdikas, Biruté. 1979. Orangutan adaptation at the Tanjung Puting Reserve. Ph.D. thesis, University of California Los Angeles.

Galdikas, Biruté. 1980, June. Living with orangutans. *National Geographic* 157(6):830–53.

Gardner, Beatrice T. and R. Allen Gardner. 1980. Two comparative psychologists look at language acquisition. In K. E. Nelson, ed. *Children's Language*, vol. 2. New York: Halsted, pp. 331–69.

Gartlan, John S. 1964. Dominance in East African monkeys. *Proceedings of the East African Academy* 2:75–79.

Gates, Henry Louis, Jr., ed. 1985. *"Race," Writing, and Difference*. Special issue of *Critical Inquiry* 12(1).

Gates, R.R. 1944. Phylogeny and classification of hominids and anthropoids. *American Journal of Physical Anthropology* 2:279–92.

Geddes, Patrick and J. Arthur Thomson. 1890. *The Evolution of Sex*. New York: Humboldt.

Giddings, Paula. 1985. *When and Where I Enter: The Impact of Black Women on Race and Sex in America*. Toronto: Bantam.

Gilligan, Carol. 1982. *In a Different Voice*. Cambridge, MA: Harvard University Press.

Gilman, Sander L. 1985. Black bodies, white bodies: Toward an iconography of female sexuality in late nineteenth-century art, medicine, and literature. *Critical Inquiry* 12(1):204–242.

Gilmore, Hugh. 1981. From Radcliffe-Brown to sociobiology: Some aspects of the rise of primatology within physical anthropology. *American Journal of Physical Anthropology* 56(4):387–92.

Goffman, Irving. 1979. *Gender Advertisements*. New York: Harper & Row.

Goldberg, Stephen. 1973. *The Inevitability of Patriarchy*. New York: William Morrow.

Goldizen, Anne Wilson. 1987. Tamarins and marmosets: Communal care of offspring. In Smuts et al. 1987:34–43.

Golub, Edward S. 1987. *Immunology, a Synthesis*. Sunderland, MA: Sinauer.

Goodall, Jane. 1962. Nest-building behavior in the free-ranging chimpanzee. *Annals of the New York Academy of Sciences* 102(2):455–67.

Goodall, Jane. 1963a. Feeding behaviour of wild chimpanzees: A preliminary report. *Symposium of the Zoological Society of London* 10:39–48.

Goodall, Jane. 1963b, August. My life among the wild chimpanzees. *National Geographic* 124(2):272–308.

Goodall, Jane. 1965a. Chimpanzees of the Gombe Stream Reserve. In DeVore 1965a:425–73.

Goodall, Jane. 1965b. New discoveries among Africa's chimpanzees. *National Geographic* 127:802–31.

Goodall, Jane. 1967a. Mother-offspring relationships in chimpanzees. In Morris 1967b:287–346.

Goodall, Jane [van Lawick]. 1967b. *My Friends the Wild Chimpanzees*. Washington, D.C.: National Geographic.

Goodall, Jane. 1968a. Behaviour of free-living chimpanzees of the Gombe Stream area. *Animal Behavior Monographs* 1:163–311.

Goodall, Jane. 1968b. Expressive movements and communication in free-ranging chimpanzees: A preliminary report. In Jay 1968:313–74.

Goodall, Jane [van Lawick]. 1971. *In the Shadow of Man*. Boston: Houghton Mifflin.

Goodall, Jane. 1973. Cultural elements in a chimpanzee community. In E. W. Menzel, ed. *Precultural Primate Behaviour*, vol 1. Basel: Karger, Fourth IPC Symposium Proceedings.

Goodall, Jane. 1977. Infant-killing and cannabalism in free-living chimpanzees. *Folia Primatologica* 28:259–82.

Goodall, Jane. 1979. Life and death at Gombe. *National Geographic* 155:592–621.

Goodall, Jane. 1984. The nature of the mother-child bond and the influence of the family on the social development of free-living chimpanzees. In Kobayashi and Brazelton 1984:47–66.

Goodall, Jane. 1986. *The Chimpanzees of Gombe: Patterns of Behavior*. Cambridge, MA: Harvard University Press.

Goodall, Jane, Adriano Bandoro, Emilie Bergmann, Curt Busse, Hilali Matama, Esilom Mpongo, Ann Pierce, and David Riss. 1979. Intercommunity interactions in the chimpanzee population of the Gombe National Park. In Hamburg and McCown 1979:13–54.

Goodman, Morris and Jack E. Cronin. 1982. Molecular anthropology: Its development and current directions. In Spencer 1982:105–46.

Gordon, Linda. 1976. *Woman's Body, Woman's Right: A Social History of Birth Control in America*. New York: Viking.

Gould, Stephen Jay. 1980. G.G. Simpson, paleontology, and the modern synthesis. In Mayr and Provine 1982:153–72.

Gould, Stephen Jay. 1981. *The Mismeasure of Man*. New York: Norton.

Gould, Stephen Jay. 1982. The Hottentot Venus. *Natural History* 91(10):20–27.

Gould, Stephen Jay. 1986. Cardboard Darwinism. *New York Review of Books* (September 25):47–54.

Gray, Chris. 1988. Postmodern war. Ph.D. qualifying essay, University of California Santa Cruz.

Greene, John C. 1981. *Science, Ideology, and World View*. Berkeley: University of California Press.

Gregory, W. K. 1934. *Man's Place among the Anthropoids*. London: Oxford University Press.

Gubar, Susan and Sandra Gilbert. 1979. *Madwoman in the Attic*.New Haven: Yale University Press.

Guggisberg, G.A. 1977. *Early Wildlife Photographers*. New York: Talpinger.

H

Hahn, Emily. 1988. *Eve and the Apes.* New York: Weidenfeld & Nicolson.

Haldane, J.B.S. 1932. *The Causes of Evolution.* London: Longmanns, Green.

Hall, Diana Long. 1973–74. Biology, sex hormones, and sexism in the 1920s. *Philosophical Forum* V: 81–96.

Hall, Diana Long and Thomas F. Glick. 1976. Endocrinology: A brief introduction. *Journal of the History of Biology* 9:229–34.

Hall, K.R.L. 1962. The sexual, agonistic and derived social behaviour patterns of the wild chacma baboon, *Papio ursinus. Proceedings of the Zoological Society of London* 13:283–27.

Hall, K.R.L. 1965. Behaviour and ecology of the wild patas monkeys, *Erythrocebus patas,* in Uganda. *Journal of the Zoological Society of London* 148:15–87.

Hall, K.R.L. and Irven DeVore. 1965. Baboon social behavior. In DeVore 1965a:53–110.

Haller, John. 1971. *Outcasts from Evolution.* Urbana, IL: Illinois University Press.

Hamburg, [Captain] David A. 1953. Psychological adaptive processes in life threatening injuries. In *Symposium on Stress,* National Research Council, Division of Medical Sciences.

Hamburg, David A. 1961. The relevance of recent evolutionary change to human stress biology. In Washburn 1961:278–88.

Hamburg, David A. 1963. Emotions in the perspective of human evolution. In P. Knapp, ed. *Expression of Emotions in Man,* New York: International Universities Press, pp. 300–17.

Hamburg, David A., ed. 1970. *Psychiatry as a Behavioral Science.* Englewood Cliffs, NJ: Prentice-Hall.

Hamburg, David A. 1972. Evolution of emotional responses: Evidence from recent research. In Duane Quiatt, ed., *Primates on Primates.* Minneapolis: Burgess, pp. 61–71.

Hamburg, David A., George V. Coelho, and John E. Adams. 1974. Coping and adaptation: Steps toward a synthesis of behavioral and social perspectives. In Coelho et al. 1974:403–40.

Hamburg, David A. and Elizabeth McCown, eds. 1979. *The Great Apes.* Menlo Park, CA: Benjamin/ Cummings.

Hamilton, G.V. 1914. A study of sexual tendencies in monkeys and baboons. *Journal of Animal Behavior* 4:295–318.

Hamilton, W.D. 1963. The evolution of altruistic behavior. *American Naturalist* 97:354–56.

Hamilton, W.D. 1964. The genetical evolution of social behavior, I and II. *Journal of Theoretical Biology* 7:1–52.

Hammonds, Evelyn. 1986. Race, sex, and AIDS: The construction of "other." *Radical America* 20(6):28–38.

Haraway, Donna J. 1976. *Crystals, Fabrics, and Fields: Metaphors of Organicism in Twentieth Century Developmental Biology.* New Haven: Yale University Press.

Haraway, Donna J. 1978a. Animal sociology and a natural economy of the body politic, Part I. A political physiology of dominance. *Signs* 4:21–36.

Haraway, Donna J. 1978b. Animal sociology and a natural economy of the body politic. Part II. The past is the contested zone: Human nature and theories of production and reproduction in primate behavior studies. *Signs* 4:37–60.

Haraway, Donna J. 1979. The biological enterprise: Sex, mind,and profit from human engineering to sociobiology. *RadicalHistory Review,* no. 20, pp. 206–37.

Haraway, Donna J. 1981. In the beginning was the word: The genesis of biological theory. *Signs* 6(3):469–81.

Haraway, Donna J. 1981–82. The high cost of information in post World War II evoltionary biology: Ergonomics, semiotics, and the sociobiology of communication systems. *Philosophical Forum* XIII(2–3):244–78.

Haraway, Donna J. 1983a. The contest for primate nature: Daughters of man the hunter in the field,

1960–80. In Mark Kann, ed. *The Future of American Democracy: Views from the Left*. Philadelphia: Temple University Press, pp. 175–207.

Haraway, Donna J. 1983b. Signs of dominance: From a physiology to a cybernetics of primate society, C.R. Carpenter, 1930–70. *Studies in History of Biology* 6:129–219.

Haraway, Donna J. 1985. A manifesto for cyborgs: Science, technology, and socialist feminism in the 1980s. *Socialist Review* 15(2): 65–108.

Haraway, Donna J. 1986. Primatology is politics by other means: Women's place is in the jungle. In Bleier 1986:77–118.

Haraway, Donna J. 1987. Geschlecht, Gender, Genre: Sexualpolitik eines Wortes. In Kornelia Hauser, ed. *Viele Orte. Uberall? Feminismus in Bewegung, Festschrift für Frigga Haug*. Berlin: Argument-Verlag, pp. 22–41.

Haraway, Donna J. 1988. Situated knowledges: The science question in feminism as a site of discourse on the privilege of partial perspective. *Feminist Studies* 14(3):575–600.

Haraway, Donna J. forthcoming a. The biopolitics of postmodern bodies: Determinations of self in immune system discourse. *Differences: A Journal of Feminist Cultural Studies* 1(1).

Haraway, Donna J. forthcoming b. *Simians, Cyborgs, and Women: The Reinvention of Nature*. London: Free Association Books.

Harcourt, Alexander H. 1979. Social relationships and determinants of ranging patterns in the mountain gorilla. Ph.D. thesis, Cambridge University.

Harcourt, Alexander H. and Dian Fossey. 1981. The Virunga gorillas: Decline of an "island" population. *African Journal of Ecology* 19:83–97.

Harding, R.S.O. 1980. Agonism, ranking, and the social order of adult male baboons. *American Journal of Physical Anthropology* 47:349–54.

Harding, R.S.O. and Geza Teleki. 1981. *The Omnivorous Primates*. New York: Columbia University Press.

Harding, Sandra. 1983. Why has the sex/gender system become visible only now? In Harding and Hintikka 1983:311–24.

Harding, Sandra. 1986. *The Science Question in Feminism*. Ithaca: Cornell University Press.

Harding, Sandra and Merrill Hintikka, eds. 1983. *Discovering Reality: Feminist Perspectives on Epistemology, Metaphysics, Methodology, and Philosophy of Science*. Dordrecht: Reidel.

Harlow, Harry. 1951. Primate learning. In Stone 1951:183–238.

Harlow, Harry. 1952. Functional organization of the brain in relation to mentation and behavior. In Milbank Memorial Fund. *The Biology of Mental Health and Disease*. New York: Hoeber, pp. 244–53.

Harlow, Harry. 1962. The heterosexual affectional system in monkeys. *American Psychologist* 17:1–9.

Harlow, Harry. 1963. An experimentalist views the emotions. In H. Knapp, ed. *The Expression of Emotion in Man*. New York: International Universities Press, pp. 254–65.

Harlow, Harry. 1977. Birth of the surrogate mother. In W.R. Klemm, ed. *Discovery Processes in Modern Biology*. Huntington, New York: Krieger.

Harlow, Harry and Margaret K. Harlow. 1965. The affectional systems. In Schrier, Harlow, and Stolnitz, Vol. II, 1965.

Harlow, Harry, Margaret K. Harlow, and Stephen J. Suomi. 1971. From thought to therapy: Lessons from a primate laboratory. *American Scientist* 659: 538–49.

Harlow, Harry and Clara Mears. 1979. *The Human Model: Primate Perspectives*. New York: Wiley.

Harlow, Harry and L.E. Moon. 1956. The effects of repeated doses of total-body X-radiation on motivation and learning in rhesus monkeys. *Journal of Comparative and Physiological Psychology* 49:60–65.

Harlow, Harry, Allen M. Schrier, and D.G. Simons. 1956. Exposure of primates to cosmic radiation above 90,000 feet. *Journal of Comparative and Physiological Psychology* 49:195–200.

Harris, Neil. 1973. *Humbug: The Art of P.T. Barnum*. Boston: Little, Brown.

Harrison, Barbara. 1963. *Orangutan*. New York: Doubleday.

Hartman, Carl. 1932. Studies on reproduction in the monkey and their bearing on gynecology and anthropology. *Bulletin of the Association for the Study of Internal Secretions* 25: 670–82.

Hartouni, Val. 1987. Personhood, membership and community: Abortion politics and the negotiation of public meanings. Ph.D. thesis, University of California Santa Cruz.

Hartsock, Nancy. 1983a. *Money, Sex and Power*. New York: Longman.

Hartsock, Nancy. 1983b. The feminist standpoint: Developing the ground for a specifically feminist historical materialism. In Harding and Hintikka 1983:283–10.

Haug, Frigga. 1986. The women's movement in West Germany. *New Left Review* 155:50–74.

Haug, Frigga et al. 1987. *Female Sexualization: A Collective Work of Memory*. London: Verso.

Hausfater, Glenn. 1975. Dominance and reproduction in baboons: a quantitative analysis. *Contributions to Primatology*, Vol. 7. Basel: Karger.

Hausfater, Glenn and Sarah Blaffer Hrdy, eds. 1984. *Infanticide: Comparative and Evolutionary Perspectives*. New York: Aldine.

Hausfater, Glenn and Barbara Skoblick. 1985. Evidence for Premenstrual Syndrome in Baboons, Amboseli National Park. Presented at the meetings of the American Primatological Society, June 4.

Hayes, Catherine. 1951. *The Ape in Our House*. New York: Harper & Row.

Hays, Samuel. 1959. *Conservation and the Gospel of Efficiency: The Progressive Conservation Movement, 1890–1920*. Cambridge, MA: Harvard University Press.

Hediger, Heini P. 1950. *Wild Animals in Captivity*. London: Butterworth.

Hegel, Georg Wilhelm Friedrich. 1979. *Phenomenology of Spirit*. Translated by A.V. Miller. New York: Oxford University Press.

Heims, Steve J. 1975. Encounter of behavioral sciences with new machine-organism analogies in the 1940's. *Journal of the History of the Behavioral Sciences* 11:368–73.

Heims, Steve J. 1980. *John von Neumann and Norbert Wiener: From Mathematics to the Technologies of Life and Death*. Cambridge, MA: M.I.T. Press.

Henderson, Lawrence J. 1935. *Pareto's General Sociology: A Physiologist's Interpretation*. Cambridge, MA: Harvard University Press.

Henry, Alice. 1982. Sex and society among the primates. A review of *The Woman that Never Evolved*. *Off Our Backs* (January):18–19.

Hern, Lafcadio. 1971 [1899]. *In Ghostly Japan*. Rutland, Vermont and Tokyo: Charles E. Tuttle & Co.

Herschberger, Ruth. 1970 [1948]. *Adam's Rib*. New York: Harper & Row.

Hess, Lilo. 1954. *Christine, the Baby Chimp*. London: Bell.

Higham, John. 1975. *Send These to Me: Jews and Other Immigrants in Urban America*. New York: Atheneum.

Hilgard, Ernest. 1965. Robert Means Yerkes. *National Academy of Sciences Memoirs* 38:385–425.

Hinde, Robert, ed. 1983. *Primate Social Relationships: An Integrated Approach*. Sunderland, MA: Sinauer.

Hinde, Robert and Thelma Rowell. 1962. Communication by postures and facial expressions in the rhesus monkey (*Macaca mulatta*). *Procedings of the Zoological Society of London* 138:1–21.

Hinde, Robert and Yvette Spencer-Booth. 1967. The behaviour of socially living rhesus monkeys in their first two and a half years. *Animal Behaviour* 15:169–96.

Hinkle, Lawrence E. 1973/74. The concept of 'stress' in the biological and social sciences. *Science, Medicine, and Man* 1:31–48.

Hiraiwa-Hasegawa, Mariko, Hasegawa Toshisada, and Nishida Toshisada. 1983. Demographic study of a large-sized unit group of chimpanzees in the Mahale Mountains, Tanzania. Mahale Mountains Chimpanzee Research Project, Ecological Report, no. 30.

Hofer, Helmut. 1962. Uber die Interpretation des Altesten Fossilen Primatengehirne. *Bibliotheca Primatologica Fasc.* 1:1–31.

Hofer, Helmut, A.H. Schultz, and D. Stark, eds. 1956. *Primatologia*, Vol. 1. Basel: Karger.

Hofstadter, Richard. 1955. *The Age of Reform*. New York: Knopf.

Holloway, Ralph, ed. 1974. *Territoriality and Xenophobia*. New York: Academic Press.

Holton, Gerald and William Blanpied, eds. 1976. *Science and Its Public: The Changing Relation*. Dordrecht: Holland.

Honoré, F. 1927. Les "singeries" de L'Institut Pasteur à Kindia et Paris. *L'Illustration, Paris* 169: 407–9.

Hooks, Bell. 1981. *Ain't I a Woman*. Boston: Southend.

Hooton, E.A. 1939. *Crime and the Man*. Cambridge, MA: Harvard University Press.

Hooton, E.A. 1942. *Man's Poor Relations*. New York: Doubleday.

Hooton, E.A. 1946. *Up from the Ape*, 2nd ed. New York: Macmillan.

Horowitz, MaryAnn C. 1976. Aristotle and women. *Journal of the History of Biology* 9:183–213.

Howard, H.E. 1920. *Territory in Bird Life*. London: Murray.

Howells, W.W. 1954. Obituary of Earnest Albert Hooton. *American Journal of Physical Anthropology* 12:445–53.

Howells, W.W., ed. 1962. *Ideas on Evolution: Selected Essays, 1949–61*. Cambridge, MA: Harvard University Press.

Hoyt, Augusta Maria. 1941. *Toto and I: A Gorilla in the Family*. Philadelphia: J.B. Lippincott.

Hrdy, Sarah Blaffer. 1977. *Langurs of Abu*. Cambridge, MA: Harvard.

Hrdy, Sarah Blaffer. 1979. The evolution of human sexuality: The latest word and the last. Review of Donald Symons, *The Evolution of Human Sexuality*. *The Quarterly Review of Biology* 54:309–14.

Hrdy, Sarah Blaffer. 1981. *The Woman That Never Evolved*. Cambridge, MA: Harvard University Press.

Hrdy, Sarah Blaffer. 1983. Heat lost. *Science 83*. October, pp. 73–8.

Hrdy, Sarah Blaffer. 1986. Empathy, polyandry, and the myth of the coy female. In Bleier 1986:119–46.

Hrdy, Sarah Blaffer and William Bennett. 1981. "Lucy's husband: What did he stand for?" *Harvard Magazine* (July–August):7–9, 46.

Hrdy, Sarah Blaffer and George C. Williams. 1983. Behavioral biology and the double standard. In Wasser 1983b:3–17.

Hubbard. Ruth. 1982. Have only men evolved? In Hubbard, Henifin, and Fried 1982:17–46

Hubbard, Ruth, M.S. Henifin, and B. Fried, eds. 1982. *Biological Woman—the Convenient Myth. A Collection of Essays and a Comprehensive Bibliography*. Cambridge, MA: Schenkman.

Huizinga, J. 1973 [1934]. The idea of history. In Fritz Stern, ed. *The Varieties of History*. New York: Vintage, pp. 290–303.

Hull, Gloria. 1985. "What is it I think she's doing anyway?" A reading of Toni Cade Bambara's *The Salt Eaters*. In Marjorie Pryse and Hortense Spillers, eds. *Conjuring: Black Women, Literary Tradition, and Fiction*. Bloomington, IN: Indiana University Press, pp. 216–32.

Hutchinson, G. Evelyn. 1948. Circular causal systems in ecology. *Annals of the New York Academy of Sciences* 50:221–46.

Hutchinson, G. Evelyn. 1959. Hommage to Santa Rosalia; or, Why are there so many kinds of animals? *American Naturalist* 93:145–59.

Hutchinson, G. Evelyn. 1960. John Farquhar Fulton (1899–1960). *Yearbook of the American Philosophical Society* 24:140–42.

Hutchinson, G. Evelyn. 1965. *The Ecological Theater and the Evolutionary Play*. New Haven: Yale University Press.

Hutchinson, G. Evelyn. 1978. *An Introduction to Population Ecology*. New Haven: Yale University Press.

Huxley, Julian. 1942. *Evolution, the Modern Synthesis*. London: Allen & Unwin.

I

Imanishi, Kinji. 1960. Social organization of subhuman primates in their natural habitat. *Current Anthropology* 1(5–6):390–405.

Imanishi, Kinji. 1961. The origin of the human family—a primatological approach. *Japanese Journal of Ethnology* 25:119–30.

Imanishi, Kinji. 1963. Social behavior in Japanese monkeys, *Macaca fuscata*. In Southwick 1963:68–81.

Irigaray, Luce. 1985. Is the subject of science sexed? *Cultural Critique* 1:73–88.

Isaac, Glynn L. 1978. The food-sharing behavior of proto-human hominids. *Scientific American* 238(4): 90–106.

Itani, Junichiro. 1954. Japanese monkeys at Takasakiyama. In Imanishi Kinji, ed. *Social Life of Animals in Japan* (in Japanese). Tokyo: Kobunsya.

Itani, Junichiro. 1958. On the acquisition and propogation of new food habits in the troop of Japanese monkeys at Takasakiyama. *Primates* 1:84–98.

Itani, Junichiro. 1963. Paternal care in the wild Japanese monkey, *Macaca fuscata*. In Southwick 1963: 91–97.

Itani, Junichiro. 1975. Twenty years with the Mount Takasaki monkeys. In Bermant and Lindberg 1975:197–249.

Itani, Junichiro and Suzuki Akira 1967. The social unit of chimpanzees. *Primates* 8:355–82.

J

Jacob, François. 1974. *The Logic of Life: A History of Heredity.* Translated by Betty E. Spillman. New York: Pantheon.

Jameson, Fredric. 1981. *The Political Unconscious: Narrative as a Socially Symbolic Act.* Ithaca: Cornell University Press.

Jay, Phyllis. 1962. The social behavior of the langur monkey. Ph.D. thesis, University of Chicago.

Jay, Phyllis. 1963a. Mother-infant relations in langurs. In Rheingold 1963:282–304.

Jay, Phyllis. 1963b. The Indian langur monkey (*Presbytis entellus*). In Southwick 1963:114–23.

Jay, Phyllis. 1963c. The female primate. In S.M. Farber and R.H.L. Wilson, eds. *Man and Civilization: The Potential of Women.* New York: McGraw-Hill, pp. 3–12.

Jay, Phyllis. 1965a. Field studies. In Schrier, Harlow, and Stolnitz, Vol. II, 1965:525–91.

Jay, Phyllis. 1965b. The common langurs of north India. In DeVore 1965a:197–249.

Jay, Phyllis, ed. 1968. *Primates: Studies in Adaptation and Variability.* New York: Holt, Rinehart & Winston.

Jepsen, G.L., Ernst Mayr, and George Gaylord Simpson, eds. 1949. *Genetics, Paleontology, and Evolution.* Princeton: Princeton University Press.

Johanson, Donald C. 1980. Early African hominid phylogenesis: A re-evaluation. In L.-K. Konigsson, ed. *Current Argument on Early Man.* New York: Pergamon, pp. 31–69.

Johanson, Donald C. and Maitland Edey. 1981. *Lucy: The Beginnings of Humankind.* New York: Simon & Schuster.

Johanson, Donald C. and Timothy White. 1979. A systematic assessment of early African hominids. *Science* 202:321–30.

Johnson, Martin. 1936. Camera safaris. In *The Complete Book of African Hall.* New York: American Museum of Natural History.

Johnson, Osa. 1940. *I Married Adventure: The Lives and Adventures of Martin and Osa Johnson.* Garden City, NY: Garden City Publishers.

Johnson, William Davison. 1958. *T.R.: Champion of the Strenuous Life.* New York: Theodore Roosevelt Association.

Jolly, Alison. 1966. *Lemur Behavior.* Chicago: Chicago University Press.

Jolly, Alison. 1972, 1985. *The Evolution of Primate Behavior*. New York: Macmillan. 1st & 2nd editions.

Jolly, Alison. 1980. *A World Like Our Own: Man and Nature in Madagascar*. New Haven: Yale University Press.

Jolly, Alison. 1987. Madagascar: A world apart. *National Geographic* 171(2):148–183.

Jolly, Alison, P. Oberle, and R. Albignac, eds. 1984. *Madagascar: Key Environments Series*. Oxford: Pergamon.

Jones, Clara B. 1978. Aspects of reproductive behavior in the mantled howler monkey (*Alouatta palliata* Gray). Ph.D. thesis, Cornell University.

Jones, Clara B. 1980. The functions of status in the mantled howler monkey, *Alouatta paliata* Gray: Intraspecific competition for group membership in a folivorous Neotropical primate. *Primates* 21:389–405.

Jones, Clara B. 1981. The evolution and socioecology of dominance in primate groups: Theoretical formulation, classification and assessment. *Primates* 22:70–83.

Jones, Clara B. 1983. Social organization of captive black howler monkeys (*Alouatta caraya*): "Social competition" and the use of non-damaging behavior. *Primates* 24:25–39.

Jordanova, Ludmilla J. 1980. Natural facts: A historical perspective on science and sexuality. In MacCormack and Strathern 1980:42–69.

Jordanova, Ludmilla J., ed. 1986. *Languages of Nature: Critical Essays on Science and Literature*. London: Free Association Books.

K

Kano, Takayoshi. 1979. A pilot study on the ecology of pygmy chimpanzees, *Pan paniscus*. In Hamburg and McCown 1979:123–35.

Kano, Takayoshi. 1982. The social group of pygmy chimpanzees (*Pan paniscus*) of Wamba. *Primates* 23:171–88.

Kano, Takayoshi and Mbangi Mulavwa. 1984. Feeding ecology of the pygmy chimpanzees (*Pan paniscus*) of Wamba. In Sussman 1984:233–74.

Kargon, Robert H. 1977. Temple to science: Cooperative research and the birth of the California Institute of Technology. *Historical Studies in the Physical Sciences* 8:3–31.

Katchadourian, H., ed. 1978. *Human Sexuality: A Comparative and Developmental Perspective*. Los Angeles: University of California Press.

Kawai, Masao. 1958. On the rank system in a natural group of Japanese monkey, I and II. *Primates* 1 (2):11–48, in Japanese with English summary.

Kawai, Masao. 1965. Newly acquired precultural behaviour of the natural troop of Japanese monkeys on Koshima Islet. *Primates* 6:1–30.

Kawai, Masao. 1969. *Nihonzaru no seitai (Life of Japanese Monkeys)*. Tokyo.

Kawamura, Syunzo. 1958. Matriarchal social ranks in the Minoo-B troop: A study of the rank system of Japanese monkeys. *Primates* 1 (2):149–156 (in Japanese).

Kawamura, Syunzo. 1963. The progress of sub-culture propogation among Japanese monkeys. In Southwick 1963: 82–90.

Kaye, Howard L. 1986. *The Social Meaning of Modern Biology: From Social Darwinism to Sociobiology*. New Haven: Yale University Press.

Keller, Evelyn Fox. 1983. *A Feeling for the Organism*. New York:Freeman.

Keller, Evelyn Fox. 1985. *Reflections on Gender and Science*. New Haven: Yale University Press.

Keller, Evelyn Fox. 1986. Problems of radical individualism in evolutionary theory. Paper presented at the meetings of the American Philosophical Association, December.

Keller, Evelyn Fox. 1987a. The gender/science system: Or, is sex to gender as nature is to science? *Hypatia* 2(3):37–50.

Keller, Evelyn Fox. 1987b. Reproduction and the central project of evolutionary theory. *Biology and Philosophy* 2:73–86.

Keller, Evelyn Fox and Christine Grontkowski. 1983. The mind's eye. In Harding and Hintikka 1983:207–24.

Kellogg, W.N. and L.A. Kellogg. 1933. *The Ape and the Child.* New York: McGraw-Hill.

Kempf, Edward J. 1917. The social and sexual behavior of infrahuman primates with some comparable facts of human behavior *Psychoanalytic Review* 4: 127–54.

Kennedy, John Michael. 1968. Philanthropy and science in New York City: The American Museum of Natural History, 1868–1968. Ph.D. thesis, Yale University.

Kevles, Bettyann. 1976. *Watching the Wild Apes.* New York: Dutton.

Kevles, Bettyann. 1986. *Females of the Species.* Cambridge, MA: Harvard University Press.

Kevles, Daniel J. 1968. Testing the army's intelligence: Psychologists and the military in World War I. *Journal of American History* 55: 565–81.

Kevles, Daniel J. 1978. *The Physicists.* New York: Knopf.

Kevles, Daniel J. 1985. *In the Name of Eugenics: Genetics and the Uses of Human Heredity.* New York: Knopf.

King, Katie. 1987. Canons without Innocence. Ph.D. thesis, University of California Santa Cruz.

King, Katie. 1988. Audre Lorde's lacquered layerings. The lesbian bar as the site of literary production. *Cultural Studies* 2(3):321–42.

Kingsland, Sharon E. 1985. *Modeling Nature: Episodes in the History of Population Ecology.* Chicago: University of Chicago Press.

Kinzey, Warren, G., ed. 1987. *The Evolution of Human Behavior: Primate Models.* Albany: State University of New York.

Kitcher, Philip. 1985. *Vaulting Ambition. Sociobiology and the Quest for Human Nature.* Cambridge, MA: M.I.T. Press.

Kleiman, Devra G. 1977a. Monogamy in mammals. *Quarterly Review of Biology* 52:39–69.

Kleiman, Devra G., ed. 1977b. *The Biology and Conservation of the Callitrichidae.* Washington: Smithsonian Institution Press.

Klopfer, Peter. 1974. *An Introduction to Animal Behavior: Ethology's First Century,* 2nd ed. Englewood Cliffs, NJ: Prentice-Hall.

Knorr-Cetina, Karin D. 1983. The ethnographic study of scientific work: Towards a constructivist interpretation of science. In Knorr-Cetina and Mulkay 1983:115–40.

Knorr-Cetina, Karin D. and Michael Mulkay, eds. 1983. *Science Observed: Perspectives on the Social Study of Science.* London: Sage.

Knutsen, F., ed. 1973. *Control of Aggression: Implications for Basic Research.* Chicago: Aldine.

Kobayashi, N. and T. Berry Brazelton, eds. 1984. *The Growing Child in Family and Society.* Tokyo: University of Tokyo Press.

Kohler, Robert. 1976. The management of science: The experience of Warren Weaver and the Rockefeller Programme on Molecular Biology. *Minerva* 14:279–306.

Köhler, Wolfgang. 1925. *The Mentality of Apes.* New York: Harcourt Brace.

Kohts, Nadie. 1923. *Untersuchungen uber die Erkenntnissfähigkeiten des Schimpansen.* Moscow: Zoopsychologischen Laboratorium des Museum Darwinianum.

Kohts, Nadie. 1935. *Infant Ape and Human Child* (in Russian, English summary). Moscow: Scientific Memoirs of the Darwin Museum, No. 3, pp. 524–91.

Kolko, Gabriel. 1977. *The Triumph of Conservatism.* New York: Free Press.

Kondo, Dorinne. 1984. If you want to know who they are . . . *The New York Times Book Review* (September 16):13–14.

Kortlandt, Adrian. 1962. Chimpanzees in the wild. *Scientific American* 206(5):128–38.

Kortlandt, Adrian and M. Kooij. 1963. Protohominid behaviour in primates. *Symposium of the Zoological Society, London* 10:61–88.

Kratochvil, C.H. 1968. Introduction. In *First Holloman Symposium on Primate Immunology and Molecular Genetics. Primates in Medicine* 1:vii–xv. Basel: Karger.

Kuhn, Annette. 1982. *Women's Pictures: Feminism and Cinema.* London: Routledge & Kegan Paul.

Kuhn, Thomas S. 1962. *The Structure of Scientific Revolutions.* Chicago: University of Chicago Press.

Kummer, Hans. 1968. Social organization of Hamadryas baboons: A field study. *Bibliotheca Primatologica* 6:1–189.

Kummer, Hans. 1971. *Primate Societies: Group Techniques of Ecological Adaptation.* Chicago: Aldine.

Kummer, Hans and F. Kurt. 1963. Social units of a free-living population of hamadryas baboons. *Folia Primatologica* 1:4–19.

Kuroda, Suehisa. 1980. Social behavior of the pygmy chimpanzees. *Primates* 21(2):181–97.

Kuroda, Suehisa. 1982. *Pygmy Chimpanzees.* Tokyo: Chikumashoko.

L

Lack, David. 1966 [1954]. *Population Studies of Birds.* Oxford: Clarendon.

Lancaster, Jane B. 1972. Play-mothering: the relations between juvenile females and young infants among free-ranging vervet monkeys. In Poirier 1972:83–104.

Lancaster, Jane B. 1973. In praise of the achieving female monkey. *Psychology Today* 7(4):30–36, 99.

Lancaster, Jane B. 1975. *Primate Behavior and the Emergence of Human Culture.* New York: Holt, Rinehart & Winston.

Lancaster, Jane B. 1979. Sex and gender in evolutionary perspective. In H.A. Katchadourian, ed. *Human Sexuality: A Comparative and Developmental Approach.* Los Angeles: University of California Press, pp. 51–80.

Lancaster, Jane B. 1984. Introduction. In Small 1984:1–12.

Lancaster, Jane and Richard Lee. 1965. The annual reproductive cycle in monkeys and apes. In DeVore 1965a:486–513.

Landau, Misia. 1981. The anthropogenic: Paleoanthropological writing as a genre of literature. Ph.D. thesis, Yale University.

Landau, Misia. 1984. Human evolution as narrative. *American Scientist* 72:362–68.

Lang, E.M. 1962. *Goma, the Baby Gorilla.* London: Victor Gollancz.

Lange, Lynda. 1983. Woman is not a rational animal: On Aristotle's biology of reproduction. In Harding and Hintikka 1983:1–15.

Laporte, Léo F. and Zihlman, Adrienne L. 1983. Plates, climate and hominoid evolution. *South African Journal of Science* 79:96–109.

Laqueur, Thomas. 1986. Orgasm, generation, and the politics of reproductive biology. *Representations* 14:1–41.

Lasch, Christopher. 1977. *Haven in a Heartless World.* New York: Basic Books.

Latour, Bruno. 1978. Observing scientists observing baboons observing. . . . Paper prepared for the Wenner Gren conference, "Baboon Field Research: Myths and Models." Unpublished manuscript.

Latour, Bruno. 1983. Give me a laboratory and I will raise the world. In Knorr-Cetina and Mulkay 1983: 141–70.

Latour, Bruno. 1987. *Science in Action: How to Follow Scientists and Engineers through Society.* Cambridge, MA: Harvard University Press.

Latour, Bruno. 1988. *The Pasteurization of France.* Translated by Alan Sheridan and John Law. Cambridge, MA: Harvard University Press.

Latour, Bruno and Shirley C. Strum. 1986. Human social origins: Please tell us another story. *Journal of Social and Biological Structures* 9: 167–87.

Latour, Bruno and Stephen Woolgar. 1979. *Laboratory Life: The Social Construction of Scientific Facts.* London: Sage.

Laughlin, William S. 1968. Hunting: An integrating biobehavior system and its evolutionary importance. In Lee and DeVore 1968:304–20.

Le Gros Clark, W.E. 1934. *Early Forerunners of Man.* London: Balleere, Tindall, and Cox.

Leavitt, Ruby Rohrlich. 1975. *Peaceable Primates and Gentle People.* New York: Harper & Row.

Lederberg, Joshua. 1983. David A. Hamburg: President-elect of AAAS. *Science* 221:431–2.

Lee, Richard. 1968. The hand to mouth existence: A note on the origin of human economy. Wenner-Gren Symposium no. 42, Social Organization and Subsistence in Primates.

Lee, Richard. 1979. *The !Kung San: Men, Women and Work in a Foraging Society.* New York: Cambridge University Press.

Lee, Richard. 1984. *The Dobe !Kung.* New York: Holt, Rinehart & Winston.

Lee, Richard. 1985. The gods must be crazy, but the producers knowwhat they are doing. *South Africa Report*, June.

Lee, Richard and Irven DeVore, eds. 1968. *Man the Hunter.* Chicago: Aldine.

Lee, Richard and Irven DeVore, eds. 1976. *Kalahari Hunter-Gatherers: Studies of the !Kung San and their Neighbors.* Cambridge, MA: Harvard University Press.

Lee, Richard and Jane B. Lancaster. 1965. The annual reproductive cycle in monkeys and apes. In DeVore 1965a:486–513.

Leibowitz, Lila. 1975. Perspectives on the evolution of sex differences. In Reiter 1975:22–35.

Leibowitz, Lila. 1978. *Females, Males, Families: A Biosocial Approach.* Belmont, CA: Duxbury.

Leibowitz, Lila. 1983. Origins of the sexual division of labor. In Lowe and Hubbard 1983:123–47.

Lewin, Roger. 1987a. *Bones of Contention: Controversies in the Search for Human Origins.* New York: Simon & Schuster.

Lewin, Roger. 1987b. My close cousin the chimpanzee. *Science* 238:273–75.

Lewontin, R.C. 1978. Adaptation. *Scientific American* 239:212–30.

Lewontin, R.C. 1979. Sociobiology as an adaptationist program. *Behavioral Science* 24:5–14.

Lewontin, R.C., Steven Rose, and Leon J. Kamin. 1984. *Not in Our Genes: Biology, Ideology, and Human Nature.* New York: Pantheon.

Lilienfeld, Robert. 1978. *The Rise of Systems Theory.* New York: Wiley.

Lindburg, Donald G. 1967. A field study of the reproductive behavior of the rhesus monkey. Ph.D. thesis, University of California Berkeley.

Lindburg, Donald G. 1971. The rhesus monkey in north India: An ecological and behavioral study. In L.A. Rosenblum, ed. *Primate Behavior*, Vol. 2. New York: Academic Press, pp. 2–106.

Lindburg, Donald G. 1975. Mate selection in the rhesus monkey, *Macaca mulatta.* Paper presented at American Association of Physical Anthropology meetings, Denver.

Linden, Eugene. 1976. *Apes, Men, and Language.* New York: Penguin.

Linden, Eugene. 1986. *Silent Partners: The Legacy of the Ape Language Experiments.* New York: Times Books.

Linton, Ralph. 1945. *The Sciences of Man in the World Crisis.* New York: Appleton-Century.

Linton, Sally. 1970. See Sally Slocum, 1975.

Lloyd, Genevieve. 1984. *The Man of Reason: "Male" and "Female" in Western Philosophy.* Minneapolis: University of Minnesota Press.

Lloyd, Elizabeth. n.d. On the primate female orgasm. Unpublished manuscript, University of California at San Diego

Long, Diana E. 1987a. Physiological identity of American sex researchers between the two world wars. *Physiology in the American Context, 1850–1940.* American Physiological Society, pp. 263–78.

Long, Diana E.   1987b. Endocrinology in American medicine. Unpublished manuscript.

Long, Diane E.   1988. Medical bibliography in the "Golden Age" of American medicine: Subject headings concerning sex in the *Index Catalogue of the Library of the Surgeon General's Office*, series I–V, 1880–1961. Paper prepared for the Wenner Gren conference, Analysis in Medical Anthropology, Lisbon, Portugal, March 5–13, 1988.

Longino, Helen and Ruth Doell.   1983. Body, bias, and behavior: A comparative analysis of reasoning in two areas of biological science. *Signs* 9(2):206–27.

Lorenz, Konrad.   1957. The concept of instinctive behavior. In Claire Schiller, translator and editor. *Instinctual Behavior: The Development of a Modern Concept*. New York: International Publishers.

Lorenz, Konrad.   1966. *On Aggression*. Translated by M.K. Wilson New York: Harcourt, Brace & World.

Lorenz, Konrad.   1971. *Studies in Human and Animal Behavior*. 2 Vols. Translated by R. Merton. Cambridge, MA: Harvard University Press.

Losos, Elizabeth.   1985. Monkey see, monkey do: Primatologists' conceptions of primate societies. Senior thesis, Harvard University.

Lotka, A.J.   1925. *Elements of Physical Biology*. Baltimore: Williams & Wilkins; reprinted as *Elements of Mathematical Biology*. New York: Dover, 1956.

Lovejoy, Arthur O.   1960 [1936]. *The Great Chain of Being: A Study of the History of an Idea*. New York: Harper Torchbooks.

Lovejoy, Owen.   1981. The origin of man. *Science* 211(4480):341–50.

Lovejoy, Owen.   1984. The natural detective. *Natural History* 93(10):24–28.

Lowe, Donald.   1982. *The History of Bourgeois Perception*. Chicago: University of Chicago Press.

Lowe, Marian and Ruth Hubbard, eds.   1983. *Woman's Nature: Rationalizations of Inequality*. New York: Pergamon.

Ludmerer, Kenneth.   1972. *Genetics and American Society*. Baltimore: Johns Hopkins University Press.

M

McCann, Carole Ruth.   1987. Race, class and gender in U.S. birth control politics. Ph.D. thesis, University of California Santa Cruz.

MacCormack, Carol and Marilyn Strathern, eds.   1980. *Nature, Culture, Gender*. Cambridge: Cambridge University Press.

McCullough, David.   1981. *Mornings on Horseback*. New York: Simon & Schuster.

McGreal, Shirley.   1981. Monkeys go to war. *Mainstream* (Winter), vol. 6.

McGrew, William.   1979. Evolutionary implications of sex differences in chimpanzee predation and tool use. In Hamburg & McCown 1979:440–63.

McGuire, M.T.   1974. The St. Kitts vervet. *Journal of Medical Primatology* 3:285–97.

McGuire, Wayne, as told to Beverly Trainer and Gina Maranto.   1987, February. "I didn't kill Dian. She was my friend," *Discover* 8(2): 28–48.

McIntosh, Robert P.   1977. Ecology since 1900. In Benjamin J. Taylor and Thurman J. White, eds. *Issues and Ideas in America*. Norman: University of Oklahoma Press.

McIntyre, Vonda.   1983. *Superluminal*. New York: Simon & Schuster.

McKay, Mary.   1980. Akeley, Mary Lee Jobe, 1978–1966. *Notable American Women: The Modern Period*. Cambridge, MA: Harvard University Press, pp. 8–10.

McKinney, W.T., Stephen Suomi, and Harry Harlow.   1971. Depression in primates. *American Journal of Psychiatry* 127:1313–20.

MacKinnon, Catharine A.   1982. Feminism, marxism, method and the state: An agenda for theory. *Signs* 7(3):515–44.

MacKinnon, Catharine A.   1987. *Feminism Unmodified: Discourses on Life and Law*. Cambridge, MA: Harvard University Press.

Macphearson, C.B. 1962. *The Political Theory of Possessive Individualism*. New York: Oxford University Press.

Malinowski, Bronislaw. 1927. *Sex and Repression in Savage Society*. London: Kegan Paul, Trench, Trubner.

Malinowski, Bronislaw. 1944. *A Scientific Theory of Culture and Other Essays*. Chapel Hill: University of North Carolina Press.

Mani, Lata. 1987. The construction of women as tradition in early nineteenth-century Bengal. *Cultural Critique*, no. 7:119–56.

Manning, Kenneth. 1983. *Black Apollo of Science*. New York: Oxford University Press.

Marais, Eugene. 1926. Baboons, hypnosis, and insanity. *Psyche* 7:104–10.

Marais, Eugene. 1939. *My Friends the Baboons*. London: Blond & Briggs.

Marais, Eugene. 1969. *The Soul of the Ape*. New York: Atheneum.

Marais, Eugene. 1980. Soul of the white ant. South Africa Radio Broadcasting.

Maran, Rita. 1987. Torture during the French-Algerian War and the role of the 'mission civiliacitrice.' Ph.D. thesis, University of California Santa Cruz.

Marler, Peter. 1976. Social organization, communication and graded signals: the chimpanzee and the gorilla. In P.G. Bateson and R.A. Hinde, eds. *Growing Points in Ethology*. New York: Cambridge University Press, pp. 239–80.

Martin, Bob and Paul H. Harvey. 1987. Human bodies of evidence. *Nature* 330:697–98.

Martin, Emily. 1987. *The Woman in the Body: A Cultural Analysis of Reproduction*. Boston: Beacon Press.

Martin, M. Kay and Barbara Voorhies. 1975. *Female of the Species*. New York: Columbia University Press.

Marx, Leo. 1964. *The Machine in the Garden*. London: Oxford University Press.

Maslow, Abraham H. 1936a. The role of dominance in the social and sexual behavior of infra-human primates. I. Observations at the Villas Park Zoo. *Journal of Genetic Psychology* 48:278–309.

Maslow, Abraham H. 1936b. The role of dominance in the social and sexual behavior of infra-human primates. III. A theory of sexual behavior of infra-human primates. *Journal of Genetic Psychology* 48:310–38.

Maslow, Abraham H. 1936c. The role of dominance in the social and sexual behavior of infra-human primates. IV. The determination of hierarchy in pairs and in a group. *Journal of Genetic Psychology* 49:161–98.

Maslow, Abraham H. 1940. Dominance-quality and social behavior in infra-human primates. *Journal of Social Psychology* 11: 313–24.

Maslow, Abraham H. 1943. The authoritarian character structure. *Journal of Social Psychology* 18:401–411.

Maslow, Abraham H. 1954. *Motivation and Personality*. New York: Harper.

Maslow, Abraham H. 1966. *The Psychology of Science: A Reconaissance*. New York: Harper & Row.

Mason, William A. and R.R. Sponholz. 1963. Behavior of rhesus monkeys raised in isolation. *Psychiatric Research* 1:1–8.

Masters, William H. and Virginia Johnson. 1966. *Human Sexual Response*. Boston: Little, Brown.

May, Mark and Leonard Doob. 1937. *Competition and Cooperation*, A report to the Sub-Committee on Competitive-Cooperative Habits, of the Committee on Personality and Culture, based on analyses of research achievement and opportunity by members of the Sub-Committee (Gordon Allport, Gardner Murphy, Mark May). New York: Social Science Research Council, Bulletin No. 25.

Maynard-Smith, J. 1964. Group selection and kin selection. *Nature* 201:1145–47.

Maynard-Smith, J. 1978. *The Evolution of Sex*. Cambridge: Cambridge University Press.

Mayr, Ernst. 1942. *Systematics and the Origin of Species*. New York: Columbia University Press.

Mayr, Ernst. 1950. Taxonomic categories in fossil hominids. *Cold Spring Harbor Symposium in Quantitative Biology* 15:109–18.

Mayr, Ernst. 1975. Epilogue on American ornithology. In Erwin Stresemann, ed. *Ornithology: From Aristotle to the Present*. Cambridge, MA: Harvard University Press, pp. 365–96.

Mayr, Ernst. 1982a. Reflections on human paleontology. In Spencer 1982:231–38.

Mayr, Ernst. 1982b. *The Growth of Biological Thought: Diversity, Evolution, and Inheritance*. Cambridge, MA: Harvard University Press.

Mayr, Ernst and William B. Provine, eds. 1980. *The Evolutionary Synthesis: Perspectives on the Unification of Biology*. Cambridge, MA: Harvard University Press.

Mead, Margaret. 1923. *Coming of Age in Samoa*. New York: Morrow.

Means, Russell. 1980. Fighting words on the future of the earth. *Mother Jones*, December, pp. 22–38.

Medawar, Peter B. 1960. *The Future of Man*. London: Methuen.

Memmi, Albert. 1967. *The Colonizer and the Colonized*. Boston: Beacon.

Merchant, Carolyn. 1980. *The Death of Nature: Women, Ecology, and the Scientific Revolution*. New York: Harper & Row.

Miller, George A. 1951. *Language and Communication*. New York: McGraw-Hill.

Miller, George A., F.M. Wiener, and S.S. Stevens. 1946. *Transmission and Reception of Sounds under Combat Conditions*. Summary Technical Report of the Division 17, Section 3, NDRC. Washington, DC: NDRC.

Miller, G.S. 1931. The primate basis of human sexual behavior *Quarterly Review of Biology* 6: 379–410.

Miller, Warren B. and Lucile F. Newman, eds. 1978. *The First Child and Family Formation*. Chapel Hill: University of North Carolina Population Center.

Mills, C. Wright. 1956. *The Power Elite*. New York: Oxford University Press.

Milton, Katie. 1980. *The Foraging Strategies of Howler Monkeys: A Study in Primate Economics*. New York: Columbia University Press.

Mitchison, Naomi. 1985 [1962]. *Memoirs of a Spacewoman*. London: The Women's Press.

Mitmann, Gregg. 1988. From the population to society. *Journal of the History of Biology* 21(2):173–94.

Mittermeier, Russell A. and Dorothy Cheney. 1987. Conservation of primates and their habitats. In Smuts et al. 1987:477–90.

Mixon, Veronica. 1979, April. Futurist woman: Octavia Butler. *Essence* 9:12, 15.

Miyade, Denzaburo. 1964. Report on the activity of the Japan-India Project in primate investigation. *Primates* 5(3–4):1–6.

Mohanty, Chandra Talpade. 1984. Under western eyes: Feminist scholarship and colonial discourse. *Boundary* 2/3 (nos. 12 & 13):333–58.

Mohnot, S.M. 1974. Ecology and behavior of the common Indian langur, *Presbytis entellus entellus* Duforisne. Ph.D. thesis, University of Jodhpur.

Mohnot, S.M. and M.L. Roonwal. 1977. *Primates of South Asia*. Cambridge, MA: Harvard University Press.

Montagu, M.F. Ashley. 1942. The genetical theory of race, and anthropological method. *American Anthropologist* 44:369–75.

Montagu, M.F. Ashley. 1945. Sociometric methods in anthropology. *Sociometry* 8:62–63.

Montagu, M.F. Ashley. 1952. *Man's Most Dangerous Myth: The Fallacy of Race*. New York: Harper.

Montagu, M.F. Ashley. 1965. *The Idea of Race*. Lincoln: University of Nebraska Press.

Montagu, M.F. Ashley, ed. 1968. *Man and Aggression*. Oxford: Oxford University Press.

Moraga, Cherrie. 1983. *Loving in the War Years*. Boston: Southend.

Moraga, Cherrie and Gloria Anzaldua, eds. 1981. *This Bridge Called My Back*. Watertown, MA: Persephone.

Morawski, Jill G.   1986. Organizing knowledge and behavior at Yale's Institute of Human Relations. *Isis* 77:219–42.

Morawski, Jill G.   1987. The troubled quest for masculinity, feminity, and androgyny. *Review of Personality and Social Psychology* 7:44–69.

Morbeck, Mary Ellen, E.H. Preuschoft, and Neil Gomberg, eds.   1979. *Environment, Behavior, and Morphology: Dynamic Interactions in Primates*. New York: Fischer.

Morell, Virginia.   1986, April. Field science and death in Africa. *Science 86*:17–21.

Moreno, J. L.   1945. Two sociometries, human and subhuman. *Sociometry* 8:64–75.

Morgan, Elaine.   1972. *The Descent of Woman*. New York: Stein & Day.

Morgan, Robin.   1984. *Sisterhood Is Global: The International Women's Movement Anthology*. Garden City, New York: Anchor.

Morris, Charles.   1938. *Foundation of the Theory of Signs*. Chicago: University of Chicago Press.

Morris, Desmond.   1962. *The Biology of Art*. London: Methuen.

Morris, Desmond.   1967a. *The Naked Ape*. New York: McGraw-Hill.

Morris, Desmond, ed.   1967b. *Primate Ethology*. London: Weidenfeld & Nicolson.

Morris, Desmond.   1969. *The Human Zoo*. New York: McGraw-Hill.

Moskowitz, Milton, Michael Katz, and Robert Levering.   1980. *Everybody's Business: The Irreverent Guide to Corporate America*. San Francisco: Harper & Row.

Moss, Cynthia.   1982 [1975]. *Portraits in the Wild*. Chicago: University of Chicago Press.

Moss, Cynthia.   1988. *Elephant Memories*. New York: Morrow.

Mowat, Farley.   1987. *Woman in the Mists: The Story of Dian Fossey and the Mountain Gorillas of Africa*. New York: Warner Books.

Moynihan, Daniel Patrick.   1965. *The Negro Family: the Case for National Action*. Washington, D.C.: U.S. Government Printing Office, Office of Policy Planning and Research of the Department of Labor.

Mulvey, Laura.   1975. Visual pleasure and narrative cinema. *Screen* 16(3):6–18.

Murdock, George.   1968 [1945]. The common denominator of cultures. In Washburn and Jay 1968:230–45.

Murrell, K. H. F.   1965. *Ergonomics*. London: Chapman & Hall.

N

Nakane, Chie.   1982. Becoming an anthropologist. In Derek Richter, ed. *Women Scientists: The Road to Liberation*. London: Macmillan, pp.45–60.

Napier, John R. and N.A. Barnicot, eds.   1963. *The Primates, Symposium of the London Zoological Society*, no. 10.

Napier, John R. and P.H. Napier.   1967. *A Handbook of Living Primates*. New York: Academic Press.

Nash, Leanne Taylor.   1978. The development of the mother-infant relationship in wild baboons (*Papio anubis*). *Animal Behaviour* 26:746–59.

Nash, Roderick, ed.   1976. *American Environment: Readings in the History of Conservation*, 2nd ed. Reading, MA: Addison-Wesley.

Nash, Roderick.   1977. The exporting and importing of nature: Nature-appreciation as a commodity, 1850–1980. *Perspectives in American History* XII: 517–60.

Nash, Roderick, ed.   1979. *Environment and the Americas: Problems and Priorities*. Melbourne, FL: Krieger.

Nash, Roderick.   1982. *Wilderness and the American Mind*, 3rd ed. New Haven: Yale University Press.

Naureckas, Jim.   1986. Signs of simian distress falling on deaf ears. *In These Times* (May 28–June 10):18.

Needham, Joseph.   1936. *Order and Life*. New Haven: Yale University Press.

Nesbit, William.   1926. *How to Hunt with the Camera*. New York: Dutton.

New Internationalist Cooperative.   1985. *Women, a World Report*. London: Oxford University Press.

Nice, Margaret M.   1937. Studies on the life history of the song sparrow, vol. I: A population study of the song sparrow. *Transactions of the Linnaean Society* 4(6): 1–247.

Nicolson, Nancy.   1982. Weaning and the development of independence in olive baboons. Ph. D. thesis, Harvard University.

Nielson, Waldemar A.   1972. *The Big Foundations*. New York: Columbia University Press.

Nishida, Toshisada.   1968. The social group of wild chimpanzees in the Mahale Mountains. *Primates* 9:167–224.

Nishida, Toshisada.   1972. Preliminary information on the pygmy chimpanzees (*Pan paniscus*) of the Congo Basin. *Primates* 13:415–25.

Nishida, Toshisada.   1979. The social structure of chimpanzees in the Mahale Mountains. In Hamburg and McCown 1979:73–121.

Nissen, H.W.   1931. A field study of the chimpanzee. *Comparative Psychology Monographs* 8:1–122.

Noble, David F.   1977. *America by Design: Science, Technology and the Rise of Corporate Capitalism*. New York: Knopf.

Nolte, Angela.   1954. Field observations on the daily routine and social behavior of common Indian monkeys, with special reference to the bonnet macaque. *Journal of the Bombay Natural History Society* 55(2):177–84.

O

Oakley, Kenneth.   1972 [1954]. Skill as a human possession. In Washburn & Dolhinow 1972:14–50.

O'Brien, Mary.   1981. *The Politics of Reproduction*. New York: Routledge & Kegan Paul.

Odum, Eugene P.   1977. The emergence of ecology as a new integrative discipline. *Science* 195: 1289–93.

Ohnuki-Tierney, Emiko.   1987. *The Monkey as Mirror: Symbolic Transformations in Japanese History and Ritual*. Princeton: Princeton University Press.

Olds, Elizabeth Fagg.   1985. *Women of the Four Winds: The Adventures of Four of America's First Women Explorers*. Boston: Houghton-Mifflin.

Oleson, Alexandra and John Voss, eds.   1979. *The Social Organization of Knowledge in Modern America, 1860–1920*. Baltimore: Johns Hopkins University Press.

Ortner, Sherry B.   1972. Is female to male as nature is to culture? *Feminist Studies* 1:5–31. Reprinted Rosaldo and Lamphere 1974:7–88.

Ortner, Sherry B. and Harriet Whitehead, eds.   1981. *Sexual Meanings: The Cultural Construction of Gender and Sexuality*. Cambridge: Cambridge University Press.

Osborn, Henry Fairfield.   1930. *Fifty-two Years of Research, Observation, and Publication*. New York: American Museum of Natural History.

P

Parascondola, John. 1971.   Organismic concepts in the thought of L.J. Henderson. *Journal of the History of Biology* 4:63–114.

Parsons, Talcott.   1970. On building social system theory. *Daedalus* 99(4):826–881.

Patterson, Francine.   1978, October. Conversations with a gorilla. *National Geographic* 154:438–65.

Patterson, Francine and Eugene Linden.   1981. *The Education of Koko*. New York: Holt, Rinehart & Winston.

Pauly, Philip.   1976. Social Pathology: American Psychologists and the Treatment of Nazi Germany. Unpublished manuscript.

Pauly, Philip.   1979. The world and all that is in it, the National Geographic Society, 1888–1918. *American Quarterly* 31(4):517–32.

Pearl, Mary. 1982. Networks of social relations among Himalayan rhesus monkeys (*Macaca mulatta*). Ph. D. thesis, Yale University.

Pearl, Raymond. 1930. Some aspects of the biology of human populations. In Cowdry 1930.

Petchesky, Rosalind. 1987. Fetal images: The power of visual culture in the politics of reproduction. *Feminist Studies* 13(2):263–92.

Pickens, Donald. 1968. *Eugenics and the Progressives*. Nashville: Vanderbilt University Press.

Piercy, Marge. 1976. *Woman on the Edge of Time*. New York: Knopf.

Pietz, William. 1983. The phonograph in Africa: International phonocentrism from Stanley to Sarnoff. Unpublished paper from the Second International Theory and Text Conference, Southampton, England.

Pilbeam, David. 1980. Major trends in human evolution. In L.-K. Konigsson, ed. *Current Argument on Early Man*. New York: Pergamon, pp. 261–85.

Poirier, Frank, ed. 1972. *Primate Socialization*. New York: Random House.

Politzer, William S. 1981. The development of genetics and population studies. *American Journal of Physical Anthropology* 56:483–89.

Porges, Irwin. 1975. *Edgar Rice Burroughs: The Man Who Created Tarzan*. Provo, Utah: Brigham Young University Press.

Porter, Charlotte. n.d. The rise to Parnassus: Henry Fairfield Osborn and the Hall of the Age of Man. Unpublished manuscript, Smithsonian Institution.

Potts, Richard. 1984. Home bases and early hominids. *American Scientist* 72:338–47.

Pratt, Mary Louise. 1985. Scratches on the face of the country; or, what Mr. Barrow saw in the land of the Bushmen. *Critical Inquiry* 12(1):119–43.

Pratt, Mary Louise. 1986, May. A reading of *The Gods Must Be Crazy*. Unpublished talk for the Group for the Critical Study of Colonial Discourse. University of California Santa Cruz.

Premack, A.J. and D. Premack. 1972. Teaching language to an ape. *Scientific American* 227:92–99.

Provine, William B. 1971. *The Origins of Theoretical Population Genetics*. Chicago: University of Chicago Press.

Pusey, Anne E. 1977. The physical and social development of wild adolescent chimpanzees. Ph.D. thesis, Stanford University.

Pusey, Anne E. 1979. Intercommunity transfer of chimpanzees in Gombe National Park. In Hamburg & McCown 1979:465–80.

R

Rabinow, Paul. 1986. Representations are social facts: Modernity and postmodernity in anthropology. In Clifford and Marcus 1986:234–61.

Radcliffe-Brown, A.R. 1937. *A Natural Science of Society*. Chicago: University of Chicago Press.

Radcliffe-Brown, A.R. 1952. *Structure and Function in Primitive Society*. New York: Free Press.

Ralls, Katherine. 1976. Mammals in which females are larger than males. *Quarterly Review of Biology* 51:245–76.

Ralls, Katherine. 1977. Sexual dimorphism in mammals: Avian models and unanswered questions. *American Naturalist* 111:917–38.

Ransom, Timothy W. 1971. Ecology and social behavior of baboons in Gombe Stream National Park. Ph.D. thesis, University of California Berkeley.

Ransom, Timothy W. and Bonnie Ransom. 1971. Adult male-infant relations among baboons (*Papio anubis*). *Folia Primatologica* 16:179–95.

Rapp, Rayna. 1987. Constructing amniocentesis: Maternal and medical discourses. Paper delivered at the meetings of the American Anthropological Association, Philadelphia.

Raven, H.C. 1933. Further adventures of Meshie: A chimpanzee that has lived most of her life in a New York suburban home. *Natural History* 33:607–17.

Redfield, Robert, ed. 1942. *Levels of Integration in Biological and Social Systems.* Lancaster, PA: Jacques Cattell Press.

Reed, Evelyn. 1975. *Woman's Evolution.* New York: Pathfinder.

Reed, James. 1978. *From Private Vice to Public Virtue: The Birth Control Movement and American Society since 1830.* New York: Basic Books.

Reite, Martin and Nancy Caine, eds. 1983. *Child Abuse: The Nonhuman Primate Data.* New York: Alan Liss.

Reiter, Rayna Rapp, ed. 1975. *Toward an Anthropology of Women.* New York: Monthly Review Press.

Reynolds, Vernon. 1967. *The Apes.* New York: Dutton.

Reynolds, Vernon and Frances Reynolds. 1965. Chimpanzees in the Budongo Forest. In DeVore 1965a:368–424.

Rheingold, Harriet, ed. 1963. *Maternal Behavior in Mammals.* New York: Wiley.

Ribnick, Rosalind. 1982. A short history of primate field studies: Old World monkeys and apes. In Spencer 1982:49–73.

Rich, Adrienne. 1980. Compulsory heterosexuality and lesbian existence. *Signs* 5(4):631–60.

Richard, Alison F. 1978. *Behavioral Variation: Case Study of a Malagasy Lemur.* Lewisburg, PA: Bucknell University Press.

Richard, Alison F. 1981. Changing assumptions in primate ecology. *American Anthropology* 83:517–33.

Richard, Alison F. 1985. *Primates in Nature.* New York: Freeman.

Richard, Alison F. 1987. Malagasy prosimians: Female dominance. In Smuts et al. 1987: 25–33.

Richards, Robert J. 1974. The innate and the learned: The evolution of Konrad Lorenz's theory of instinct. *Philosophy of the Social Sciences* 4: 111–33.

Riley, Denise. 1983. *War in the Nursery: Theories of the Child and Mother.* London: Virago.

Rioch, David and Edwin A. Weinstein, eds. 1964. *Disorders of Communication.* Baltimore: William & Wilkins.

Riopelle, Arthur. 1967, March. "Snowflake," the world's first white gorilla. *National Geographic* 131(3):443–48.

Riopelle, Arthur. 1970, October. Growing up with Snowflake. *National Geographic* 138(4):491–502.

Robinson, John G. and Charles H. Janson. 1987. Capuchins, squirrel monkeys, and atelines: Socioecological convergence with Old World primates. In Smuts et al. 1987:69–82.

Robinson, John G., Patricia C. Wright, and Warren G. Kinzey. 1987. Monogamous cebids and their relatives: Intergroup calls and spacing. In Smuts et al. 1987:44–53.

Roe, Anne and George Gaylord Simpson, eds. 1958. *Behavior and Evolution.* New Haven: Yale University Press.

Rosaldo, Michelle Z. 1980. The use and abuse of anthropology: Reflections on feminism and cross-cultural understanding. *Signs* 5(3):389–417.

Rosaldo, Michelle Z. and Louise Lamphere, eds. 1974. *Woman, Culture, and Society.* Palo Alto: Stanford University Press.

Rose, Hilary. 1983. Hand, brain, and heart: A feminist epistemology for the natural sciences. *Signs* 9:73–90.

Rosenberg, Rosalind. 1982. *Beyond Separate Spheres: Intellectual Roots of Modern Feminism.* New Haven: Yale University Press.

Rosenfelt, Deborah and Judith Stacey. 1987. Second thoughts on the second wave. *Feminist Studies* 13(2):341–62.

Rossiter, Margaret. 1982. *Women Scientists in America: Struggles and Strategies to 1914.* Baltimore: Johns Hopkins University Press.

Rothmann, M. 1912. Uber die Errichtung einer Station zur psychologischen und hirnphysiologischen Erforschung der Menschenaffen. *Berlin klinische Wschr.* 49:1981–85.

Rowell, Thelma. 1966. Forest living baboons in Uganda. *Journal of Zoology* 149:344–64.

Rowell, Thelma. 1972. *The Social Behaviour of Monkeys.* Middlesex, England: Penguin Books.

Rowell, Thelma. 1974. The concept of dominance. *Behavioral Biology* 11:131–54.

Rowell, Thelma. 1984. Introduction. In Small 1984:13–16.

Rowell, Thelma. 1988. Monkey business. *Natural History* 97(1):58–60.

Rowland, Guy. 1964. The effects of total social isolation upon learning and social behavior in the rhesus monkeys. Ph.D. thesis, University of Wisconsin.

Rubin, Gayle. 1975. The traffic in women: Notes on the 'political economy' of sex. In Reiter 1975:157–210.

Ruch, Theodore C. 1941. *Bibliographia Primatologica.* Introduction by J. Fulton. Springfield and Baltimore: Charles C. Thomas.

Ruddick, Sara. 1980. Maternal thinking. *Feminist Studies* 6(3): 343–67.

Rumbaugh, D.M., ed. 1977. *Language Learning by a Chimpanzee: The Lana Project.* New York: Academic Press.

Russ, Joanna. 1983. *How to Suppress Women's Writing.* Austin: University of Texas Press.

Russett, Cynthia. 1966. *The Concept of Equilibrium in American Social Thought.* New Haven: Yale University Press.

S

Sade, Donald S. 1965. Some aspects of parent-offspring and sibling relations in a group of rhesus monkeys, with a discussion of grooming. *American Journal of Physical Anthropology* 23(1):1–17.

Sade, Donald S. 1967. Determinants of dominance in a group of free-ranging rhesus monkeys. In S.A. Altmann 1967:99–114.

Said, Edward. 1978. *Orientalism.* New York: Pantheon.

Salvaggio, Ruth. 1986. Octavia E. Butler. In Marleen Barr, Ruth Salvaggio, and Richard Law. *Suzy McKee Charnas/Octavia Butler/Joan D. Vinge.* Mercer Island, WA: Starmont House.

Sarich, Vincent M. 1971. A molecular approach to the study of human origins. In Phyllis Dolhinow and Vincent M. Sarich, eds. *Background for Man.* Boston: Little, Brown, pp. 60–81.

Sarich, Vincent M. and J.E. Cronin. 1976. Molecular systematics of the primates. In M. Goodman and R.E. Tashian, eds. *Molecular Anthropology.* New York: Plenum, pp. 141–70.

Sarich, Vincent M. and A.C. Wilson. 1967. Immunological time scale for hominid evolution. *Science* 158:1200–03.

Savage-Rumbaugh, E. Sue. 1986. *Ape Language from Conditioned Response to Symbol.* London: Oxford University Press.

Savage-Rumbaugh, E. Sue, D.M. Rumbaugh, and S. Boysen. 1978. Symbolic communication between two chimpanzees (*Pan troglodytes*). *Science* 201:641–44.

Sayers, Janet. 1982. *Biological Politics: Feminist and Anti-Feminist Perspectives.* London: Tavistock.

Schaller, George B. 1963. *The Mountain Gorilla: Ecology and Behavior.* Chicago: University of Chicago Press.

Schaller, George B. 1964. *The Year of the Gorilla.* Chicago: University of Chicago Press.

Schaller, George B. 1967. *The Deer and the Tiger: A Study of Wildlife in India.* Chicago: University of Chicago Press.

Schaller, George B. 1972. *The Serengeti Lion.* Chicago: University of Chicago Press.

Schaller, George B., Hu Jinchu, Pan Wenshi, and Zhu Jing. 1984. *The Giant Pandas of Woolong.* Chicago: University of Chicago Press.

Schiebinger, Londa. 1986. Skeletons in the closet: The first illustrations of the female skeleton in eighteenth-century anatomy. *Representations* 14:42–82.

Schiebinger, Londa. 1987. The history and philosophy of women in science: A review essay. *Signs* 12(2):305–32.

Schjelderup-Ebbe, Thorlief. 1922. Beitrage zur Sozialpsychologie des Haushuhns. *Zeitschrift zur Psychologie* 88:225–52.

Schneider, David. 1984. *A Critique of the Study of Kinship*. Ann Arbor: University of Michigan Press.

Schneirla, T.C. 1950. The relationship between observation and experimentation in the field study of behavior. *Annals of the New York Academy of Sciences* 51:1022–44.

Schrier, Allen M., H.F. Harlow, and F. Stolnitz, eds. 1965. *The Behavior of Nonhuman Primates*, I & II. New York: Academic Press.

Schultz, A.H. 1969. *The Life of Primates*. London: Weidenfeld & Nicolson.

Schultz, A.H. 1971. The rise of primatology in the twentieth century. *Proceedings of the Third International Congress of Primatology* (Zurich, 1970) 1:2–15.

Scott, Joan. 1986. Gender: A useful category of historical analysis. *The American Historical Review* 91(5):1053–75.

Scott, J.P. 1950. Methodology and techniques for the study of animal societies. *Annals of the New York Academy of Sciences* 51 (Art. 6):1004.

Seager, Joni and Ann Olson. 1986. *Women in the World Atlas*. New York: Simon & Schuster.

Seay, Billy M. and Harry Harlow. 1966. Mothering in motherless mother monkeys. *British Journal of Social Psychiatry* 1:63–69.

Sebeok, Thomas. 1967. Discussion of communication processes. In S. A. Altmann 1967:363–70.

Sebeok, Thomas and Ann Umiker Sebeok, eds. 1980. *Speaking of Apes*. New York: Plenum.

Selye, Hans. 1956. *The Stress of Life*. New York: McGraw-Hill.

Senn, P.R. 1966. What is behavioral science?—Note toward a history. *Journal of the History of the Behavioral Sciences* 1:107–203.

Seyfarth, Robert M. 1976. Social relationships among adult female baboons. *Animal Behaviour* 24:917–38.

Shannon, Claude E. and Warren Weaver. 1949. *The Mathematical Theory of Communication*. Urbana: University of Illinois Press.

Shapiro, Harry L. 1981. Earnest A. Hooton, 1887–1954, *in memoriam cum amore*. *American Journal of Physical Anthropology* 56:431–34.

Sheldon, Racoona. 1985. Morality meat. In Jen Green and Sarah Lefanu, eds. *Despatches from the Frontiers of the Female Mind*. London: The Women's Press.

Shelford, Victor E. 1913. *Animal Communities in Temperate America*. Chicago: University of Chicago Press.

Sherfy, Mary Jane. 1966. *The Nature and Evolution of Female Sexuality*. New York: Random House.

Shelley, Mary. 1818. *Frankenstein*. London: Lackington, Hughes, Harding, Mavor, and Jones.

Shostack, Marjorie. 1981. *Nisa. The Life and Words of a !Kung Woman*. Cambridge, MA: Harvard University Press.

Shoumatoff, Alex. 1986, September. The fatal obsession of Dian Fossey. *Vanity Fair* 49(9):84–90, 130–38.

Siddiqi, M. Farooq and Charles H. Southwick. 1977. Population trends of rhesus monkeys in Aligarh district. In M.R.N. Prasad and T C. Anand-Kumar, eds. *Use of Nonhuman Primates in Biomedical Research*. New Delhi: Indian National Sciences Academy, pp. 15–23.

Silk, Joan. 1978. Patterns of food-sharing among mother and infant chimpanzees at Gombe National Park, Tanzania. *Folia Primatologica* 29(2):129–41.

Simonds, Paul. 1963. Ecology of bonnet macaques. Ph.D. thesis, University of California Berkeley.

Simpson, George Gaylord. 1944. *Tempo and Mode in Evolution*. New York: Columbia University Press.

Simpson, George Gaylord. 1953. *The Major Features of Evolution*. New York: Columbia University Press.

Simpson, George Gaylord. 1958. "The study of evolution: Methods and present status of theory" and "Behavior and evolution". In Roe and Simpson 1958: 7–26, 507–36.

Skybreak, Ardea. 1984. *Of Primeval Steps and Future Leaps. An Essay on the Emergence of Human Beings, the Source of Women's Oppression, and the Road to Emancipation*. Chicago: Banner Press.

Slocum, Sally. 1975. Woman the gatherer: Male bias in anthropology. In Reiter 1975:36–50. Originally published as Sally Linton. 1971. In Sue-Ellen Jacobs, ed. *Women in Perspective. A Guide for Cross Cultural Studies*. Urbana: University of Illinois Press, pp. 9–21.

Small, Meredith, ed. 1984. *Female Primates: Studies by Women Primatologists*. New York: Allan Liss.

Smith, Robert J. 1983. *Japanese Society: Tradition, Self, and the Social Order*. Cambridge: Cambridge University Press.

Smith-Rosenberg, Carroll and Charles Rosenberg. 1984. The female animal: Medical and biological views of woman and her role in nineteenth-century America. In Judith Walzer Leavitt, ed. *Women and Health in America*. Madison, WI: University of Wisconsin Press, pp. 12–27.

Smuts, Barbara B. 1982. Special Relationships between adult male and female olive baboons (*Papio anubis*). Ph.D. thesis, Stanford University.

Smuts, Barbara B. 1983. Sisterhood is powerful: Aggression, competition, and cooperation in nonhuman primate societies. Manuscript for D. Benton and P.F. Brian, eds. *The Aggressive Female*. Announced but not published, Montreal: Eden.

Smuts, Barbara B. 1985. *Sex and Friendship in Baboons*. Chicago: Aldine.

Smuts, Barbara B. 1987a. Sexual competition and mate choice. In Smuts et al. 1987:385–99.

Smuts, Barbara B. 1987b. Gender, aggression, and influence. In Smuts et al. 1987:400–12.

Smuts, Barbara B., Dorothy L. Cheney, Robert M. Seyfarth, Richard W. Wrangham, and Thomas T. Struhsaker, eds. 1987. *Primate Societies*. Chicago: University of Chicago Press.

Snitow, Ann, Christine Stansell, and Sharon Thompson, eds. 1983. *Powers of Desire: The Politics of Sexuality*. New York: Monthly Review Press.

Sober, E. and Richard Lewontin. 1982. Artifact, cause, and genic selection. *Philosophy of Science* 47:157–80.

Sofia, Zoe. 1984. Exterminating fetuses: Abortion, disarmament and the sexo-semiotics of extra-terrestrialism. *Diacritics* 14(2):47–59. See Zoe Sofoulis.

Sofoulis, Zoe. 1988. Through the lumen: Frankenstein and the optics of re-origination. Ph.D. thesis, University of California Santa Cruz.

Sohn-Rethel, Alfred. 1978. *Intellectual and Manual Labor*. London: Macmillan.

Sontag, Susan. 1977. *On Photography*. New York: Delta.

Southwick, Charles, ed. 1963. *Primate Social Behavior*. Princeton: Van Nostrand.

Southwick, Charles H. and M.A. Beg. 1961. Social behavior of rhesus monkeys in a temple habitat in northern India. *American Zoologist* 1(3):262.

Southwick, Charles H., Thomas Richie, Henry Taylor, H. Jane Teas, and M. Farooq Siddiqi. 1980. Rhesus monkey populations in India and Nepal: Patterns of growth, decline, and natural regulation. In Cohen, Malpass, and Klein 1980:151–70.

Southwick, Charles H. and M. Farooq Siddiqi. 1985, February. Rhesus monkeys' fall from grace. *Natural History* 94(2):63–70.

Spencer, Frank. 1981. The rise of academic physical anthropology in the United States, 1880–1980. *American Journal of Physical Anthropology* 56:353–364.

Spencer, Frank, ed. and intro. 1982. *A History of American Physical Anthropology, 1930–1980*. New York: Academic Press.

Spencer, Herbert. 1852. A theory of population, deduced from the genral law of animal fertility. *Westminster Review*, n.s. 1:468–501.

Spencer-Booth, Yvette and Robert Hinde. 1971. Effects of 6 days separation from mother on 18–32 week old rhesus monkeys. *Animal Behaviour* 19:174–91.

Sperling, Susan. 1988. *Animal Liberators: Research and Morality*. Berekely: University of California Press.

Spillers, Hortense. 1987, Summer. Mama's baby, papa's maybe: An American grammar book. *Diacritics* 17(2):65–81.

Spivak, Gayatri Chakravorty. 1985. Three women's texts and a critique of imperialism. *Critical Inquiry* 12(1):243–61.

Spivak, Gayatri Chakravorty. 1986. Imperialism and sexual difference. *Oxford Literary Review* 8(1,2):225–40.

Stack, Carol. 1974. *All Our Kin: Strategies for Survival in a Black Community*. New York: Harper & Row.

Stacey, Judith. 1987. Sexism by a subtler name? Postindustrial conditions and postfeminist consciousness. *Socialist Review* 17(6):7–30.

Starr, Paul. 1982. *The Social Transformation of American Medicine*. New York: Basic Books.

Stein, D.M. 1981. The nature and function of social interactions between infant and adult male yellow baboons (*Papio cynocephalus*). Ph.D. thesis, University of Chicago.

Stepan, Nancy. 1982. *The Idea of Race in Science: Great Britain, 1800–1960*. Hamden, CN: Shoe String Press.

Stewart, Kelly J. 1981. Social development of wild mountain gorillas. Ph.D. thesis, University of Cambridge.

Stewart, Kelly J. 1988. Chronicles of the Pumphouse Gang. Review of *Almost Human*. *New York Times Book Review* (January 10): 14.

Stocking, George W. 1968. *Race, Culture, and Evolution: Essays in the History of Anthropology*. New York: Free Press.

Stocking, Goerge W. 1979. *Anthropology at Chicago: Tradition, Discipline, Department*. Chicago: University of Chicago.

Stone, Calvin P., ed. 1951. *Comparative Psychology*, 3rd ed. New York: Prentice-Hall.

Strathern, Marilyn. 1980. No nature, no culture: The Hagan case. In Strathern and MacCormmach 1980:174–222.

Strathern, Marilyn. 1986a. John Locke's servant and the hausboi from Hagen: Some thoughts on domestic labour. *Critical Philosophy* (Sydney) 2:21–48.

Strathern, Marilyn. 1986b. An awkward relationship: The case of feminism and anthropology. *Signs* 12:276–92.

Strathern, Marilyn. 1987. Out of context: The persuasive fictions of anthropology. *Current Anthropology* 28:251–81.

Strathern, Marilyn. 1988. *The Gender of the Gift*. Los Angeles: University of California Press.

Struhsaker, Thomas. 1967. Ecology of vervet monkeys (*Cercopithecus aethiops*) in the Masai-Amboseli Game Reserve, Kenya. *Ecology* 48:891–904.

Struhsaker, Thomas. 1971. Social behaviour of mother and infant vervet monkeys (*Cercopithecus aethiops*). *Animal Behaviour* 19:233–50.

Struhsaker, Thomas. 1981. Forest and primate conservation in East Africa. *African Journal of Ecology* 19:99–114.

Strum, Shirley. 1975. Life with the pumphouse gang. New insights into baboon behavior. *National Geographic* 147(5):672–91.

Strum, Shirley. 1976. Predatory behavior of olive baboons (*Papio anubis*) at Gilgil, Kenya. Ph.D. thesis, University of California Berkeley.

Strum, Shirley. 1978. Dominance Hierarchy and Social Organization: Strong or Weak Inference? Unpublished manuscript for the Wenner Gren Conference on Baboon Field Research: Myths and Models.

Strum, Shirley. 1982. Agonistic dominance in male baboons: An alternate view. *International Journal of Primatology* 3:175–202.

Strum, Shirley. 1983a. Use of females by male olive baboons (*Papio anubis*). *American Journal of Primatology* 5:93–109.

Strum, Shirley. 1983b. Why males use infants. In Taub 1983:146–85.

Strum, Shirley. 1987a. The Pumphouse Gang moves to a strange new land. *National Geographic* 172(5):676–90.

Strum, Shirley. 1987b. *Almost Human: A Journey into the World of Baboons.* New York: Random House.

Strum, Shirley and William Mitchell. 1987. Baboon models and muddles. In Kinzey 1987:87–104.

Sugiyama, Yukimaru, K. Yoshiba, and M.W. Parthasarathy. 1965. On the social change of Hanuman langurs (*Presbytis entellus*). *Primates* 6(3–4):381–429.

Suomi, Stephen and Harry Harlow. 1969. Apparatus conceptualization for psychopathological research in monkeys. *Behavior Research Methods and Instrumentation* 1:225–39.

Suomi, Stephen and Helen A. LeRoy. 1982. *In memoriam*: Harry F. Harlow (1905–81). *American Journal of Primatology* 2:319–42.

Suomi, Stephen J. and Chris Ripp. 1983. A history of the motherless monkey mothering at the University of Wisconsin Primate Laboratory. In Reite and Caine 1983:49–78.

Sussman, Randall L, ed. 1979. *Primate Ecology: Problem Oriented Field Studies.* New York: Wiley.

Sussman, Randall L., ed. 1984. *The Pygmy Chimpanzee: Evolutionary Biology and Behavior.* New York: Plenum.

Symons, Donald. 1979. *The Evolution of Human Sexuality.* New York: Oxford University Press.

T

Tanaka J. 1980. *The San Hunter Gatherers of the Kalahari: A Study of Ecological Anthropology.* Tokyo: University Press.

Tanner, Nancy. 1981. *On Becoming Human.* Cambridge: Cambridge University Press.

Tanner, Nancy. 1987. The chimpanzee model revisited and the gathering hypothesis. In Kinzey 1987:3–27.

Tanner, Nancy and Adrienne Zihlman. 1976. Women in evolution. Part 1. Innovation and selection in human origins. *Signs* 1:585–608.

Taub, D.M., ed. 1983. *Primate Paternalism.* New York: Van Nostrand.

Taylor, Joshua. 1979. Malvina Hoffman. *American Art and Antiques* 2 (July/August):96–103.

Taylor, Peter. 1987. Technocratic optimism, H.T. Odum and the partial transformation of ecological metaphor after World War II. Paper presented at the meeetings of the History of Science Society, Pittsburg.

Teleki, Geza. 1973. *The Predatory Behavior of Wild Chimpanzees.* Lewisburg: Bucknell University Press.

Temerlin, Maurice. 1975. *Lucy: Growing Up Human.* Palo Alto, CA: Science and Behavior Books.

Terborgh, John. 1983. *Five New World Primates: A Study in Comparative Ecology.* Princeton: Princeton University Press.

Terrace, Herbert S. 1979. *Nim, a Chimpanzee Who Learned Sign Language.* New York: Knopf.

Thorpe, William. 1979. *The Origins and Rise of Ethology.* London: Heinemann.

Tiger, Lionel and Robin Fox. 1971. *The Imperial Animal.* New York: Holt, Rinehart & Winston.

Tiger, Lionel and Heather Fowler, eds. 1978. *Female Hierarchies.* Chicago: Beresford.

Tiptree, James, Jr. 1973. *Ten Thousand Light Years from Home.* New York: Ace.

Tiptree, James, Jr. 1975. *Warm Worlds and Otherwise.* Introduction by Robert Silverberg. New York: Ballantine.

Tiptree, James, Jr. 1978a. *Star Songs of an Old Primate.* New York: Ballantine.

Tiptree, James, Jr.   1978b. *Up the Walls of the World*. New York: Medallion.

Tiptree, James, Jr.   1981. *Out ᵭf the Everywhere, and Other Extraordinary Visions*. New York: Ballantine.

Tiptree, James, Jr.   1985. *Brightness Falls from the Sky*. New York: Tom Doharty Associates.

Tooby, John and Irven DeVore.   1987. The reconstruction of hominid behavioral evolution through strategic modeling. In Kinzey 1987:183–237.

Tournier, Michel.   1972. *Vendredi ou les limbes du Pacifique*. Paris: Gallimard.

Travis, Carol, ed.   1973.*The Female Experience*. Del Mar, CA: Communications Research Machines, Inc.

Traweek, Sharon.   1988. *Beamtimes and Lifetimes: The World of High Energy Physics*. Cambridge, MA: Harvard University Press.

Trinh, T. Minh-ha.   1988. Not you/like you: Post-colonial women and the interlocking questions of identity and difference. *Inscriptions* (nos. 3–4): 71–77.

Trivers, Robert.   1972. Parental investment and sexual selection.In Campbell 1972:136–79.

Tutin, Carolyn E.G.   1975. Sexual behaviour and mating patterns in a community of wild chimpanzees (*Pan troglodytes schweinfurthii*). Ph.D. thesis, University of Edinburgh.

U

United Nations.   1949 [1948]. *Universal Declaration of Human Rights*. Washington, D.C.: U.S. Department of State.

United Nations.   1975. *UN Plan of Action for the Decade for Women: Equality, Development, and Peace*. New York.

United Nations.   1980. *Report of the World Conference for the Decade for Women*. Copenhagen.

United Nations.   1985. *Report of the World Conference to Review and Appraise the Achievements of the UN Decade for Women* (A/CONF.116/28, July 1985). New York.

United Nations Educational, Scientific and Cultural Organization.   1952. *The Race Concept: Results of an Inquiry*. Paris: UNESCO.

V

Vance, Carol, ed.   1984. *Pleasure and Danger: Exploring Female Sexuality*. Boston: Routledge & Kegan Paul.

Varley, John.   1978. *The Persistence of Vision*. New York: Dell.

Veit, Peter.   1983. A life among gorillas. *Natural History* 92(8):66–69.

Vessels, Jane.   1985. Koko's kitten. *National Geographic* 167(1):110–13.

Virilio, Paul and Sylvere Lotringer.   1983. *Pure War*. Translated by Mark Polizotti. New York: Semiotext(e).

W

de Waal, Frans.   1982. *Chimpanzee Politics: Power and Sex among Apes*. New York: Harper & Row.

Waddington, Conrad Hal, ed.   1970. *Toward a Theoretical Biology*. Vol. 3. Edinburgh: Edinburgh University Press.

Waddington, Conrad Hal.   1977. *Tools for Thought*. London: J. Cape.

van Wagenen, Gertrude.   1972. Vital statistics from a breeding colony: Reproduction and pregnancy in *Macaca mulatta*. *Journal of Medical Primatology* 1:3–28.

van Wagenen, Gertrude and Sophie B.D. Aberle.   1931. Menstruation in *Pithecus (Macacus) rhesus* following bilateral and unilateral ovariectomy performed early in the cycle. *American Journal of Physiology* 99:271–278.

Washburn, Sherwood L.   1946a. The effect of facial paralysis on the growth of the skull of rat and rabbit. *The Anatomical Record* 94(2): 163–68.

Washburn, Sherwood L. 1946b. The effect of removal of the zygomatic arch in the rat. *Journal of Mammalology* 27(2):169–72.

Washburn, Sherwood L. 1951a. The analysis of primate evolution with particular reference to the origin of man. *Cold Spring Harbor Symposia on Quantitative Biology* 15:67–78. In Howells 1962:154–171.

Washburn, Sherwood L. 1951b. The new physical anthropology. *Transactions of the New York Academy of Sciences*, series II, 13(7):298–304. Reprinted in J.D. Jennings and E.A. Hoebel, eds. 1966. *Readings in Anthropology*, 2nd ed. New York: McGraw-Hill, pp. 75–81.

Washburn, Sherwood L. 1952. The strategy of physical anthropology. In A.L. Kroeber, ed. *Anthropology Today*. Chicago:University of Chicago Press, pp. 714–27.

Washburn, Sherwood L., ed. 1961. *The Social Life of Eartly Man*. Viking Fund Publication in Anthropology, No. 31. New York: Wenner-Gren Foundation for Anthropological Research and Chicago: Aldine.

Washburn, Sherwood L., ed. 1963a. *Classification and Human Evolution*. Viking Fund Publication in Anthropology, No. 37. New York: Wenner-Gren Foundation for Anthropological Research and Chicago: Aldine.

Washburn, Sherwood L. 1963b. The curriculum in physical anthropology. In D.G. Mandelbaum, G.W. Lasker, and E.M. Albert, eds. *The Teaching of Anthropology*. Berkeley: University of California Press, pp. 39–47.

Washburn. Sherwood L. 1977. Field study of primate behavior. In Geoffrey Bourne, ed. *Progress in Ape Research*. New York: Academic Press, pp. 231–42.

Washburn, Sherwood L. 1978. Human behavior and the behavior of other animals. *American Psychologist* 33(5):405–18.

Washburn, Sherwood L. 1982. Fifty years of studies on human evolution. *Bulletin of the American Academy of Sciences* XXXV(4):25–39.

Washburn, Sherwood L. 1983. Evolution of a teacher. *Annual Review of Anthropology* 12:1–24.

Washburn, Sherwood L. and Virginia Avis. 1958. Evolution of human behavior. In Roe and Simpson 1958:421–36.

Washburn, Sherwood L. and S.R. Detwiler. 1943. An experiment bearing on the problem of physical anthropology. *American Journal of Physical Anthropology*, n.s., I(2):171–90.

Washburn, Sherwood L. and Irven DeVore. 1961. Social behavior of baboons and early man. In Washburn 1961:91–105.

Washburn, Sherwood L. and Phyllis Dolhinow, eds. 1972. *Perspectives in Human Evolution*, Vol. 2. New York: Holt, Rinehart & Winston.

Washburn, Sherwood L. and David A. Hamburg. 1965. The study of primate behavior. In DeVore 1965a:1–13.

Washburn, Sherwood L., David A. Hamburg, and Naomi Bishop. 1974. Social adaptation in non-human primates. In Coelho et al. 1974:3–12.

Washburn, Sherwood L. and Phyllis Jay, eds. 1968. *Perspectives on Human Evolution*, Vol. 1. New York: Holt, Rinehart & Winston.

Washburn, Sherwood L. and C.S. Lancaster. 1968. The evolution of hunting. In Lee and Irven DeVore 1968:293–303.

Washburn, Sherwood L. and S. Ranieri. 1981. Who brought home the bacon? Review of *Woman the Gatherer*, ed. Frances Dahlberg. *New York Review of Books*, September 24.

Wasser, Samuel K. 1983a. Reproductive competition and cooperation among female yellow baboons. In Wasser 1983:350–90.

Wasser, Samuel K., ed. 1983b. *Social Behavior of Female Vertebrates*. New York: Academic Press.

Weaver, Kenneth F. 1961, May. Countdown for space. *National Geographic* 119(5):702–34.

Weiner, J.S., F.P. Oakley, and W.E. Le Gros Clark. 1953. The solution of the Piltdown problem. *Bulletin of the British Museum of Natural History, Geology* 2(3):141–56.

Weinstein, James. 1969. *The Corporate Ideal in the Liberal State, 1900–1918*. Boston: Beacon.

Werskey, Gary. 1979. *The Visible College*. New York: Holt, Rinehart & Winston.

West, Candance and D.H. Zimmermann. 1987. Doing gender. *Gender and Society* 1(2):125–51.

Western, J. David and Shirley Strum. 1983. Sex, kinship, and the evolution of social manipulation. *Ethology and Sociobiology* 4:19–28.

Westing, Arthur. 1980. *Warfare in a Fragile World: Military Impact on the Human Environment*. Sweden: Stockholm International Peace Research Institute.

White, Hayden. 1987. *The Content of the Form*. Baltimore: Johns Hopkins University Press.

Whiting, Beatrice B. and John W.M. Whiting. 1975. *Children of Six Cultures: Psycho-Cultural Analysis*. Cambridge, MA: Harvard University Press.

Whitman, Charles Otis. 1919. The behavior of pigeons. *Carnegie Institution of Washington Publications* 257:1–161.

Wiebe, Robert. 1966. *The Search for Order, 1877–1920*. New York: Hill & Wang.

Wiener, Norbert. 1948. *Cybernetics, or Control and Communication in the Animal and the Machine*. New York: Wiley.

Williams, George C. 1966. *Adaptation and Natural Selection*. Princeton: Princeton University Press.

Williams, George C. 1975. *Sex and Evolution*. Princeton: Princeton University Press.

Williams, Sarah. 1987. American identities and Banana Republic bodies. Paper presented to the American Anthropological Association, Chicago, Nov. 22, 1987.

Williams, Sherley Anne. 1986, March. On Octavia Butler. *Ms.* 14(9):70–72.

Wilson, E.O. 1968. The ergonomics of caste in social insects. *American Naturalist* 102:41–66.

Wilson, E.O. 1975. *Sociobiology: The New Synthesis*. Cambridge:Harvard University Press.

Wilson, E.O. 1978. *On Human Nature*. Cambridge: Harvard University Press.

Windle, William F. 1980. The Cayo Santiago primate colony. *Science* 209:1486–91.

Wittenberger, J.F. 1980. *Animal Social Behavior*. Boston: Duxbury.

Wittgenstein, Ludwig. 1953. *Philosophical Investigations*. Oxford: Blackwell.

Wittig, Monique. 1981. One is not born a woman. *Feminist Issues* 2:47–54.

Wolfe, Linda D. 1984. Japanese macaque female sexual behavior: A comparison of Arashiyama East and West. In Small 1984:141–57.

Wolfe, Tom. 1969. *Pump House Gang*. New York: Bantam.

Wolpoff, M.H. 1971. Competitive exclusion among lower Pleistocene hominids: The single species hypothesis. *Man* 6(4):601–14.

Wollstonecraft, Mary. 1792. *Vindication of the Rights of Woman*. London.

Wrangham, Richard. 1974. Artificial feeding of chimpanzees and baboons in their natural habitats. *Animal Behaviour* 22:83–93.

Wrangham, Richard. 1975. The behavioural ecology of chimpanzees in Gombe National Park, Tanzania. Ph.D. thesis, Cambridge University.

Wrangham, Richard. 1979a. On the evolution of ape social systems. *Social Science Information* 18(3):335–69.

Wrangham, Richard W. 1979b. Sex differences in chimpanzee dispersion. In Hamburg and McCown 1979:481–89.

Wrangham, Richard W. 1980. An ecological model of female-bonded primate groups. *Behaviour* 75:269–99.

Wrangham, Richard W. 1987. The significance of African apes for reconstructing human social evolution. In Kinzey 1987:51–71.

Wrangham, Richard W. and Barbara B. Smuts. 1980. Sex differences in the behavioral ecology of chimpanzees in Gombe National Park. *Journal of Reproduction and Fertility*, Supplement, 28:13–31.

Wright, Patricia C. 1984. Biparental care in *Aotus trivirgatus* and *Callicebus moloch*. In Small 1984:59–76.

Wright, Patricia C. 1988. Lemurs lost and found. *Natural History* 97(7):56–60.

Wright, Susan. 1986. Recombinant DNA technology and its social transformation, 1972–82. *Osiris*, 2nd series 2:303–60.

Wynne-Edwards, V.C. 1962. *Animal Dispersion in Relation to Social Behaviour*. New York: Hafner.

Y

Yerkes, Robert M. 1916. The mental life of monkeys and apes. *Behavior Monographs* 3:1–145.

Yerkes, Robert M. 1922. What is personnel research? *Journal of Personnel Research* 1:56–63.

Yerkes, Robert M. 1925. *Almost Human*. New York: Century.

Yerkes, Robert M. 1927a. A program of anthropoid research. *American Journal of Psychology* 39:181–99.

Yerkes, Robert M. 1927b. The mind of a gorilla. Parts I & II. *Genetic Psychology Monographs* 2:1–193; 375–551.

Yerkes, Robert M. 1928. The mind of a gorilla. Part III. *Comparative Psychology Monographs* 5:1–92.

Yerkes, Robert M. 1930. Mental evolution in primates. In Cowdry 1930:115–38.

Yerkes, Robert M. 1932a. Robert Mearns Yerkes, psychobiologist. In Carl Murchison, ed. *A History of Psychology in Autobiography*. Worcester: Clark University Press, pp. 381–407.

Yerkes, Robert M. 1932b. Yale Laboratories of Comparative Psychobiology. *Comparative Psychology Monographs* 8:1–33.

Yerkes, Robert M. 1935–36. The significance of chimpanzee culture for biological research. *Harvey Lectures* 31:57–73.

Yerkes, Robert M. 1937. Primate cooperation and intelligence. *American Journal of Psychology* 50:254–70.

Yerkes, Robert M. 1939a. Sexual behavior in the chimpanzee. *Human Biology* 11:78–111.

Yerkes, Robert M. 1939b. Social dominance and sexual status in the chimpamzee. *The Quarterly Review of Biology* 14:115–36.

Yerkes, Robert M. 1943. *Chimpanzees: A Laboratory Colony*. New Haven: Yale University Press.

Yerkes, Robert M. 1963. Creating a chimpanzee community. *The Yale Journal of Biology and Medicine* 36:205–23.

Yerkes, Robert M. 1969 [1920]. How psychology happened into the war. In R. M. Yerkes, ed. *The New World of Science*. Freeport, New York: Books for Libraries Press, p. 358.

Yerkes, Robert M., J. W. Bridges, and R. S. Hardwick. 1915. *A Point Scale for Measuring Mental Ability*. Baltimore: Warwick & York.

Yerkes, Robert M. and Ada W. Yerkes. 1929. *The Great Apes*. New Haven: Yale University Press.

Yerkes, Robert M. and Ada W. Yerkes. 1934. The comparative psychopathology of infrahuman primates. In Madison Bentley and E. V. Cowdry, eds. *The Problem of Mental Disorder*. New York: McGraw-Hill, pp. 327–38.

Yerkes, Robert M. and Ada W. Yerkes. 1935. Social behavior in infrahuman primates. In Carl Murchison, ed. *A Handbook of Social Psychology*. Worcester: Clark University Press, pp. 973–1033.

Young, Robert M. 1970. *Mind, Brain, and Adaptation in the Nineteenth Century*. London: Oxford University Press.

Young, Robert M. 1977. Science *is* social relations. *Radical Science Journal* 5:65–129.

Young, Robert M. 1981. The naturalization of value systems in the human sciences. In *Science and Belief: Darwin to Einstein*, 7 vols, Block VI: *Problems in the Biological and Human Sciences*. Milton Keynes: Open University Press, pp. 63–110.

Young, Robert M. 1985a. Is nature a labour process? In Les Levidow and R.M. Young, eds. *Science*,

Technology and the Labour Process: Marxist Studies, Vol. 2. London: Free Association Books, pp. 206–32.

Young, Robert M. 1985b. *Darwin's Metaphor: Nature's Place in Victorian Culture*. London: Cambridge University Press.

Yoxen, Edward J. 1977. The social impact of molecular biology. Ph.D. thesis, Cambridge University.

Yoxen, Edward J. 1981. Life as a productive force: Capitalising the science and technology of molecular biology. In Les Levidow and R.M. Young, eds. *Science, Technology, and the Labor Process: Marxist Studies*, Vol. 1. London: CSE Books, pp. 66–122.

Yoxen, Edward J. 1983. *The Gene Business: Who Should Control Biotechnology?* New York: Harper & Row.

Z

Zaki, Hoda M. 1988. Fantasies of difference. *The Women's Review of Books* V(4):13–14.

Zihlman, Adrienne. 1967. Human locomotion: A reappraisal of the functional and anatomical evidence. Ph.D. thesis, University of California Berkeley.

Zihlman, Adrienne. 1974. Review of *Sexual Selection and the Descent of Man, 1871–1971*. *American Anthropologist* 76:475–78.

Zihlman, Adrienne. 1976. Sexual dimorphism and its behavioral implications in early hominids. In Phillip V. Tobias and Yves Coppens, eds. *Les plus anciens hominidés*. Paris: Centre National de la Recherche Scientifique, pp. 268–93.

Zihlman, Adrienne. 1978a. Women and evolution, Part 2. Subsistence and social organization among early hominids. *Signs* 4:4–20.

Zihlman, Adrienne. 1978b. Motherhood in transition: From ape to human. In Miller and Newman 1978:35–50.

Zihlman, Adrienne. 1978c. Interpretations of early hominid locomotion. In C.J. Jolly, ed. *Early Hominids of Africa*. New York: St. Martin's Press, pp. 361–77.

Zihlman, Adrienne. 1979. Pygmy chimpanzee morphology and the interpretation of early hominids. *South African Journal of Science* 75(4):165–68.

Zihlman, Adrienne. 1981. Women as shapers of human adaptation. In Dahlberg 1981:75–120.

Zihlman, Adrienne. 1982a. *The Human Evolution Coloring Book*. New York: Barnes and Noble.

Zihlman, Adrienne. 1982b. What happened to woman the gatherer? Paper presented at the meetings of the American Anthropological Association, Washington, D.C.

Zihlman, Adrienne. 1982c. Sexual dimorphism in *Homo erectus*. In *L'Homo erectus et la place de l'Homme de Tautavel parmi les hominides fossiles*. Premier Congrés International de Paleontologie Humaine. Prépublication vol. 2:949–70. Paris: CNRS.

Zihlman, Adrienne. 1983a. A behavioral reconstruction of *Australopithecus*. In Kathleen J. Reichs, ed. *Hominid Origins: Inquiries Past and Present*. Washington, D.C.: University Press of America, pp. 207–38.

Zihlman, Adrienne. 1983b. Dietary divergence of apes and early hominids. Paper for Wenner Gren symposium on Food Preferences and Aversions, Oct. 23–30.

Zihlman, Adrienne. 1984. Pygmy chimps, people and the pundits. *New Scientist* 104:364–77.

Zihlman, Adrienne. 1985a. Gathering stories for hunting human nature. *Feminist Studies* 11:364–77.

Zihlman, Adrienne. 1985b. Sex, sexes and sexism in human origins. Keynote address, American Association of Physical Anthropologists, Knoxville, 12 April. *Yearbook of Physical Anthropology* 30(1987):11–19.

Zihlman, Adrienne. 1985c. Who lost estrus, and where did it go? Paper presented at December meetings of the American Anthropology Association, Washington, D.C.

Zihlman, Adrienne. 1985d. *Australopithecus afarensis*, two sexes or two species? In P.V. Tobias, ed. *Hominid Evolution: Past, Present and Future*. From Taung 60 International Symposium. New York: Liss, pp. 213–20.

Zihlman, Adrienne and Douglas Cramer. 1976. Human locomotion. *Natural History* 85(1):64–69.

Zihlman, Adrienne and Douglas Cramer. 1978. Skeletal differences between pygmy (*Pan paniscus*) and common chimpanzees (*Pan troglodytes*). *Folia Primatologica* 29(2): 86–94.

Zihlman, Adrienne, J. E. Cronin, Douglas Cramer, and Vincent M. Sarich. 1978. Pygmy chimpanzee as a possible prototype for the common ancestor of humans, chimpanzees and gorillas. *Nature* (London) 275:744–46.

Zihlman, Adrienne and Jerold Lowenstein. 1983. A few words with Ruby. *New Scientist* (14 April):81–83.

Zihlman, Adrienne and Nancy Tanner. 1978. Gathering and the hominid adaptation. In Tiger and Fowler 1978:163–94.

Zuckerman, Solly. 1932. *The Social Life of Monkeys and Apes.* London: Routledge & Kegan Paul.

Zuckerman, Solly. 1933. *Functional Affinities of Man, Monkeys, and Apes.* New York: Harcourt Brace & Co.

Zuckerman, Solly. 1963. Concluding Remarks by the Chairman. *Symposium of the London Zoological Society,* no. 10:119–22.

Zuckerman, Solly. 1978. *From Apes to Warlords. The Autobiography of Solly Zuckerman, 1904–46.* New York: Harper & Row.

Zumpe, Doris and R.P. Michael. 1968. The clutching reaction and orgasm in the female rhesus monkey (*Macaca Mulatta*). *Journal of Endocrinology* 40:117–23.

INDEX